Ceramic- and Carbon-matrix Composites

Soviet Advanced Composites Technology Series

Series editors: J. N. Fridlyander, Russian Academy of Sciences, Moscow, Russia
I. H. Marshall, University of Paisley, Paisley, UK

This series forms a unique record of research, development and application of composite materials and components in the former Soviet Union. The material presented in each volume, much of it previously unpublished and classified until recently, gives the reader a detailed insight into the theory and methodology employed, and the results achieved, by the Soviet Union's top scientists and engineers in relation to this versatile class of materials.

Titles in the series

1. Composite Manufacturing Technology
   Editors: A. G. Bratukhin and V. S. Bogolyubov
2. Ceramic- and Carbon-matrix Composites
   Editor: V. I. Trefilov
3. Metal Matrix Composites
   Editor: J. N. Fridlyander
4. Polymer Matrix Composites
   Editor: R. E. Shalin
5. Fibre Science and Technology
   Editor: V. I. Kostikov
6. Composite Materials in Aerospace Design
   Editors: I. F. Obraztsov and G. E. Lozino-Lozinski

Ceramic- and Carbon-matrix Composites

Edited by

Academician V. I. Trefilov
Institute for Problems of Materials Science, Kiev, Ukraine

 Springer-Science+Business Media, B.V.

First edition 1995
© 1995 Springer Science+Business Media Dordrecht
Originally published by Chapman & Hall in 1995
Softcover reprint of the hardcover 1st edition 1995
Typeset in Times 10/12 pt by Pure Tech Corporation, Pondicherry, India

ISBN 978-94-010-4558-2   ISBN 978-94-011-1280-2 (eBook)
DOI 10.1007/978-94-011-1280-2

Apart from any fair dealing for the purposes of research or private study, or criticism or review, as permitted under the UK Copyright Designs and Patents Act, 1988, this publication may not be reproduced, stored, or transmitted, in any form or by any means, without the prior permission in writing of the publishers, or in the case of reprographic reproduction only in accordance with the terms of the licences issued by the Copyright Licensing Agency in the UK, or in accordance with the terms of licences issued by the appropriate Reproduction Rights Organization outside the UK. Enquiries concerning reproduction outside the terms stated here should be sent to the publishers at the London address printed on this page.

The publisher makes no representation, express or implied, with regard to the accuracy of the information contained in this book and cannot accept any legal responsibility or liability for any errors or omissions that may be made.

A catalogue record for this book is available from the British Library

Library of Congress Catalog Card Number: 94-72025

Contents

Contributors	viii
Introduction: State-of-the-art, problems and prospects for ceramic- and carbon-matrix composites *V. I. Trefilov, E. L. Shvedkov*	xi
References	xxii

1 Ceramic-matrix composites — 1
  1.1 Structural materials — 3
    1.1.1 Theoretical fundamentals (*B. A. Galanov, O.N. Grigoriev, Yu. V. Milman, V. I. Trefilov*) — 3
    1.1.2 Technology (*G. G. Gnesin*) — 29
    1.1.3 Machining techniques (*A. V. Bochko*) — 53
    1.1.4 Joining (brazing) of ceramic materials (*Y. V. Naidich*) — 73
    1.1.5 Mechanical properties (*S. M. Barinov*) — 95
    1.1.6 Engineering (*A. G. Romashin, A. D. Burovov, A. A. Postnikov*) — 115
  1.2 Functional materials — 158
    1.2.1 Ceramic composites for electronics (*V. V. Skorokhod, V. A. Dubok*) — 158
    1.2.2 Ferroelectrics, piezoelectrics, and high-temperature superconductors (*M. D. Glinchuk*) — 216
  References

2 Glass ceramic-based composites — 238
(*A. E. Rutkovskij, P. D. Sarkisov, A. A. Ivashin and V. V. Budov*) — 255
  2.1 Glass ceramic-based materials reinforced with metallic fibres and nets — 256
    2.1.1 Moulding of materials — 260

    2.1.2 Thermomechanical properties of metallic
Net-reinforced glass ceramics   264
    2.1.3 Investigation of interaction at matrix–fibre interface   265
2.2 Ceramic fibre-reinforced glass ceramic composites   266
    2.2.1 Reinforced glass ceramics   266
    2.2.2 Mechanical and corrosion properties of reinforced glass ceramics after one-sided radiant heating   269
    2.2.3 Chemical stability of glass ceramic-based composites at elevated temperatures   270
    2.2.4 Structure of composites   270
2.3 Particle-filled glass ceramic-based materials   272
    2.3.1 Filled glass ceramics   272
    2.3.2 Composites with nonmetallic particles   277
    2.3.3 Impact of microstructure of glass-ceramic matrices on properties of composites   279
References   283

3 Carbon-Based Composites   286
(V. I. Kostikov)

3.1 Theoretical fundamentals of development of carbon-based composites   287
    3.1.1 Structure of materials   289
    3.1.2 Mechanics of carbon–carbon composites   297
    3.1.3 Physicochemical features of carbon–carbon composites   303
3.2 Technology of product manufacture from carbon-based materials   330
    3.2.1 Winding of preforms   331
    3.2.2 Lay-up   337
    3.2.3 Weaving of three-dimensional structures   337
    3.2.4 Braiding   340
    3.2.5 Assembling of multidirectional skeletons   344
    3.2.6 Manufacturing technology for two-dimensional composites   348
    3.2.7 Manufacturing technology for three- and four-dimensional composites   352
3.3 Structure and properties of composites   364
    3.3.1 Properties and application of two-dimensional composites   364
    3.3.2 Friction-purpose carbon–carbon composites   385
    3.3.3 Three- and four-dimensional composites   399
3.4 Carbon- and carbide-matrix composites   414

3.4.1	Requirements for Structural Constituents of three-dimensionally reinforced high-temperature heat-resistant composites	415
3.4.2	Manufacturing technology for carbon fibre–refractory carbide composites	418
3.4.3	Mechanical properties of carbon fibre–silicon carbide materials	421
3.4.4	Heat resistance of carbon–carbide matrix materials	429
References		432
Index		438

Contributors

S. M. Barinov
IMET
Moscow
Russia

A. V. Bochko
Institute for Problems of Materials Science
Kiev
Ukraine

V. V. Budov
Institute for Problems of Materials Science
Kiev
Ukraine

A. D. Burovov
ONPO
Obninsk
Russia

V. A. Dubok
Institute for Problems of Materials Science
Kiev
Ukraine

B. A. Galanov
Institute for Problems of Materials Science
Kiev
Ukraine

M. D. Glinchuk
Institute for Problems of Materials Science
Kiev
Ukraine

Contributors

G. G. Gnesin
Institute for Problems of Materials Science
Kiev
Ukraine

O. N. Grigoriev
Institute for Problems of Materials Science
Kiev
Ukraine

A. A. Ivashin
Institute for Problems of Materials Science
Kiev
Ukraine

V. I. Kostikov
Graphit Scientific Research Institute
Moscow
Russia

Yu. V. Milman
Institute for Problems of Materials Science
Kiev
Ukraine

Yu. V. Naidich
Institute for Problems of Materials Science
Kiev
Ukraine

A. A. Postnikov
ONPO
Obninsk
Russia

A. G. Romashin
ONPO
Obninsk
Russia

A. E. Rutkovskij
Institute for Problems of Materials Science
Kiev
Ukraine

P. D. Sarkisov
MHTI
Moscow
Russia

E. L. Shvedkov
Institute for Problems of Materials Science
Kiev
Ukraine

V. V. Skorokhod
Institute for Problems of Materials Science
Kiev
Ukraine

V. I. Trefilov
Institute for Problems of Materials Science
Kiev
Ukraine

Introduction: State-of-the-art, problems and prospects for ceramic- and carbon-matrix composites

V. I. Trefilov and E. L. Shvedkov

We will begin with definitions, since discussion of any problem must be preceded by a strict definition of the subject of discussion. Unfortunately, the discord in definitions and classifications reigning in the international technical literature, not only prevents a clear understanding of what is meant in one or other case by terms 'composite material' or 'ceramics', but also often results in quite incorrect evaluations of real situations. The latter occurs especially often in research-on-research or economic studies.

The above may be illustrated by the fact that such an authoritative journal as *Chemical Abstracts* includes graphite materials, diamonds and glasses in the section on ceramics. *The Engineering Materials Handbook*, a fundamental six-volume reference book on structural materials, contains excellent definitions of composite material, ceramics, matrix, filler and reinforcing material, which are above all criticism [1]. Accordingly, we will employ in this volume the basic concepts whose definitions are as follows [2, 3].

Composite materials – artificially created materials consisting of two or more phases integrated with one another or having formed during manufacturing stage (but not resulting from a subsequent processing of the material). There exist quite a number of natural heterophase materials (wood, animal bone, etc.) which exhibit all of the characteristics for classing them with composite materials, but are not actually ones, as composite materials are artificially manufactured to meet predetermined demands, using the properties of starting materials. A not less important attribute of composite materials, distinguishing them from a number of other materials, such as steels or cast irons, is the existence in a finished material of those phases from which it has been manufactured or which have formed during its manufacture rather than phases that have appeared at the after-treatment of the material. The latter occurs, e.g., in producing certain heterogeneous structures of ferrous alloys (spheroidal-graphite cast iron, etc.). The properties of a composite material result either from the adding-up of properties of its constituents or (more often) from their mutual improvement due to synergy [4].

Matrix (of a material) – a continuous constituent of a composite, accommodating its other constituents. This term is often applied to the constituent that occupies the greatest part of the material volume, but this condition is neither necessary nor sufficient; for example, a tough matrix in hard alloys may occupy no more than 10% of the volume. The matrix is often called a binder, which is not quite correct since the binder is not necessarily a continuous constituent of a material.

Fillers – substances added to composite materials to facilitate their processing, to impart certain required properties, and also, to reduce their cost. These materials include reinforcing fibres, felt, bundles, meshes, fabrics, whiskers and particles.

Reinforcing materials – high-strength and/or high-modulus materials added to other materials for strength. They include reinforcing fibres, felt, bundles, meshes, fabrics, whiskers and particles.

Ceramic – a consolidated polycrystalline material based on compounds of nonmetals of groups III–VI of the periodic table, with one another and/or with metals, produced by means of processes ensuring mass transfer that results in bonding of the constituents. The term covers solid bodies, films, coatings and fibres (but not powders and granules). Starting compounds include borides, carbides, nitrides, oxides, silicides, phosphides, chalcogenides and their complex compounds, both natural and synthetic. The production processes include all types of sintering, hot pressing, as well as gas-thermal or vapour-phase coating deposition and impregnation techniques. These processes, together with any preform moulding, are commonly united under the somewhat loose heading 'ceramic technology'.

What is the cause of the confusion and vagueness in definitions of the terms relating to composites and particularly to ceramics? Most likely, it stems from the fact that the field of ceramic composites is now among the most dynamic and vigorously advancing fields of scientific activity (in this respect the carbon-matrix composites may be classed with relatively 'quiet' fields of science and technology [5, 6]).

New ceramics (or 'advanced ceramics' [USA] or 'fine ceramics' [Japan and Germany]) and hence ceramic composites, since in the vast majority

Introduction

of cases a new ceramic is just a composite, bursting into engineering and technology in the 1960s, brought about a real revolution in materials science. They also became, in the opinion of, e.g., Japanese experts, 'the third industrial material' (after metal and polymers), and resulted in the onset of a 'new Stone Age'. The daring assertion is justified in that in the near future not a single important scientific or engineering development, in any way serious, can be achieved without resorting to new ceramics.

The key position of new ceramics among new technologies (being an outcome of high technologies, itself, it brings about a profusion of higher ones!) has led to a 'ceramic boom' in all developed countries and a 'fight over ceramics' between technological world leaders, the USA and Japan.

Direct comparisons and estimates of expenditure of research in various countries, investments in the industry and amounts of production in terms of cost are difficult, not only because of floating currency exchange rates or incomplete information, but also due to the aforementioned terminological 'disorder', thus, one estimate will include glasses, another, materials with a binder, still another, diamonds, graphites, etc. and yet another, solely refractory compound-based materials. The notion on the advance of efforts in the field of new ceramics and composite materials can nonetheless be given by the following data.

In 1988, expenditure on the production of such materials in the USA amounted to US $2 billion, while in 2000 it may rise to US $20 billion (for the actual product, i.e., materials and parts, excluding the machinery) [7]; it is expected that in the USA the annual consumption of new ceramics will, by 2000, reach a level of US $4–5 billion [8] and the scope of their market, according to various predictions, varies from US $1–5 billion in 2000 [7] to $7 billion, already in 1995 [9]; for comparison, the world market for the same periods is predicted to be from $9 billion to $33 billion respectively; a still more optimistic prediction of the world market for new ceramics in 2000, of US $50 billion (!), has been made [10]. The new ceramics market growth rates are striking. Whereas the average annual growth in the USA between 1985 and 1990 amounted to 16% (25% for structural ceramics), right up to 1995, the market in the USA will grow by at least 10% annually [9], and in Japan, by at least 11% [11] (primarily due to functional ceramics, which in 2000 will account for 70–80% of the market). It is interesting that Japan, already a recognized leader in the field of functional ceramics [12], now seems to be also outstripping the USA in the field of structural ceramics [13]; thus, it is expected that by 2000, Japan will supply about 40% of the world market of ceramics for heat engines, while the share of the USA will be as little as 25%.

In general it should be noted that stakes in the 'fight between giants' for the new ceramics market are exceedingly high. Thus, experts from the

Department of Energy of the USA believe [9] that if the USA commands the market of new ceramics the gross national product of their country will by 2000 rise by US $28.2 billion, and the number of jobs, by about 250 000. The assertion by experts from the Charles Rivers Assoc., Inc., is still more daring: whoever conquers the world ceramics market, will be the conqueror of the overall world market!

All this accounts for copious investment by the governments of the USA and Japan countries in the R&D of new ceramics. The annual investment in this field in the USA is not less than $100 million (50% by the government, 50% by private companies) [9]. Thus, in 1986, government contracts with the United States Advanced Ceramics Association (USACA) amounted to US $43.6 million, while the provision by private companies for R&D was US $152.9 million [14] (which was at least 23% of the ceramic product sales). Expenses are growing continuously. According to various data, the contribution to R&D efforts in Japan is much greater. In view of the prodigious scientific intensity of this field, the inference drawn by authors of the analytical review [9], that Japan is probably winning the fight over ceramics, does not look like an exaggeration. This is a very important conclusion notwithstanding the fact that the share (but not the role!) of ceramic composites in the family of composites is fairly modest [15]: even with allowance for a vigorous growth of their market (over 19% annually) in 2000, ceramic composites will account for less than 9% of the whole market of composite materials; for comparison, the annual growth of the market of polymer-matrix composites is about 10%, and of metal-matrix ones, over 22%.

Compared with the above, in the former USSR the situation in the field of composites generally, and of ceramic composites in particular looks rather gloomy [12, 14, 16, 17]. With the exception of military and space technologies, studies on ceramic and carbon composites can be counted on the fingers of one hand. In the field of new ceramics in general, there still exist more or less coordinated R&D efforts (a constituent congress of the Ceramic Association of the USSR was held in December 1989), conducted primarily at three research centres (in St. Petersberg, Moscow and Kiev), and a special financing of this work by the government (the 'Motoceram' inter-industry programme was at last approved in 1989), but in the field of ceramic and carbon composites very little is being done both in the Academy of Sciences of the USSR and in sectoral institutes, where foreign studies and results are primarily repeated and reproduced. Thus, studies of the possibility of production of coreless silicon carbide fibres, manufactured abroad in enormous quantities, are hardly conducted; fibres of carbon, boron and silicon carbide, generally of a fairly high quality, are only manufactured on a small scale for experiments, which makes them very costly. Civilian sectors of engineering (where the situation in the automotive industry is perhaps slightly better) have today practically no raw materials, no equipment, no commercial technologies, no product inspection procedures, no experts, no data bases and no banks for manufacture of ceramic and carbon composites. Causes of this situation include absence of relevant information and lack of knowledge by designers and production engineers of what composite materials are, what their properties and features are, and what contribution they can make to engineering.

Let us, if even briefly, answer these questions.

Modern engineering technology experiences an obvious demand for radically new materials, capable of performing (with a reasonable economic efficiency) utterly new functions (such as in power lasers, in sheathing of space shuttles, in superconducting devices, etc.); for a drastic improvement of properties of already-existing materials (such as through new techniques for their processing or for application of coatings); and for upgrading the efficiency of operation of facilities being both already used and designed (such as increasing the working temperature in heat engines).

All these, and many other demands are most fully met by capabilities of new ceramic (including composite) materials, which offer practically unlimited combinations of properties, capacities and functional characteristics.

First of all, why just ceramic composites? This is because they, as well as other composite materials, allow the designer and production engineer not only to 'design' a material with predetermined properties [18] (development of ceramic composites is essentially structural design), but also to create 'smart' materials which adapt themselves to operating conditions [16, 17]. But, in contrast to other composites, ceramics are light, exhibit the best relation of strength, hardness and toughness, are stable in the most severe environments (in corrosive media, at high temperatures), their manufacture components are practically unlimited, their fabrication technologies (even already existing ones) are diverse and versatile, and the potential cost of products made from ceramic composites is exceedingly low. Thus, although coming last regarding the scopes of market and consumption in the family of composite materials as noted above, ceramic composites are nevertheless already being successfully introduced into many key sectors of engineering and technology, and their use is growing rapidly. The role of the military and space technologies in the growth is far from being a major one; as predicted in [15], the average annual growth until 2000 in the latter sectors will amount to 5.5%; for comparison, for ceramic composites for cutting tools it will be 22%, for wear-resistant parts, 20.2%, for engines, 38.5%, and for power engineering, 40.8%.

It is interesting to note that the whole 'ceramic boom' was initiated not by the functional or, say, the nuclear ceramics, but by the structural high-temperature ceramics (the breakthrough in the field of high-temperature superconducting ceramics occurred later). So, speaking of ceramic composites or of new ceramics in general, structural high-temperature ceramics are usually implied [19, 20]. It is noteworthy that carbon–carbon composite materials also started their way into modern engineering from high-temperature applications, such as the thermal protection of hypersonic aircraft and spacecraft [21–23]. Later, these composites found use in the manufacture of parts and assemblies of gas turbine [24] or liquid-fuel rocket engines [25] and recently, of disc brake [26], electric heaters, HIP moulds, nuclear reactor tubes, etc. [27]. But this is not all. Speaking of ceramic or carbon–carbon composites, one often means ceramic- (or carbon-) matrix materials reinforced with whiskers or fibres rather than with particles, since both theory and practice indicate that the maximum increase of the material toughness and a controlled crack growth are most efficiently provided just by elongated rather than equiaxial inclusions [28].

The development of ceramic composites followed an ordinary pathway, typical for nearly all novel materials [29]. After it was clear that the properties of a ceramic matrix could be radically altered by incorporating certain structural elements, the work was advanced slowly and sluggishly by a few enthusiasts (driving force: curiosity; attitude of the scientific community: condescending tolerance; attitude of financing organizations: restraint). A striking event was needed for the 'boom' to occur; such an event was probably the publicity demonstration of an all-ceramic engine by Japanese experts. Then, for all those involved in the ceramic technology not to get disappointed, some real results were needed, not something delicate and exotic (such as a turbine wheel), but obvious and useful (such as a cutting tool or bearing). Numerous such elements have now appeared; efforts in the field of new ceramics and ceramic composites became more ordered; it seems that all those involved in the ceramic technology would not be disappointed.

A full utilization of all potentialities of new ceramics, however, needs the elimination of their present substantial shortcomings, such as high cost, brittleness, low reliability, poor reproducibility of properties and poor machinability.

The shortcomings can be overcome only on the basis of new developments in theory and technology, as well as with the establishment of a total special nondestructive inspection of products. In contrast to, e.g., powder metallurgy or traditional ceramic technology, which in their essence are a craft or art, the new ceramic technology is an exceedingly science-intensive field so that the theory plays a governing part in eliminating the shortcomings.

As regards the theory, an active advance of coordinated efforts in the field of chemistry, physical chemistry and solid-state physics is needed, first of all for all types of new ceramics, although it is clear that various specific types of new ceramics will call for different emphases in this triad, addition of other branches of science and different degrees of detail in the work [30].

Studying the long-term behaviour of properties is essential for all new ceramics types (creep resistance and long-term strength for structural ceramics, swelling and radiation stability for nuclear ceramics, etc.). For the mechanoceramics, which, as mentioned above, are the 'key' type of new ceramics, however, the first and foremost problem is to control their tendency to sudden, catastrophic, 'unwarned' fracture. This means that development of mechanoceramics for commercial use requires upgrading their reliability, elaboration of the theory of fracture under a multiaxial dynamic loading, and creation eventually of 'tough' ceramics [28, 31, 32].

The development of fibre-reinforced ceramic composites has initiated active research into theoretical problems of the physicochemical interaction between the reinforcement and the matrix. This is due to the fact that the toughness of a composite can be increased through the deflection of cracks not more than twice. The whole remaining effect of reinforcement is due to 'pulling' the fibres out of the matrix, where the interphase interaction plays the major role [9, 13, 30, 32].

The above suffices to conclude that the traditional 'formulatory' approach to technology does not work for the new ceramics. The new ceramic technology must rely on a strong theoretical foundation, the more so as it is the governing factor for producing new materials and products. Moreover, a number of basic defects of ceramics, such as pores, cracks, voids, inclusions, inhomogeneity zones, coarse grains, etc., can be eliminated solely by elaborate production techniques.

For convenience of presentation, we will divide the ceramic technology into stages of preparation of starting powders, their consolidation, aftertreatment and inspection, and will discuss the stages successively.

The vast majority of ceramic production engineers believe that production of high-quality new ceramics calls for 'ideal' powders, i.e., pure controlled-composition, submicronic, spherical, monodisperse, non-agglomerating powders [29, 31, 33]. This set of requirements contains at least two slightly unclear items; first, monodispersity: if a higher degree of recrystallization at high gradients of dispersity of blends is excluded, then specially selected polydisperse blends can, in a number of cases, yield a much higher particle-packing density, which is particularly important for 'wet' moulding methods [31]; secondly, non-agglomerability: a successful use of so-called 'controllable soft' agglomerates is known [34, 35], though more often for monophase systems, the requirement for homogeneity, excluding any agglomeration, is the first and foremost one for polyphase systems. If the two unclear points are put aside, one can assert with confidence that numerous methods for producing such powders exist at present: deposition from the vapour phase, co-deposition from solutions, reactive spray drying, plasma-chemical synthesis, solution–gel process, self-propagating high-temperature synthesis, laser synthesis, as well as others [10, 14, 31, 33, 34]. These methods are capable of yielding refractory compound powders with particle sizes down to 0.2–0.3 nm (boehmite powder with a particle size of 0.1 nm and a purity of 99.999% has been reported). A purity of 'five nines' is already attainable and efforts to attain 'nine nines' are under way.

Among the moulding methods it is difficult to distinguish the most popular one; both 'dry' and 'wet' techniques are being developed equally actively [34, 36]. An obvious preference among the former is given to the isostatic pressing (cold and hot), and among the latter, to the injection moulding and sol-gel method [9, 14, 37]. The developments in the isostatic pressing are aimed primarily at solving purely technical problems (upgrading the safety at higher process parameters, provision of continuity of the process, its automation, strict control of the cycle, etc.); in the injection moulding, efforts are concentrated on rheologic properties of the particle carrier, methods for its maximum saturation with the powder, techniques for its effective removal without contaminating and impairing the strength of the material, etc.; the work on sol–gel techniques involves active efforts of chemists and physicochemists on the search for working media, where conversion to alcoholate ones is characteristic, and for mould materials, such as dialysis membrane-type ones, etc. [7, 9, 10, 14, 20].

Sintering processes are becoming more energy-intensive (higher pressures, lower temperatures), while the process parameters are strictly controlled or even preset (programmed sintering) using microprocessors. Quite new sintering types, such as a plasma sintering [38], microwave-heating sintering [39] and others are appearing.

An ever-widening application is being gained by 'combined' consolidation methods, where moulding is combined with sintering, and sometimes, an *in situ* synthesis of a compound with moulding and sintering. Such methods include impact moulding (which is capable of yielding a compact density of 98.6% at 320 MPa even from coarse, on the order of 10 μm, SiC powders) [40, 41]; sintering under high pressure (HPS process) [42]; isostatic pressing of preforms being sintered (sinter-HIP process); and various types of consolidation by means of the self-propagating high-temperature synthesis (SHS): SHS – hot pressing, used in the USSR; SHS – hot isostatic pressing, used in Japan; and SHS – hot rolling, used in the USA [43–45].

A very attractive new technique for manufacture of ceramic composites (in particular silicon carbide fibre-reinforced ones) is the so-called chemical vapour impregnation (CVI) process, where a fibrous skeleton is impregnated with the matrix material from a vapour phase. This technique yields a dense (up to 80% of theoretical density) preform in as little as 24 hours instead of weeks at ordinary impregnation methods. SiC–SiC and SiC–Si_3N_4 composites have been successfully produced by this method [46, 47].

To complete this brief survey of the processes of consolidation of new ceramics and ceramic composites, two technological principles which are already clearly defining themselves have to be emphasized: first, primary attention should be given to properties of starting materials [48] and to their processing before high-temperature treatment [20], as basic defects in the material originate just at these stages: secondly, when choosing between the traditional ('physical') and the new ('chemical') ceramic technology, the preference should be perhaps given to the latter [31, 49] as only it makes it possible to attain the highest system homogenization and to minimize defects, which is of the utmost importance for ceramic composites. It is predicted [7] that over 50% of the new ceramics will be produced only by such methods by 2000.

However technically important the consolidation processes are, an equal part is played by the subsequent processing of ceramics and their inspection: in the total cost of a ceramic product, the cost of starting materials and consolidation accounts for as little as 11% (for metals, 43%), whereas the processing accounts for 38% (for metals, 43%), and the inspection, for 51% (for metals, 14%).

The principal methods of the aftertreatment of ceramics include thermal treatment, dimensional machining and product surface modification.

The thermal treatment is carried out to optimize the material structure and to improve mechanical properties of the material. Appropriately selected thermal treatment conditions provide for crystallization of the intergranular glass phase, characteristic for some ceramics, and thereby increase significantly (by 20–30%) the hardness and fracture toughness [34].

Practically all ceramic materials (except a few specially developed so-called 'machinable ceramics') are very hard to machine. Because of this, the consolidation processes which are capable of yielding 'near-net-shape' products are highly rated in ceramic technology. Even such products, however, call for surface finishing, at least to attain the maximum possible surface quality; this is extremely important for ceramics, as the minutest surface defects – irregularities, scratches, etc. – may become stress concentrators, the 'weak links' taken into account by the Weibull statistics.

The dimensional machining of current-conducting ceramics is successfully effected by electric discharge and high-output electrochemical methods, while nonconducting ones are machined by diamond, laser and ultrasonic techniques (which can obviously also be used for conducting ceramics). Attempts to increase the electrical conductivity of a ceramic material by small additions to its basic composition (e.g., of silicon nitride) to a level which makes it possible for electric discharge or electrochemical machining have been undertaken recently and proved successful in some cases [54].

The finishing of ceramic product surfaces is most often carried out by diamond wheels, although a successful use of ordinary abrasives, such as green silicon carbide, to obtain mirror-like defect-free surfaces has been reported. A fine chemical and electrochemical treatment is also used for this purpose. For modification of surfaces, use is made of various electrophysical methods, such as laser machining, ion implantation, etc. The most universal and powerful method for the aftertreatment of ceramics is, of course, deposition of coatings on them; this procedure for carbon–carbon composite materials is indispensable, as otherwise a high-rate oxidation of the material at high working temperatures would occur; however, the deposition of appropriate, ordinarily ceramic, coatings provides the highest strength to mass ratio presently achieved right up to 2270 K [18] and reduces many times the mass loss at 1500–1600 K [51]. The deposition of protective and functional coatings makes it possible to solve a number of technological and economic problems at once: to heal surface defects; to impart new surface properties to a product or to improve existing properties; and to combine, when required, cheapness or special properties of the substrate with scarcity of material or required properties of a thin working layer. The deposition of coatings on ceramic materials often yields striking results, increasing the durability and reliability of products by orders of magnitude [7].

The concept of ceramics fracture at a 'weak' link necessitates a total (not a sampling!) nondestructive evaluation of ceramic products [14, 34]. Actual sizes of critical defects in new ceramics require a very high resolution of the inspection method used: not worse than 100 µm (better, 10–20 µm). Of numerous existing nondestructive evaluation methods (vibrational, luminescent, electrical, thermal, radiographic, X-ray, optical, acoustic, etc. [52–54]) the most extensively used for new ceramics is at present, the X-ray and ultrasonic flaw detection. The former, with the use of sharp-focus tubes and microprocessors, makes possible 90-fold magnifications and detection of defects sizing on the order of 50 µm. (Striking reports [54] on detection of defects with a size on the order of 2 µm have appeared recently.) The latter, with the use of a sharply focused sonic wave of a high (over 300 MHz) frequency, provides a resolution of 0.08–1 µm over the thickness, and of 1.5–5 µm over the section of a defect. Active developments of various advanced acoustic methods – acoustic and laser-acoustic microscopy, US echo-pulse method, etc. – are under way [55]. The appearance can be expected, in the near future, of such simple and reliable nondestructive inspection methods for ceramics that would make a computerized flaw detector as routine an inspection tool as, say, a hardness tester.

Introduction

When the above-described shortcomings of ceramics and ceramic composites (first and foremost, their inadequate reliability and reproducibility of properties [31]) are overcome (which will probably occur during the next few years) and the ceramic production techniques which seem exotic today become ordinary, the diversity of commercially used ceramic materials will become so high that their users (designers, production engineers) will be hardly able or, perhaps, unable at all to orient themselves in this immense world without dependable guides, such as monographs, reference books (which are so far scarce, encyclopaedias, explanatory dictionaries, glossaries, thesauri, including those similar to the one being now published in the USSR [3]. At the same time, this new world of new objects and technologies cannot exist without new data processing methods. Today these include automated information systems and data banks, tomorrow these will be knowledge banks based on expert systems and artificial intelligence. Regretfully, only a few information systems in the world are at present actually working in this field, mainly bibliographic ones, and even those are often incorporated into the context of universal databases such as INSPEC [56]. Specialized systems on new ceramics and ceramic composites can be literally counted on the fingers of one hand, for example, the Chemical Abstracts database incorporated in the DIALOG world information service [57]; the Silica international documental system of information on ceramics and glass [58]; and the DOFIN system, commercially operated in the USSR [59]. Such a situation is today obviously abnormal. An indispensable prerequisite for a regular and effective advance of the ceramic technology and development of the market of new ceramics and ceramic composites is the establishment of a global network of computerized specialized information services. It seems that a good start Western Europe has been made by the Eureka program [17]. Having been implemented, such programs will make it possible not only to optimize 'horizontal' connections (developer – production engineer – designer – user), but also to arrange a normal training of specialists in this field, whose lack is so acutely experienced in the USSR [17].

This very brief survey of the situation in the field of new ceramics and ceramic composites can be summarized as follows: if scientific, technological, organizational and informational problems are solved in the near future, then ceramic composites will change from tomorrow's to today's materials, and an optimistic prediction by Japanese experts (development of materials with angström-sized structural components by 2000; of 'defect-free' ceramics by 2005; of an all-ceramic computer by 2001; and of an all-ceramic engine by 2007 [50]) will become reality.

We hope that this book will, even if to a small extent, fill the above-mentioned information gap for the time being.

REFERENCES

1. *Engineering Materials Handbook* (1987) Vol. 1, Composites. Met. Park, ASM Internat., 1987.
2. Shvedkov E. L., Kovensky I. I. and Kutsenok T. (1987) Definition and interpretation of the term 'ceramics'. *Science of Sintering*, **19**(1), 39–47.
3. Shvedkov E. L., Kovensky I. I., Denisenko E. T. and Zyrin A. V. (1991) Glossary. *New Ceramics Reference Book*. Naukova Dumka, Kiev.
4. Piatti J. (ed.) (1978) *Advances in Composite Materials*. Ispra Establishment: Applied Science Publications, New York.
5. Sigalas J. (1988) Composite materials – the tailormade solution. *South African Mechanical Engineers*, **38**(8), 486–90.
6. Kostikov V. I. Carbon–carbon composites. (1989) *Zhurnal Vsesojuznogo Obshchestva im. D. I. Mendeleeva*, 1989, **34**(5), 492–501.
7. *Advanced Materials by Design.* (1988) Congress of the US Office of Technology Assessment. US Government Printing Office, Washington, DC.
8. Birchall J. D. (1988) High strength ceramics: problems and possibilities. *J. Phy. and Chem. Solids*, **49**(8), 859–62.
9. *Fiber-reinforced ceramics: advanced materials for todays needs*. Emerging Technologies, No. 22. Technical Insights, Inc., Fort See, 1987.
10. Tretjakov Ju.D. and Metlin Ju.G. (1987) *Ceramics: material of the Future*. Znanie, Moscow.
11. Okudo Hiroshi (1989) Future uses of ceramics. *Toraj Reports* (Japan), **34**(2), 152–5.
12. Reh H. (1989) Wie realistisch sind Marktprognozen für Technische Keramik? *Keramik Zeitschrift*, **41**(3), 176–82.
13. What's ahead for ceramics in heat engines? *Metal Progress*, **134**(2), 11, 12, 14.
14. Shvedkov E. L. and Judin A. G. (1990) *Ceramic Materials: State-of-the-Art in R&D, Future Prospects*, MTsNTI, Moscow.
15. Abraham T., Bryant R. W. and Mooney P. J. (1988) The prospects for advanced polymer-, metal- and ceramic-matrix composites. *Journal of Metals*, **40**(11), 46–8.
16. Fridljander I. N., Bratukhin A. G. and Shalin R. E. (1989) *Engineering without Steel (Composite Materials)*. Sovetskaja Rossija, Moscow.
17. Fridljander I. N., Bratukhin A. G. and Shalin R. E. (1989) *On Composites*. Znanie, Moscow, 1989.
18. Prevo K. M. (1989) Fiber-reinforced ceramics: new opportunities for composite materials. *American Ceramic Society Bulletin*, **68**(2), 395–400.
19. Lewis C. F. (1988) Ceramic matrix composites: the ultimate materials dream. *Materials Engineering*, **105**(9), 41–5.
20. *Ceramic Technology for Advanced Heat Engines*. (1987) National Academy Press, Washington, DC.
21. Hermes, Preventing nose and leading edges from melting. *Revue Aerospatiale*, **54**, 44.
22. Aerospace materials for year 2000. *Engineering Materials and Design*, **32**(8), 26, 29, 31.
23. The unique capabilities of carbon–carbon composites. *Materials Engineering*, **106**(4), 27–31.
24. Munson M.C. (1988) Sixth International Symposium on Superalloys. *Journal of Metals*, 1988, **40**(5), 8–9.
25. Melchior A. Poulignen M. F. and Soler E. (1987) Thermostructural composite materials for liquid propellant rocket engines. *AIAA Papers*, No. 2119, pp. 1–11.

26. Pat. 2616779 France, CO4B 35/52, F16D 69/02. Procédé de Fabrication d'une Pièce Nottamment de Disque de Frein. J. -L. Chareire and J. Salem. Publ. 23 Dec. 1988.
27. Sumo Kendzo. (1988) Carbon–carbon composites. *Cast., Forg. and Heat Treat.* (Japan), **41**(2), 29–32.
28. Buljan S. -T., Pasto A. E. and Kim H. J. (1989) Ceramic whisker and particulate-composites: properties, reliability, and applications. *American Ceramic Society Bulletin*, **68**(2), 387–94.
29. Warren R. (1987) Ceramic-matrix composites. *Composites*, **18**(2), 86–7.
30. Davidge R. W. (1987) Fibre-reinforced ceramics. *Composites*, **18**(2), 92–8.
31. Greil P. (1989) Opportunities and limits in engineering ceramics. *Powder Metallurgy International*, **21**(2), 40–2, 44–6.
32. Shvedkov E. L. (1987) 'Tough' ceramics abroad (theoretical prerequisites). Preprint No. 7. IPM AN USSR, Kiev.
33. Mengelle C. (1968) Principales conclusions des jounées d'étude sur: 'Le présent et le futur des céramiques dans les industries méchaniques'. *Matériaux Mécanique Electricité*, **426**, 7–17.
34. Shvedkov E. L. (1987) 'Tough' ceramics abroad (practical introduction). Preprint No. 8. IPM AN USSR, Kiev.
35. Lungberg R., Nyberg B., Willicader K. *et al.* (1987) Processing of whisker-reinforced ceramics. *Composites*, **18**(2), 125–7.
36. Pivinskij J. E. (1989) Structural ceramics and problems of their technology. In *Chemistry and Technology of Silicate and Refractory Compounds*. Leningrad, pp. 109–25.
37. Livage J. (1988–9) Les procédes sol-gel, un noveau mode d'élaboration des matériaux céramiques. *Courrier de CNRS*, **71**, 29–30.
38. Plasma sintering of ceramics and ceramic composites. *Futurtech.*, 1989, **81**.
39. Sutton W. H. (1989) Microwave processing of ceramic materials. *American Ceramic Society Bulletin*, **68**(2), 376–86.
40. Shock consolidation of coarse SiC powder. (1985) *J. of American Ceramic Society*, **68**(12), 322–4.
41. Gourdin W. H. and Neinland S.L. (1985) Dynamic compaction of aluminium nitride powder. *J. of American Ceramic Society*, **68**(12), 674–9.
42. High technology ceramics and ceramic composites: focus of three years laboratory study. *CI News*, **1**(3), 19.
43. Abramovici R. (1985) Composite ceramics in powder or sintered form obtained by alumothermal reactions. *Material Science and Engineering*, **71**, 313–20.
44. Cutler R. A., Vircor A. V. and Holty B. (1985) Synthesis and densification of oxide–carbide composites. *Ceramic Engineering and Science Proceedings*, **6**(7–8), 715–28.
45. Osamu Y., Yeshikari M. and Mitsue K. (1985) High pressure self-combustion sintering of silicon carbide. *American Ceramic Society Bulletin*, **64**(2), 319–21.
46. Pitzer E. and Godow R. (1986) Fiber reinforced silicon carbide. *American Ceramic Society Bulletin*, **65**(2), 326–35.
47. Chiang Y. -M., Haggerty J. S., Messner R. P. *et al.* (1988) Reaction-based processing methods for ceramic-matrix composites. *American Ceramic Society Bulletin*, **68**(2), 420–28.
48. Mechalsky I. I. (1989) Engineering research needs of advanced ceramics and ceramic-matrix composites. *American Ceramic Society Bulletin*, **68**(2), 367–75.

49. Mehrabian R. (1988) New pathways to processing composites. *High Temperature/High Performance Composites*. Proceedings of the Symposium held in Rend (Nev.), 5–7 April 1988. Pittsburgh (Pa.), 1988, pp. 3–21.
50. Firestone R. F. (1988) Guide des méthodes nouvelles d'usinage des céramiques et des verres. *Industrie Céramique*, **11**, 776–7.
51. Hsu S. E. Improvement and Property Analysis for 3D C/C Composites. Atmam '87: Int. Conf. Anal. and Test. Methodol. of Adv. Mater., Montreal, 26–28 Aug. 1987, pp. 1–9.
52. Misuharu S. and Geruo K. (1988) Ceramics investigation methods. Nondestructive evaluation techniques. *Ceramics* (Japan), **23**(12), 1168–77.
53. Reynolds W. N. (1988) The inspection of engineering ceramics. *Non-Destructive Testing*. Proceedings of the 4th European Conference in London, 12–13 Sept. 1987. Vol. 3, Oxford, pp. 1688–90.
54. Firestone R. F. (1988) NDE improving reliability of advanced ceramics. *American Ceramic Society Bulletin*, **68**(6), 1177–8, 1186.
55. Generazio E. R., Roth D. J. and Backlini G. I. (1988) Acoustic testing of subtle porosity variations in ceramics. *Materials Evaluation*, **46**(10), 1338–43.
56. *Search Guide to Material Science and Metallurgy in the INSPEC Database* (1988) Piscataway, IEEE.
57. *DIALOG Database Catalog* (1989) Dialog Information Services, Inc.
58. Müller-Lorentz M. and Willmann G. (1988) Konstruiren mit Keramik und Glass-Beschafftung von Informationen. *Sprechsaal*, **121**(10), 934–40.
59. Shvedkov E. L. (1987) Experience in development and operation of problem-oriented automatic information retrieval system (as exemplified by powder metallurgy). *Information Bulletin of GSSSD*, **16–17**, 17–19.
60. Mykio H. (1988) Future of New Materials. *Ceramics* (Japan), **23**(1), 11–12.

FURTHER READING

Jamet J. F. (1988) Ceramic–ceramic composites for use at high temperature. *New Materials and Their Applications*. Proceedings of Institute of Physics Conference, Warwick, 22–25 Sept. 1987. Bristol. 1988, pp. 63–75.

Low-cost ceramic lube good to 1300+ °F. *High-Tech. Materials Alert*, 1989, **6**(4), 2.

Olander D. R. (1989) I ceramici nella technologia della fusione nucleare. *Ceramurgia*, **19**(1), 17–24.

Deroin P. (1989) Les céramiques composites à la conquête de l'extrême. *Science et Technologie*, **15**, 26–7.

Rice R. W. (1984) Mechanically reliable ceramics. Needs and opportunities to understand and control fracture. *Journal of Physics and Chemistry of Solids*, **45**(10), 1033–50.

Evans A. G. (1985) Engineering property requirements for high performance ceramics. *Material Science and Engineering*, **71**, 3–21.

1
Ceramic-matrix composites

Nomenclature

ΔT_{cr}	critical temperature difference
t^*	characteristic deformation temperature
T_m	melting temperature
T_b	cold brittleness temperature
T_b^m	monocrystal cold brittleness temperature
T_b^l, T_b^u	lower and upper cold brittleness temperatures respectively
T_b^t, T_b^i	cold brittleness temperatures at transcrystalline and intercrystalline fracture respectively
t_r	recrystallization temperature
S	activation entropy
H	activation enthalpy
U	dislocation movement activation energy
V	activation volume
γ_0	theoretical value of surface energy of fracture
γ_{ef}	effective surface energy of fracture
γ	density
ε_p^*	plastic strain corresponding to stress σ_s
$\dot{\varepsilon}$	strain rate
δ	plasticity
v_0	frequency factor
N	number of elements being activated
E	Young's modulus
G_c	critical rate of elastic energy release
ΔG_c	increment of G_c
HV	Vickers hardness
σ	effective stress in existing slip system
σ_a	applied stress

Ceramic-matrix composites

σ_i	long-range internal stresses
σ_s	yield stress, flow stress
σ_{max}	theoretical tensile strength
τ_{max}	theoretical shear strength
σ_f^p	polycrystal fracture stress
$\Delta\sigma$	strain hardening
σ_f	fracture stress
σ_{ij}^0	fast-oscillating components of tensor of stresses in composite
$\tilde{\sigma}_{ij}^0$	components of tensor of mean stresses
$\tilde{\sigma}_{ij}^s$	averaged (over cell Y) stresses in sth constituent of composite ($\tilde{\sigma}_{ij}^0 = \sum_s \tilde{\sigma}_{ij}^s$)
σ_{ij}^r	components of tensor of residual (internal) stresses
$\tilde{\sigma}_{ij}^r = \sigma^{rs} \delta_{ij}$	residual stresses in sth constituent of composite, averaged (over cell Y) ($\sum_s \tilde{\sigma}_{ij}^{sr} = 0$)
K_M (M = I, II, III)	stress intensity factor (M = I, II, III for normal-fracture (opening-mode), cross-shear (sliding-mode) and longitudinal shear (tearing-mode) cracks respectively)
$\tilde{K}_M$	factor of intensity of stresses $\tilde{\sigma}_{ij}^0$ (mean intensity factor)
K_M^s	factor of intensity of stresses $\tilde{\sigma}_{ij}^s (K_M^s = \alpha_s \tilde{K}_M)$
K_{Ic}^s	fracture toughness of sth constituent of composite
$\tilde{K}_{Ic}$	fracture toughness of composite of given composition
ΔK_{Ic}	increment of fracture toughness K_{Ic}
a_{ijlm}	components of tensor of elastic moduli
a_{ijlm}^h	components of tensor of effective elastic moduli
a_{ijlm}^{hs}	components of tensor of effective elastic moduli of sth constituent of composite ($a_{ijlm}^h = \sum_s a_{ijlm}^{hs}$)
k	Boltzmann's constant
$\tilde{G}, \tilde{v}$	effective shear modulus and Poisson's ratio
$\tilde{G}_s, \tilde{v}_s$	effective (averaged) shear modulus and Poisson's ratio of sth constituent of composite ($\tilde{G} = \sum_s \tilde{G}_s$)
G_s, v_s	shear modulus and Poisson's ratio of sth constituent of composite
b_{ijlm}^h	tensor of elastic compliances, corresponding to tensor a_{ijlm}^h
$\alpha_s = \tilde{G}_s \tilde{G}^{-1}$	factor determining the fraction of factor $\tilde{K}_M$ of intensity of stresses $\tilde{\sigma}_{ij}^0$, which falls at factor K_M^s of intensity of stresses $\tilde{\sigma}_{ij}^s (\sum \alpha_s = 1)$
$\tilde{e}_{ij}^0$	components of tensor of mean strains
$|Y|$	volume of periodicity cell (element) Y

$|Y_s|$	volume of region $Y_s \subset Y$ containing sth constituent of composite
C	size of crack (defect)
ω	measure of microscopic damage
$\alpha = \dfrac{U}{kT_m}$	measure of covalent component's contribution to interatomic bond
δ_{ij}	Kronecker delta
r	fibre radius
L	fibre length
v_f	volume fraction of fibre in matrix
τ_i	friction resistance at matrix–fibre interface

1.1 STRUCTURAL MATERIALS

1.1.1 Theoretical fundamentals
B. A. Galanov, O. N. Grigoriev, Yu. V. Milman and V. I. Trefilov

Extensive efforts on developing heterophase ceramic composites have been conducted in the world over the last 15 years. These materials have a ceramic matrix based on oxides, carbides, nitrides, etc. with inclusions of whiskers and fibres of silicon carbide and nitride, aluminium nitride, platelet and isometric grains of carbides, borides, nitrides and their more complex refractory compounds.

The interest in the composites stems from the possibility of attaining states that offer a set of mechanical and performance properties unattainable with single-phase materials. In particular, when used as structural materials, heterophase composites exhibit an effective operation of the processes of energy dissipation at fracture, discovered recently in ceramics (pulling of fibres out of the matrix, microcracking, phase transformation, crack branching, etc.), with a corresponding steep increase in the strength and fracture toughness, improvement of performance characteristics (abrasive and wear resistance, thermal stability, reliability, working temperature range).

Table 1.1 Some properties of modern ceramic-matrix composites

Strength at 20 °C (MPa)	2000–2500
Strength at 1400 °C (MPa)	500
Fracture toughness at 20 °C (MPa m$^{1/2}$)	> 15
Weibull modulus m	20
Thermal shock resistance (critical temperature difference ΔT_{cr}) (°C)	1000
Maximum working temperature in oxidizing environment (°C)	1600

Ceramic materials based on sialons, zirconia, silicon carbide, etc., whose properties are presented in Table 1.1, have been prepared on a laboratory scale in various laboratories over the world; their pilot- and commercial-scale production is expected to begin in the mid-1990s.

The materials under consideration have gained a wide application in various industries, in particular as:

1. heat- and erosion-resistant, refractory engineering ceramics for metallurgy and chemical engineering;
2. structural ceramics for engines;
3. wear-resistant ceramics for working elements of milling equipment, cultivating machines, etc.;
4. impact-resistant materials.

As a result of the diversity and complexity of the electronic structure of the refractory materials used for heterophase ceramics, they exhibit rare combinations of properties inherent in both covalent (high atomization energy, hardness, refractoriness, elastic moduli, chemical stability) and metallic crystals at a relative simplicity of structures, such as those of NaCl-type ionic crystals.

A straightforward calculation of properties of the compounds from the 'first principles' is at present difficult; only qualitative correlations between electronic structure details and, e.g., mechanical properties can be determined. Due to this, interpretation of properties of the refractory compounds from features of their electronic structure has been up to now based on numerous models [1–5]. It is believed that the set of properties of most practically important refractory compounds is determined by the dominating influence of the covalent component in interatomic bonds.

Table 1.2 Physical properties of diamond-structure refractory compounds

Material	Melting temperature T_m (°C)	Vickers hardness HV (GPa)	Young's modulus E (GPa)	Density γ (g/cm^3)
Cubic boron nitride	3900	80.0	800–1000	3.51
Silicon carbide	2830	35.0	500	3.17
Silicon nitride	1900	33.4	280	3.18
Aluminium nitride	2400	12.5	380	3.12

According to the character of the electronic structure, the refractory compounds under consideration are expediently divided into two large groups. The first group consists of diamond-like covalent crystals, such as silicon carbide SiC, silicon nitride Si_3N_4, boron carbide B_4C and boron nitride BN. Covalent refractory compounds with a diamond-like lattice (including also diamond) are characterized by a spatial tetrahedral distribution of interatomic bonds formed by overlap of sp^3 orbitals of atoms. These crystals exhibit exceedingly high elastic properties (Young's modulus $E = 400–500$ GPa) and a high hardness, approaching in some cases that of diamond ($HV \approx 80$ GPa for BN), in combination with a low density ($\gamma \approx 3$ g/cm^3) (Table 1.2).

The second, more numerous, group of the refractory compounds consists of interstitial solutions formed by light elements with a small atomic radius and transition metals. This results in formation of carbides, nitrides, oxides, borides and other compounds. As indicated by the analysis conducted in [5], interatomic bonds in these compounds exhibit a complete filling of the outer electronic shell of the nonmetal with d, s electrons of the metal up to the s^2p^6 configurations, the overlap of p orbitals of metal and nonmetal ions in the shortest directions between the metal and nonmetal giving rise to a strong covalent component of the interatomic bond. At an excess of valence electrons needed for creation of s^2p^6 configuration, this results in a metallic bond in the metal sublattice which has an electronic conductivity. Melting points of the refractory compounds can exceed those of the transition metals incorporated in the compounds; the maximum values are exhibited by hafnium and tantalum carbides (≈ 4000 °C) [5].

Along with a high hardness, all the refractory compounds, both diamond-like and interstitial solutions, are characterized by the existence of a broad temperature range, from 0 K to $(0.4–0.8)T_m$, where the crystals exhibit a low plasticity and the fracture is of a brittle or quasi-brittle character. A steep drop of the yield strength (and of hardness) with increasing test temperature, the ductile–brittle transition and the existence of crystalline structure imperfections – all results in a rather complex character of the effect exerted by external factors (first of all temperature) on the set of mechanical properties of the refractory compounds. The mechanical properties, mechanism of plastic deformation, fracture and recovery of properties at annealing in a broad temperature range were previously studied for every individual refractory compound but the advance in recent years of physical concepts of the mechanism of deformation and fracture of covalent and partly covalent crystals, in particular of the concept of characteristic deformation with temperature, made it possible to generalize numerous experimental results, to introduce a quantitative estimation of the contribution of the covalent component to the interatomic bond, and to analyse the temperature effect on the behaviour under load of a large group of crystalline materials from common positions [6–10].

1.1.1.1 Features of crystalline structure of refractory compounds

Crystalline structures of diamond-like refractory compounds based on principles of different sequences of packing of identical layers are similar.

Their characteristic feature is the existence of numerous polytypes with high values of periods along the c-axis, which have also been found for diamond and BN.

Refractory compounds based on transition metals (interstitial phases) crystallize into a wide variety of crystalline structures, most often into cubic lattice structures of NaCl, CaF_2, ZnS and CaB_6 types as well as hexagonal and tetragonal lattice ones.

Compositions of the compounds correspond approximately to formulas MeX, Me_2X, Me_4X, MeX_2 (where X is a nonmetal), but most of the interstitial phases are substances of a varying composition with broad homogeneity regions. The nature of the broad homogeneity regions was discussed in [11] within the framework of notions of stability of electronic configurations [12]. The existence of broad homogeneity regions is accompanied by appearance of structural vacancies in the nonmetal sublattice with formation of ordered structures. Some possible types of ordered interstitial structures for f.c.c. and b.c.c. Bravais lattices as well as more complex crystalline structures of carbides are presented in [2, 13, 14].

Dislocation structures of covalent crystals and interstitial solutions with diamond, wurtzite and sphalerite lattices as well as of compounds

Table 1.3 Slip systems of some refractory compounds

Compound	Structure type	Slip system	Comment
Diamond	Diamond	$\{111\} \langle 110 \rangle$	
BN, SiC (3C)	Sphalerite	$\{111\} \langle 110 \rangle$	
BN, SiC (2H)	Wurtzite	$\{0001\} \langle 11\bar{2}0 \rangle$	
		$\{10\bar{1}0\} \langle 11\bar{2}0 \rangle$	
TiC	NaCl	$\{111\} \langle 110 \rangle$	$T = 1100\text{--}1900$ K
ZrC	NaCl	$\{110\} \langle 110 \rangle$	
		$\{111\} \langle 110 \rangle$	$T = 1350\text{--}2300$ K
		$\{100\} \langle 110 \rangle$	
TaCl	NaCl	$\{111\} \langle 110 \rangle$	
		$\{001\} \langle 110 \rangle$	$T > 1500$ K
		$\{001\} \langle 010 \rangle$	
		$\{011\} \langle 010 \rangle$	
MgO	NaCl	$\{110\} \langle 110 \rangle$	
		$\{100\} \langle 110 \rangle$	
UO_2	CaF_2	$\{100\} \langle 110 \rangle$	
		$\{110\} \langle 110 \rangle$	
WC	WC (hexagonal syngeny)	$\{1\bar{1}00\} \langle 11\bar{2}0 \rangle$	
		$\{1\bar{1}00\} \langle 0001 \rangle$	
		$\{01\bar{1}0\} \langle \bar{2}113 \rangle$	
TiB_2	AlB_2 (hexagonal syngeny)	$\{0001\} \langle 11\bar{2}0 \rangle$	
		$\{10\bar{1}0\} \langle 11\bar{2}0 \rangle$	

Structural materials

with structures of NaCl, CaF_2, etc., have been studied to a considerable extent and generalized by now. Features of the structure of dislocations and of their slip with allowance for the polyatomicity of materials, directionality of bonds, and requirement for electroneutrality are discussed, e.g., in [4]. Some characteristics of dislocations and slip systems of the refractory compounds are presented in Table 1.3.

Note that a considerable body of recent data evidences a substantial splitting of dislocations with formation of stacking faults in covalent crystals [15]; processes of phase transition, associated with an ordered arrangement of stacking faults, with formation of multilayered polytypes are observed at plastic deformation, particularly at a concentrated loading [16].

Polycrystals of heterophase ceramics exhibit a broad set of structural parameters (number of various types of structural components; morphology and size of grains; composition, structural state and thickness of intergranular interlayers of foreign phases; porosity characteristics; texture; characteristics of fracture of reinforcing elements, etc.). The structure of internal interfaces and the existence of amorphous grain-boundary interlayers (Figure 1.1), which often determine the set of mechanical properties of a ceramic material over a broad temperature range, is of particular importance.

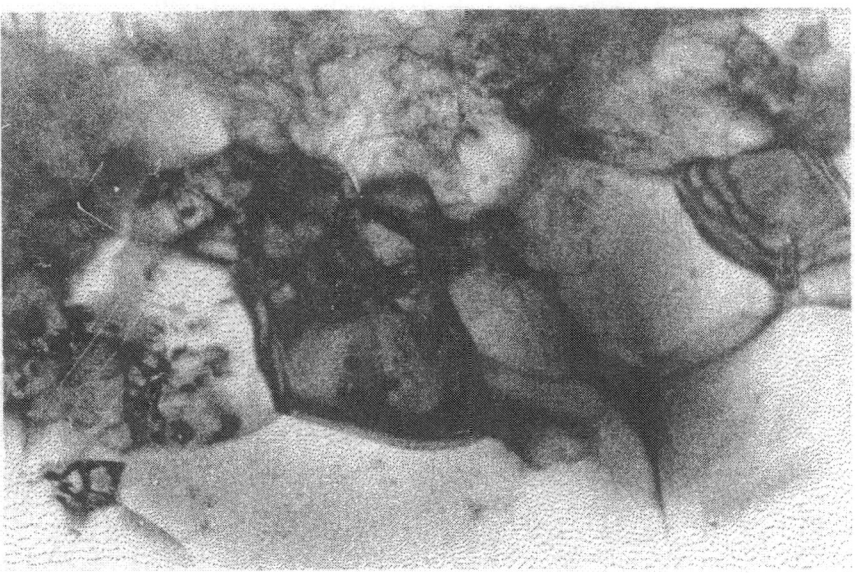

Figure 1.1 Transmission electron micrograph of Si_3N_4–2%Al_2O_3–5%Y_2O_3 ceramic having amorphous grain-boundary interlayers, × 20 000.

1.1.1.2 Physical concepts of temperature dependence of yield strength

Crystals of refractory compounds exhibit the highest values of theoretical tensile (σ_{max}) and shear (τ_{max}) strength; they are the highest for diamond (σ_{max} = 210 GPa; τ_{max} = 125 GPa [17]). Both σ_{max} and τ_{max} decline with decreasing covalent component of the interatomic bond as well as with increasing temperature; τ_{max} declines more rapidly, so that the σ_{max}/τ_{max} ratio grows, which, according to the cold brittleness criterion [18], reduces the brittleness. At high temperatures the crystals are plastic enough.

Based on theoretical and experimental studies [19–29] within the scope of a general thermoactivation analysis [30–33], basic regularities of variation of mechanical properties of the materials with temperature have been formulated in [9, 34, 35, 36], where several different approaches have been presented.

Proceeding from the Arrhenius equation for the strain rate $\dot{\varepsilon}$, Conrad with co-workers [37–42] derived the expression

$$\dot{\varepsilon} = v\exp\left(\frac{-(H - V\sigma)}{kT}\right), \quad v = v_0 N \exp(S/k), \quad (1.1)$$

where v_0 is the frequency factor, N the number of elements being activated, S the activation entropy, H the activation enthalpy, often called activation energy, $\sigma = \sigma_a - \sigma_i$ the effective stress in the existing slip system, the difference between the applied stress σ_a and the resistance to the movement of dislocations from long-range internal fields of stresses σ_i, and V the activation volume.

However, this equation, especially at low $\dot{\varepsilon}$ and σ, is too rough an approximation, and therefore it was suggested in [6, 9, 43] to consider the rate of strain in the direction of action of external forces as the result of the difference between forward and backward thermoactivated dislocation jumps. Using this approach, the authors of [9] derived the following expression describing the temperature dependence of the yield strength σ_s corresponding to some small plastic strain rate $\dot{\varepsilon}_p^*$, one and the same for all temperatures:

$$\sigma_s = C\dot{\varepsilon}\frac{1}{(m + n + 1)}\exp\left(-\frac{H}{kT}\right), \quad (1.2)$$

where C is a constant of the material; $\dot{\varepsilon}$ is the strain rate; H is the experimental (apparent) activation energy; m and n are exponents in the power dependence of the velocity V of individual dislocations and of their density N on stresses:

$$V = V_0(\sigma/\sigma_0)^m \exp(-U/kT), \quad V \sim \sigma^m$$
$$N = (N_0 + N_1\varepsilon^p)(\sigma/\sigma_0)^n, \quad N \sim \sigma^n.$$

Structural materials

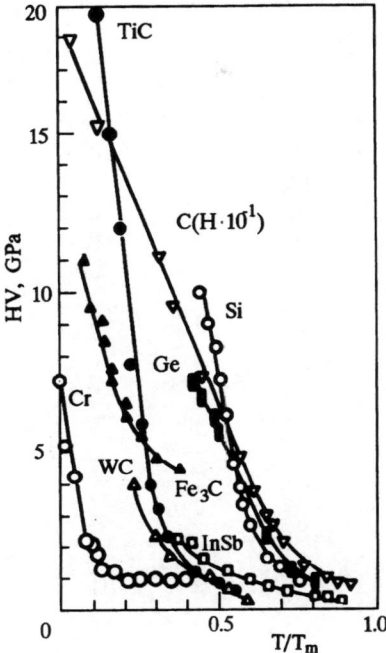

Figure 1.2 Temperature dependence of hardness for some covalent crystals.

The dislocation movement activation energy (U) and the apparent activation energy (H) are related as follows:

$$H = U/(m + n + 1).$$

Experimental data and calculations [6, 9] indicate that at high temperatures $m + n + 1 = 3$, the high temperature range being determined by the condition $\sinh[V\sigma_s/(kT)] \approx V\sigma_s/(kT)$. For low temperatures, where $\sinh[V\sigma_s/(kT)] \approx 0.5 \exp[V\sigma_s/(kT)]$,

$$\sigma_s = (U - kT \ln M/\dot{\varepsilon})/V, \tag{1.3}$$

where M is a quantity little dependent on temperature.

Experimental data (Figure 1.2) is in a good agreement with expressions (1.2) and (1.3), since the dependence $\sigma_s(T)$ at low temperatures is linear and at higher temperatures exponential. Typical values of the activation energy U for a number of compounds are presented in Table 1.4.

The theory of movement of dislocations in crystals with high Peierls barrier was further refined and elaborated in greater detail by Petukhov and Pokrovsky [44] and Seeger [45].

Table 1.4 Activation energy U, parameters α, t^* and primary recrystallization temperature t_r for covalent crystals

Material	U (eV)	T_m (K)	α (rel. units)	t^*	T^* (K)	t_r (rel. units)
C (diamond)	2.60	2000	15.0	0.85	1700	–
Ge	1.60	1210	15.3	0.83	990	0.89
Si	2.20	1680	15.1	0.82	1370	0.84
InSb	0.70	800	10.1	0.67	530	–
Al_2O_3	1.90	2300	9.7	0.65	1320	–
ZrC	3.12	3800	9.5	0.65	2455	–
TiC	2.55	3520	8.4	0.61	2160	0.51–0.60
WC	1.50	3160	5.49	0.49	1770	0.52–0.55
NbC	1.80	3670	5.7	0.48	1770	0.53
Fe_3C	0.78	1830	4.9	0.46	840	–
TiB_2	1.13	3220	4.06	0.42	1340	0.51–0.53
HfB_2	0.80	3200	2.9	0.35	1130	–
ZrB_2	0.60	3310	2.09	0.29	970	0.47
Be	0.40	1560	2.98	0.35	550	–
Cr	0.20	2200	1.05	0.21	440	0.3–0.4
Mo	0.19	2880	0.77	0.17	490	0.3–0.4
W	0.49	3670	0.54	0.25	910	0.3–0.4
V	0.18	2225	0.94	0.19	420	0.3–0.4
Nb	0.24	2740	0.85	0.18	490	0.3–0.4
Ta	0.30	3270	1.06	0.20	650	0.3–0.4

The most important parameter for estimating the effect of temperature on mechanical properties of refractory compounds is the characteristic deformation temperature t^* [6, 8]. It is defined as the temperature at which the resistance exerted by the crystal lattice to the movement of dislocations becomes substantial and brings about a steep rise of the flow stress when the temperature is lowered below t^*. In some crystals (e.g., in metals with a close-packed lattice) the Peierls–Nabarro stress exerts no substantial effect on the mobility of dislocations over the whole temperature range above 0 K. These crystals exhibit no steep rise of the flow stress with lowering temperature, and it can be conventionally regarded that for them the whole temperature range is above the characteristic deformation temperature.

For refractory compounds, however, as also for other covalent and partly covalent crystals, the characteristic deformation temperature is substantially above 0 K and determines to a great extent the behaviour of the crystal in deformation and annealing.

The temperature t^* can be determined from curves shown in Figure 1.2 or from the dependence $\sigma_s(T)$ as the temperature corresponding to a steep rise of σ_s or HV. It can also be found from the relation [6, 8]

$$t^* = T^*/T_\mathrm{m} \approx 0.22\,\alpha^{1/2},$$

where $\alpha = U/(kT_\mathrm{m})$; T_m is the melting temperature (K).

The kT_m value may be considered as a characteristic of the interatomic bond force, while U is the height of the potential barrier to be overcome by dislocations in the course of their thermally activated movement. It follows that $U/(kT_\mathrm{m})$ characterizes the rigidity of the crystal lattice with respect to the movement of dislocations. These considerations and the analysis conducted in [6, 8] allow α to be treated as the measure of contribution of the covalent component to the interatomic bond. Since α determines the t^* value, the same applies also to the characteristic deformation temperature t^*. The highest values, $\alpha \approx 15$, $t^* \approx 0.85$, are exhibited by crystals with a purely covalent interatomic bond (see Table 1.4). For carbides of transition metals, $t^* = 0.5$–0.7, $\alpha = 2$–3.5.

The parameters α and t^* for carbides of transition metals of group IVA (ZrC, TiC) are higher than for NbC and WC, which agrees with modern concepts of a greater covalent component in the former carbide group.

The parameters α and t^* for borides of transition metals are somewhat lower than for carbides (see Table 1.4). For b.c.c. metals the parameters are the minimum in the crystal group under consideration where the interatomic bond has a covalent component.

Above t^*, practically all high-purity covalent crystals can be plastically deformed with high reductions and their fracture is usually ductile, whereas below this temperature there commonly occurs a quasi-brittle or brittle cleavage fracture. An easy formation of a cellular or fragmented dislocation structure is observed only above t^*, whereas formation of randomly oriented dislocation structures is typical below this temperature. Conditions for the recovery of mechanical properties also change at this temperature.

The temperature t^* is not the cold brittleness temperature T_b, but the possibility of brittle fracture appears below t^* for crystals which are plastic above t^*. In other words, the condition $T_\mathrm{b} \leq t^*$ is always satisfied.

The use of the characteristic deformation temperature as a parameter made it possible to generalize the concepts of the hot, warm and cold deformation, known for metals, to covalent crystals, refractory compounds [8, 46, 48] (Figure 1.3). The character of deformation is here estimated from structural features: characteristics of the granular structure, dislocation substructure and also the internal stress level.

A hot deformation is to be carried out above the material recrystallization temperature t_r, from which one can obtain an equiaxial granular structure, free from dislocations and internal stresses. The recrystallization can here proceed both directly in the course of the shape change (dynamic recrystallization) and in the course of the material reheating

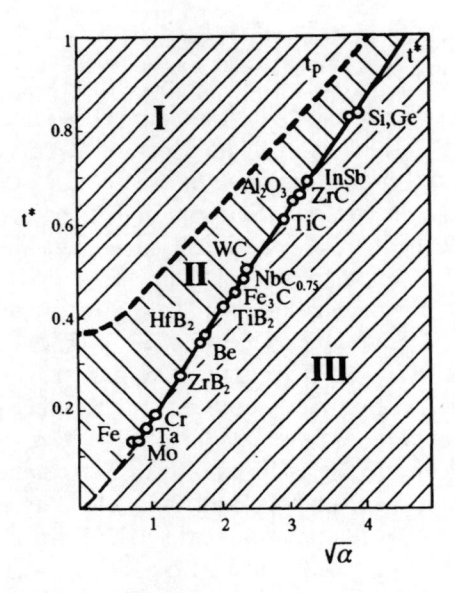

Figure 1.3 Hot (I), warm (II) and cold (III) deformation regions for covalent and partly covalent crystals.

between reductions (static recrystallization). The temperature of recrystallization in the course of deformation turns out to be much higher than under static conditions at the material annealing after a cold or a warm deformation. Since $t_r > t^*$, even under static conditions, the hot deformation range (region I in Figure 1.3) is the smaller, the greater the covalent component in the interatomic bond, i.e., the higher the parameters α and t^*. As shown in [8, 46], the temperature t^* is a natural boundary between regions of a warm deformation (II in Figure 1.3) and a cold deformation (III in Figure 1.3). A warm deformation results in an unequiaxial granular structure in the crystal, and a cellular or fragmented dislocation structure within grains. The level of internal stresses and strain hardening are in this case higher, but fracture of high-purity single crystals is generally still of a ductile character. After a cold deformation, as after a warm one, the granular structure is of an unequiaxial character, but the formation of a cellular structure is impeded and, even after much greater reductions (if allowed by the plasticity), a randomly distributed structure of dislocations is formed. The internal stress level is, in this case, exceedingly high.

The concept of the characteristic deformation temperature and summarization of available literature data made it possible to suggest a generalized scheme of variation of mechanical properties of covalent and partly covalent crystals with temperature. Such a scheme for carbides of transition metals is shown in Figure 1.4. For other refractory compounds

Structural materials

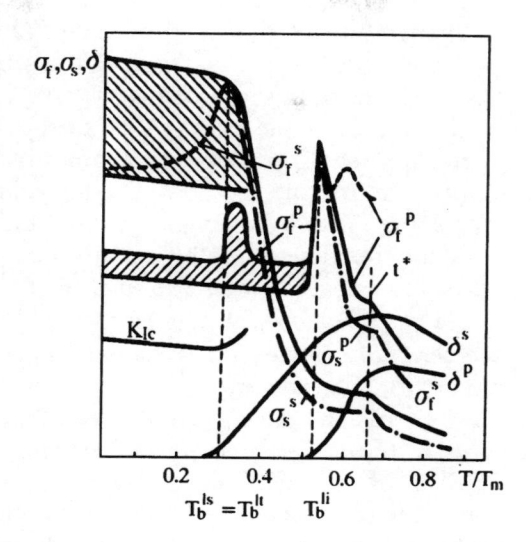

Figure 1.4 Generalized scheme of temperature dependence of mechanical properties for transition metal carbides: σ_s and σ_f – yield and fracture stress respectively; δ – plasticity before fracture; T_b – cold brittleness temperature; the superscripts ^{ls}, ^{lt} and ^{li} refer to lowest stress, lowest transcrytalline and lowest intercrystalline respectively; m and p – denote mono- and polycrystals respectively.

the scheme remains essentially unchanged, but its basic reference points (temperatures of cold brittleness of crystals in various structural states and t^*) shift along the temperature scale.

As seen from Figure 1.4, at low temperatures, in the brittle fracture region, the strength of perfect monocrystals is higher than that of polycrystals. The fracture stress σ_f can vary substantially with the structural state and chemical composition; this is represented in Figure 1.4 by σ_f scatter regions. According to the established concepts, it is shown in Figure 1.4 that the cold brittleness temperature T_b of a polycrystal depends on the fracture character (transcrystalline, T_b^t, or intercrystalline, T_b^i), $T_b^i > T_b^t$, whereas for monocrystals $T_b^m \approx T_b^t$. In the low-temperature region of cold brittleness, there ordinarily occurs a purely brittle fracture, for which a linear decrease in the fracture toughness of material with increasing temperature is typical [48]. The effective fracture energy γ_{ef} turns out here to be close to the theoretical estimate, $2\gamma_0$, which agrees with concepts of a brittle character of fracture.

Monocrystals above the temperature $T_b^m \approx 0.3 T_m$ generally exhibit a steady strength decrease, stemming from a decline in the resistance to plastic deformation and decrease in the flow stress in compliance with expressions (1.2) and (1.3), linear at low temperatures and exponential at higher ones. At a temperature of $(0.6-0.7)T_m$, the $\sigma_f^m(T)$ and $\sigma_s^m(T)$ curves often have a knee, which is due to a change – increasing role of diffusion processes – in the mechanism of deformation.

The temperature dependence of the polycrystal fracture stress σ_f^p has as a rule peaks [47, 50]. A steep σ_f^p rise and the highest peak occur in the region of ductile–brittle transition, i.e., near the temperature T_b^i (see Figure 1.4). The steep σ_f^p rise with increasing temperature stems here from the increase of the fracture toughness and the resulting decline in the role of existing defects as stress concentrators. The further σ_f^p decrease with increasing temperature is associated with the flow stress decline.

At a further temperature increase, the fracture is preceded by a macroscopic plastic deformation so that $\sigma_f^p = \sigma_s^p + \Delta\sigma$, where σ_s^p is the flow stress and $\Delta\sigma$ the strain hardening. As the temperature rises, σ_s^p declines while $\Delta\sigma$ grows. In most cases σ_s^p decreases more rapidly than $\Delta\sigma$ grows, and therefore σ_f^p declines steadily with increasing temperature. If, however, plasticity in the crystal grows steeply until the fracture, then a steep σ rise with temperature can occur, outstripping the σ_s^p decrease; this case is represented by another maximum of the $\sigma_f^p(T)$ curve (indicated by a broken line in Figure 1.4). Such a case was observed, e.g., for UO_2 in [49]. Still another peak of the $\sigma_f(T)$ curve for polycrystals can occur at a temperature $T \approx 0.3T_m$, i.e., at the temperature T_b^t. As shown in [47] for ZrC as an example, this peak occurs if the fracture is transcrystalline, but is absent at an intercrystalline fracture. Figure 1.4 also shows conventionally the variation of plasticity δ in mono- and polycrystalline states. The plasticity appears at the corresponding cold brittleness temperature, increases with temperature, and then, in the ductile fracture region, can stabilize or even decline.

Let us also present as an example the results of studying the temperature-rate dependence of the ultimate bending strength of diamond [51]. The yield stress σ_s (critical shear stress τ) was calculated from the load P_s at the point of loading diagram departure from linearity, which allowed the use of the relation $\dot{\varepsilon} \approx \dot{\sigma}$, valid at the elastic loading stage, for determining the experimental (apparent) values of the activation energy $H = -kT^2(\partial\tau/\partial T)_{\dot{\varepsilon}}/(\partial\tau/\partial\ln\dot{\varepsilon})_T$ and activation volume $V = kT/(\partial\tau/\partial\ln\dot{\varepsilon})_T$, where k is the Boltzmann constant, T the temperature (K), $\dot{\varepsilon}$ the plastic strain rate and $\tau = 0.408\sigma_s$ (0.408 is the Schmidt factor for the loading scheme used).

The measured values of the activation energy, 2.6–0.5 eV, and activation volume (Figure 1.5) show a good accord with the data obtained from the temperature dependence of hardness (Table 1.4).

The temperature dependence of strength of diamond is shown in Figure 1.5. More than 20 specimens were tested at room temperature over the indicated loading-rate range. At high temperatures, the points represent the data averaged over 6–8 specimens in determining the

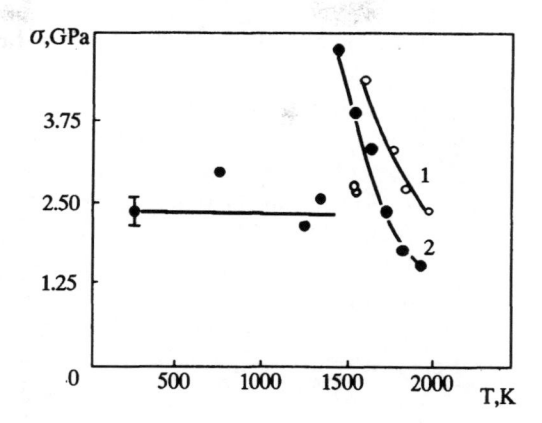

Figure 1.5 Temperature dependence of fracture stress ($T < T_b$) and yield stress ($T > T_b$) of diamond at loading rates of (1) 2.5 MPa/s and (2) 25 MPa/s.

$\tau = f(\dot{\tau})$ dependence for every temperature. In accordance with the general scheme (Figure 1.4), the temperature dependence of strength exhibits a steep fracture stress rise at the cold brittleness temperature T_b^1, the latter varying significantly with the loading rate. Such a behaviour is typical of materials with a high loading-rate sensitivity of the yield stress, which is quantitatively expressed by the parameter m^* in the equation $\dot{\varepsilon} = \text{const } \sigma^{m^*}$ and is higher, the lower the m^* value. Diamond featured low m^* values over the studied temperature range (Figure 1.6), which apparently determined a high cold brittleness temperature detected at the bending test by applying a constant load to diamond plates [6]. In the present study, diamond also fractured at a linear part of the loading

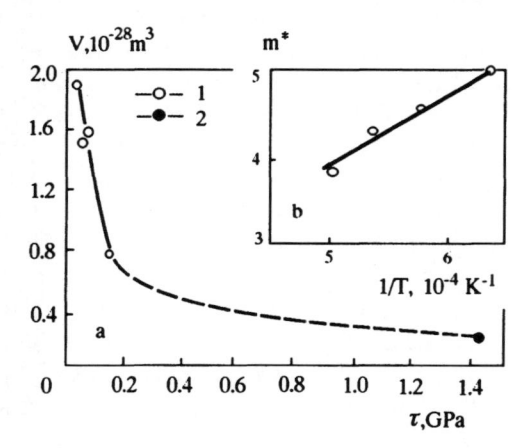

Figure 1.6 Diamond plastic deformation process parameters: (a) dependence of activation volume V on critical shear stress τ according to (1) bending and (2) microindentation test data; (b) temperature dependence of $m^* = \Delta \ln \dot{\varepsilon}/\Delta \ln \tau$.

curve without a residual plastic deformation at a temperature of 1900 K when the loading rate exceeded 250 MPa/s; when the loading rate had been reduced by two orders of magnitude, diamond specimens were bent by 4° at 1600 K and by about 30° at 2100 K without fracture. The analysis of the loading-rate dependence of T_b^1 within the scope of the A. F. Ioffe's classical scheme indicates that at the observed character of the $m^*(T)$ dependence the value of $T_b^1 \approx 1600$ K at $\sigma = 2.5$ MPa/s is close to the lower limit of the cold brittleness temperature. It is characteristic that this value turns out to be close to the T_b^1 value obtained from the temperature dependence of the number of cracks at the hardness indentation [52].

The high strain-rate sensitivity of the yield stress governs also the diamond fracture character: as in [6], diamond specimens fractured by cleavage over the whole temperature range even after a considerable preceding deformation. As indicated by an analysis [6], this phenomenon stems from the fact that at a low m^* value, there occurs a very high sensitivity of mechanical properties to stress concentrators. At the crack tip, where the local stress is much higher, with low m^* the rate of plastic strain ε is little increased, and the stress relaxation has no time to proceed as the stress grows up to the critical value corresponding to the brittle fracture stress.

High values of the activation energy, obtained in [53], may result from the procedure of its determination; activation energy values close to those presented in Table 1.4 can be obtained from the same experimental data with allowance for the quantity $(m + n + 1)$ in equation (1.2).

It should be pointed out that if diamond or a refractory compound is in a metastable state, then the temperature of a high-rate phase transition to a stable state can, from the standpoint of mechanical properties, play the role of the melting temperature. This brings about an intense weakening of metastable structures with increasing temperatures and restricts the possibilities of their use at high temperatures.

It should also be noted that, because of a low value of the parameter m^*, which characterizes the strain-rate dependence on the stress, diamond, as well as other covalent crystals, exhibits a high sensitivity of critical shear stress to the strain rate. In processes of fracture, this leads to a brittle extension of cracks after a plastic flow and to a strong dependence of the brittle–ductile transition temperature T_b on the strain rate, which substantially restricts the temperature region and the limiting rate of a plastic deformation of refractory compounds with low m^* values; a low m^* value of said crystals is directly determined by a low value of the activation volume V of the plastic deformation process [6] and in the final analysis by a sharp directionality and rigidity of interatomic bonds.

Studying the temperature dependence of mechanical properties of complex multiphase materials is of special interest. Mechanical properties of a self-bonded silicon carbide were studied in [54]. This material has a continuous carbide skeleton of primary silicon carbide, surrounded by an envelope of secondary silicon carbide that forms directly during the preparation of the material as a result of the silicon–carbon interaction. Apart from the carbide phase, the material contains 10–20% free silicon in the form of inclusions dispersed in the carbide skeleton. The study [54] disclosed a fairly complex character of the temperature dependence of the fracture stress, $\sigma_f(T)$: a maximum of σ_f within 600–800 °C, its decrease at 900–1000 °C and a further steep drop at temperatures over 1400 °C.

This $\sigma_f(T)$ character was explained on the basis of temperature dependences of mechanical properties of two constituent phases, SiC and Si. It was shown that the account of their features makes the explanation possible. Thus, the maximum strength within 600–800 °C is associated with an increase in the silicon fracture toughness; the strength decline at 900–1000 °C results from a decrease of the flow stress (and strength) as well as an anomalous microbrittleness of silicon carbide within this temperature range; and a further steep drop of the strength above 1400 °C stems from reaching the melting temperature of the silicon phase. In the general case, the temperature effect on mechanical properties of multiphase systems of covalent crystals can be explained on the basis of the temperature dependence of mechanical properties of each constituent phase; the mechanical properties of every phase can be here described by the generalized scheme of Figure 1.4 with reference points t^* and T_b for this phase.

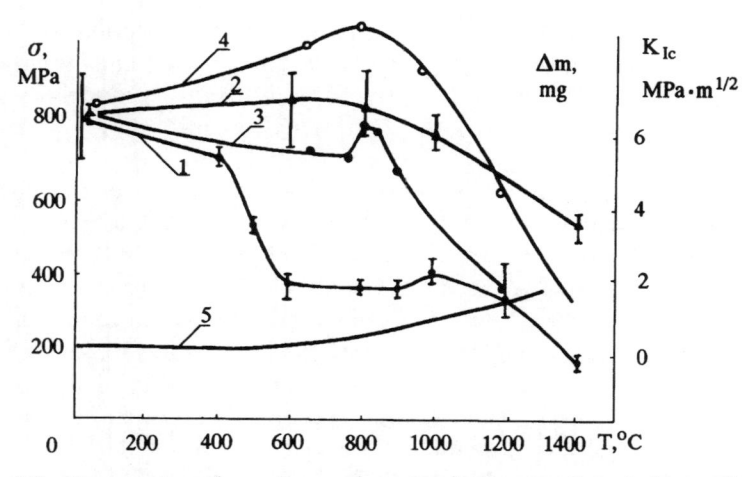

Figure 1.7 Temperature dependence of mechanical properties of silicon nitride-based materials Si_3N_4–Y_2O_3–Al_2O_3 (1, 2, 3, 5) and Si_3N_4–Y_2O_3–Al_2O_3–TiN (4) at tests in air (1, 4, 5) and in vacuum (2, 3); strength (1, 3, 4); crack resistance (2); and thermogravimetric curve (5).

To conclude, let us note that in the high-temperature region the temperature dependences of strength of heterophase ceramics containing glass-phase interlayers at grain boundaries depend very little on properties of the crystalline constituents and can be fully determined by processes of the grain-boundary slip on the interlayers. Besides, the character of temperature dependences of mechanical properties is significantly affected by processes of chemical interaction of the material with the environment. As an example, Figure 1.7 shows these dependences for hot-pressed silicon-nitride-based materials [55]. As shown by the electron microscope and X-ray studies, for all the examined materials and test conditions, there occurred a grain-boundary slip, while effects of plastic deformation of Si_3N_4 grains themselves were not detected. The weakening of the material in the moderate temperature region (≈ 500 °C) resulted from the conditions of tests in the air with oxidation; it can be suppressed by an appropriate selection of the composition of grain-boundary phases and introduction of additional crystalline phases (TiN, serving as a precipitation-hardening and refining additive).

1.1.1.3 Criteria of fracture of ceramic-matrix composites

The analysis of dependence of properties of heterophase composite ceramic materials on their composition indicates that the change-over from single-phase to heterophase states is generally accompanied by a substantial improvement of the set of their mechanical and service properties. Very important for the insight into the nature of formation of the properties are mechanico-mathematical models. The most important characteristic of structural ceramics is the fracture toughness, which is greatly dependent on the microstructure, composition and manufacturing technology of structural ceramics, and therefore models of their fracture toughness, which represent this dependence, are of prime importance.

One such model [56] is discussed below. It generalizes the classical fracture mechanics' force fracture criterion to the case of macroscopically homogeneous isotropic and linearly elastic heterophase materials with a periodic microstructure. A material with a periodic structure implies a macroscopic volume (sample) of a material, composed of a periodically repeating element (cell) εY[57]. The microstructure means that $\delta L^{-1} \ll 1$, where δ is the characteristic dimension of the periodic cell εY and L is the characteristic macroscopic dimension of the material sample. This makes it possible to treat the sample material as a continuum and to use asymptotic methods of averaging [57] for deriving its effective characteristics as well as local stress and strain fields. The theory of such methods, developed by now, makes it possible to trace the dependence of macroproperties of the medium on its microproperties.

Structural materials

As known [57], macroscopic properties of a composite material sample are described by the following averaged equation of state

$$\tilde{\sigma}_{ij}^0 = a_{ijkh}^h \tilde{e}_{kh}^0, \tag{1.4}$$

where the fourth-rank tensor

$$a_{ijkh}^h = \{a_{ijlm}[\delta_{lk}\delta_{mh} + e_{lmy}(W^{kh})]\}^{\sim}$$

is the tensor of effective elastic moduli. The designations and phase boundary conditions adopted here and hereinafter are the same as in [57]. The symbol ~ denotes 'operator of the mean value'

$$\tilde{\varphi} = |Y|^{-1} \int_Y \varphi(y) \, dy, \tag{1.5}$$

which puts in correspondence to any Y-periodic function $\varphi(Y)$ its mean value $\tilde{\varphi}$ ($|Y|$ is the volume of the parallelepiped (cell) Y with edges y_i^0 [57]. Functions $a_{ijlm}(x\varepsilon^{-1})$, which determine elastic moduli of phases, are εY-periodic in x (or Y-periodic in $y = x\varepsilon^{-1}$), where the period is a parallelepiped with edges εy_i^0.

Along with the operator (1.5) we introduce an 's-operator of the mean value' of the formula

$$\overset{s}{\tilde{\varphi}} = |Y|^{-1} \int_Y \varphi(y) \eta_s(y) \, dy,$$

where $\eta_s(y)$ is the characteristic function of the set $Y_s \subset Y$, i.e., a function equal to 1 at $y \in Y_s$ and equal to 0 at the addition of Y_s to Y. The region Y_s is that part of the cell Y which is occupied by the sth constituent of the composite. It is obvious that

$$\tilde{\varphi} = \sum_{s=1}^N \overset{s}{\tilde{\varphi}}, \qquad Y = \sum_{s=1}^N Y_s,$$

where N is the total number of constituents of the composite. In view of this, the tensor a_{ijkh}^h and mean stresses $\tilde{\sigma}_{ij}^0$ can be written as sums

$$a_{ijkh}^h = \sum_{s=1}^N a_{ijkh}^{hs}; \quad \tilde{\sigma}_{ij}^0 = \sum_{s=1}^N \tilde{\varphi}_{ij}^s;$$

$$a_{ijkh}^{hs} = \{a_{ijlm}[\delta_{lk}\delta_{mh} + e_{lmy}(W^{kh})]\}^{\overset{s}{\sim}};$$

$$\tilde{\varphi}_{ij}^s = a_{ijkh}^{hs} \tilde{e}_{kh}^0.$$

In the model discussed below, it is assumed that $\forall s$ the tensors a_{ijkh}^{hs} are presented in the following form:

$$a_{ijkh}^{hs} = \frac{2\tilde{G}_s \tilde{v}_s}{1 - 2\tilde{v}_s} \delta_{ij}\delta_{kh} + \tilde{G}_s(\delta_{ik}\delta_{jh} + \delta_{ih}\delta_{jk}).$$

Coefficients $\tilde{G}_s$ and $\tilde{v}_s$ are respectively called the effective averaged shear modulus and Poisson's ratio of the sth constituent of the composite. The respective actual coefficients are denoted by G_s, v_s. Then the following formulae are valid for the effective shear modulus $\tilde{G}$ and Poisson's ratio $\tilde{v}$ of the composite:

$$\tilde{G} = \sum_{s=1}^{N} \tilde{G}_s; \quad \sum_{s=1}^{N} \frac{2\tilde{G}_s \tilde{v}_s}{1 - 2\tilde{v}_s} = \frac{2\tilde{G}\tilde{v}}{1 - 2\tilde{v}},$$

and mean stresses $\tilde{\sigma}_{ij}^s$ in the sth constituent are written as

$$\tilde{\sigma}_{ij}^s = \frac{\tilde{G}_s}{\tilde{G}} (\tilde{\sigma}_{ij}^0 + \beta_s \tilde{\sigma}_{ll}^0 \delta_{ij}); \quad \beta_s = \frac{\tilde{v}_s(1 - 2\tilde{v})}{(1 - 2\tilde{v}_s)(1 + \tilde{v})} - \frac{\tilde{v}}{1 + \tilde{v}}.$$

One should distinguish here stresses $\tilde{\sigma}_{ij}^s$ from actual mean stresses in the sth constituent, which are equal to $|Y||Y_s|^{-1}\tilde{\sigma}_{ij}^s$ ($|Y_s|$ is the volume of the region Y_s).

Thus, it is assumed that a macroscopically linearly elastic homogeneous and isotropic composite material is an additive mixture of homogeneous and isotropic elastic materials with elastic moduli tensors a_{ijkh}^{hs}. The mathematical model of the composite material is a continuum whose every point contains all the constituents of the composite. Such a model allows the introduction of factors $\tilde{K}_M$ of the intensity of stresses $\tilde{\sigma}_{ij}^0$ of the composite and factors K_M^s of the intensity of stresses $\tilde{\sigma}_{ij}^s$, which are interrelated as follows:

$$K_M^s = \alpha_s \tilde{K}_M; \quad \tilde{K}_M = \sum_{s=1}^{N} K_M^s; \quad \alpha_s = \tilde{G}_s \tilde{G}^{-1} \leq 1 \quad (1.6)$$

$$\left(\sum_{s=1}^{N} \alpha_s = 1; \quad M = \text{I}, \text{ II}, \text{ III} \right).$$

In formulae (1.6), $M = \text{I}$ corresponds to opening mode crack (mode I); $M = \text{II}$, to sliding mode crack (mode II); and $M = \text{III}$, to tearing mode crack (mode III). Based on results of study [57], fast-oscillating stresses σ_{ij}^0 at the crack tip (see Figure 1.8) are presented (with an accuracy to within $0(\varepsilon)$) as

$$\sigma_{ij}^0 = \frac{\tilde{K}_M}{\sqrt{2\pi r}} \{a_{ijlm}[\delta_{lk}\delta_{mh} + e_{lmy}(W^{kh})]\} b_{khrs}^h f_{rs}(\tau),$$

$$r = \sqrt{x_1^2 + x_2^2}, \quad x_1 = r\cos\theta, \quad x_2 = r\sin\theta,$$

where b_{khrs}^h is the tensor of elastic compliances, corresponding to the tensor a_{ijkh}^h.

The above-described model of a composite material allows the introduction of a force criterion of its fracture. It is regarded that the initial heterophase material has fractured if even one of its constituents has fractured. In other words, the fracture also implies a change of the starting material with some initial composition to a heterophase material

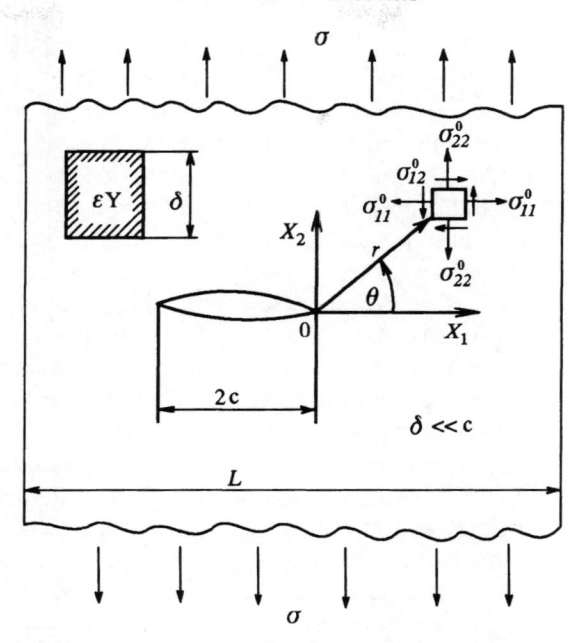

Figure 1.8 Crack that developed in specimen under nominal stresses σ.

with another composition. Because of that, even one constituent (phase) of the starting material has ceased to resist the fracture due to its cracking (fracture). Therefore, if we denote the critical value of K_I^s as K_{Ic}^s (i.e., K_{Ic}^s is the fracture toughness of the material of the sth component, determined either experimentally on specimens of a pure sth component or from reference books containing already available experimental results), it is natural to assume (in accordance with formulae (1.6)) that the fracture of the starting composite will occur when $\tilde{K}_I$ has reached its limiting value $\tilde{K}_{Ic} = \min_s(\alpha_s^{-1} K_{Ic}^s)$, i.e., on condition that

$$\tilde{K}_I = \tilde{K}_{Ic} \equiv \min_s(\alpha_s^{-1} K_{Ic}^s). \tag{1.7}$$

The factor $\tilde{K}_{Ic}$ can be called the fracture toughness of a composite or the toughness of transformation of the starting composite into another one. The value $s = s_c$, corresponding to the minimum in (1.7) ($\tilde{K}_{Ic} = \alpha_{s_c}^{-1} K_{Ic}^{s_c}$), indicates those constituents of the composite from which its fracture or transformation of the starting composite with some initial composition into a composite with another composition begins due to the fact the starting composite's constituent with the $s = s_c$ has ceased to resist the fracture because of its cracking (fracture). It is obvious that at this transformation, the minimum in (1.7) can occur for several s values, which corresponds to a simultaneous cracking of several composite constituents corresponding to the s values. Several such transformations can occur before a complete fracture of the starting composite material; it is proposed to use the criterion (1.7) at each of them.

If there is no complete fracture, then it is more correct to speak of a change to a microcracked material rather than to a material with another composition. When using the criterion (1.7), one has to determine the effective moduli $\tilde{G}$ and $\tilde{G}_s$ for the material with cracks (or with pores if the cracked constituents are treated as pores) at each such transition. This is known [57], however, to be a very difficult task, and the moduli $\tilde{G}$ and $\tilde{G}_s$ sometimes do not exist since averaged equations of states for microcracked (or porous) materials are, generally speaking, nonlinear (due to a one-sided contact of crack edges or pores). Because of this, it is proposed to determine the moduli $\tilde{G}$ and $\tilde{G}_s$ from linear approximations of said nonlinear averaged equations of states.

The approximate approach under consideration to estimating the fracture toughness of a composite, consisting in the determination of the moduli $\tilde{G}$ and $\tilde{G}_s$ as well as the use of the criterion (1.7) at each of the above-mentioned changes or transitions, is regarded to be justified if no substantial qualitative change in the stressed state at the macrocrack tip occurs from transition to transition. It is obvious that the accuracy of the criterion (1.7) declines from transition to transition.

If the fulfilment (even an approximate) of the inequality

$$|\overset{s}{e}_{lmy}(W^{kh})| \ll \delta_{lk}\delta_{mh}|Y_s||Y|^{-1}$$

is possible, then an approximate formula for $\tilde{K}_{Ic}$ results:

$$\tilde{K}_{Ic} = \min_s \left(\frac{\sum_{s=1}^{N} G_s|Y_s|}{G_s|Y_s|} K_{Ic}^s \right).$$

For porous or microcracked materials it can be set in (1.7) that $\alpha_s = 2E_s(\tilde{E} + E_s)^{-1} \geqslant 1$, where $\tilde{E} = E_s(1 - \omega)$ and ω is the measure of the degree of microscopic damage of a defect-free (i.e., free from pores and cracks) material with the Young's modulus E_s.

At a combined deformation the following criterion can be used:

$$\tilde{K}_I^2 + \tilde{K}_{II}^2 + \frac{\tilde{K}_{III}^2}{1 - \tilde{v}} = \tilde{K}_{Ic}^2,$$

where $\tilde{K}_{Ic}$ is determined by formula (1.7).

As is known, in the absence of actions of external forces, materials may be acted upon by the field of their inherent internal (residual) stresses resulting from their manufacturing technologies. We will denote the stresses by σ_{ij}^r and assume them to be Y-periodic in $y = \varepsilon^{-1}x$.

It is obvious that

$$\tilde{\sigma}_{ij}^r = \sum_{s=1}^{N} \overset{s}{\tilde{\sigma}}_{ij}^r = 0.$$

As regards mean stresses, $\overset{s}{\tilde{\sigma}}_{ij}^r$ in the sth constituent of the composite, we will assume them to have the structure $\overset{s}{\tilde{\sigma}}_{ij}^r = \sigma^{rs}\delta_{ij}$, where σ^{rs} is a constant depending on s alone.

The actual mean residual stresses in the sth constituent are equal to $|Y||Y_s|^{-1}\sigma^{rs}\delta_{ij}$.

The following formulas were derived in [56] for the factor $\tilde{K}_I$ of intensity of stresses $\tilde{\sigma}_{ij}^0$ and fracture toughness $\tilde{K}_{Ic}$ for a specimen with a crack (see Figure 1.8):

$$\tilde{K}_I = \alpha_s^{-1} K_I^s - F\sigma^{rs}\sqrt{C}; \quad \tilde{K}_{Ic} = \min_s (\alpha_s^{-1} K_{Ic}^s - F\sigma^{rs}\sqrt{C}); \qquad (1.8)$$

they show that residual stresses can both increase and reduce the fracture toughness. They also indicate a diversity of possible versions of fractures of the material specimen (the dimensionless factor F in (1.8) depends solely on the shape of the specimen with a crack).

The proposed approach to determination of the fracture toughness (formulae 1.7 and 1.8) makes it possible to state and to solve problems of optimizing the composite structure, consisting in maximization of $\tilde{K}_{Ic}$. In particular, the optimum composition of a composite can be found from the criterion:

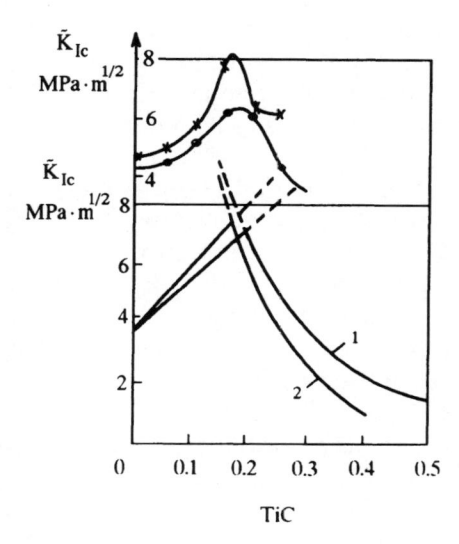

Figure 1.9 Dependence of predicted and measured fracture toughness $\tilde{K}_{Ic}$ in sialon–TiC for: (1) $\sigma^{rs} \neq 0$, $C = 50\,\mu\text{m}$; (2) $\sigma^{rs} \neq 0$, $C = 100\,\mu\text{m}$.

$$\tilde{K}_{\text{Ic}}^{\text{opt}} = \max_{|Y_s|} \min_{s} (\alpha_s^{-1} K_{\text{Ic}}^s - F\sigma^{rs}\sqrt{C}). \tag{1.9}$$

Figure 1.9 shows the $\tilde{K}_{\text{Ic}}$ dependence on $V_2 = |Y_2| \, |Y|^{-1}$ for the two-phase Si_3N_4–TiC composite, calculated from the formula (1.8) at $\alpha_s = G_s |Y_s| \, (\sum_{s=1}^{2} G_s |Y_s|)^{-1}$ (the region Y_1 is occupied by Si_3N_4; $G_1 = 92$ GPa; $G_2 = 197$ GPa; $K_{\text{Ic}}^1 = K_{\text{Ic}}^2 \approx 4$ MPa m$^{1/2}$). The optimum value $\tilde{K}_{\text{Ic}}^{\text{opt}} \approx 7$ MPa m$^{1/2}$ corresponds to $V_2 = 1 - V_1 \approx 0.2$. The figure also shows experimental $\tilde{K}_{\text{Ic}}$ values, which agree fairly well with the calculated ones.

1.1.1.4 Micromechanisms of fracture

A wide variety of fracture micromechanisms operate in refractory compounds, prone to brittle fracture at low temperatures, as the temperature is increased. In accordance with [58–59], basic types of the fracture micromechanisms can be characterized by the fracture toughness and structural character of fracture. Brittle, quasi-brittle, and ductile fracture are distinguished with respect to the fracture toughness; transcrystalline and intercrystalline fracture, with respect to the structural character. These fracture toughness and structural characteristics may occur in any combination, which makes possible six basic fracture types. It is to be pointed out at once that an intercrystalline fracture can occur not only in polycrystals, but also in monocrystals in a case where the crack is extending on low-angle grain boundaries which can form both in the course of monocrystal growth and in the course of plastic deformation preceding the fracture.

If the intercrystalline fracture for monocrystals of refractory compounds is excluded, then transition from a ductile to a quasi-brittle and then to a brittle fracture should occur as the temperature decreases. This change of fracture micromechanisms with decreasing temperature results in a phenomenon that is most important for the materials under consideration: cold brittleness. During measurement of mechanical properties, the cold brittleness shows up as a steep decrease of the macroscopic plasticity (practically down to zero) with decreasing test temperature. When studying the b.c.c. metals, A. F. Ioffe [60] came to the conclusion that the cold brittleness is the consequence of a strong temperature dependence of the flow stress σ_s. For refractory compounds, where the $\sigma_s(T)$ dependence is still stronger and shows up at higher temperatures (corresponding to t^*), the cold brittleness is of the same nature, but manifests itself over a much wider temperature range. When the cold brittleness phenomenon is studied experimentally, the ductile–brittle transition temperature T_b is generally determined. It is expedient to determine the lower and the upper cold brittleness temperature, T_b^l and T_b^u. T_b^l corresponds to the temperature of exhibition of plastic deformation symptoms and can be found from the Ioffe's condition: $\sigma_f(T_b^l) = \sigma_s(T_b^l)$.

The T_b^u value corresponds to the transition to the ductile fracture mechanism and for pure monocrystals is generally about the same as the characteristic deformation temperature or somewhat lower. Values of the lower cold brittleness temperature T_b^l and the position of the temperature t^* are presented in the scheme in Figure 1.4.

Investigation into the fracture micromechanisms in the low-temperature region of brittle fracture is of prime importance for solving the problem of upgrading the fracture toughness of ceramic-matrix composites. Numerous ingenious research efforts dealing with mechanisms of the transformation toughening, microcracking, crack front bending, nonplanar crack propagation between obstacles of various forms, processes of crack propagation in fibre- and whisker-reinforced ceramic composites have been conducted; they are summarized, e.g., in reviews [61–64]. Not dwelling upon a specific description of the above-listed mechanisms which are well enough known, we will only note that their efficiency as to upgrading the fracture toughness depends on the level and character of distribution of the microstress and microstrain fields. Corresponding σ^r or ε^r values are directly used in practically all model calculations of the fracture toughness, and therefore the problem of optimizing the structural state and composition of a composite can be reduced to the problem of optimizing the fields of internal elastic stresses along with optimizing the grain-boundary strength.

In some instances the internal stress fields not only affect the efficiency of a certain mechanism, but also are responsible for a change of fracture mechanism. It is with the high level of elastic internal stresses in heterophase polycrystals based on high-pressure boron nitride phases that the phenomenon of an explosive self-sustaining fracture, detected in the polycrystals, was associated in [65], whereas single-phase polycrystals without nonuniform internal stresses fractured at propagation of solitary cracks opening by a simple cleavage.

Two basic factors that result in retarding the crack formation and upgrading the fracture toughness and strength are generally distinguished in brittle ceramic materials: 'screening' of the crack tip by compressive stresses in the material and interaction of the crack front with structural elements (disperse particles of the second phase, tough interlayers, etc.). Most important for fibre-reinforced ceramic materials is the energy-intensive interaction between the crack and reinforcement, resulting in 'pull-out' of fibres from the matrix. For the crack tip to be screened, irreversible microstructural changes, occurring at a rate not less than the crack propagation velocity, are needed. The screening can be effected by two main mechanisms: through phase transformations with a volume increase and through formation of a microcracked structure of the material. Both calculations and experimental evidence (with the use of the phase transformation in ZrO_2 and HfO_2) show the former mechanism to be more effective. It involves a martensitic-type phase transformation from a tetragonal (T) into a monoclinic (M) modification, proceeding at 1440 K and accompanied by a volume increase, which is of 3–7% depending on the presence of impurities and the state of the material [66]. The T-phase can be stabilized at room temperature by doping ZrO_2 with calcia, magnesia, yttria, ceria or oxides of other lanthanides. Particles of such a partly stabilized zirconia in the form of T-phase are distributed throughout a rigid matrix (Al_2O_3, Y_2O_3, etc.). Tensile stresses arising in the crack tip bring about transition of the particles to the M-phase; this involves absorption of a considerable energy in the 'excitation zone' [67] while great compressive stresses arise in the vicinity of the crack tip and arrest the crack.

The increase of the fracture toughness of a ceramic material in this case can be estimated from the McMicking–Evans formula [68]:

$$\Delta K_{Ic} = 0.3\, E\sigma V_f W^{1/2},$$

where E is the elastic modulus of the matrix, σ the stress arising at the T → M phase transition, V_f the volumetric content of ZrO_2 particles in the matrix, and W the 'excitation zone' width.

It is essential that the mechanism under consideration of upgrading the fracture toughness through phase transformation increases both the fracture toughness and the strength of the material, whereas formation of a microcracked structure at the crack tip only results in growth of the effective energy of fracture and fracture toughness, while the strength may even decline.

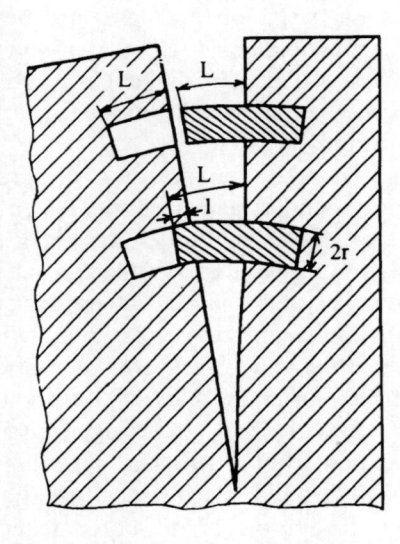

Figure 1.10 Fibre pull-out of rigid matrix. Protruding lengths of fibres and holes remain in fracture surface.

The use of the mechanism of crack retardation through the martensitic transformation of ZrO_2 at an adequate amount of the T-phase and relatively high matrix characteristics yielded materials with a high K_{Ic} level (up to 15 MPa m$^{1/2}$) along with a high strength, which even resulted in the appearance of a term 'ceramic steel'.

In the case of pulling a fibre out of the matrix (Figure 1.10), the force pulling out a fibre is $F = \tau_i 2\pi r l$, where r is the fibre radius, l is the length of the fibre not yet pulled out, and τ_i the frictional resistance at the matrix–fibre interface [69].

Then the work needed to pull N fibres of length L (on one side of the crack plane) out of the matrix is

$$W_b = N \int_0^L \tau_i 2\pi r l \, dl = \pi N \tau_i r l^2.$$

The number of fibres per flat disc-shaped crack of $2c$ in diameter will be

$$N = \frac{V_f c^2}{r^2},$$

where V_f is the volume fraction of fibres in the matrix.

Then the work required to pull out fibres as a function of the crack size is

$$W_b = \pi V_f \tau_i r k^2 c^2,$$

where $k = L/r$ is the geometric factor of the fibre.

Study [69] analysed the balance of energy at the crack extension. It took into account the elastic energy released at the crack growth,

$$W_{el} = -\frac{8(1-v^2)\sigma^2 c^3}{3E},$$

as well as the energy needed to form two new crack surfaces,

$$W_n = \pi c^2 G_c,$$

where G_c is the critical rate of release of the elastic energy of the matrix.

Then the total energy of the material with the crack is $W = W_b + W_{el} + W_n$.

Based on the condition $dW/dc = 0$, first introduced by Griffith, the stress at which the crack growth occurs is found:

$$\sigma_c = \frac{\pi E(G_c + V_f \tau_i r k^2)}{4c(1-v^2)}. \tag{1.10}$$

From (1.10) it is seen that

$$\Delta G_c = V_f \tau_i r k^2, \tag{1.11}$$

i.e., ΔG_c grows with increasing stress τ_i needed to pull a fibre out of the matrix. At the crack extension it is necessary, however, that a fibre be pulled out of the matrix rather than be broken by the crack front. If the fibre strength is σ_f, then the force to break it is $F_f = \sigma_f \pi r^2$. For a fibre to be pulled out, but not broken, it is needed that $F_f > F$, i.e., $\sigma_f r > 2\tau_i l$, or, approximately,

$$\sigma_f > \tau_i k. \tag{1.12}$$

Thus, the friction force τ_i, i.e., the adhesive strength at the interface, should not exceed the value of σ_f/k. From this standpoint it is desirable to use fibres with a high strength σ_f, while the k and τ_i values have to be controlled in accordance with (1.12). The operation of the fibre pull-out mechanism is indicated by the presence of pits and protruding lengths of fibres at the fracture surface (Figure 1.10). Having substituted (1.12) into (1.11), we obtain the maximum possible increment

$$\Delta G_{c\max} = V_f \sigma_f L. \tag{1.13}$$

Let us estimate $\Delta G_{c\max}$ at $V_f = 0.1$, $\sigma_f = 1000$ MPa and $L = 25\,\mu\text{m} = 2.5 \times 10^{-5}$ m; the result is $\Delta G_{c\max} = 0.025$ MPa m $= 2.5 \times 10^4$ J m^{-2}. Neglecting the value of G_c of the starting matrix as compared with $\Delta G_{c\max}$, at $E = 4 \times 10^5$ MPa we find $K_{Ic} = \sqrt{E\Delta G_{Ic\max}} = 100$ MPa m$^{1/2}$, which greatly exceeds the actually attainable values for materials not reinforced with fibres. At the same time an exceedingly high value, $K_{Ic} > 25$ MPa m$^{1/2}$, has been attained for fibre-reinforced materials (SiC–SiC or C–SiC) [70] and, as seen from the above estimate, a further K_{Ic} increase can be expected.

It is to be pointed out, however, that in fibre-reinforced materials the above-discussed mechanism is effective only for cracks whose front is perpendicular to the direction of fibres, whereas in unindirectionally reinforced material, a crack can freely extend parallel to fibres, not crossing them. This results in a strong anisotropy of K_{Ic} and strength of fibre-reinforced ceramic materials.

Note also that the reserve for K_{Ic} increase is to a great extent determined by the fibre strength. When deriving the relations (1.10) and (1.11) it was assumed that a disc-shaped flat crack of radius r has a smooth circular front in the form of circumference. The crack front, however, can get fixed at separate points and continue to move, bending out between them. Thus, some critical density of crack front fixation points, at which the bending-out of the front between the points becomes energetically disadvantageous, is needed. Study [71] analysed the movement of the front of a coin-shaped flat crack of radius c, fixed by disperse particles of the second phase, spaced at distance d from one another. It showed that the bending-out of the crack front between the fixation points increases the critical energy release rate G_c by

$$\Delta G_c = \frac{2c}{3d} G_c. \quad (1.14)$$

This relation can also be applied to the case of crack front bending-out between points of fixation by reinforcing fibres. For such a crack front movement mechanism to become energetically disadvantageous and the fracture process to be governed by the pulling of fibres out of the matrix, it is needed, as seen from (1.14), that $c/d \gg 1$. If the existence of cracks with a radius $c = 10\,\mu m$ in the material is tolerable, then it is needed that $d < 1-2\,\mu m$. Knowing the volume fraction of second-phase fibres and their spacing d, one can determine the fibre radius from the obvious relation

$$r = d\sqrt{V_f/\pi} \quad (1.15)$$

At $V_f = 0.1$ and $d = 1\,\mu m$ we find from (1.15) that $r = 0.18\,\mu m$. At a greater spacing of fibres (and hence at a greater r value) the crack will grow to a size exceeding $10\,\mu m$ before the mechanism of crack front bending-out between reinforcing fibres becomes energetically disadvantageous. In this case, although the crack resistance in testing macroscopic specimens with a long crack front will be high, the strength σ_f will decrease since $\sigma_f \sim c^{-1/2}$.

This example of dissimilar behaviours of the strength and fracture toughness of two-phase materials is not a unique one, but rather, most probable. Thus, a study of materials where the phase transformation in ZrO_2 is employed to increase the fracture toughness demonstrated that a nonmonotonic (along a curve with a maximum) variation of the strength involves a fracture toughness growth. At a high fracture toughness the strength turns out to be substantially below its maximum value and, of course, depends on the defect sizes at which the fracture toughness increases and the crack retardation mechanism becomes effective.

Modern ceramic materials employ, as a rule, a combination of several fracture toughness increase mechanisms, the main of which are reduction of the effective size of a structural element (grain), toughening by phase transformation, precipitation hardening, matrix reinforcement with fibres, initiation of crack branching process, and creation of a microcracked structure.

1.1.2 Technology
G. G. Gnesin

Ceramic-matrix composite material include various structural elements, each of which performs certain functions. A combination of the structural elements, arranged in a definite preprogrammed pattern and at the optimum quantitative ratio, provides the required set of properties and predetermined structural state of a composite is determined first of all by the quality (i.e., composition, structure and properties) of starting components and conditions of their treatment at individual stages of the manufacturing process.

It is to be noted that the specifics of the technology of manufacturing products from composite materials is determined not only by the requirements for their properties and performance characteristics, but also by the configuration and dimensions of the products as well as by the economic efficiency of one or other process.

Thus, the technology of ceramic-matrix composites is a set of various procedures involving dispersion and homogenization of heterophase blends; addition of reinforcing or modifying phases; moulding and structurization of the composite material; and a surface working of products. A strict observation of the specified manufacturing process parameters at every stage provides fabrication of high-quality products from composites.

1.1.2.1 Classification of ceramic-matrix composites

The class of ceramic-matrix composite materials includes polyphase materials containing, along with the primary (matrix) phase, inclusion phases having an equiaxial or an unequiaxial configuration. Equiaxial inclusions can be either of a regular spherical or polyhedral shape or have irregular outlines. Unequiaxial phases are of an acicular, platelet or fibrous form and can also be present as grain-boundary interlayers.

In ceramics, the structural states characteristic of composite materials are created by a statistically uniform or oriented distribution of these constituents; we will conventionally call them 'modifying phases'. Included in these can also be a 'void phase', i.e. porosity, which most substantially affects most physical properties of ceramics.

The character of action and role of the modifying phases in composite formation processes have been fully studied [1, 4]. Intensive studies aimed at determining correlations among the structure, composition and properties of composites ascertained the following basic tendencies in the effect of modifying phases on properties of composites.

1. Uniformly distributed inclusions of stronger equiaxial monocrystal particles in a polycrystalline ceramic matrix tend to increase the hardness, strength and fracture toughness of the material on condition of an optimum ratio between the modifying and the matrix phase [3].
2. Addition of optimum amounts of high-modulus fibres and whiskers to a ceramic matrix tends to increase the dynamic strength, fracture toughness and thermal shock resistance of the composite [4].
3. Addition of conducting inclusions to a dielectric matrix and, vice versa, of insulating inclusions to a conducting matrix brings about changes

in electrophysical properties, described by the percolation theory [5, 6].
4. 'Void phase', i.e., porosity, uniformly distributed in a ceramic matrix, substantially affects thermophysical and mechanical properties as well as the thermal shock resistance [7].
5. Preprogrammed oriented distribution of the modifying phase in a ceramic matrix results in a material with an anisotropic or a varying character of properties in certain directions.

It should be pointed out that the above tendencies only schematically characterize the role of modifying phases in the formation of composite materials, since several types of modifying phases can be used in a material at a time, and special studies on optimizing the properties of such a composite material would be needed.

Ceramic-matrix composite materials are polyphase systems having a set of structural imperfections that determine their thermodynamic non-equilibrium. Various structural defects are formed at all stages of the composite material manufacturing process. Their formation is associated with kinetic and thermodynamic parameters of the synthesis, dispersion and blending of starting components. These procedures involve formation of a more developed surface, microcracks, adsorbed layers and films. Concentration of such defects is closely related to the energetics of the process. Thus, the use of high-energy actions (such as energy of an explosion) as well as surfactants gives rise to microdistortions of crystal lattices of the phases contained in the composite. In the course of a high-temperature formation of the material structure, the concentration of defects declines due to recrystallization, reduction of oxide films and diffusion tending to reduce gradients of impurity atom concentration. At the same time the 'heredity', determined by the existence of such stable structural defects as dislocations or impurity atoms, persists in a composite.

In designing ceramic-matrix composites, one should take into consideration not only the role of hereditary defects, but also the capabilities of the technology, which allows a purposeful incorporation of new types of imperfections in the form of modifying phases.

Incorporation of modifying phases can be aimed either at activating the material structurization processes or at obtaining a qualitatively new combination or level of properties, unattainable in single-phase materials. Thus, incorporation into a silicon nitride-based ceramic matrix of equiaxial inclusions based on some metal-like refractory compounds, such as titanium nitride or carbide, whose thermal expansion coefficient exceeds that of the primary phase, generates in the material a field of structural stresses that prevent the extension of a main crack. A composite having such a heterophase structure features a higher fracture toughness, defined by the critical stress concentration factor (K_{Ic}) [8].

Reinforcing phases in the form of fibres or whiskers most substantially improve mechanical properties of a material. Propagation of main cracks in fibre- or whisker-reinforced composites is impeded due to higher energy expenditures for fracturing the material because of the need for pulling the reinforcing element out of the ceramic matrix [1].

The most important characteristic of a composite affecting its mechanical, thermo- and electrophysical properties as well as the reactivity with respect to the environment, is porosity. Powder technology makes it possible to produce materials with various porosity levels (from monolithic structures to foam compositions) and various morphologic features (open, closed and directional pore systems). Spherical pores, profiled channels and cellular structures can be formed. Porosity reduces the mass of products made from composites, imparts heat-insulating properties to them, and allows ceramics to be used as filters and catalyst carriers [9].

Methods for solving the problems of using composite materials for products to be operated under complex conditions of multifactor external actions include the manufacture of materials with a layered structure, with a gradual variation of the phase composition according to a predetermined law, and with coatings. The modern technology makes it possible to combine ceramics with metals and other materials by brazing, diffusion welding, laser and plasma techniques so as to attain the required macrostructure in products.

Table 1.5 Classification of structural states of ceramic-matrix composite materials

Characteristic of structural state	Properties
Uniformly distributed equiaxial inclusions	Higher fracture toughness and strength, controllable electrophysical properties
Randomly distributed short fibres or whiskers	Higher strength, fracture toughness, thermal shock resistance and high-temperature strength
Continuous fibres	Higher high-temperature strength, thermal shock resistance and fracture toughness
Equiaxial pores	Lower density and thermal conductivity, high thermal shock resistance
Oriented nonequiaxial pores	Anisotropy of thermophysical properties, higher thermal shock resistance
Oriented channels	Anisotropic permeability, higher thermal shock resistance
Layered ceramic–ceramic or ceramic–metal structure	Anisotropy of mechanical, thermo- and electrophysical properties, higher strength and fracture toughness
Ceramic with coating	Higher corrosion and erosion resistance, control of surface properties

Structural materials 33

Products where dissimilar materials are combined, each of which performs a certain function, may be classed as 'macrocomposites'.

The diversity of structural elements allows the designing of structural states either responsible for the provision of a predetermined set of properties of a material or assisting its rational manufacture (see Table 1.5).

The designing of a composite material should take into account the geometry and service conditions of a specific product as well as its compatibility with other materials and environment. Because of this, the overall process of manufacturing a product from a composite should include not only methods of incorporation of modifying phases, moulding and high-temperature treatment, but also methods of applying coatings and joining dissimilar materials from which the product consists.

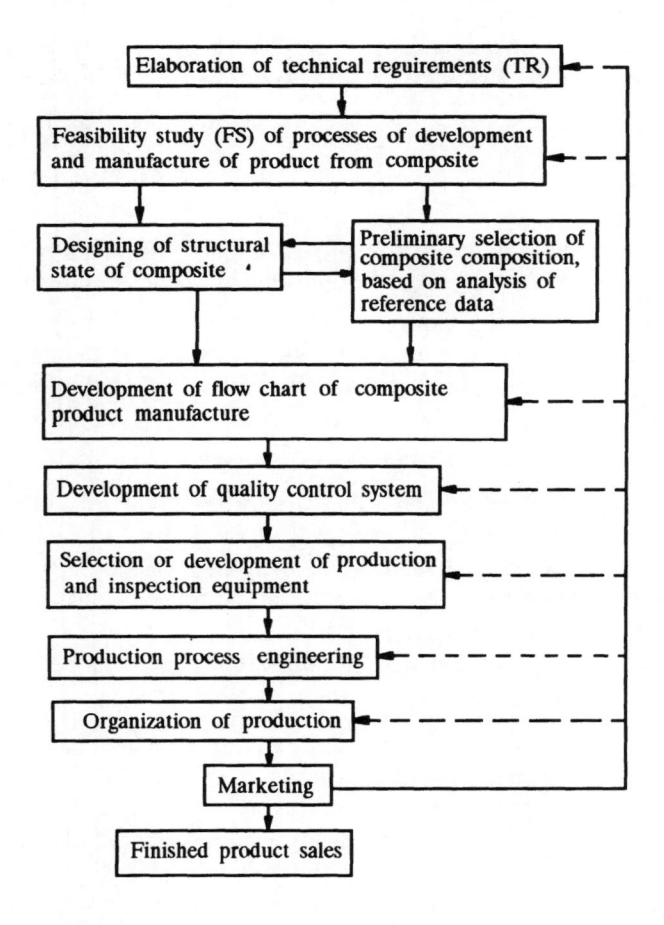

Figure 1.11 The organization of work for development and manufacture of products from new composite materials.

Thus, the designing of composites and manufacture of products from them is to be treated as an integral process, including not only the manufacture of a material, but also the 'technological assembly' of a product. A successful solution of the problem of obtaining products from composites with higher performance characteristics is possible only through joint efforts by experts in the fields of the physics of solids and physical chemistry, production engineers and designers. Each of the stages of the technological assembly can include both the use of already known procedures and equipment, and development of new techniques. The process of designing new materials should therefore be considered as a system of research efforts and engineering-organizational measures, providing an economically founded technology. A generalized diagram illustrating the sequence of stages in the manufacture of products from composites is shown in Figure 1.11

As follows from the diagram, the optimum way of manufacturing a product from composites involves the account of the following factors:

- economic efficiency of development of a new material as compared with traditional methods;
- possibility of fulfilling the technical requirements with the available raw stock, semi-products and equipment;
- need for developing new processing plants, automation and inspection means with the aim of engineering the whole manufacturing process.

The special role of marketing is to be emphasized, as its results affect all stages of the work organization system, because the changing and rising market demands spur the improvement of the technology.

1.1.2.2 Manufacturing processes for ceramic-matrix composites

The manufacture of ceramic-matrix composites involves numerous and diverse processing steps based on chemical reactions as well as on various physical processes such as dispersion, blending, moulding, melting and crystallization, condensation from plasma and gaseous phases, sintering, machining, thermomechanical treatment, etc.

Ceramic-matrix composite manufacturing processes can be conventionally classed into the following groups:

- technology where the starting components are chemical compounds based on elements that in dissociation or reduction in a gaseous phase produce refractory compounds in the form of structural components of the material of the predetermined composition and in the required combination (chemical technology);
- the starting components are powders, whiskers, fibres of refractory compounds, from which the composite is formed without melting of the primary phase (powder technology);

Structural materials

- the starting components are melted and arranged by a crystallization process, which promotes formation of specific composite structures (melting technology);
- technology involving a high-rate heating and intense mass transport of the components (in particular, under the action of electromagnetic fields), heated up to a solid, a liquid or a gaseous state, to the condensation zone where the material structure is formed (condensation technology).

Chemical technology methods are employed in manufacture of composite materials with an oriented (layered) structure, formed as a result of gas-transport reactions on hot substrates. The reactions result in polycrystalline layers. Varying the deposition temperature, gas mixture flow velocity, component ratio and gas pressure in the reaction zone makes it possible to produce composites differing in the crystallite size, porosity, thickness of alternating layers, their orientation and other properties. Gas-transport chemical reactions with participation of volatile or gaseous halides, hydrocarbons, organoelements and other compounds result in the formation of refractory metal or ceramic layers on graphite surfaces heated up to the preset temperature. This technology is widely used to produce high-modulus ceramic fibres intended for reinforcement of composite materials [10]. Gas-transport reactions result in deposition of thin high-density layers of silicon carbide, boron carbide, boron on tungsten or carbon filaments heated by electric current [11]. A continuous process of the deposition from a gas phase on a moving tungsten or carbon filament can be arranged at commercial production of high-modulus fibres.

Thus, the chemical technology yields composite materials in the form of single- or multilayered coatings. The gas-phase (pyrolytic) materials feature a high-density, fine-grained, uniform structure. They exhibit the

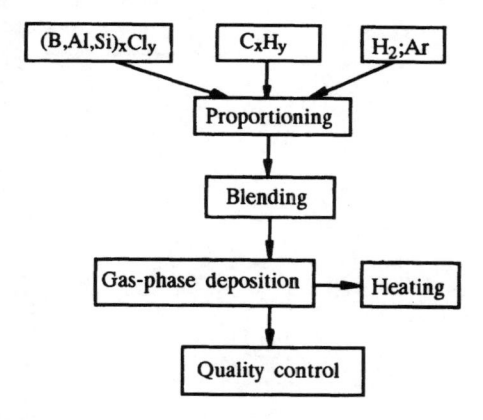

Figure 1.12 The manufacture of ceramic-matrix composite materials by chemical deposition from gaseous phase.

36 *Ceramic-matrix composites*

highest mechanical, thermo- and electrophysical properties. The flow diagram of basic steps in manufacture of gas-phase composite materials is shown in Figure 1.12

A low output of the process and its restrictions as to manufacture of thick-layered products result in a comparatively narrow scope of applications of pyrolytic composites.

Much broader capabilities are offered by powder technology, which makes it possible to produce numerous diverse ceramic-matrix composites and products from them with predetermined structure, composition and properties. Powder technology involves the use of classified powders, whiskers and/or fibres of refractory compounds, their blending with required production binders, activators, modifying additives, followed by

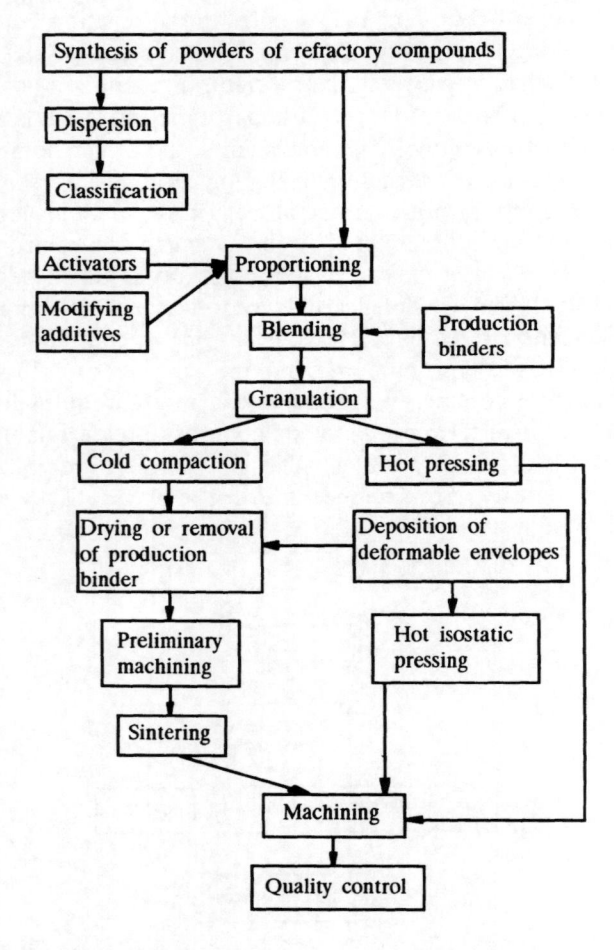

Figure 1.13 The manufacture of ceramic-matrix composite materials by powder technology.

moulding (shaping) and high-temperature treatment (sintering). The sequence of steps in manufacture of composites by the powder technology is shown in Figure 1.13.

Basic versions of the powder technology differ in the specifics of the processes of synthesis, dispersion and homogenization of constituents, moulding conditions and the possibility of combining the moulding and sintering steps (hot pressing or hot isostatic pressing). The optimum sequence of manufacturing steps is selected depending on the purpose and configuration of the product. The variety of moulding methods stems from the need for producing preforms whose configuration is close to that of the finished product. The cost-effectiveness and output of the process should be taken into account here. Figure 1.14 shows the classification of basic methods of moulding of products from composites obtained from ceramic powders.

As seen from Figure 1.14, the moulding methods employed in powder technology include intermittent and continuous processes, static and dynamic methods, and methods of moulding without the application of pressure (slip casting). The diversity of moulding methods extends the possibilities for manufacturing a broad assortment of products. Thus, to manufacture complex-configuration products it is recommended to use such moulding methods as a hot casting of thermoplastic slips, injection, hydrostatic or vibratory-pulse moulding followed by sintering or hot isostatic pressing. Lengthy products (bars, tubes, complex-profile guides) are moulded by continuous methods (extrusion).

The final step in powder technology for ceramic-matrix composites is sintering: thermal treatment of compacted preforms to impart them with

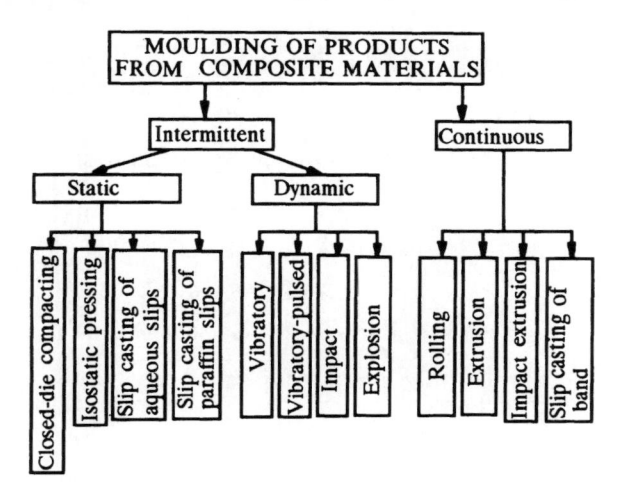

Figure 1.14 Classification of methods of cold moulding of products from ceramic-matrix composite materials.

predetermined physicomechanical properties and performance characteristics. High-density composites are produced by processes which combine the compaction and sintering steps (hot pressing or hot isostatic pressing). These processes are most expedient in a mass production of cutting inserts from composite ceramics [3, 8, 12] and of structural elements [13].

1.1.2.3 Formation of composites: processes and equipment

let us discuss the formation of composites in the sequence of manufacturing steps, shown in Figure 1.12 and 1.13, beginning with the production of powders and fibres and right up to completion of structurization in sintering or hot pressing.

The manufacture of composites involves the following most important processes:

- synthesis of refractory-compound powders and fibres;
- dispersion and blending of powders;
- cold moulding;
- sintering and hot pressing.

(a) *Synthesis and preparation of powders and fibres*
The powder and fibre production methods can be conventionally divided into physicochemical and mechanical. Physicochemical methods include various synthetic techniques, such as:

- direct formation of refractory compounds from elements, for example:

$$Si + C \rightarrow SiC,$$
$$3Si + 2N_2 \rightarrow Si_3N_4,$$
$$2Al + N_2 \rightarrow 2AlN;$$

- reduction of oxides with carbon, metals or ammonia:

$$SiO_2 + 3C \rightarrow SiC + 2CO,$$
$$2B_2O_3 + 7C \rightarrow B_4C + 6CO,$$
$$2B_2O_3 + 6Mg + C \rightarrow B_4C + 6MgO,$$
$$3SiO_2 + 4NH_3 \rightarrow Si_3N_4 + 6H_2O;$$

- reduction of halides in a gaseous phase:

$$BCl_3 + NH_3 \rightarrow BN + 3HCl,$$
$$4BCl_3 + CCl_4 + 8H_2 \rightarrow B_4C + 16HCl,$$
$$3SiCl_4 + 4NH_3 \rightarrow Si_3N_4 + 12HCl,$$
$$SiCl_4 + CH_4 \rightarrow SiC + 4HCl.$$

These chemical reactions can be carried out in various processing plants that provide the required thermal activation (heating).

Methods for synthesizing refractory compound powders in high-temperature electric furnaces have been developed in the greatest detail. Both resistance and induction furnaces, where accomplishment of the above-listed reactions is possible, are widely used for this purpose at present as they make it possible to control temperature–time parameters of the process over broad ranges and to optimize the synthesis conditions so as to meet the demands placed upon the powders. Disadvantages of the 'furnace' synthesis processes include their high power consumption, the possibility of contamination of the products being synthesized with impurities from the furnace lining and formation of fairly strong cakes requiring a subsequent crushing and grinding.

The use of a low-temperature plasma speeds up the synthesis reactions and makes it possible to produce ultradispersed high-activity powders [14]. For this purpose, d.c. and a.c. arc plasma generators are used, as well as high-frequency plasma generators. The introduction into the plasma jet of solid, dispersed as well as the gaseous components, results in their intense heating, melting, evaporation and partial ionization, which promotes a rapid and complete proceeding of reactions. A high velocity of components in the plasma jet allows the accomplishment of hardening of reaction products, resulting in formation of submicronic particles with a high concentration of crystal lattice defects. Depending on the parameters of the plasma chemical synthesis, the process can yield either equiaxial (spherical or polyhedral) or acicular particles. Just such a variety in the morphology of particles was observed, e.g., in ultradisperse silicon nitride particles produced at interaction between silicon and nitrogen in a low-temperature plasma [14]. The plasma chemical synthesis offers a high output and the possibility of producing high-activity powders. The main disadvantage of this method is the possibility of contamination of sythesized products with atmospheric components.

The self-propagating high-temperature synthesis (SHS), developed in the USSR, which has gained a wide application, relies on the use of exothermal reactions of formation of some refractory compounds [15]. Thermal initiation with the aid of an electric resistor brings about a steep temperature rise in the reaction zone and the chemical combustion front advances throughout the reagents. SHS is carried out in special pressure-tight reactors where an elevated gas pressure can be created to ensure completeness of the reaction. The SHS method features a low power consumption, a high output and simplicity of the required equipment. Its limitations consist in the possibility of synthesizing only those refractory compounds whose formation involves a significant exothermal effect. The purity of SHS products is determined by the quality of starting components, as the material of the reactor, whose walls are cold, does not contaminate the products. However, the high temperatures in the combustion zone result in very strong cakes that have to be crushed and ground, and hence the powders can get contaminated from the wear of walls or from the grinding units.

The purest powders of refractory compounds are produced by the gas-phase synthesis, where reactions between gaseous components yield products with a low impurity content since starting substances can be subjected to a profound purification [16].

Starting products for the gas-phase synthesis are halogen or organic derivatives of appropriate metals. Synthesized powders feature a high dispersity (0.01–0.4 µm) and a monofractionality. Drawbacks of this method include the complexity of controlling the multistage synthesis process, the complexity of the equipment needed, the toxicity and explosion hazard of starting substances and reaction products.

New ceramic powder synthesis methods have been developed recently to increase the dispersity and upgrade the characteristics of the powders as well as to reduce the power consumption. The methods can be divided into three main groups: production of powders from solutions or colloidal systems (sol–gel process); condensation from a vapour phase; and decomposition of salts in a solid phase [17].

An advantage of the solution method is the possibility of controlling the solution compositions over the broadest range as well as a high accuracy of controlling the introduction and dispersion of low concentrations of additives.

One of variations of the solution method, which provides for obtaining ultradispersed oxide powders, is hydrothermal synthesis, involving superheating of supersaturated aqueous solutions at an elevated pressure. The hydrothermal synthesis includes steps of solvent removal through evaporation, solid phase precipitation and solvent extraction through its filtering-out, burnout or freezing-out. Methods of deposition with filtration yield, as a rule, aggregated and multicomponent powders. The adverse effect of powder aggregation is reduced by the use of surfactants [18]. Employment of dilute solution allows a precision doping of ceramic powders with small amounts of additives (down to $3 \times 10^{-4}\%$) [19].

The sol–gel process utilizes the transition from a liquid colloidal state to a solid one. The process involves condensation of sol into gel, ageing of gel, its drying, moulding and sintering. The sol–gel process includes hydrolysis of salts, followed by condensation of sols. It can yield powders, fibres, coatings, and also ceramic-matrix composites.

Two varieties of sol–gel processes are known. One is employed to produce ceramic powders from solutions of alkoxides and other organometallic compounds. The other involves preparation of aqueous colloidal dispersions (sols) of hydroxides, which are converted into gel by dehydratation and then into oxide by calcination [20].

Nontraditional ceramic powder synthesis methods include processes involving an action on a vapour or gaseous phase by the plasma or laser energy which gives rise to physicochemical phenomena that promote formation of ultradisperse particles of refractory compounds [21,22].

The action of the laser emission relies on the resonance between the frequency of the monochromatic emission of a CO_2 laser and the absorption band of the IR spectrum of the reagent gas (such as silane SiH_4, ethylene C_2H_2, ammonia NH_3, employed in the synthesis of Si_3N_4 and SiC). Reactions proceed at exceedingly high rates at temperatures over 800 °C, are initiated in fractions of a millisecond, and stop abruptly, resulting in monodisperse powder particles [23].

In contrast to physicochemical methods, a mechanical production of powders is based on the use of a chemically synthesized starting product without changing its composition, by crushing, grinding and attrition of solids.

It should be noted that the division of powder production methods into the two groups is conventional. Physicochemical processes play an important part in mechanical methods. Thus, a high-temperature grinding involves concurrent comminution of components being dispersed and their tribochemical interaction [24]. Similar processes occur at an intense grinding in the blending of oxygen-free refractory compounds in oxygen-containing atmospheres, which results in a partial oxidation of the materials being dispersed.

An important requirement for all the powder and fibre production methods is the provision of the required dispersity (for fibres, of the diameter and length), morphology and composition. In solid-phase methods, this is attained mainly through the temperature–time parameters, which should, on the one hand, ensure the completeness of the reaction and, on the other, not bring about an excessive growth and aggregation of particles.

(b) *Moulding*

Moulding processes play an important part in production of ceramic-matrix composites. The basic feature of compacting multicomponent ceramic powder blends is the absence of plasticity in refractory compound powders and fibres. Taking this fact into account, the following main characteristics should be provided in moulding:

- predetermined level of density (porosity);
- maximally uniform density (or predetermined pattern of density variation);
- shape and dimensions of the product with allowance for dimensional changes at subsequent production steps;
- preservation of reinforcing elements (whiskers, fibres);

- adequate strength of the moulded preform, ensuring its intactness at subsequent productions steps.

To improve the mouldability of ceramic blends, they are plasticized with the aid of production binders, which upgrade the rheologic properties of blends being compacted and impart a certain strength to moulded raw preforms. The powder technology of ceramic-matrix composites employs moulding methods which can be classed into static methods, dynamic (high-speed) methods and pressureless moulding.

Investigations into processes of moulding plasticized ceramic blends demonstrated that compaction with the use of static methods requires application of external pressures not over 50–60 MPa; higher static pressures result in a subsequent fracture of raw compacts under the action of elastic after-effect forces.

The moulding media which can be used in the compaction of ceramic blends includes solids (cold compaction in metallic dies or hot pressing in graphite dies); elastomers, such as rubbers and polyurethane (elastostatic pressing); liquids (hydrostatic, hydrodynamic, pulsed electrohydraulic pressing); and electromagnetic fields (magnetopulsed pressing).

Static and dynamic pressing methods differ in the following characteristics: loading rate; maximum loading of movable elements of the equipment; and duration of the pressure action. Static methods feature low loading and deformation rates, measured in centimetres per second, while dynamic processes involve high loading rates, amounting to tens and hundreds of metres per second, moulding element accelerations over 10 G, and extremely short deformation times (milli- and microseconds).

The widest use has been found by a static pressing in steel dies on hydraulic and mechanical presses. It is extensively employed in mass fabrication of products of a simple geometric shape and a small height. To increase the process output, use is made of multicavity dies, automatic blend metres, mechanized die opening, ejection and transportation of compacts. Robot-manipulators which perform the whole set of procedures in the pressing of preforms are employed at mass production.

Lengthy constant-section products are moulded by extrusion or rolling. In contrast to hydraulic and mechanical presses, where the force is directed vertically and operation is cyclic, in extrusion plants the plasticized mass is extruded horizontally and the process can be continuous. The configuration of the product being moulded is determined by the shape of the working orifice of the extrusion die, which allows the extrusion of round and shaped tubes, as well as rods of various cross-sectional shapes. The extrusion method also makes possible the production of multilayered composites as well as preforms from textured composites and materials with an oriented porosity. In extrusion plants, force to the mass being moulded is transmitted by a piston or screw. Plants with a vacuum treatment are employed to improve the mouldability and elimina-

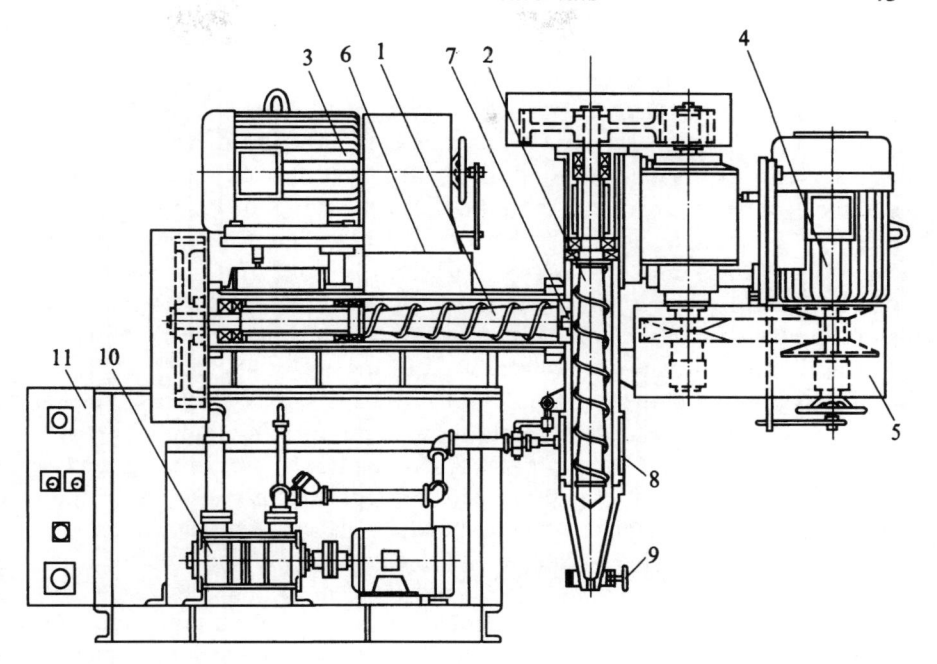

Figure 1.15 Vacuum extruder: (1) feed screw; (2) pressing screw; (3) feed screw drive; (4) pressing screw drive; (5) stepless control belt drive; (6) feed trough; (7) guide plate; (8) cooled pressing cylinder; (9) mandrel with die; (10) vacuum unit; (11) control desk.

tion of air bubbles in the mass. A vacuum extruder is schematically shown in Figure 1.15.

Sheets, strips and plates can be produced from plasticized ceramic blends by rolling, which offers a high output and ensures a uniform density of long preforms.

Complex-shaped products are manufactured by slip casting [25] and injection moulding. Two processes are employed for moulding composite-material products: casting of aqueous slips into porous moulds and hot casting of thermoplastic slips. In the former process, based on aqueous suspensions, ammonium alginate, ammonia, caustic soda as well as polyvinyl alcohol, carboxyl cellulose and soluble glass are added to stabilize the suspensions and improve their fluidity. Capillary forces as well as vacuum filtration promote compaction and a uniform distribution of powder particles, reinforcing elements and pores.

In the 'hot' casting of thermoplastic slips, a mass is used consisting of a blend of powders and fibres of a predetermined composition and a binder based on paraffin and wax. A heated and thoroughly stirred paraffin slip is subjected to vacuum treatment and cast under pressure into metallic moulds. The injection moulding involves a forced injection

44 Ceramic-matrix composites

of a vacuum-treated slip. After cooling and solidification, the preforms are heat-treated to remove the binder. To prevent shrinkage and distortion, a programmed heating is needed, whose conditions are selected with allowance for the mass, dimensions and configuration of the preform. The slip casting and injection moulding are used to produce thin-walled complex-shaped elements (turbine blades, rotor discs of turbines, crucibles, etc.). Since the forces acting on the mass being moulded in slip casting processes are very small, such brittle reinforcing elements as whiskers and fibres remain intact in the preform. A disadvantage of the slip technology is the impossibility of producing higher-density preforms. The minimum porosities for slip-cast preforms are of 45–55%.

The method of hydrostatic (isostatic) pressing is free from this drawback, as a uniform compression of a powder preform by liquid can yield a higher density level and a uniform distribution of residual porosity. The hydrostatic pressing employs elastic or plastically deformable capsules wherein the blend to be moulded is placed. Before sealing the capsules it is preferable to subject the blend to a vacuum treatment so that the gas present in pores does not prevent the compaction. The capsule configuration should match that of the part being moulded with allowance for the shrinkage. The blend sealed in the capsule is placed into a high-pressure container filled with a liquid (water, glycerin, mineral oil). The liquid pressure in the container is increased so that the powder in the capsule gets uniformly compressed. The rational range of pressures used is of 100–500 MPa. Advantages of the hydrostatic pressing are a high (75–85%) and a uniform density of the moulded product. Rubber, latex and polyurethane are used as the material for capsules. A wide use of this method is prevented by its relatively low output as compared with the static pressing in steel dies, as well as by difficulties in producing preforms with a high accuracy of geometric dimensions. A

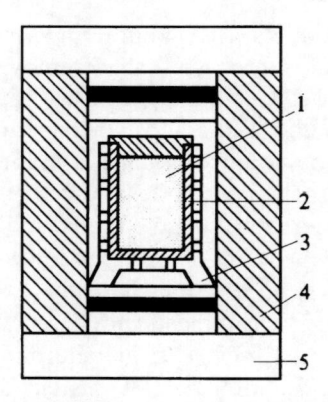

Figure 1.16 Hydrostatic pressing plant: (1) powder preform; (2) deformable capsule; (3) high-pressure container; (4) casing; (5) plug.

Structural materials 45

diagram of a hydrostatic pressing plant is shown in Figure 1.16. Plants with an outside high-pressure source are generally used for the hydrostatic pressing. Such a plant consists of a working chamber and a high-pressure generator, interconnected by piping. Hydraulic pumps are used as high-pressure generators (at pressures up to 500 MPa), hydraulic compressors (up to 600 MPa) and hydraulic intensifiers (up to 1200 MPa). The hydrostatic pressing can yield preforms with any height–diameter or height–thickness ratio. The overall dimensions of the preform being pressed are limited only by the dimensions of the working chamber.

One of varieties of the isostatic moulding is the elastostatic pressing. In this case the composite material blend to be moulded is placed into a thick-walled elastic capsule that fills the press mould cavity. Here, the elastic capsule is the deforming medium which creates in the material being compacted such conditions that approach the pattern of a uniform cubic compression (see Figure 1.17). The material of the elastic capsule creating the isostatic compaction conditions should take and retain the preset geometric shape, and offer the required elasticity and strength. Rubber, rubber vulcanizate and polyurethane are used for this purpose since they feature practically no residual deformation. Elastomers are used for mass moulding of complex-shape parts (sphere, thin-walled bushings, etc.) using an ordinary pressing equipment capable of developing a pressure of 500–1000 MPa in the die. The pressure transmission through elastic media makes it possible to create practically any patterns of the stressed-strained state in the material being pressed and to reach high degrees of compaction of complex-shape elements.

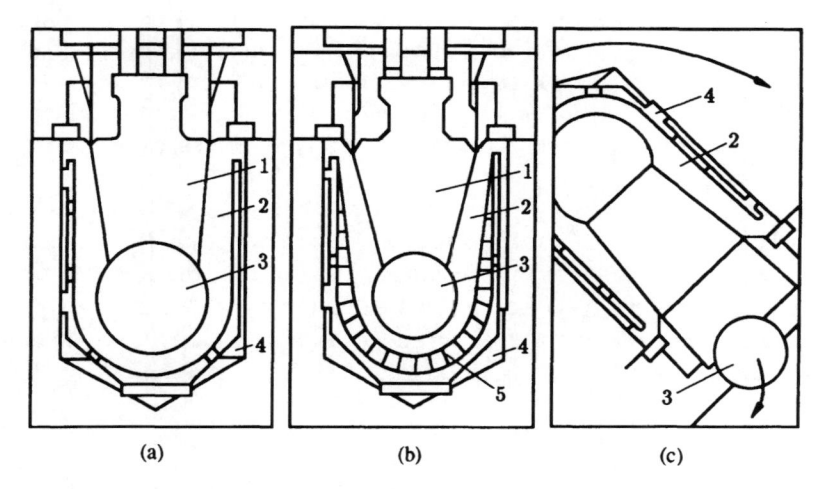

Figure 1.17 Plant for elastostatic pressing of spherical preforms (a) initial stage; (b) compacting stage; (c) preform ejection stage; (1) elastic punch; (2) elastic die; (3) preform being compacted; (4) housing; (5) working liquid.

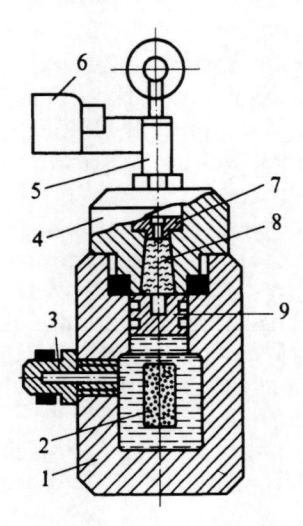

Figure 1.18 Hydrodynamic pressing plant: (1) housing; (2) preform being compacted; (3) pressure sensor; (4) closing screw; (5) firing pin mechanism; (6) electromagnetic triggering unit; (7) percussion cap; (8) gunpowder charge; (9) piston acting on working liquid.

Consider now dynamic moulding methods in accordance with the classification shown in Figure 1.14. They can be classed into two groups: (1) methods of a one-time high-speed action on the powder being moulded and (2) methods of a repeated loading at predetermined frequency and amplitude of deformation [26].

The high-speed one-time action methods are distinguished according to the type of energy carrier used; the variety of energy carriers makes it possible to control the compaction parameters over a wide range. Thus, the use of gunpowder as the energy carrier provides for a high-speed pressing in metallic dies. The dynamic pressing yields a high enough uniformity of the porosity distribution.

Hydrodynamic pressing plants employ a liquid as the energy transmission medium. The powder charge explosion energy is transmitted to a piston that compresses the working liquid wherein the preform being pressed is placed in an elastic capsule (see Figure 1.18). A family of hydrodynamic pressing plants with a working chamber diameter up to 450 mm and height up to 1200 mm, operating at pressures of 400–1200 MPa, have been developed [27].

In gas-dynamic moulding plants the gunpowder gas pressure acts directly on the material being pressed. A plant intended for the gas-dynamic pressing at pressures of 350–700 MPa is shown in Figure 1.19.

The use of high explosives (HE) provides for a pulse with a pressure of 1–2 GPa, which is of interest for moulding low-plasticity ceramic

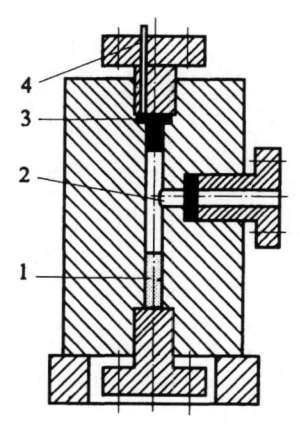

Figure 1.19 Gas-dynamic pressing plant: (1) preform being compacted; (2) safety rod which adjusts maximum pressure; (3) gunpowder charge; (4) firing pin.

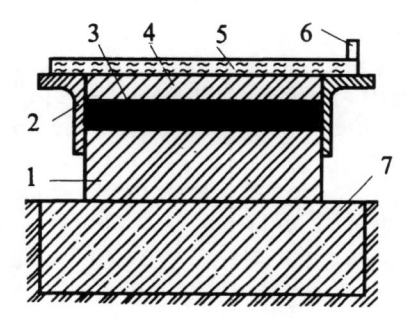

Figure 1.20 Explosive pressing device: (1) supporting plate; (2) moulding frame; (3) powder being compacted; (4) punch; (5) high explosive charge; (6) electric detonator; (7) concrete base.

powder blends. The pressure arising at the HE explosion compacts powder preforms by planar or travelling shock waves. One version of the explosive pressing of flat powder preforms is shown in Figure 1.20. This device yields preforms with a uniformly distributed porosity at a relative density over 90%.

The use of various vibratory moulding versions is very attractive for nonplastic ceramic powder blends, which make it possible to obtain large complex-shaped products with a strictly uniform porosity distribution throughout the product. The compaction occurs through redistribution and a closer packing of particles with retention of their shape and practically without fracture, which is particularly important in the pressing of blends containing unequiaxial reinforcing elements (whiskers, fibres).

The vibratory moulding of powders is carried out with the use of vibroexciters of various types, each having its working frequency range. The working frequencies of main types of vibrators employed in the technology of composite materials are presented in Table 1.6.

Table 1.6 Working frequencies of vibroexciters

Vibroexciter type	*Frequency* (s^{-1})
Mechanical centrifugal	3–100
Mechanical eccentric	2–300
Pneumatic	1–40
Hydraulic	1–500
Electromagnetic and electrodynamic	5–3000
Electric-discharge hydraulic	1–60

The vibroimpact moulding consists of acting on the powder being compacted by low-frequency impact pulses. The process is carried out on special disbalance sets (see Figure 1.21) where vibration of the working platform that carries the press tooling and the clamping system is accompanied by impacts against limiters so that the powder is subjected to a pulsed load from the punch. Phases of interaction between the die with the powder being compacted and the punch can be varied by varying the clamping force so as to attain their counter-impact at which the compaction is most efficient. A substantial drawback of vibroimpact sets is that impacts of a massive platform result in great dynamic loads on the disbalance mechanism and foundation.

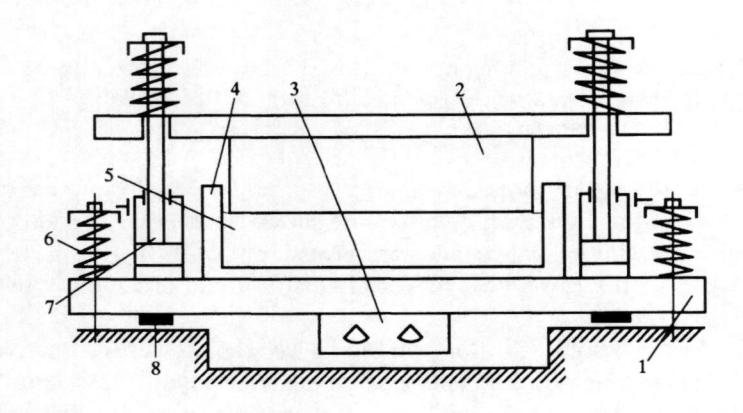

Figure 1.21 Vibroimpact moulding device: (1) supporting platform; (2) punch; (3) vibrator; (4) die; (5) powder being compacted; (6) springs; (7) clamping cylinders; (8) elastic gaskets.

Structural materials 49

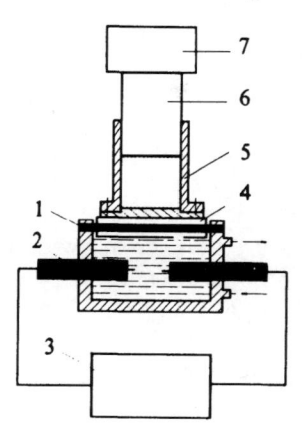

Figure 1.22 Plant with electric-discharge vibration exciter for vibropulsed compaction: (1) rubber diaphragm; (2) electrodes; (3) power unit; (4) lower punch; (5) die with powder being compacted; (6) upper die; (7) weight.

Among the vibroexciter types listed in Table 1.5, the widest use has been found by mechanical, hydraulic and electric-discharge devices.

A variety of the vibropulsed moulding is a vibratory moulding with the use of the electrohydraulic effect [28]. Plants for this process are based on electric-discharge exciters (see Figure 1.22) where pulses of oscillation of the tooling that acts on the powder being compacted are generated by the electric discharge in liquid. In view of the considerable power of the vibroexciter (up to 1 kJ in a pulse), such plants are useful in moulding large products from nonplastic ceramic powders. It is recommended to use concurrently small (up to 45 MPa) static pressures in the process. The electric control circuit allows control of the power and pulse repetition frequency so as to select the required moulding conditions.

It was shown in [28] that the vibropulsed moulding of nonplastic ceramic powders yields the maximum compaction when low-frequency vibrations with elevated amplitudes are used. It was found that the coefficient of side pressure in the vibropulsed moulding is 1.5–2.0 times as high as that in a static pressing in rigid metallic dies, which makes the results of compaction by this method close to the results of a hydrostatic pressing. It was also found that the required external forces in the vibropulsed moulding are nearly an order of magnitude lower than in the static pressing; this makes possible the manufacture of complex-shaped and larger-sized products at a low power of the pressing equipment.

The above-described composite moulding processes allow the selection of the optimum technology, depending on the operating conditions, required properties and configuration of products as well as on the cost-effectiveness, to ensure the predetermined level and stability of properties and of other parameters of a product made from composite materials under commercial production conditions.

(c) *Sintering*

Ceramic-matrix composites are sintered in special electric furnaces with a controlled atmosphere. Composite materials based on oxide phases are sintered in oxygen-containing atmospheres, such as air, whereas materials containing nonoxide refractory compounds or metallic phases have to be sintered in neutral atmospheres (argon, nitrogen) or in a vacuum. It should be noted that composites based on nonmetallic nitrides with a covalent chemical bond (Si_3N_4, AlN, BN) may dissociate at temperatures over 1800–50 °C. To suppress this undesirable phenomenon, sintering in a nitrogen atmosphere under pressure (10–100 MPa) is recommended.

Specialized high-temperature plants for sintering various products from ceramic-matrix composites have been developed: compression furnaces with an elevated atmosphere pressure, vacuum furnaces, hot pressing presses, and hot isostatic pressing plants.

Compression furnaces with an elevated pressure of nitrogen in the working space allow an intensified process of reaction and activated sintering of nitride ceramics-based composite materials, since a high nitrogen pressure, suppressing dissociation of nitrides, makes it possible to raise the process temperature and thereby intensify the compaction.

The hot pressing method is used to produce nonporous ceramic-matrix composites. It is most efficient to produce high-density polyphase composite materials based on nonmetallic nitrides with activating additives [29, 30]. This method combines the moulding and the sintering, for which purpose the hot pressing plants have both pressing and electric heating devices. Designs of the hot pressing plants vary depending on the heating method, pressure application character, need for a protective atmosphere or vacuum in the working space, dimensions or configuration of the product being moulded, and on the level of temperatures needed to compact a given material.

The pressing tools for the hot pressing are as a rule made from high-density structural graphite, which allows the process to be run at pressures up to 250 MPa. Owing to a high electrical conductivity of graphite, the heating can be carried out by a direct current flow through the punches, through the die, through an external heating screen or by eddy currents induced in graphite by an external high-frequency field. Various versions of heating in the hot pressing are schematically shown in Figure 1.23.

The material being pressed is protected from oxidation either directly by the die material (graphite) or by carrying out the hot pressing in a vacuum chamber or a chamber filled with an inert gas. A hot pressing plant is fitted with devices for measuring and controlling the pressure

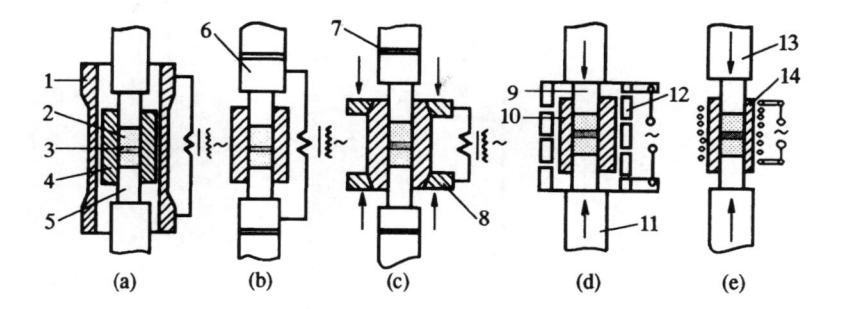

Figure 1.23 Die heating in hot pressing: (a) indirect resistance heating; (b) direct resistance heating with current feed to punches; (c) direct resistance heating with current feed to die; (d) induction heating of current-conducting graphite die; (e) induction heating of current-conducting powder in dielectric (ceramic) die. (1) heater; (2) powder being compacted; (3) sintered compact; (4) die; (5, 6, 9) punches; (7) insulation; (8) graphite or copper (water-cooled) contact; (10) graphite die; (11, 13) ceramic punches; (12) inductor; (14) ceramic die.

and temperature, determining the shrinkage and programming the conditions in time.

The main shortcoming of the hot pressing process is a nonuniform distribution of the thermal field and pressure throughout the volume of the die in producing complex-shaped elements.

The hot isostatic pressing (HIP) provides for compaction of products from ceramic-matrix composites at high temperatures under a uniform compression [31]. It is accomplished in gas autoclaves, high-pressure vessels where electric heaters are mounted. The autoclave contains a system of screens and thermal insulation which precludes heating of its walls. At a gasostatic pressing the energy-transmitting medium is an inert gas which is heated in resistance furnaces with a tungsten, molybdenum or graphite heater.

Before the hot isostatic pressing, the material to be compacted is enclosed in a deformable 'can' made from nickel, molybdenum, borate or silicate glass, or quartz. Screening coatings are deposited on the inside and outside of a can to reduce its interaction with the material being pressed and with the gas atmosphere. Glass or quartz coatings, whose softening temperature corresponds to the HIP temperature, are formed by applying a slip to the surface of the preform, which is followed by drying, roasting and flame polishing, in the course of which, there occur healing of pores and a final sealing of the can or capsule. Glass capsules can be prepared beforehand and then filled with the material to be pressed, vacuum-treated and sealed.

On completion of the HIP process, the containers with finished products are cooled outside the gasostat. Capsules for complex-shaped

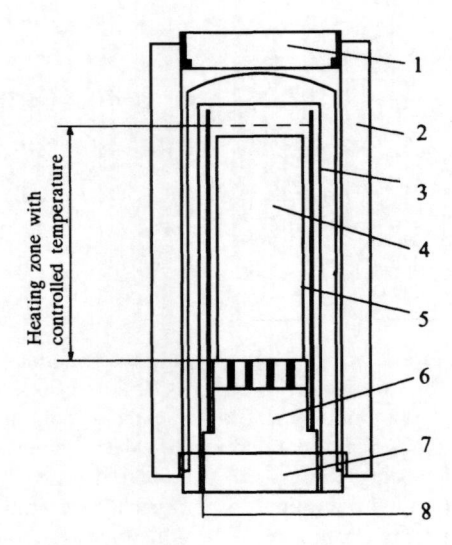

Figure 1.24 Working zone of hot isostatic pressing plant: (1) end-cover; (2) high-pressure container, strengthened by wire winding; (3) heat-insulating furnace casing; (4) heaters; (5) preform being compacted; (6) support for preform in furnace and bottom heat insulation; (7) thermocouple inlet; (8) electric power inlet.

products are made by applying a slip to a wax pattern which is then melted off after drying.

A HIP version without the use of capsules can be employed for processing presintered preforms having no open porosity; in this case, the HIP yields a density close to the theoretical one. The HIP process makes it possible to manufacture large thin-walled complex-shaped products from ceramic-matrix composites.

The working zone of a hot isostatic pressing plant is schematically shown in Figure 1.24.

The above-discussed processes and production equipment types make it possible to select the most rational way to manufacture products from ceramic-matrix composites, proceeding from such basic factors as the configuration and dimensions of the product, composition and structure of the material, its properties determined by service conditions, economic characteristics and the amount of output of the products. Ceramic-matrix composites find the most diverse fields of application since the principles of designing and processes for manufacturing polyphase composite ceramics open up possibilities for the creation of function-oriented structural states which provide the required set and level of properties and performance characteristics.

1.1.3 Machining techniques
A. V. Bochko

The terms 'machinability of ceramic composites' and 'machinability of metals', may be said to characterize a material property or quality which can be distinctly established and measured to determine the ability of a material to undergo cutting. In actual fact, however, there exists no such clearly defined cutting conditions which would in a unique fashion correspond to these terms. In practice, the machinability of a material can be expressed in terms of the number of parts produced per unit time, cost of machining and quality of the machined surface.

For metals and alloys, this complex process is most often examined on the basis of their composition, structure, heat treatment and properties. Such an approach is, to a considerable extent, also valid in dealing with the machinability of composite ceramic materials. In this case, the machinability of materials can be evaluated in terms of the tool life; maximum material removal rate (at the standard minimum tool life); cutting forces or power consumed; surface roughness and chip shape (under standard cutting conditions).

All the above machinability criteria are as far as possible taken into account in examining every group of materials being machined [1].

Distinctive features of ceramic materials, first of all their high brittleness and hardness, predetermine the selection of their machining methods and techniques. In view of these features, nontraditional machining methods, such as electrophysical, laser and others, involving no coarse mechanical action on a material, would seem to be preferable here, but too great an amount of a material to be machined, low machining rates and complexity of the equipment have all prevented them until recently from ousting the traditional machining to any considerable extent.

Abrasive machining processes have, up to the present, been predominant among ceramic-material product manufacturing and finishing operations. Many properties of ceramic materials are sensitive to the surface state, and therefore grinding and finishing processes occupy an important place among the factors governing the quality of finished ceramic parts. Thus, the sensitivity of the strength of brittle ceramic materials to inherent cracks exhibits itself in that the mechanical strength of a finished product depends on the character of the machining.

A high hardness of ceramic materials predetermines the necessity for using diamond tools with metallic and ceramic bonds in cutting and coarse grinding, and with an organic bond in finishing and polishing operations.

The cutting by a single diamond grain, often employed in modelling the grinding processes, may be considered as an independent process of a precision machining of ceramic materials or ceramic coatings by cutting tools of polycrystalline synthetic diamonds, synthesized or sintered at high pressures and temperatures.

The machining of ultrahard ceramic materials based on diamond, cubic and wurtzite boron nitride is in turn a separate, fairly complex task. Basic regularities involved in machining of these materials coincide in many respects with conclusions drawn on the basis of studying the machinability of oxide, oxide-carbide and nitride ceramics.

1.1.3.1 Machinability of ceramic-matrix composites

The machining of ceramic materials covers a broad scope of processes, from cutting and coarse grinding to finishing and polishing. It involves a mandatory removal of material, which in the course of the removal is subjected primarily to a brittle fracture and a plastic deformation. Generalizing, it can be noted that the part played by fracture in the material removal process rises with intensification of the abrasive machining process, i.e., with increasing applied loads, abrasive grain size, cutting fluid feed rate, etc. The mechanism of abrasive removal of relatively tough ceramic materials is comparable with the mechanism of plastic flow in the machining of metals.

On the whole, the problem of machining brittle nonmetallic materials is a complex, multifaced one and remains, so far, unsolved in a rigorously scientific respect [2].

According to [3], a higher brittleness of many nonmetallic materials, such as glasses and glass ceramics, is characterized by a great difference between the tensile and the compressive strength ($\sigma_t/\sigma_c = 0.04$–0.05). At the same time these materials exhibit a practically ideal elasticity at $\sigma_c = 2.4$ GPa. A similar feature is also exhibited by oxide ceramics ($\sigma_t/\sigma_c = 0.08$) and leucosapphire ($\sigma_t/\sigma_c = 0.035$).

The mechanism of the abrasive diamond machining of ceramic materials is based on regularities common for brittle nonmetallic materials, which were determined by studies of microcutting of optical glass and ceramics by a single diamond grain and of the mutual microfracturing of the tool and the material being machined.

In the general opinion of researchers [2–4], the process of action of an individual fixed grain on the surface of the material being machined is a microcutting-scratching and depends on properties of the material and on conditions of action of the grain on it. In the action on a material that has plastic properties, microcutting occurs with separation of fracture products in the form of chips, whereas a microscratching process with separation of fracture products in the form of minutest particles and fragments predominates in the action on brittle materials. There is, however, no distinct boundary between the manifestations of the two processes. In various studies the process of microcutting by a single diamond grain or indentor was in most cases evaluated by the a_z/ρ or h_z/ρ ratio, where h_z is the cutting depth, or by the P_z/P_y ratio.

Although processes of microcutting of brittle nonmetallic materials, characterized by the compressive strength, have much in common with those of hard brittle engineering materials, characterized by the tensile strength, they cannot be treated as equivalent to the latter. The P_z/P_y ratio has the maximum value of 0.65–0.70 in microcutting of soft and ductile metals and decreases to 0.15, approaching the friction mode conditions, in microcutting of steels as their hardness, brittleness and the cutting speed increase; it varies inversely with the ratio of the microhardness of a material to its tensile strength, H_μ/σ_t.

In the microcutting of, e.g., quartz, optical and borosilicate glass, glass ceramics, single-crystalline silicon and germanium, the P_z/P_y ratio is within 0.66–0.74 and rises rather than declines with increasing σ_c. In the machining of a broad class of such nonmetallic brittle materials as ceramics, glass, glass ceramics, etc., a diamond grain can form a groove and disperse the material in the volume of the groove by nearly any part of the grain rather than only by the sharp projection, while an abrasive grain cannot do this, having no hardness and strength required for the cutting-scratching of such materials. The dispersing capability of fixed grains is three to four orders of magnitude higher than that of free grains [4, 5]. The surface roughness in the former case has a regularly directional, and in the latter case, a uniform dead structure. In grinding by a tool with fixed grains, as the grain size increases, the depth of the fractured layer increases at a slower rate than in the case of a free abrasive and is less than in the latter case.

In the process of an abrasive machining of glass, when its surface layer is being microscopically fractured by points of abrasive grains, regions of compressive stress σ_c arise under them, and regions of tensile stress σ_t, under the latter. Only an elastic deformation appears at $\sigma_c = 1$–2 GPa. When the value of σ_t under the grain point reaches the glass yield strength σ_y, a brittle fracture occurs in macroregions above 100 μm in size with a simultaneous display of plastic phenomena in microregions. Thus, the penetrating action of the hard point of a grain gives rise to a plastic deformation, while internal tensile stresses are the cause of destruction phenomena. A viscous flow of glass and plastic deformation in regions less than 100 μm in size appear at $\sigma_c = 5$–6 GPa. The brittle fracture under the grain points determines the production efficiency of removal of the allowance in the glass grinding, while microdeformations are only accompanying.

The existence of a kind of boundary in the character of proceeding of these processes is associated with the grain size and, accordingly, with geometric parameters of the diamond powder used in the tool. At a grain size 40/28–28/20 separate traces of a brittle fracture still take place. Grains of size 20/14 and less, leave separate traces of a plastic deformation. A plastic shear occurs at a grain size 10/7, and in the presence of a

Table 1.7 Physicomechanical properties of some brittle nonmetallic materials and characteristics of their microcutting by single fixed diamond grain

Material	Microhardness H (GPa)	Elastic modulus E (GPa)	σ_t/σ_c	Relative machinability coefficient	Cut-off thickness (μm)	P_z/P_y
Optical glass	5.65	71.6	0.220	1.00	–	0.66
Quartz glass, transparent, especially pure	9.5–11.5	104.0	0.090	0.20	25	0.74
Borosilicate industrial glass	6.43	74.0	–	1.15	30	0.69
Glass ceramics	5.40–9.67	60.0–130.0	0.240–0.100	0.50	30	0.69
Single-crystalline silicon	7.95	174.0	–	0.62	30	0.67
Single-crystalline germanium	8.60	141.0	–	0.70	30	0.69
Optical ceramics	3.00	85.5	0.066–0.125	1.60	–	–
Oxide ceramics	19.7–21.0	43.2	0.080	–	20–25	0.71
Corundum (ruby, sapphire)	22.8–29.0	46.9	0.035	–	–	0.31
Natural diamond	100	1145.0	10.800	–	–	–

cutting fluid, the process proceeds with low values of the friction coefficient and material removal rate [6].

Establishing a similarity between glass and ceramic machining processes is assisted by comparing physicomechanical properties of various kinds of these materials and of diamond as the machining material (Table 1.7). The plasticity of glass and its ability to form chips increases with decreasing relative hardness, estimated from the grinding-off rate. Thin sections of chips being cut from brittle materials become plastic to a certain extent as the temperature in the machining zone increases. In the microscratching of most brittle nonmetallic materials, the chip formation and chipping-off are usually observed under the action of P_z ahead of the grain (Figure 1.25a). For silicon and germanium the chip-formation is mainly determined by P_y and occurs under the diamond grain (Figure 1.25b), and the chip separation behind the moving grain, as a result of release of the elastic strain energy of deeper elements.

Complex processes of microcutting and microfracturing on the scale of a tool consisting of a large number of grains can be treated as stochastic ones. The group average-statistical behaviour of cutting grains in them is highly dependent on the structure of the diamond tool itself. As

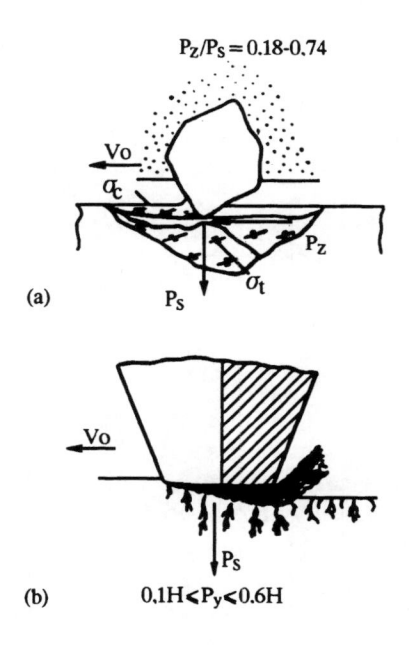

Figure 1.25 Schematic of grinding wheel's diamond grain penetration into material being machined under action of P_y as well as chip formation and separation under action of P_z: (a) for most of brittle materials; (b) for germanium and silicon (after [6]).

known, the factors determining the structure of its working layer are the number of grains and the character of their distribution in the volume of the diamond-containing layer as well as the ratio of the contents of diamond grains, bond and pores. However, even well-known regularities of the real process of grinding by diamond-micropowder tools have so far not been successfully approximated by accurate enough relations. Moreover, effecting such a process with predetermined conditions of deformation of the material being machined, i.e., controlling it, is only possible at a high precision level of the machine tool–accessories–tool–workpiece system, which takes place in operation on precision edge-tool lathes.

To substantiate the above, consider a study of the finish diamond turning and diamond grinding of germanium, silicon and silicon carbide, conducted at the Precision Machining Center of the North Carolina State University (USA) [7].

The turning was carried out on a 'Paul' lathe having a rigidity of 9 MN/m in the direction normal to the surface. The tool point radius was 3.18 mm; the relief angle, 1°; the spindle rotation speed, 1000 rpm; the cutting speed, 2.5 m/s; the depth of cut, 0.46, 1.1 and 2.3 μm; and the feed, 2.1, 4.2 and 5.9 μm/rev.

The grinding was carried out on a 'Pegasus' machine by a cup grinding wheel with 4–8 μm diamond grains, a concentration of 3.3 carat/cm^3, with a metallic and plastic bond. The machine had 10 cm diameter air-pressure bearings. A stepless variation of the rotation speed within 0–5000 rpm was provided by a d.c. motor. The depth of cut was of 30 μm, and the feed, of 0.25–25 mm/rev. Cutting fluid was water, oil, ethanol or dry grinding.

The study ascertained that the studied ceramic materials can be machined under plastic deformation conditions with the use of a small depth of cut and a rigid machine tool.

1.1.3.2 Machining of oxide ceramics

The primary machining of silicon dioxide-based ceramic materials is usually carried out by abrasive wheels, and the final machining by diamond wheels. Owing to high hardness and thermal conductivity of diamond, the diamond grinding is the most efficient method for machining of ceramic materials. Most suitable for this purpose are synthetic diamonds with a large surface of grains and a skeleton structure, which result in a good holding of grains in the bond and a permanent renewal of cutting edges because of their brittle fracture in the course of machining [5].

Wheels of the 14A1 type of synthetic diamonds AS4 of grain size 200/160 and a 100% concentration are recommended for grinding of ceramic materials. Silicon dioxide-based materials, such as Kersil, Niasit, etc., as well as quartz glass should be ground by wheels with metallic bonds M1 (Cu–Sn) and MK (Cu–Sn–Al$_2$O$_3$); heavy flint glasses, by wheels with bonds MK and M52 (Cu–Sn–WC–Co); and borosilicate glasses, by wheels with bonds M52 and M1. The surface of a ceramic part after machining is a set of roughnesses resulting from the chipping-out of separate sections, i.e., the predominating mechanism of the chip formation is a brittle fracture, which gives rise to a cyclic action of cutting forces and temperatures on the tool in the course of grinding. A low thermal conductivity of silicon dioxide-based materials results in high temperatures in the machining zone. The temperature rises most rapidly when the depth of cut is increased (at $t = 0.5$ mm the chip temperature is 150–200 °C). In grinding quartz glass without a cutting fluid the temperature in the grinding zone is of 675 °C and more, and therefore an abundant feed of cutting fluid, a 1% aqueous solution of soda ash or, more effective, a 2% aqueous solution of extradiol, should be used in machining such glasses; the solutions have a high heat-removal capability, but affect properties of porous materials.

The greatest influence on the process of grinding ceramic materials is exerted by the diamond concentration and the wheel bond; the cross-feed and depth of cut in grinding exert a lesser effect.

The following conditions are recommended for a surface grinding of SiO$_2$-based ceramics by diamond wheels: $V_w = 30$ m/s; $S_{cr} = 0.9$–1.5 mm/stroke; $S_{long} = 8$–12 m/min; $t = 0.3$–0.6 mm. For an external cylindrical grinding: $V_w = 30$–40 m/s; $V_{work} = 60$–80 m/min; $S_{cr} = 1.8$–3.0 mm/rev; $t = 0.6$–1.2 mm.

The wear of a diamond wheel in the surface grinding occurs on the edge and proceeds nonuniformly: at a higher rate in the initial period of grinding, and then stabilizes. At an external grinding of quartz glass by a 200 mm diameter, 14A1 diamond wheel, the wear rate is the highest during the first 60 min of grinding and then during 210 min, decreases to a constant, much lower (2.5-fold) value. After the end of the stable wear period the wear rate rises steeply.

Alumina-based mineral ceramics, such as TsM 332, are widely used as a structural material in the instrument engineering and also as cutting inserts in the metal machining.

This material after its firing has a high hardness (HRC 89–92), and therefore parts made from it are machined in two stages, before and after the firing. Before the firing the ceramic is a brittle material requiring no great cutting forces, so that a hard alloy or a diamond tool with great rake and relief angles is used at this stage. The final machining of high-precision ceramic parts after the firing is carried out mainly by grinding and finishing.

A similar production technique is employed in the hard alloy technology. A preform, slightly presintered, is machined by tools made of polycrystalline diamonds, cubic or wurtzite boron nitride and then, after the final sintering, is calibrated by cutting tools only of a polycrystalline diamond, such as carbonado.

A correct selection of the tool material and geometry and of the optimum machining conditions is very important in the machining of fine ceramics [8]. In particular, the hardness of the tool should be more than four times as high as that of tools used in the cutting of most traditional materials. Thus, the highest durability in terms of the flank in the cutting of fine zirconia ceramics is exhibited by a single-crystalline diamond tool, but it is costly, and therefore a sintered diamond tool is often used. In the latter case, cutting at speeds of 10–40 m/min is accompanied by appearance during the first 5–20 min of a 0.3 mm wide wear band on the tool flank. A higher wear is also observed in cutting fine Al_2O_3-based ceramics. Because of this, it is usually recommended to carry out the cutting of fine ceramics discretely at small feeds and depths of cut. A cutting fluid should be used to prevent a temperature rise in the working zone; the tool life can be increased 30-fold through selecting the optimum cutting fluid composition alone.

Experiments on a round cutting by a single-crystalline diamond tool were conducted [9] with the aim of studying the cutting of sialon or zirconia under conditions similar to grinding. The depth of cut was here very small, but reached a maximum at one point. The slip of the cutting edge of the diamond, normal cutting force, cut profile, material removal, etc., were studied. Three different regions, an elastic, a plastic and a cutting one, were identified within the cut.

1.1.3.3 Machining of oxide-carbide tool ceramics

The doping of Al_2O_3-based ceramics with refractory metal carbides (WC, TiC, NbC, etc.) as well as the use of a hot pressing instead of a cold pressing, followed by sintering, resulted in production of cutting ceramics with higher strength properties, including bending strength (600–700 MPa). Such ceramics have been called oxide–carbide (mixed) or 'black' ceramics.

A further improvement of the cutting properties, based on improvement of strength characteristics, has been attained through the use of silicon nitride as the base of composite tool materials.

The bending strength of such composites has reached 900–1200 MPa, which allowed a change-over not only from finishing operations, typical for white ceramics, but also from semifinish ones, typical for oxide–carbide ceramics, to semifinish and roughing operations in the machining of steels and cast irons over a broad hardness range (30–50 HRC).

The comparison of physicomechanical properties and compositions of various grades of tool ceramics (Table 1.8) manufactured in different countries discloses a great similarity in physicomechanical properties

Table 1.8 Physicomechanical properties of main tool ceramics grades

Grade	Composition (% mass)		Colour and production method	Density ($10^{-3} \times kg/m^3$)	Hardness (HRC)	Bending strength (MPa)	Grain size (μm)	Origin
I. Oxide ceramics								
TsM 332	Al_2O_3 MgO	99 1	White, cold pressing	3.85–3.90	91	300–350	4	USSR
VO-13	Al_2O_3	99	White, cold pressing	3.92–3.95	92	450–500	3–4	USSR
VSh-75	Al_2O_3	99	Black-grey, hot pressing	3.98	91–92	500	3	USSR
ONT-2 (Kortinit)	$Al_2O_3 + TiN$		Dark-brown, hot pressing	4.39	90–92	640	—	USSR
SN56	Al_2O_3	99.7	White, cold pressing	3.91	$HV = 24$ GPa	550	2.8	Germany, Feldmülle
SN60	Al_2O_3 ZrO_2	90 10	White, cold pressing	3.97	$HV = 20$ GPa	440	3	Germany, Feldmülle
SN80	Al_2O_3 ZrO_2	80 20	White, cold pressing	4.16	$HV = 20$ GPa	500	—	Germany, Feldmülle
W80	Al_2O_3	99	White, hot isostatic pressing	3.97–3.98	$HV = 24$ GPa	700–800	1.2–1.6	Japan, Sumitomo Electric
Widalox G	Al_2O_3 ZrO_2	95 5	White, hot isostatic pressing	4.02	$HV = 17$ GPa	700	2	Germany, Widia Krupp
GC620	$Al_2O_3 + ZrO_2$		White, cold pressing	3.97	$HV = 16$ GPa	—	2–3	Sweden, Sandvik Koromant
KO60	Al_2O_3	99.9	White, cold pressing		93.5	700–770		USA, Kennametal

Table 1.8 (continued)

Grade	Composition (% mass)		Colour and production method	Density ($10^{-3} \times kg/m^3$)	Hardness (HRC)	Bending strength (MPa)	Grain size (μm)	Origin
II. Oxide–carbide (mixed) ceramics								
VOK-60	Al$_2$O$_3$	60	Black, hot pressing	4.20	94	600	2–3	USSR
	TiC	40						
V-3	Al$_2$O$_3$	60	Black, hot pressing	4.30	93	550–650	2–3	USSR
	TiC	40						
VOK-63	Al$_2$O$_3$	60	Black, hot pressing	4.20–4.30	94	650–700	2–3	USSR
	TiC	40						
VOK-71	Al$_2$O$_3$	60	Black, hot pressing	4.20–4.30	94	650–700	2–3	USSR
	TiC	40						
SHT-1	Al$_2$O$_3$ + TiC		Black, hot pressing	4.28	HV = 30 GPa	650	1.8	Germany, Feldmülle
SH-1	Al$_2$O$_3$	60	Black, hot pressing	4.30	HV = 25 GPa	380	2	Germany, Feldmülle
	TiC	40						
SH-20	Al$_2$O$_3$	80	Black, hot pressing	4.28	HV = 21 GPa	400	2	Germany, Feldmülle
	TiC	20						
HC-2	Al$_2$O$_3$ + carbides		Black, hot pressing	4.30	94.5	700–800		Japan, Nippon Technical Ceramics
MC2	Al$_2$O$_3$ + TiC		Black, hot pressing	4.30	HV = 20 GPa	500	2	Germany, Karl Hertel
NB90	Al$_2$O$_3$ + TiC		Black, hot isostatic pressing	4.30–4.35	HV = 30 GPa	950	0.8–1.2	Japan, Sumitomo Electric

Table 1.8 (continued)

Grade	Composition (% mass)	Colour and production method	Density ($10^{-3} \times kg/m^3$)	Hardness (HRC)	Bending strength (MPa)	Grain size (μm)	Origin
NB90M	$Si_3N_4 + TiC$	Black, hot isostatic pressing	4.35–4.40	$HV = 29$ GPa	900	0.8–1.2	Japan, Sumitomo Electric
CC650	$Al_2O_3 + TiN + TiC + ZrO_2$	Dark-brown, hot pressing	4.27	$HV = 18$ GPa	400–500	—	Sweden, Sandvik Koromant
KO90	$Al_2O_3 + TiC$	Black, hot pressing	—	95	910–940	—	USA, Kennametal
III. Nitride (Si_3N_4-based) ceramics							
Silinit-R	$Si_3N_4 + Al_2O_3$ and other additions	Brown, hot pressing	3.8–4.0	$HV = 18$ GPa	700	—	USSR
Silinit-R1	$Si_3N_4 + Al_2O_3$	Black, hot pressing	3.3–3.4	$HV = 20$ GPa	850	—	USSR
SL100	Si_3N_4	Violet	—	$HV = 21$ GPa	—	—	Germany, Feldmülle
SP4	$Si_3N_4 + Y_2O_3 + ZrO_2 + Al_2O_3$ with wear resistant coating	Hot isostatic pressing	—	—	—	—	Japan, Nippon Technical Ceramics
SX4	$Si_3N_4 + Y_2O_3 + ZrO_2 + Al_2O_3$	Hot isostatic pressing	—	—	—	—	Japan, Nippon Technical Ceramics

Table 1.8 (continued)

Grade	Composition (% mass)	Colour and production method	Density ($10^{-3} \times$ kg/m^3)	Hardness (HRC)	Bending strength (MPa)	Grain size (µm)	Origin
CC680	Si$_3$N$_4$ + Al$_2$O$_3$ + Y$_2$O$_3$	Grey	3.17	HV = 14.5 GPa	–	–	Sweden, Sandvik Koromant
Kion-2000	Si$_3$N$_4$ + Al$_2$O$_3$	Black, hot pressing	–	HV = 16 GPa	1200	–	USA, Kennametal
Kion-3000	Si$_3$N$_4$ + Al$_2$O$_3$	–	–	–	–	–	USA, Kennametal
S-8	Si$_3$N$_4$ + Y$_2$O$_3$ + Al$_2$O$_3$	Black, hot pressing	–	89–91	833–914	–	USA, Ford Motor
Quantum-5000	Si$_3$N$_4$ + IrO$_2$ + Al$_2$O$_3$ + TiC	Black, hot pressing	3.19	93.5	703	–	USA, Chemical Metallurgical

both within each of the three groups (oxide, oxide–carbide and nitride ceramics) as well as among these groups as a whole.

The latter fact allows the generalities revealed in studying the machinability of brittle nonmetallic materials, such as optical and quartz glass, corundum, silicon, germanium, sintered SiO_2-based ceramics, to be extended to new ceramic-types of interest, developed on the basis of Al_2O_3 with additions of carbides, nitrides and oxides of transition metals as well as on the basis of Si_3N_4 with additions of Al_2O_3, TiN, TiC, etc.

Comparison of the machinability of the cutting ceramics and of tungsten-containing hard alloys, such as VK8 (92% WC, 8% Co), whose machinability by diamond tools has been well studied by now, shows that similar behaviour in the diamond grinding process is exhibited by tungsten-containing hard alloys with an addition of titanium carbide, such as T15K6 (79% WC, 15% TiC and 6% Co), or by tungsten-free titanium carbide-based hard alloys, such as TN-50, TN-20, KNT-16. etc. [11] (the numeral indicates the amount of bond, TiC being the balance).

This observation is also substantiated by the comparison of physicomechanical properties of such a typical representative of oxide–carbide ceramics such as VOK-60 and of several hard alloy grades (Table 1.9).

Table 1.9 Physicomechanical properties of VOK-60 and hard alloys

Material	Density ($10^{-3} \times kg/m^3$)	Hardness (HRC)	Bending strength (GPa)	Linear expansion coefficient ($10^6 \times deg^{-1}$)	Grain size (μm)
VOK-60*	4.2–4.3	92–4	0.60–0.67	6.0–6.2	2–3
T15K6	11.1–11.6	90	1.10–1.15	5.2	1–2
TN-50	6.0	87	1.20	–	1–2
TN-20	5.5	89.5	1.00	7.1	1–2
KTS-2M	5.8	90	1.10	8.2	1–2
KNT-16	5.8	89	1.10	8.87	1.2–1.8
VK-8	14.4–14.8	90	1.60	5.0	–

* The composition is given in Table 1.8.

Tungsten-free hard alloys are inferior to standard tungsten-containing ones in strength characteristics, have lower crack resistance and thermal conductivity, a higher thermal expansion coefficient, which renders them more sensitive to thermal loads. However, their scaling resistance is much higher than that of, e.g., alloy T15K6 and the temperature of the beginning of adhesion to steel in vacuum (10^{-3} mmHg) is 200–250 °C higher that that for this alloy.

Tungsten-free hard alloys have a lower coefficient of friction on steel. Thus, in friction on annealed steel 40 at a load of 400 N the friction coefficient for alloys VK-8, T15K6, TN-20 and KNT-16 amounted to

0.34, 0.63, 0.14 and 0.24 respectively. Due to this, the cutting force and chip shrinkage in metal turning by tungsten-free hard alloys are substantially lower than for VK group alloys.

The microhardness of tungsten-free hard alloys up to 300 °C is higher than that of alloy T15K6, but at higher temperatures the pattern reverses. In view of this, measures for a high-rate heat removal from the cutting zone by means of an abundant cooling and shortening of the time of the cutting tool contact with the workpiece should be taken in machining tungsten-free hard alloys [5, 10].

The working capacity of diamond wheels with various bonds in the face grinding of ceramic VOK-60 by the simplest scheme was studied on a model 3V642 universal tool-grinding machine in machining cutting inserts of size $5.5 \times 12 \times 12$, $7 \times 14 \times 18$ and $14 \times 14 \times 18$ mm [12]. Wheels 12A2–45° of size $150 \times 10 \times 3$ mm of synthetic diamonds AS4 and AS6 (GOST 9206–80) of grain size 100/80 and a 100% concentration were used. The grinding conditions were as follows: wheel speed, 15 m/s; longitudinal feed, 0.3 m/min; and cross-feed, 0.25 mm/double stroke.

The machining of inserts of VOK-60 and other tool materials by wheel AS6 100/80–B11–2–100% (grinding output, 525 mm^3/min; cutting fluid, 3% aqueous solution of Na_2CO_3) yielded the results presented in Table 1.10. This shows the diamond grinding of ceramic VOK-60 to exhibit some distinctive features, namely, at the same output the effective grinding power for VOK-60 is the lowest, while the relative diamond consumption is high enough, at about the same level as in grinding of alloy TN-50.

Table 1.10 Machining inserts of VOK-60 and other materials

Material	Composition	Effective grinding power (kW)	Relative diamond consumption (mg/g)
VOK-60	Al_2O_3 + TiC	0.25	2.10
VK-8	WC + Co	0.90	0.70
T15K6	WC + TiC + Co	0.70	0.80
TN-50	TiC + Ni + Mo	0.70	2.20
R6M5*	Fe + W + Mo	1.20	17.10

* High-speed steel.

The effect of the area of contact of the wheel's working layer with the material being machined on the characteristics of grinding of ceramic VOK-60 was studied to explain the obtained data. The contact areas in the experiments were of 52 and 140 mm^2, and the contact lengths (i.e., the insert thickness), 5.2 and 14.0 mm respectively. A solution of $NaNO_3$ (5%) and $NaNO_2$ (0.5%) was used as the cutting fluid. It was ascertained (Table 1.11) that a nearly threefold increase of the contact area brings about only a 20–25% increase in the wear of wheels. This indicates that no physicochemical interaction of diamond with ceramics seems to occur, while a high relative diamond consumption, comparable with that in grinding of alloy TN-50, may be ascribed to an intense abrading action of the sludge on the wheel bond. This assumption is supported by a low effective power of grinding of VOK-60 and a decrease in the diamond consumption when wheels with metallic bond MO4 (Cu–Al–Zn–Si), whose hardness exceeds that of organic bonds, are used.

Table 1.11 Characteristics of grinding of ceramic VOK-60 by wheels with organic and metallic bonds

Wheel bond	Contact area (mm^2)	Grinding output (mm^2/min)	Relative diamond consumption (mg/g)	Effective grinding power (kW)	Surface roughness R_a (μm)
Organic bond B11-2	52	400	2.00	0.17	0.21
	140	1050	2.50	0.70	0.17
Metallic bond MO4	52	400	0.39	0.25	0.42
	140	1050	0.47	1.15	0.40

When these wheels are used, the height of micro-irregularities of the machined surface increases, although the diamond consumption decreases considerably. Besides, the grinding by wheels with metallic bond MO4, despite a high output (1000 mm^3/min), brings about chippings on side surfaces of inserts because of higher force and thermal intensity of the process over those in the machining by organic bond tools.

The working capacity of wheels of diamonds AS4 and AS6 with ceramic and organic bonds was compared in the course of the study with an aqueous solution of $NaCO_3$ used as the cutting fluid. The highest grinding power was attained for wheels with ceramic bond K5 (glass bond + Al_2O_3), but the diamond consumption in this case was higher and the roughness of the machined surface (0.42 μm) was greater than in grinding by wheels with organic bond B11-2 (0.21 μm).

The experiments demonstrated also a higher working capacity of wheels with experimental bond B11-2 over that of wheels with serially-produced bonds 01 and BP2. It was ascertained that in wheels with bond B11-2, it is more preferable to use diamonds AS4 (less strong, brittle, with a tendency to self-sharpening).

The grinding of ceramic VOK-60 by wheels of diamonds AS4 with a glass, a ceramic and an aggregated-metal coating was also studied. Wheels with an aggregated-metal coating on diamond grains, providing first of all their better holding in the bond, were found to be more effective.

Thus, wheels of diamonds AS4 (or similar) with an aggregated-metal coating and organic bond B11-2 (or its analogues) can be recommended for grinding of mineral ceramics of the VOK-60 type.

Figure 1.26 Electron micrographs (at varying magnifications) of surfaces of ceramic plates after working on diamond faceplate: (a) Si$_3$N$_4$; (b) Silinit-R; (c) AlN; (d) B$_4$C; (e) SiC.

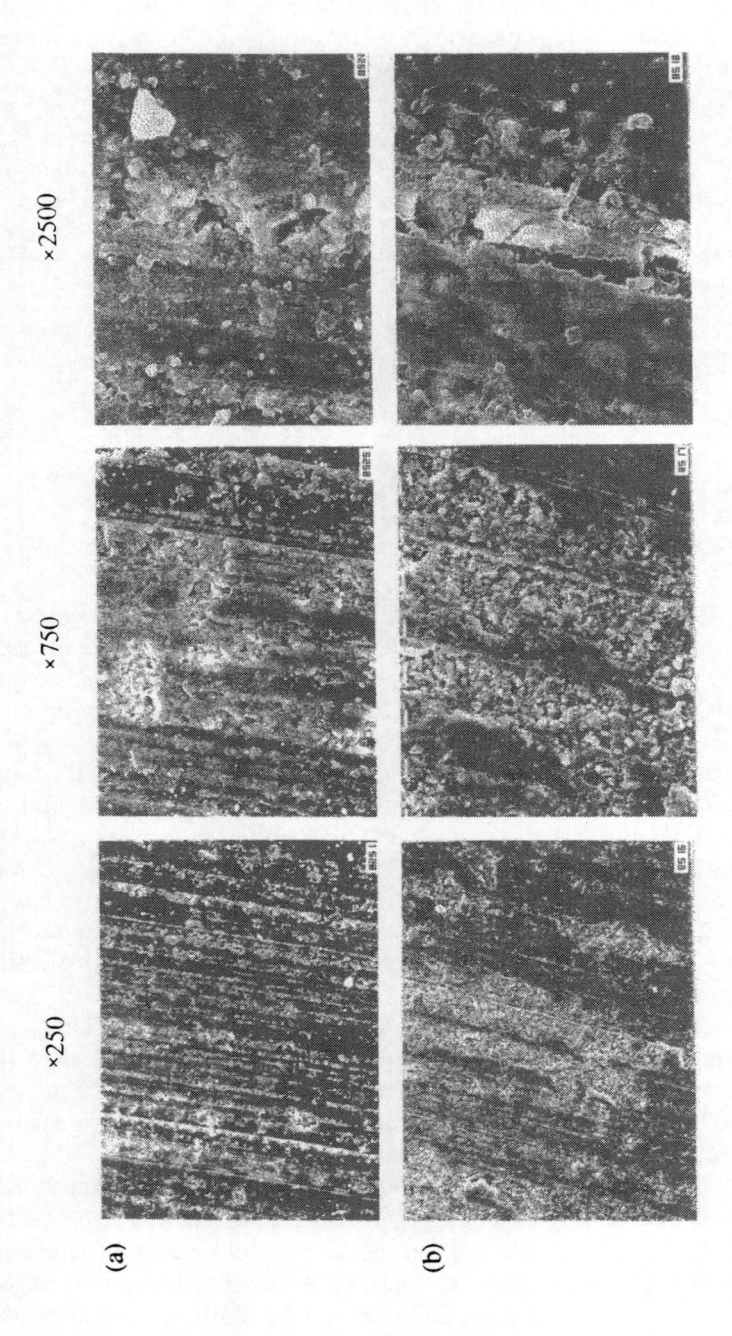

1.1.3.4 Machining of nonoxide ceramics

The need for increasing the strength, reliability and durability of various parts functioning under conditions of high temperatures and an abrasive wear, compels, more and more often, the proposal of nonoxide ceramic-based wear-resistant materials for these applications.

Pressing and sintering or hot pressing of AlN, B_4C, SiC and Si_3N_4 powders in a pure state fail to ensure the required diffusion processes and an adequate strength of sintered products, and therefore such composite materials as $SiC-Si_3N_4$, $SiC-AlN$, B_4C-AlN, Si_3N_4-AlN, and the like, where AlN is most often used as the binder, have gained a wide application.

A diamond working of SiC, B_4C, Si_3N_4, AlN and Silinit-R on a flat diamond faceplate 6A2T 400 × 3AS6 40/28 50% with water by hand, conducted by A. M. Kovalchenko, Y. G. Gogotsi and O. V. Ivashchenko of the Institute for Materials Science Problems, Academy of Sciences of the Ukraine, demonstrated the existence of both a brittle and a ductile fracture (Figure 1.26).

The roughness of the worked surface of these materials, measured by the model 201 profilograph-profilometer made by the 'Kalibr' works, is presented in Figure 1.27.

The profilograms show the roughness level, which is nearly identical for three of the five worked materials, with separate single peaks for SiC, B_4C and AlN. The roughness level for Si_3N_4 and Silinit-R is somewhat lower; there are separate peaks for both materials, but the amplitude for Silinit-R is higher than for Si_3N_4, B_4C and AlN.

As the AlN fraction in the SiC-AlN system is increased, the mean grain size grows due to development of an accumulative recrystallization in this component, whose possibility is determined by a high diffusion activity of aluminium nitride at 1800 °C.

Mechanical properties of the system improve as the SiC content in it is increased to 60% mass. The loss of high strength by the ceramics with a higher SiC content is caused by formation of low-strength SiC-SiC intergranular contacts. The best wear resistance is exhibited by the material with 50% mass SiC [13].

An electron microscope study of friction surfaces after the wear test of such ceramics demonstrated the formation of deep friction scratches on the surface at a high AlN content. A high degree of their distortion evidences a rapid wear of the material because of engagement of projections resulting from spalling and chipping-off of particles. The higher the SiC content, the finer the friction scratches; in a material with 50% Si they are already practically invisible, and the surface acquires a finely dispersed structure.

An electron microscope analysis of the structure of tool ceramics [14] disclosed that they can be regarded as a precipitation-hardened material.

(c)

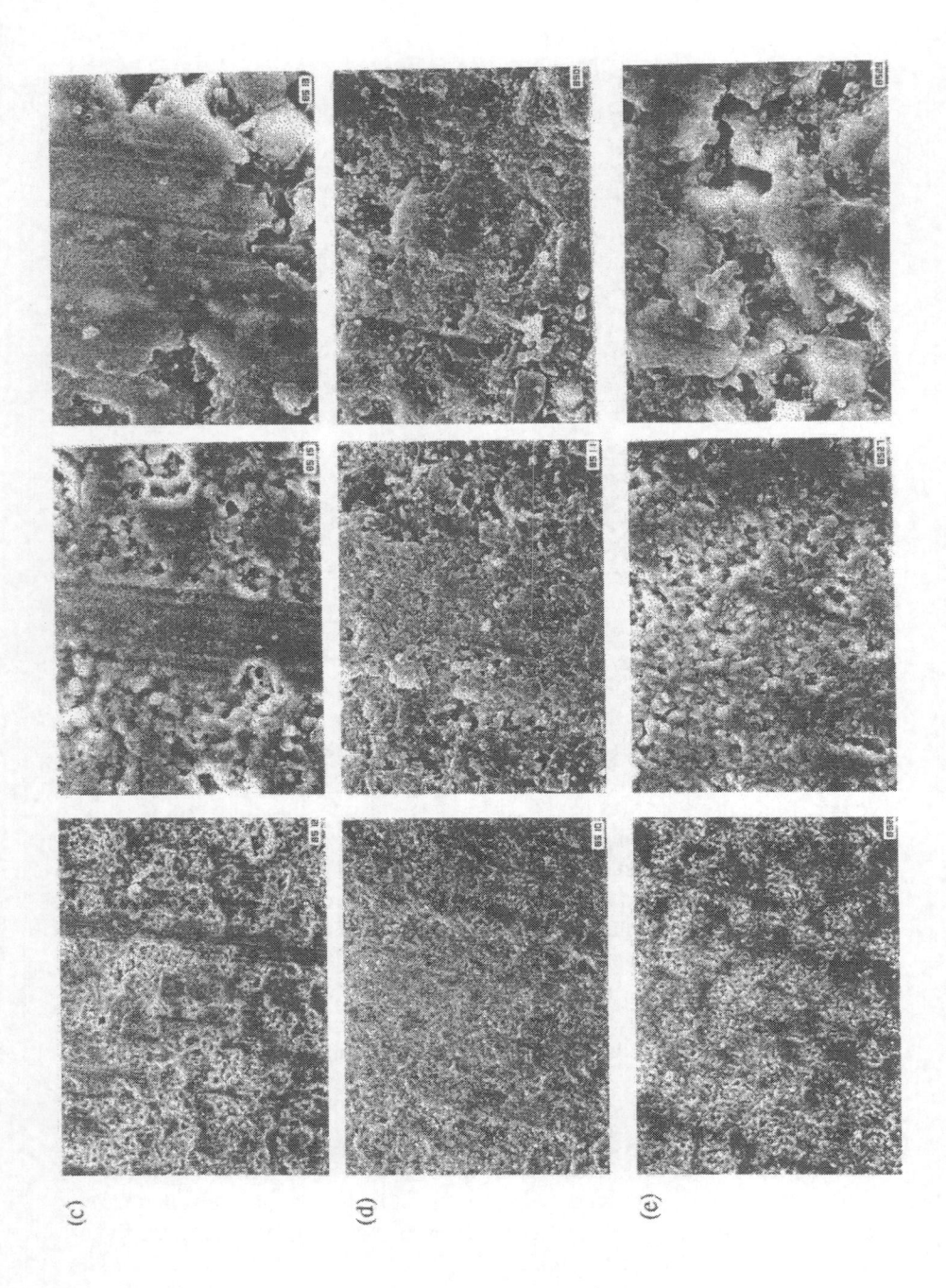

(d)

(e)

Structural materials

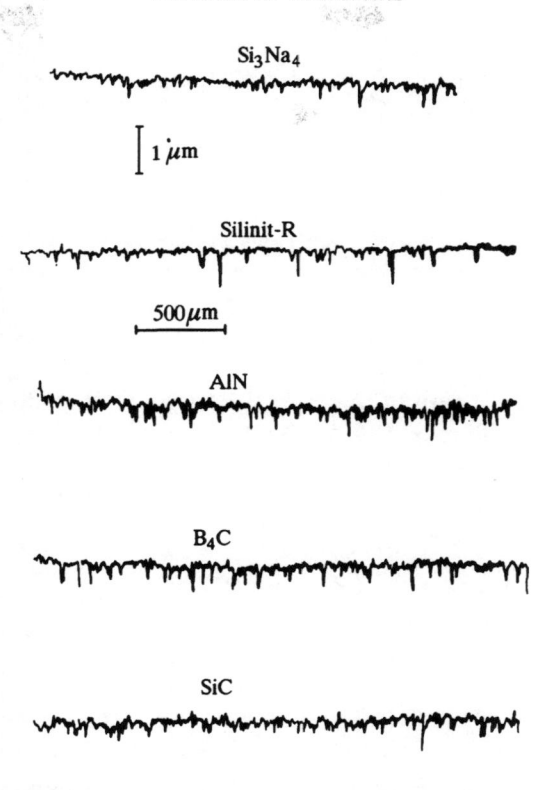

Figure 1.27 Profilograms of surfaces of ceramic plates of Si_3N_4, Silinit-R, AlN, B_4C and SiC after working on diamond faceplate.

The ceramic matrix, depending on the ceramic grade, is reinforced with particles of TiC, TiN, WC, Mo_2C, ZrO_2, etc. In ceramic TsM332, the size of Al_2O_3 grains reaches 10 µm, and of MgO grains, fractions of a micron. In ceramic VO13, the Al_2O_3 particle sizes are not over 3 µm. In oxide–carbide ceramics the Al_2O_3 grain sizes reach 2–5 µm, and those of TiC are within 0.2 to 2.7 µm. The size of reinforcing TiN particles in Kortinit is of 0.1–0.5 µm. The content of reinforcing particles in ceramics varies; for example, in oxide–carbide ceramics (VOK-60) it amounts to 20%. After the diamond machining, the surface layer of ceramics has cracks to a depth of down to 2–5 µm and high internal stresses exist in the bulk. The porosity of oxide ceramics is over 2%, while in oxide– carbide it is less than 2%. Pores in oxide ceramics are situated at grain boundaries and have an acute-angle shape, which produces an additional stress concentration in the bulk, whereas pores in oxide–carbide and nitride ceramics are of a rounded shape. Fracture stresses in ceramic TsM332 are of 30–35 MPa, and in VO-13, of 50–55 MPa. Sizes of critical defects reach 100 and 200 µm respectively.

In ceramics TsM332 and VO-13, the surface density of dislocations in individual grains after the diamond machining is of $10^6\,\text{cm}^{-2}$; the grain boundaries and particles in the bulk of inserts are free from dislocations. Since oxide–carbide and nitride ceramics are produced by a hot pressing, individual Al_2O_3 and TiC grains become fragmented, and the dislocation density can rise to $10^{14}\,\text{cm}^{-2}$.

Thus, the main causes of a high wear resistance of oxide–carbide and nitride ceramics, superior to that of oxide ceramics, are their fine-grained structure as well as a substructural and precipitation mechanism of hardening, which imposes additional requirements on diamond machining conditions.

An experimental study of parameters of machining of SiC, Si_3N_4 and Al_2O_3 by special diamond and borazon tools [15] demonstrated the feasibility of finishing of coarse-machined new ceramics by selecting the cutting speed and pressure on the tool and bond, with the result that finished surfaces (with practically no waviness and a low roughness) are obtained at a high productivity.

The author of [15] managed to provide a defect-free surface structure, excluding the appearance of burns, recrystallization zones, etc. He developed and effected the procedure of a six-step finish machining of the ceramics on one and the same machine tool without rearrangements of the workpiece, which is essential to ensure the accuracy of dimensions. The experiments demonstrated that it is more economical to machine Al_2O_3 by a diamond tool; Si_3N_4, by a cubic boron nitride tool; and SiC, equally efficiently by either of them.

An investigation into the surface state at a diamond grinding of fine ceramics, conducted by Japanese researchers [16], showed the appearance of grinding defects to be considerably affected by the diamond grinding conditions and the grinding tool type used.

The grinding was carried out on a precision grinding machine by series SD 150, SD 80, and SD 400 diamond wheels under the following conditions: peripheral grinding speed, 1600 m/min; table feed rate, 1.4 m/min; depth of cut, 15 µm; cutting fluid, JIS W2-2. The state of the surface layers of fine ceramics after the diamond grinding was examined for cracks and other defects; the sub-layer of the thin ceramics was studied.

A layer with minutest cracks was found to form under the above conditions, which in Si_3N_4- and Al_2O_3-based fine ceramics produced by hot isostatic pressing was considerably greater than in ZrO_2- and SiC-based fine ceramics. Using diamond wheels with smaller diamond grain sizes reduced the layer with cracks in Si_3N_4- and Al_2O_3-based ceramics; grinding these ceramics by metallic-bond wheels resulted in a greater thickness of the defective layer with cracks than that in grinding by wheels with a bakelite–rubber bond. Crack nucleation zones were found in the sub-layer of the fine ceramics, whose number and density depended on the type of ceramics and of the diamond wheels.

Considering the modern state and prospects for advance of the processes of grinding and polishing of hard-to-machine materials, including new grades of these ceramics, one cannot but agree with the conclusions made in [17] about the equipment needed for this purpose. Its author considers diamond wheels and special multipurpose grinding machines where the wheel-truing and an automatic change of wheels from the machine's tool magazine are provided to be most promising for this purpose. Also attractive are grinding machines with combined grinding processes: with an electrolytic and an electric-discharge grinding; with superposition of ultrasonic vibrations; and with the use of modern electric-discharge methods of the wheel-truing in an automatic mode. A further wider application for a precision, submicronic-accuracy, machining of fine ceramics will be found by a magnetic grinding, chemicomechanical polishing and combined working methods.

Along with the above-presented most widely used methods of the diamond machining of ceramic materials, an ever-increasing importance is being gained by radically new combined methods, where plasma, electron beam, laser, and other working techniques are combined with the machining. Such an approach makes it possible to combine roughing and finishing operations, increases the accuracy and output of the working, but at the same time creates new difficulties, associated with the effect of various new factors on the surface roughness and the appearance of residual stresses in the surface layer. The analysis of these complex processes should become the subject of a separate detailed study.

1.1.4 Joining (brazing) of ceramic materials
Yu. V. Naidich

Nomenclature

h	capillary rise
r	capillary radius
σ_{lv}	surface tension of melt
θ	contact angle of wetting
ρ	density
W_A	work of adhesion
$F(x)$	force of attraction between surfaces of 10^{-4} m^2 in area
a	equilibrium (initial) distance between surfaces
x_k	radius of action of interatomic forces
χ	linear thermal expansion coefficient
l	linear dimension

D	diameter of disc-shaped brazed joint
$\Delta G^0 = \Delta G^{0'} - \Delta G^{0''}$	difference of free energies of formation of solid phase oxide and liquid metal
$\Delta \bar{H}$	partial enthalpy of dissolution

Ceramic materials (inorganic nonmetallic materials based on oxides, nitrides, carbides, and other refractory compounds and composites) are often used in engineering in combination with metallic structures, which raises the problem of joining ceramic parts with metallic elements.

Detachable joints, such as threaded or flanged (bolted) ones with sealing elastic (organic) gaskets are difficult to make for ceramic materials or are bulky, heavy, unreliable, with a low working temperature.

Adhesive joints also have grave disadvantages, in destruction of organic adhesives, gas release from them, high brittleness of inorganic glass and glass-ceramic cements, etc.

In view of high demands placed on the quality of a joint, such as a high strength, leak tightness, thermal stability, small size of the assembly, it is attractive to produce such joints by metallurgical techniques, namely by brazing or welding.

However, characteristic properties of nonmetallic inorganic (ceramic) materials – high brittleness, low thermal shock resistance, considerable difference between thermal expansion coefficients of the materials and metals, and, especially, chemical inertness and nonwettability by ordinary brazing alloys – prevent the application of brazing alloys and joining methods, used in brazing of metals, which calls for developing special brazing alloys and techniques for joining ceramic materials.

Problems of joining nonmetallic materials (first of all, oxide ceramics) were dealt with in many studies conducted in 1930–60s, whose results were discussed in a number of reviews [1–4].

The research in this field was spurred by the advance, in particular, in the electronic engineering industry, namely, the replacement of glass in electron tubes with ceramic materials, offering higher strength, dielectric and thermal properties, first with those based on Al_2O_3–MgO–SiO_2 oxide systems and then with those with an alumina content as high as 90–95%. It should be noted that the developed so-called molybdenum–manganese technology of depositing a metal coating (metallization) on ceramics, followed by brazing with ordinary brazing alloys, turned out to be fairly advanced for joining the latter materials.

Quite a number of new nonmetallic materials have been developed and put into use, especially in recent years: ceramic materials based on pure corundum, including an especially dense and optical transparent as well as a single-crystalline one, on magnesia single crystals; crystalline and optically perfect quartz glass, and radioparent materials; and, finally, new high-strength structural ceramic materials of silicon nitride and

Structural materials

carbide, ultrahard tool materials based on cubic boron nitride, as well as various composites.

The development of new materials calls for working out corresponding techniques for joining elements made of them with one another and with metallic structures.

The present section discusses the problems of joining (brazing) non-metallic inorganic materials, taking into account modern achievements in physicochemistry, materials science, heterophase systems and surface phenomena.

In view of the diversity of nonmetallic (ceramic) materials used today and promised for the future, and since their contact with metallic substrates is dealt with, it is useful to class the materials into two groups as follows.

(a) Ceramic materials based on compounds with predominantly an ionic interatomic bond: oxides, glasses and silicates, glass ceramics, halides and other salt-like refractory compounds, sulphides, selenides, etc. Although single-crystal oxides or single crystals of other compounds as well as glasses are formally not covered by the term 'ceramic materials', these nonmetallic materials in essence, from the standpoint of physicochemical and a number of physicomechanical properties which are important for their joining with metals, are little different from ceramic materials proper (except a possible anisotropy of characteristics) and will be further discussed jointly with the latter.

(b) Ceramic materials based on substances and compounds with covalent interatomic bonds: silicon nitride and carbide, boron carbide, hexagonal and cubic boron nitride, polycrystalline diamond materials, etc.

In its character, the ionic bond is more different from the metallic one than is the covalent bond; substances forming the basis for group (a) materials generally have a greater heat of formation, are more chemically inert, and are worse wetted with molten metals; materials of group (b) are, as a whole, easier to wet with metallic liquids and, as will be shown below, a wider assortment of alloy compositions and methods for attaining a small contact angle of wetting of the surface of 'covalent' materials are available.

The thermal expansion coefficient of group (a) materials is usually higher than for group (b) ones (except quartz glass, some other glasses and glass ceramics).

A contact between two bodies with the greatest possible true contact area at a given geometric area of the contact boundary (plane) can be attained two ways; first, one of the bodies is in a liquid state, and the liquid phase at a good wetting fills irregularities and roughnesses on the solid body surface; secondly, one of the bodies is in a solid, but highly plastic state and deforms under load, filling the solid body surface roughnesses. It follows that two ceramic joining methods are feasible:

1. brazing with the use of a molten brazing alloy;
2. solid-phase 'brazing', based on deformation of a metallic part or gasket under pressure.

The principal problems to be solved in developing a process of joining by the above methods are:

1. provision of the required (high) degree of wetting of the surfaces being brazed together with the brazing alloy and of a high adhesion in the joint;
2. elimination or reduction of stresses in the joint zone, caused by the difference of thermal expansion coefficients of the materials being joined at the temperature change after the joint cools, or at temperature variations in the course of service of the product, or by some other causes.

Let us discuss the methods for solving the above problems as applied to each of the above-specified classes of joining methods.

1.1.4.1 Wetting with metallic melts. Adhesion

Theoretically, for the brazing alloy to flow into the gap in brazing, the angle of contact (θ) should only be less than 90°, which follows from a well-known relation:

$$h = \frac{2\sigma_{lv}\cos\theta}{r\rho g} > 0, \quad (1.16)$$

$\cos\theta > 0$; $\theta < 90°$; σ_{lv} is the surface tension of the melt and h is the height of liquid rise in a capillary of a radius r. Practically, to attain a quality brazed weld, θ must be small enough; as shown by experience, it should be within 0–20°. Unfortunately, nonmetallic materials generally do not interact with metals and are poorly wetted with standard metallic brazing melts.

Metals, on the contrary, are well wetted with metallic melts (nonwettability in intermetallic systems is caused by presence of contaminations on contact surfaces [5, 6]). Due to this, an obvious method for attaining a high degree of wetting of nonmetallic solids with metallic melts is the deposit of a metallic layer on the surface of the solids (metallization). There are various methods for the metallization of nonmetals (glass, ceramics, etc.): thermal spraying and sputtering of a metal, chemical deposition, application of special metallizing pastes, followed by annealing, etc. A number of problems should be solved here, such as providing the required film thickness to ensure the specified high wetting degree, attaining a high adhesion of the film to the substrate material, etc.

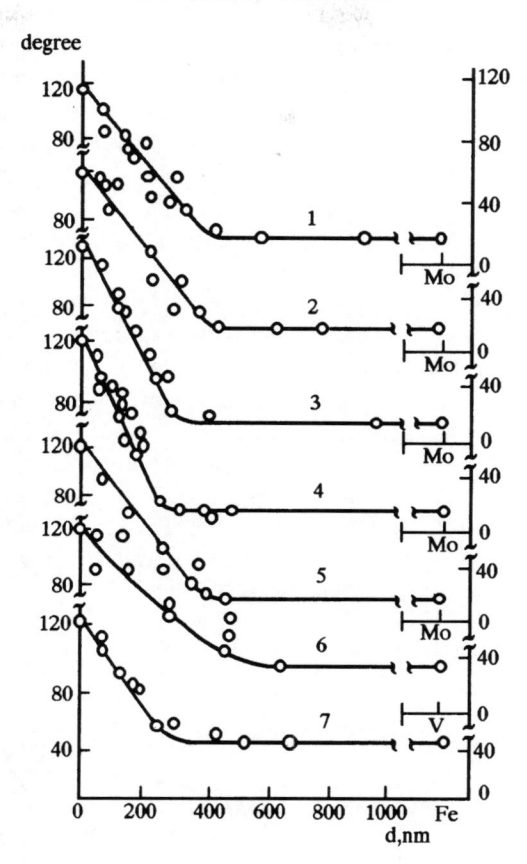

Figure 1.28 Contact angles of wetting of nonmetallic materials by metallic melts as a function of thickness of metallic film deposited on surface of materials: (1) Al_2O_3–Mo–Cu, 1150 °C; (2) SiO_2–Mo–Cu, 1150 °C; (3) Al_2O_3–Mo–Ag, 1000 °C; (4) SiO_2–Mo–Sn, 900 °C; (5) SiO_2–Mo–Sn, 900 °C (pre-annealing of film at 1150 °C); (6) SiO_2–V–Sn, 900 °C; (7) SiO_2–Fe–Pb, 700 °C.

Figure 1.28 shows the degree of wetting of a number of materials with metals as a function of the thickness of the refractory metal film deposited by the electron beam evaporation [7, 8]. This data indicates that already at thicknesses of 10^{-2} μm an island structure of the film changes to a continuous one and the angle of contact decreases to values characteristic for a given monolithic refractory metal of the film. Thermal spraying methods are employed (to a limited extent) for practical purposes of joining glasses and some ceramics with metals.

An extensive application in the ceramic–metal brazing practice has been gained by the method of 'molybdenum–manganese metallization' of ceramics, consisting in burning-in a metallizing layer on the surface of a nonmetallic material. The metallizing blend consists of molybdenum and manganese powders, and special additives. The method was proposed in the 1930s [9], has been greatly improved later on, and is at present one of the few methods used on a commercial scale. It is described in detail in [1–4] and will not be discussed in this section; it is applicable only to a certain narrow type of group (a) ceramics, namely alumina-based ceramics with a definite amount of a glass phase.

A high degree of wetting of a nonmetallic material can be attained by the use of special adhesion-active alloys.

The high-temperature capillarity, adhesion and wettability in 'liquid metal–nonmetallic solid' systems have been studied in a sufficient detail [5, 6]. (More than a thousand of such contact systems have been studied.) At the same time, the wetting of a number of types of solids has been hardly studied, if at all. Results of studying the adhesion and wettability with liquid metals of both traditional materials (oxides such as quartz glass, single and polycrystals of alumina, and magnesia) and little investigated materials, such as oxides of rare-earth elements (REEO), oxygen-free ionic compounds (some halides and chalcogenides), silicon nitride and carbide, etc., are discussed below.

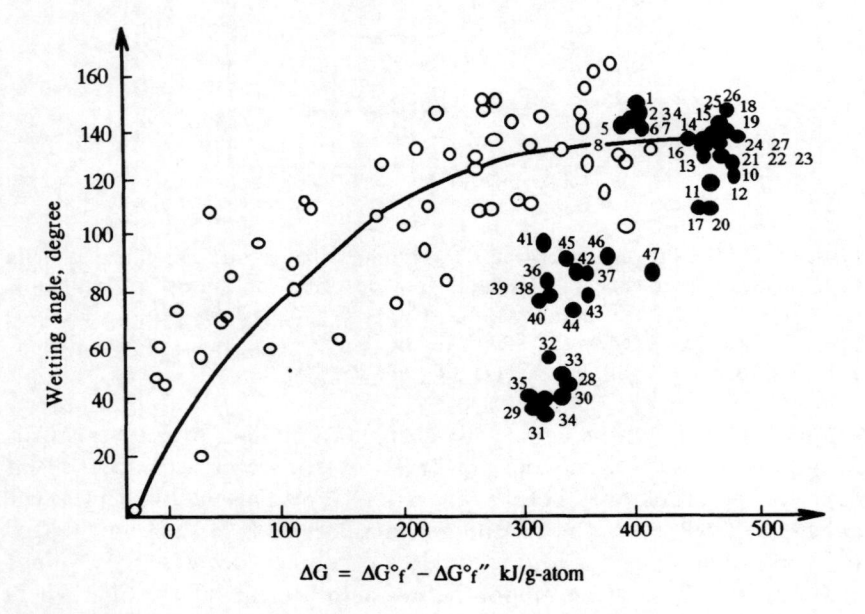

Figure 1.29 Correlation between wetting degree (angle of contact) and difference of free energies of formation ($\Delta G°$) of liquid phase metal oxide ($\Delta G°'$) and solid phase oxide ($\Delta G°''$): o contact systems presented in [5, 6]; • REEO systems with melts: Cu (1–9); Ag (10–18); Au (19–27); Ge (28–36); Sn (37–45). The REEO sequence in all systems in accordance with the indicated increasing numbers is as follows: Y_2O_3; Gd_2O_3; Tb_2O_3; Dy_2O_3; Ho_2O_3; Er_2O_3; Tu_2O_3; Yb_2O_3; Lu_2O_3.

Structural materials

The analysis of experimental evidence, in particular of that obtained recently, based on the chemical theory of wetting [5, 6, 10–13], which establishes the relation between the angle of contact and the intensity of chemical reaction, the differences of free energies of formation, at the contact interface (see Figure 1.29) substantiates the basic concept of this theory, but at the same time indicates the need to take into account additional factors: the degree of 'metallicity' of products of the reaction at the phase interface and the liquid metal's ability to form varying-valence compounds with the substrate element, the contribution of the energy of interaction between the liquid metal and the cationic sub-lattice of the oxide, the thermodynamic activity of the component having been added in the metallic melt, etc. The obtained results allow the phenomena of capillarity and wettability at a high temperature to be controlled and utilized for practical purposes of brazing ceramic materials.

It is convenient to use concentration dependences of the contact angle of wetting (wetting diagrams), which make it possible to select alloys with the required capillary characteristics. Such diagrams for a number of systems are presented below.

Consider the wettability with metals of the group (a) materials (based on ionic compounds). As shown by many studies, metals having a low affinity for oxygen (copper, tin, nickel, etc.) do not wet refractory oxides of s and p elements. The same behaviour is also exhibited by metals with

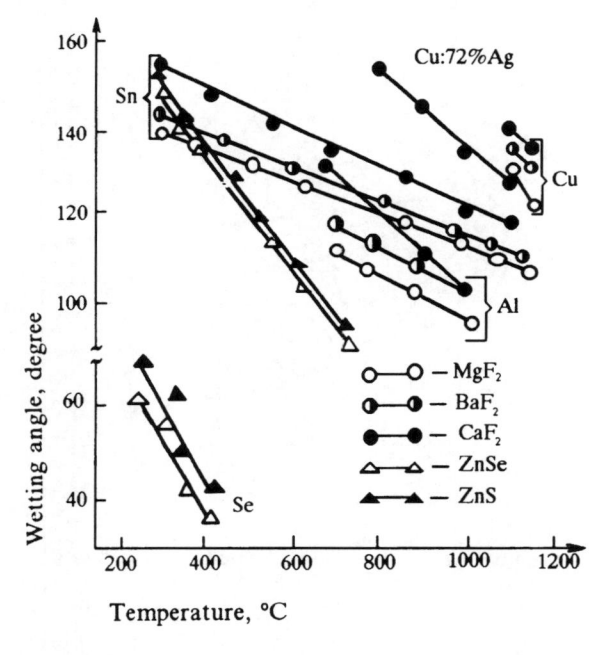

Figure 1.30 Wettability of oxygen-free ionic compounds by melts.

respect to refractory halides and chalcogenides of these elements (Figure 1.30). According to the existing theory, the surface of such phases is, due to the large size of anion, formed by negative ions of the corresponding nonmetal (oxygen, sulphur, etc.). The wettability in these systems can be increased by adding to the liquid metal:

- elements featuring a high chemical affinity for the anion (oxygen), titanium is used most often for this;
- elements featuring a high affinity for the electron, such as oxygen, sulphur, chlorine, etc.

Figure 1.31 shows data on wettability of oxides with a number of titanium-containing melts [14, 15]. Titanium drastically reduces the contact angle of wetting, but its action varies from system to system. The work of adhesion was shown [14] to be much greater in cases where the chemical reaction results in the formation at the phase interface of a transition layer of titanium monoxide (TiO), a compound having a metallic conductivity and a metallic interatomic bond, as compared with that in systems where Ti_2O_3, a compound with a considerable share of ionic bonds, is formed.

The mechanism of adhesion bonds between a metal and an ionic compound can be presented as follows: a boundary atom of the liquid-phase metal (in the case of a high chemical activity) gives off the valence electron to a solid-phase anion (vacancies at these levels result from a partial thermal dissociation of bonds in the solid). The metallic ion thus

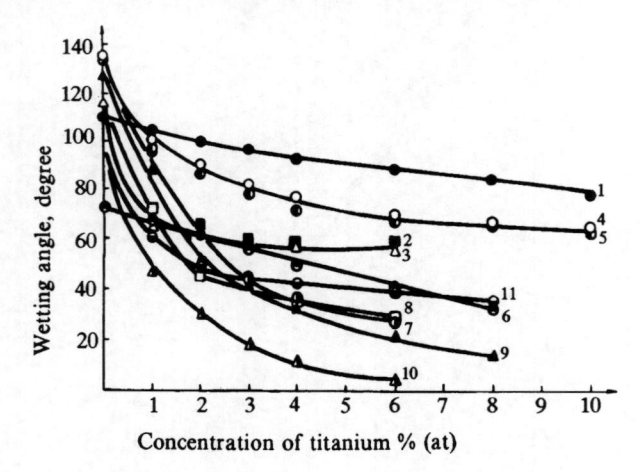

Figure 1.31 Wettability of oxides by titanium-containing melts:
(1) Al_2O_3–Ni–Ti, 1500 °C; (2) Al_2O_3–Ga–Ti, 1050 °C; (3) SiO_2–Ga–Ti, 1050 °C; (4) Al_2O_3–Au–Ti, 1100 °C; (5) Al_2O_3–Ni–Mo–Ti (58:42), 1500 °C; (6) $TiO_{0.86}$–Cu–Ti, 1150 °C; (7) MgO–Cu–Ti, 1150 °C; (8) SiO_2–Cu–Ti, 1150 °C; (9) Al_2O_3–Cu–Ti, 1150 °C; (10) SiO_2–Sn–Ti, 1150 °C; (11) Al_2O_3–Sn–Ti, 1150 °C.

formed gets bonded by an ionic bond to an anion of the oxide, thereby completing the lattice of the latter. The same ion remains bonded by a metallic bond to deeper atoms of the liquid-phase metal. The latter bond is weakened because of pulling of some or all valence electrons on the boundary atom to the anion of the oxide. The resulting bond between the liquid metal and the ionic compound is determined by a weak link in this chain, namely the energy of the ionic bond at a low affinity of the metal for oxygen or the metallic bond between the metal ion and metal atom for elements with a high affinity. The latter bond will be strong in the case where some of the valence electrons remain on the boundary metallic atom (ion), i.e., when an intermediate-valence ion is formed and is stable. This is easy to attain with transition metals, in particular titanium, where s, p and d electronic levels are situated close to one another and can hybridize. Transition elements with a high chemical affinity for a nonmetal (especially titanium) are the most widely used adhesion-active additions to brazing alloys in the wetting and brazing of ionic compounds.

Apart from titanium, additives such as zirconium, hafnium and some others can be used for doping to increase the wettability of group (a) materials by alloys based on copper, tin, nickel, gold and other nonactive metals. In practice, however, the use of these elements is impeded by their very high oxidability and the difficulty of attaining a clean (without oxide films) surface of the liquid alloy. The use of titanium and other chemically active transition materials is also effective for upgrading the wettability of sulphides and selenides [16, 17]. Contact angles of the zinc sulphide (ZnS) wetting with tin and its alloys with active additives (0.5 at. %) are as follows [16].

Table 1.12 Wetting angles of zinc sulphate with active additives to tin and its alloys

	Sn	Ti	Zr	Hf	Nb	V
Wetting angles(°)	109	52	50	45	62	70

At a contact of titanium-containing melts with halides (flourides), the formation of low-melting or even gaseous titanium fluorides at the interface should be taken into consideration; the wettability in such systems is unstable, dewetting being observed [18].

Figure 1.32 presents another method for doping the metal melt to increase its wettability, namely by addition of electronegative elements: oxygen for oxides; sulphur, selenium for sulphides and selenides [5, 6, 17]. When dissolving in the liquid metal, the electronegative elements create electron attraction (drain) sites, giving rise to positive metallic ions which get bonded to the electronegative surface of oxides or sulphides and selenides.

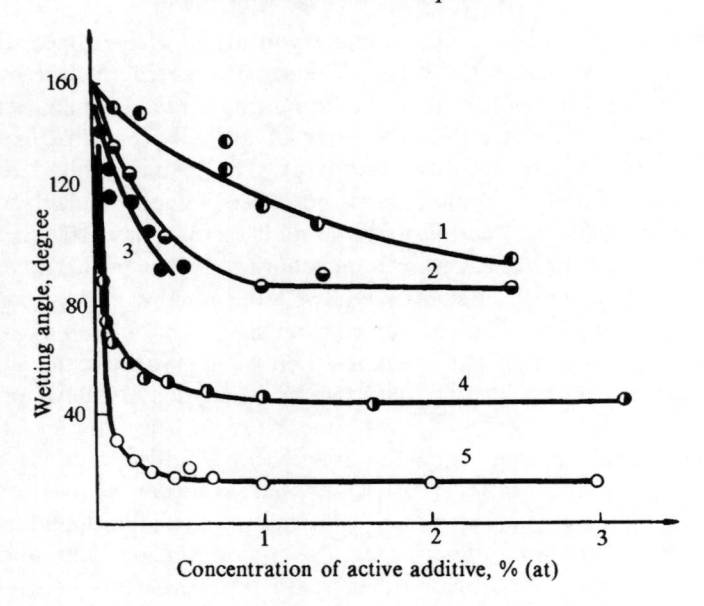

Figure 1.32 Interphase activity of nonmetals in wetting of ionic compounds by metals: (1) MgO–(Cu–O), 1150 °C; (2) Al_2O_3–(Cu–O), 1150 °C; (3) Al_2O_3–(Ni–O), 1500 °C; (4) ZnS–(Sn–Se), 700 °C; (5) ZnSe–(Zn–S), 700 °C.

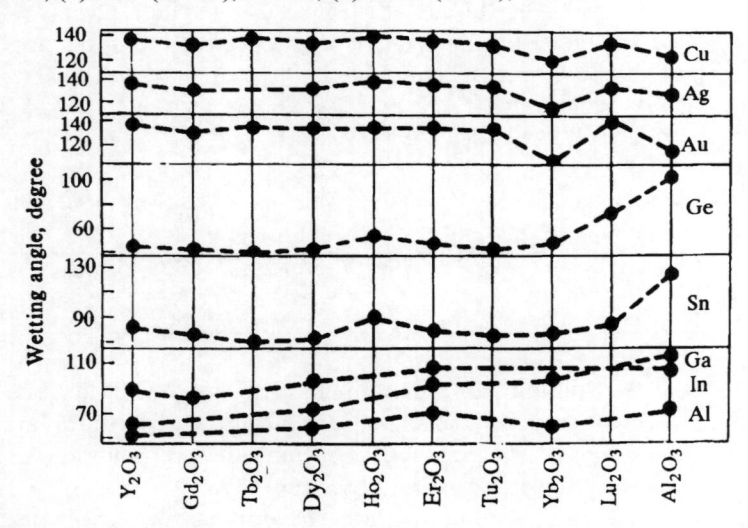

Figure 1.33 Wettability of rare earth element oxides by metals. $T = 1423$ K.

The wettability of rare earth metal oxides has a number of specific features [19]. REEO exhibit high formation energies, about 1.2 times as high as that for alumina, and therefore their poorer wetting with metallic melts as a whole should be expected; this is indeed the case for copper, silver and gold (Figure 1.33). However, a number of metals (in particular Sn, Ge), wet REEO much better than alumina; the data on contact angles of wetting are indicated below the generalized curve in Figure 1.29. This can be accounted for as follows. Due to a large size of ion ($R_{AC}{}^{+3}/R_O{}^{-2} = 0.4$), the surface of a 'traditional' oxide can be practically considered as consisting of oxygen anions, and therefore the interaction between the liquid-phase metal and cation is insignificant. High values of the ionic radius of metal are typical for REE so that the metal-to-oxygen ionic radii ratio is greater, about 0.6–0.8.

Thus, the share of cations on the REEO surface rises; at the same time, rare earth elements exhibit high values of the energy of interaction with many metals, in particular with Sn and Ge. Partial enthalpies of dissolution of some REE in germanium, tin and copper ($-\Delta \bar{H}$, kJ/mol) are as in Table 1.13.

Table 1.13 Partial enthalpies of dissociation (kJ/mol) of some REE in various metals

Metal	Rare earth metal		
	Y	Gd	Er
Ge	270	240	213
Sn	154	211	–
Cu	98	–	92

Liquid copper, although it 'geometrically' can interact with cations of a rare earth oxide, does not form these bonds because of a low chemical affinity, and a non-wetting occurs in the system.

A titanium doping of non-wetting metals reduces the angle of contact also in systems with rare earth element oxides. The data is presented in Table 1.14.

Table 1.14 Dependence of contact angles in liquid copper–titanium alloy – REE oxide system on Ti concentration

Oxide	Ti (at. %)						
	1	5	10	20	30	40	50
	Angle of contact (°)						
Y_2O_3	135	91	75	68	65	62	59
Gd_2O_3	125	98	79	67	64	60	57
Er_2O_3	132	96	81	69	66	63	58

Covalent refractory compounds (group (b) materials) are first, silicon carbide and nitride, and boron nitride (cubic and hexagonal), most practically important ones which have a high enough enthalpy of formation ($\Delta H^0_{SiC} = 62.8$ kJ/mol; $\Delta H^0_{298}(\frac{1}{4} Si_3N_4) = 251$ kJ/mol) are not wetted with a number of metals having a low chemical affinity for carbon, nitrogen

84   *Ceramic-matrix composites*

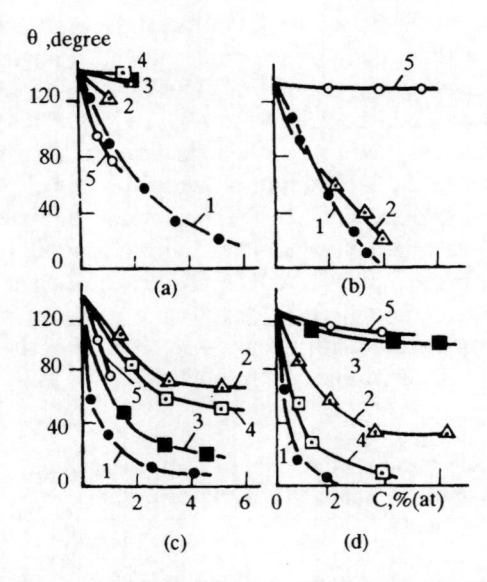

Figure 1.34 Wettability of silicon nitride by liquid alloys based on (a) copper; (b) gold; (c) tin; (d) germanium: (1) Ti; (2) V; (3) Nb; (4) Ta; (5) Cr; $T = 1150\ °C$.

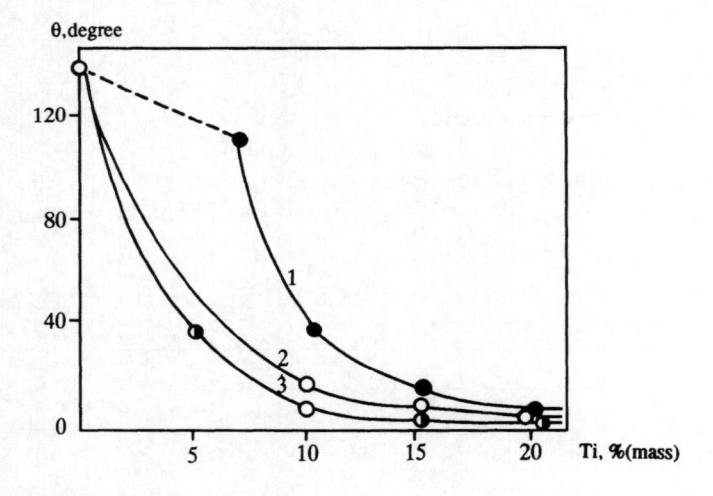

Figure 1.35 Effect of titanium on cubic boron nitride wettability by liquid metals: (1) Cu–Ti; (2) (Cu + 20 mass % Sn)–Ti; (3) (72% Cu + 18% Sn + 100% Pb)–Ti.

and silicon [5, 6, 20]. Some data on the contact angles of wetting are presented in Figure 1.34 and 1.35 and Tables 1.15 and 1.16. The observed large contact angles (90°) and a low work of adhesion ($10^{-1}\ J/m^2$) in these systems are due to a physical interaction: van der Waals forces.

Table 1.15 Contact angles of wetting in metal–silicon carbide system

Metal	T (°C)	Contact angle (°)
Ag	1100	128
Au	1150	138
Ga	800	118
In	800	130
Ge	1050	113
Cu	1100–1250	Interaction, formation of intermediate phases, wetting
Al	1100	34
Ni	1460 (short-time holding)	65
Si	1480	36

Table 1.16 Wettability in boron nitride–metal system (in vacuum)

Metal	T (°C)	Contact angle (°)	
		cubic modification	hexagonal modification
Cu	1100	137	146
Ag	1000	146	140
Au	1100	145	–
Ga	1100	130	–
In	1000	110	136
Ge	1100	138	139
Sn	1100	137	150
Si	1500	95	110
Al	1100	60	90
Co	1500	–	35
Ni	1500	–	75
B	2200	–	133

A high wettability degree should be expected in systems with metals having a considerable chemical affinity for components of the compounds under consideration (nitrogen, carbon, silicon), which is close to the thermodynamic strength (free energy of formation) of the compound.

Figure 1.34 shows diagrams of wettability of a silicon nitride ceramic (95% Si_3N_4) with metallic melts [20]. The highest adhesion activity is exhibited by titanium; a close one, by tantalum; they are followed in the order of decreasing activity by vanadium, niobium and chromium. This series corresponds to decreasing Gibbs' energy of formation of corresponding nitrides.

Ceramic materials are often composites with a metallic or a ceramic matrix. Studying the wettability of such materials in developing the brazing technology represents an independent problem, which can be formulated as determining the wettability of a composite with known wettabilities of its constituent phases. Such studies were conducted for

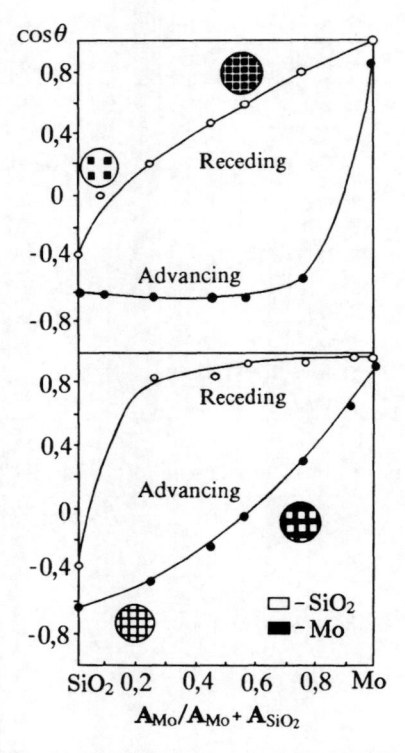

Figure 1.36 Wettability of heterogeneous glass ceramic-metal (Mo) surface as a function of component area ratio (model structure; metal islands in the form of squares with 10 μm sides on glass ceramic field). Tin, $T = 900$ °C: ● advancing contact angles; ○ receding contact angles.

two-phase structures [21]. The contact angle of wetting as a function of the ratio between areas of differently wet table surface portions (for model systems with an ordered arrangement and a uniform shape of the portions, these are squares of a metallic phase, of the order of 10 μm in size) on the oxide field is presented in Figure 1.36. A considerable difference between advancing and receding angles (wetting hysteresis), the effect of the shape and continuity of the wetted and non-wetted phases (a negative and a positive pattern) and of other factors are observed.

Figure 1.37 presents data on the wetting with metals of real graphite-metal, diamond–metal composites and of a composite ceramic material, silicon nitride–alumina (with the addition of yttria) [22]

1.1.4.2 Mechanical strength of adhesive contact

A high wetting degree determines a high capillarity of the system consisting of a metallic brazing melt and of the surfaces being brazed

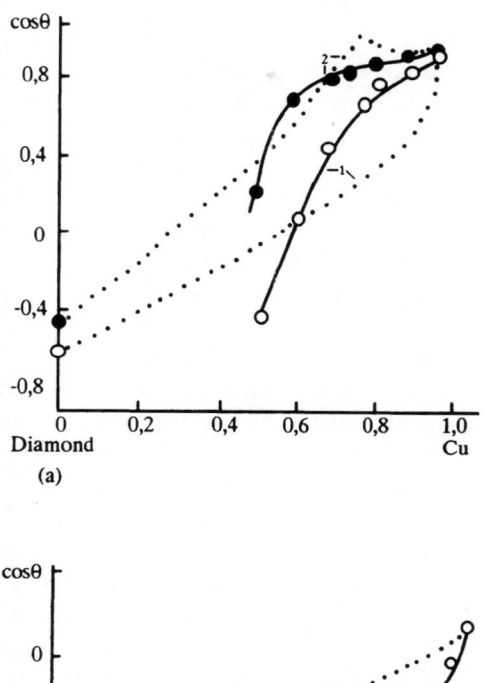

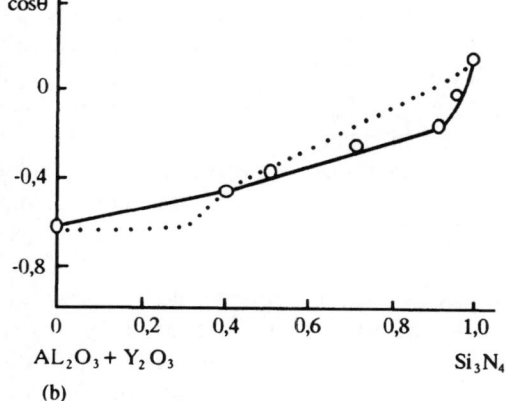

Figure 1.37 Wettability of real composites: silicon nitride–alumina; Cu–Ga–Cr alloy. $T = 1150$ °C.

together, which define the brazing gap, i.e., drawing the melt into the gap and holding it there, a high adhesion of the liquid phase to the solid one. In other words, it determines the very possibility of effecting the joining process, of forming a uniform continuous leak-tight brazing seam. A high wettability and a high thermodynamic adhesion of the liquid to the solid phase are necessary (but not sufficient) conditions for a high mechanical strength of the joint. Apart from physicomechanical factors (stress) which reduce the strength properties of a brazed joint (they will be discussed in the next section), the adhesion of a solidified brazing alloy may depend on the properties formed at an interphase reaction of new phases and on the thickness of interphase interlayers.

The formation of thick layers of intermediate chemical compounds, which generally are brittle, have a thermal expansion coefficient and a specific volume differing from those of the nonmetallic base material. They can be porous themselves or lead to the porosity of the nearest layers of the base material because of the Kirkendall–Frenkel effect, all of which can take place under strenuous conditions of the joint formation, namely high temperatures and prolonged holdings; this all exerts, generally speaking, an adverse effect on the strength properties and leak-tightness of the joint.

The bond strength in a crystallized metal–solid body surface system (a structurally sensitive quantity) is dependent on many factors: structural imperfections, defects and dislocations, and stresses in the alloy and nonmetallic material regions adjoining the contact.

What is the relation between the work of adhesion of a liquid alloy to a solid surface and the mechanical strength properties of the joint after solidification of the alloy?

For an idealized equilibrium system, where a perfectly smooth surface of a solid is brazed to a metal which has crystallized into a defect-free structure, neglecting the nonsimultaneity of breakaway of metal atoms from the solid surface and the residual stresses in the zone of the joint at a reversible isothermal disturbance of the contact, the work of adhesion is related to the mechanical strength of the joint as follows:

$$W_A = \int_a^\infty F(x)dx \approx F(x_k)x_k, \qquad (1.17)$$

where $F(x)$ is the force of attraction between 10^{-4} m^2-area surfaces being separated at a given distance between them (x); a is the equilibrium (initial) distance between the surfaces; and x_k is the radius of action of interatomic forces (about 10^{-10} m). The maximum value of the force of attraction between the surfaces, $F(x_k)$, is the theoretical breaking (rupture) strength of the contact. As for solids in general, actual contact strength values are much below theoretical ones. Nevertheless, at a change from non-wetting to wetting the work of adhesion of molten metals (substances with a high surface tension) alters very strongly, by an order of magnitude; at such intervals of great change, there exists a correlation, observed by many researchers, between the work of adhesion of a liquid alloy to a solid and the mechanical strength of the solidified alloy contact. For example, a copper–silver alloy forms on the surface of a cubic boron nitride-based ceramic material at a contact angle of 130°; the work of adhesion is of 25×10^{-2} J/m^2; and the strength of the solidified alloy contact with boron nitride is very low (less than 1 MPa). When the adhesion has risen to 200×10^{-2} J/m^2 as the wettability has

Structural materials

increased to a contact angle of 10° (titanium doping of the alloy), the contact strength rises to 100 MPa. Thus, high values of the adhesion of the contacting phases are always preferable for a strong brazed seam.

When the adhesive contact is formed without melting of one of the phases (by a plastic deformation of one of them under the applied pressure), the contact strength depends on many factors: chemical affinity of the substances, temperature–time conditions of formation of the joint, magnitude of the applied pressure, etc. Many of the dependences have been studied.

Copper is traditionally used for joining ceramic oxide materials with metals. The technology of joining alumina-based ceramics with copper was described in [4] and is used in practice. Drawbacks of this technology lie in a relatively low adhesion of copper to oxide materials (to increase it, either special oxygen-containing atmospheres are used or the copper part is pre-oxidized under special conditions). The use of aluminium, a more plastic and much more chemically active metal, is promising in this case. Systematic studies of adhesion characteristics of a contact formed by the solid-phase method with pressure application between aluminium and oxide and nitride ceramic materials were conducted [23, 24]; some of their results are presented in Figure 1.38 and 1.39. Contact strengths

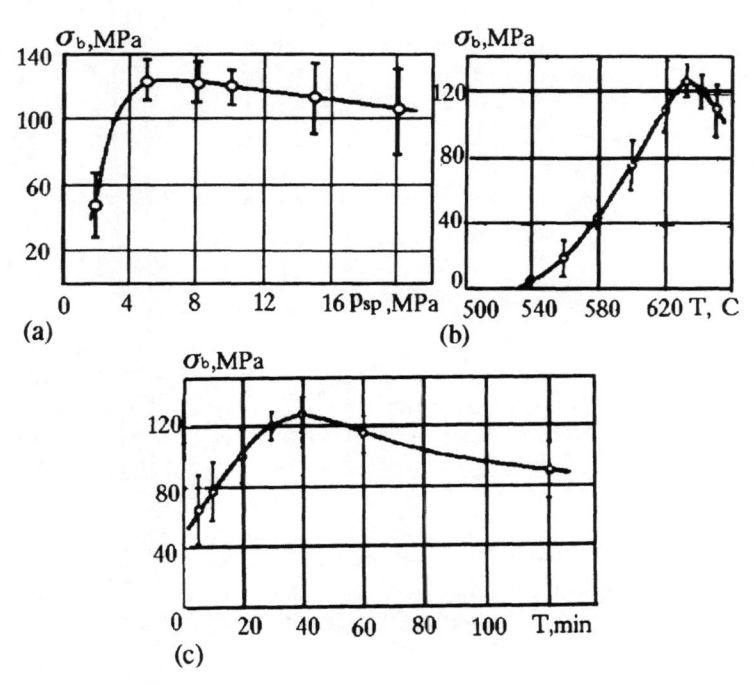

Figure 1.38 Bending strength of corundum–corundum joints through aluminium gasket as a function of: (a) pressure ($T = 630\,°C$; $\tau = 40$ min); (b) temperature ($P = 6$ MPa; $\tau = 40$ min) (c) isothermal holding time ($P = 6$ MPa; $T = 630\,°C$).

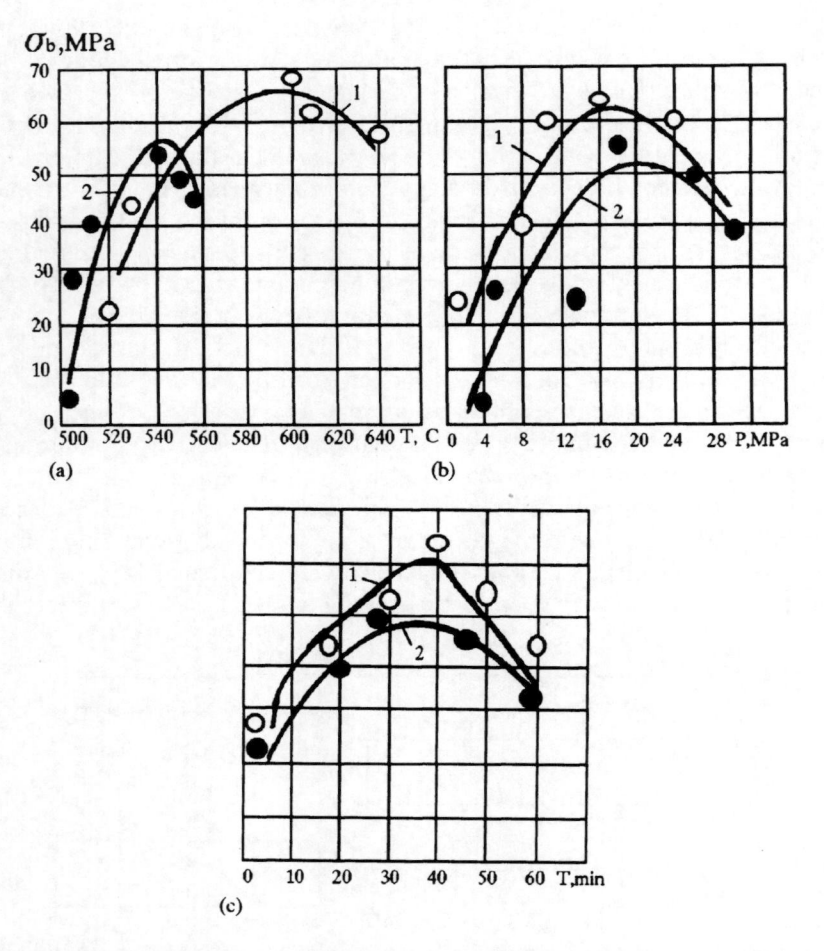

Figure 1.39 Strength of nonoxide ceramics–aluminium joints produced by solid-phase method: (1) silicon nitride–aluminium; (2) silicon carbide–aluminium: (a) effect of joint formation temperature at $P = 16$ MPa for (1) and 18 MPa for (2), $\tau = 33$ min; (b) effect of pressure at $T = 570\,°C$ for (1) and $530\,°C$ for (2), $\tau = 33$ min; (c) effect of holding time under pressure and temperature $P = 16$ MPa, $T = 570\,°C$ for (1), $P = 18$ MPa, $T = 530\,°C$ for (2).

up to 100–150 MPa were obtained for oxide materials (sapphire, quartz glass, silicon–nitride ceramics, cubic boron nitride), which is quite sufficient for practical purposes in joining (brazing) of materials.

1.1.4.3 Stresses in brazed joint zone and methods for their elimination

The sources of stresses in the brazed joint zone are temperature changes at different thermal expansion coefficients of the materials being joined

Structural materials

and brazing alloy; crystallization stresses, arising in solidification of the brazing alloy act as well, but are generally low.

The magnitude and character (compressive or tensile, tangential or normal) of stresses and their distribution throughout the parts form an independent field of research and calculations [4]. General methods for reducing or removing the stresses will be discussed below.

1. Selection of materials with close thermal expansion coefficients. A tolerable difference between thermal expansion coefficients for joining brittle ceramic materials is of a few $10^{-6}\,°C^{-1}$ units.

   A rough equalization of thermal expansions of the parts being joined can also be attained as follows [25]. Combining the values of expansion coefficients of the three brazed joint components (ceramic, metal and brazing alloy), dimensions of the parts and the gap width, we obtain for a simplest linear scheme (see Figure 1.40):

$$\alpha_c L_c + 2\alpha_b \delta_b = \alpha_m L_m \quad \text{at} \quad \alpha_b > \alpha_m > \alpha_c, \quad (1.18)$$

where α, L and δ are respectively the thermal expansion coefficients, dimensions of the parts being joined and the gap width; subscripts c, m and b denote the ceramic, metal and brazing alloy respectively.

For the case of a disc-shaped or annular brazed joint of a ceramic with a metallic case a rough (radial) equalization of temperature-induced changes in dimensions of the parts being joined will take place at the following gap width:

$$\delta = \frac{D}{2}\left(\frac{\alpha_m - \alpha_c}{\alpha_b - \alpha_m}\right)\sqrt{(1 - \alpha_b \Delta T)(1 + \alpha_m \Delta T)}. \quad (1.19)$$

This expression takes into account the correction for a change in the gap cross-sectional area at a temperature change ΔT.

2. Utilization of the deformability of the metallic material being joined, of the brazing alloy or of special stress-relaxing gaskets. A classical

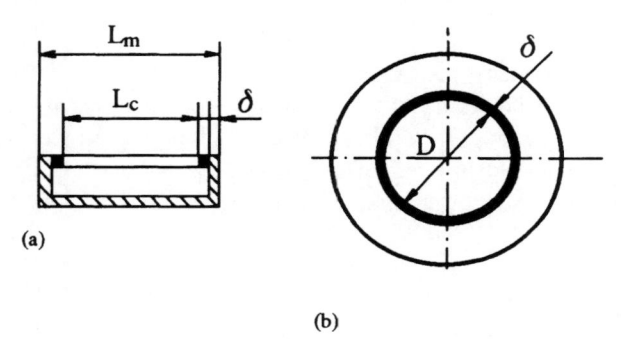

Figure 1.40 Linear-end face (a) and disk-shaped (b) ceramic-metal joint.

example of this joint type is a glass–copper joint (thin, knife-like section of the copper part, joined with glass, deforms easily).

Satisfactory results for massive brazed joints have been obtained with the use of soft brazing solders with a yield strength of 1–5 MPa at a sufficient gap width.

The use of especially highly deformable intermediate gaskets – highly porous, wire, felt, corrugated ones – is also of interest. The use of felt element-gaskets involves certain difficulties because of their easy impregnation with a molten brazing alloy which after crystallization turns the gaskets into a monolith so that they fully or partly lose their deformability. A random distribution of fibres in felt means that not all of them become ultimately brazed to both parts being joined, which reduces the strength of the joint.

A gasket material formed from periodic ordered elements – cylindrical wire springs placed close to one another – has been shown to be useful. In such gaskets every spring element, in the course of brazing gets brazed to the surfaces of both parts. The optimum material and ratios of dimensions, wire and spring diameters and pitch of the helix were investigated in [26]. Springs of 1 mm diameter from 0.1 mm thick wire are most often used. The brazed joint strength can reach several tens of MPa. This technique makes it possible to produce a large-area (0.01 m^2) brazed joint of materials with considerably differing thermal expansion coefficients, in particular a Si_3N_4-steel joint.

Gaskets whose thermal expansion coefficient is intermediate between those of the materials being joined can be used; a composite consisting of constituents with different thermal expansion coefficients can be employed for this purpose. So-called expansion compensators are also used, which are additional elements with a thermal expansion coefficient close to that of the ceramic material and more massive; they take up the loads and stresses and unload the basic 'weak' part.

1.1.4.4 Joining methods and brazed products

Apart from the general requirements placed on brazing alloys, such as a high enough melting temperature, needed from the service conditions, adequate strength properties, etc., a brazing alloy should wet the surfaces being joined well and have a certain level of plasticity. Alloys of nickel, copper, soft metals and many other metals can be used as the base of brazing alloys.

A high wettability is attained (apart from the metallization of ceramics) by doping the basic material of the brazing alloy with electropositive chemically active elements (brazing alloys of the first type) or with electronegative elements (brazing alloys of the second type). Oxide materials and materials based on oxygen-free ionic compounds (group

(a)) can be brazed with the use of brazing alloys of both types, some of which are presented below.

(a) *Brazing alloys of the first type*
These are based on copper and its alloys with silver, tin, gallium, germanium, as well as on gold and nickel, doped with titanium. The working temperature for products brazed with these alloys can be as high as 700–1300 °C.

This type also includes alloys of low-melting metals (lead, tin, indium) with titanium. The working temperature for products brazed with these alloys covers a range from cryogenic temperatures to 150–300 °C. The process of brazing of materials with the aid of these brazing alloys is conducted in vacuum. Titanium is added either beforehand (by preparing an alloy of the required concentration) or by a preliminary application of a paste of titanium powder or titanium hydride to the surfaces to be brazed together. In some technologies, titanium gets into the brazing alloy melt in the course of brazing through a partial dissolution of a titanium part.

(b) *Brazing alloys of the second type*
These are alloys based on copper, silver and low-melting metals, doped with nonmetals, such as oxygen, sulphur and selenium. In particular, an Ag–Cu–O alloy containing 5–10% copper and several per cent oxygen is used to braze oxide materials.

The brazing process is conducted in the air, whose oxygen saturates the brazing alloy and serves as a wetting additive. This technique is called a metal–oxygen technology; it is applied for joining ceramic parts with each other or with parts made of noble metals (silver, platinum).

The brazing can be carried out in two versions. The first version consists of a direct joining of ceramic oxide materials with the aid of a silver–copper (5–10%) brazing alloy prepared beforehand, while the second version involves the application of copper oxide paste to the ceramic surfaces being joined and burning the paste into the surface, after which the parts are brazed together with silver.

The 'solid-phase brazing' process with the use of aluminium is conducted in vacuum. Pressure is applied with the aid of a special hydraulic or mechanical system.

Figures 1.41–1.43 show brazed products of ceramic materials and glass, made with the use of various brazing alloys and techniques.

The use of ceramic joining (brazing) techniques makes it possible to develop radically new designs of assemblies and parts for various purposes, upgrade the performance characteristics of products and devices, and greatly expand the fields of application of nonmetallic refractory (ceramic) materials and composites.

Figure 1.41 Disk-shaped brazed joints of quartz glass windows with metal cases for vacuum engineering, optics and electronics.

Figure 1.42 Brazed joints of single-crystalline optically transparent alumina with metal cases for high-temperature optics in ultrahigh vacuum engineering, produced by solid-phase technology (the largest part in the figure), with working temperatures up to 600 °C, and brazed together with adhesion-active molten brazing alloy (smaller-size parts), with working temperature of 700 °C.

l_c	critical crack length
l_{db}	fibre–matrix bond disruption zone length
m	statistical distribution parameter
n	exponent of kinetic equation
r	whisker cross-section radius
r_c	critical dimension of dissipative zone
r_d	linear dimension of dissipative zone
R	fibre cross-section radius
t	time
V	volume fraction
w_ρ	energy dissipation density
W	work by external forces
W_d	work performed in dissipative process
W_e	work of elastic deformation in crack length increase
α	factor
β	factor
γ	specific surface energy
γ_f	specific work of fibre fracture
γ_m	specific work of matrix fracture
γ_i	specific energy of fibre–matrix interface
γ_F	specific work of material fracture
γ_1	specific work of crack extension initiation
ε	strain
ε_{ij}^T	strain tensor
ε_c	critical strain
σ	stress
σ_0	statistical distribution parameter
σ_b	strength
σ_c	critical stress
σ_T	flow stress
υ	Poisson's ratio
τ	shear stress at fibre–matrix interface
ρ	density of distribution of microcracks
ρ_0	density of distribution of structural units

Prospects for a wide application of ceramics as a structural material depend on the success in solving the problem of improving their reliability and stability [1]. The reliability represents the ability of a material or structure to retain the carrying capacity. One of main aspects of the reliability is a failure-free service. Failure, as applied to brittle ceramic materials, is a brittle fracture which necessitates stopping the operation. Structures and materials which retain a considerable proportion of service properties even on reaching the limiting state are considered as survivable. The survivance of ceramic materials depends on the kinetics

Structural materials

Figure 1.43 Ceramic cutting elements of cubic boron nitride, brazed into holders and disk saws for woodworking.

1.1.5 Mechanical properties
S. M. Barinov

Nomenclature

A_e	work of elastic deformation
E	modulus of normal elasticity
E_c	modulus of elasticity of composite
E_f	modulus of elasticity of fibre
E_m	modulus of elasticity of matrix
G	shear modulus
G_c	fracture toughness; rate of release of work of deformation
G_m	fracture toughness of matrix
h	dissipative zone width
k_1, k_2, k_3	local stress intensity factors
K_c	crack resistance; critical stress intensity factor
K_{Ic}	crack resistance in mode I (opening-mode) fracture
K_0	crack start stress intensity factor
K_1, K_2	factors
l	crack length

of crack propagation. Thus, there are two groups of problems to be solved in developing structural ceramics: (a) attaining a high level of mechanical properties at short-time static or dynamic loads; (b) ensuring a high durability at the action of static or cyclic loads. Of course, the operational reliability of a ceramic material is to a great extent dependent on the correctness of selecting the level of service loads in accordance with the specified safety factors. The latter should be specified not only with an account of loading conditions (static, cyclic, dynamic, type of stressed state), but also with allowance for properties of the material, namely its tendency to a delayed fracture, statistical scatter of strength and fracture energy values [2, 3].

The principal drawback of ceramic materials is their brittleness, which results in a high sensitivity of mechanical properties to various structural stress concentrators and, accordingly, in a low reproducibility of properties. Brittleness is an intrinsic property of ceramics, stemming from the specifics of interatomic interaction; the density of active dislocations in ceramic polycrystals is several orders of magnitude less than in metals with a high symmetry of the crystal lattice, and efforts on developing plastic ceramic materials are hopeless. Due to this, the basic course of the research is aimed at increasing the resistance to brittle fracture. The progress in this field is possible through developing composite ceramic materials, whose structure provides for a higher dissipation of work of external forces in the course of fracture [4].

1.1.5.1 Crack resistance

Brittle fracture at loads that slowly vary in time is a multistage process, including initiation of a crack from an existing defect at some threshold value of the stress intensity factor K_0, its controlled growth to the critical value l_c, and a further supercritical extension as the critical stress intensity factor K_c has been reached. The interval $\Delta K = K_c - K_0$ as a function of the crack length increment Δl is described by the R-curve of resistance to the crack propagation

$$\Delta K = f(\Delta l). \tag{1.20}$$

A higher crack resistance of ceramic materials is always associated with rising R-curves [5]. It is obvious that a higher service reliability is offered by materials having not only high K_c values, but also a high energy of fracture (Figure 1.44), which is characterized by the value

$$A = A_e + \int_0^{l_c} K^2(\Delta l)\,\mathrm{d}(\Delta l)/2E, \tag{1.21}$$

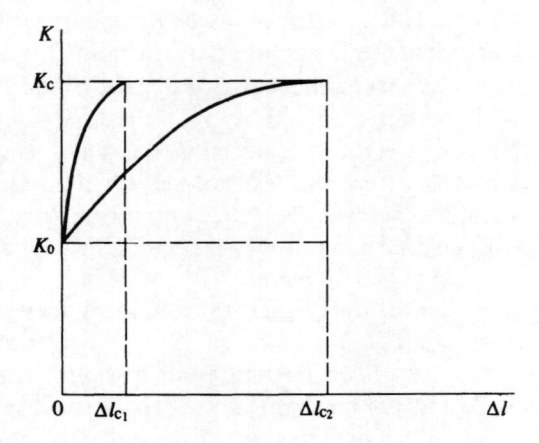

Figure 1.44 K–Δl diagrams for materials having identical K_c values, but different fracture energies.

where $A_e = K_0^2/2E$ is the elastic energy stored by the crack initiation moment and E the modulus of normal elasticity.

The K_0 value is determined first of all by such factors as the grain size, structure of phase boundaries and strength of the bond between structural components, and also the size, morphology and mechanical properties of structural defects that initiate the fracture. For many ceramic materials the subcritical crack growth is small and $K_0 \ll K_c$, and therefore those factors that affect the threshold value of the stress intensity factor determine also the strength of a material. The state of a high strength and a relatively high crack resistance is attained in materials with highly disperse or ultradisperse grains and with a specially organized structure of grain boundaries, e.g., of grains interconnected through second phases. A higher crack resistance of ultradisperse structures is ascribed to small sizes of structural defects and an effective crack retardation at grain boundaries at its growth to the critical length [6].

However, a high crack resistance, approaching that of metallic materials, together with a high strength can be attained only on condition that dissipative processes, comparable in the magnitude of expenditure with the work of plastic deformation in the crack tip region, occur in a ceramic material at its fracture. This can only be attained in composites.

By structural features, ceramic composites can be classed into:

1. composites reinforced with discrete particles;
2. composites with interpenetrating skeletons of two or more structural components;
3. composites reinforced with discrete fibres or whiskers;

Structural materials

4. composites reinforced with continuous fibres (unidirectional, crosswise flat or three-dimensional reinforcement);
5. composites with layered structures.

One or other processes of energy dissipation in the course of fracture can be provided depending on the structure, chemical nature of constituents and state of the bond between them. The following processes, initiated by the stress field in the crack tip zone, are effective to upgrade the crack resistance:

- polymorphic transformations of metastable phases;
- plastic deformation of disperse structural components;
- microcracking and crack branching;
- reflection of crack from phase boundaries and its reorientation with respect to the direction of action of the tensile component of stresses;
- pull-out of fibres from the matrix in overcoming the elastic counter-action and friction forces that prevent the crack opening.

The occurrence (simultaneous or successive) of several above-indicated dissipation processes is typical for most composites. Thus, fracture of composites with an Al_2O_3 matrix reinforced with SiC whiskers involves overcoming of the forces that oppose the crack opening both through the pull-out of whiskers and as a result of the intergranular friction in the matrix. Processes of microcracking of the matrix, fracture and pull-out of fibres occur in turn at fracture of composites unidirectionally reinforced with continuous fibres. It is therefore generally adopted to develop theories of fracture of ceramic composites not on the basis of their structure, but of the dominating type of mechanism of dissipation of the work of external forces. A micromechanical modelling of fracture makes it possible to predict limiting values of the crack resistance as well as the shape of the R-curve of resistance to the crack propagation.

1.1.5.2 Theoretical fundamentals of prediction of mechanical properties

Methods of the micromechanical modelling of ceramic composites rely on an assumption of the existence of a structural cell, whose equation of state is known and whose periodic repetition can confidently represent the structure of the whole material. It is also assumed that macroscopic characteristics of a composite can be described in terms of the mechanics of a quasi-homogeneous (continuous) medium, excluding some material volume near the crack tip within the dissipative zone. The consistency of such an approach is confirmed by numerous experimental data on condition that the size of the dissipative zone substantially exceeds the characteristic dimension of the structural cell. Optimization of the phase composition of heterogeneous ceramic materials in terms of the macroscopic force criterion of fracture in a medium with oscillating internal stresses [7] may serve as an example.

Predicting the limits of variation of mechanical properties is either based on assumptions of the additivity of stress intensity factors or proceeds from the energetic concept of the fracture mechanics.

(a) *Polymorphic transformations*
Typical materials that exhibit higher crack resistance and strength as a result of polymorphic transformations are composites containing disperse particles of tetragonal ZrO_2. In the simplest approximation the influence of the polymorphic transformation is related to the dilatational effect of the transformation, which is responsible for hysteresis and irreversible residual deformation.

In terms of linear fracture mechanics, the crack resistance increase effect corresponds to the difference ΔK between the stress intensity factors outside the polymorphic transformation zone (K_m) and within that part of the zone, where the polymorphic transformation has been complete (K_t) [8]:

$$\Delta K = K_t - K_m. \qquad (1.22)$$

In the physical meaning, ΔK characterizes the shielding action of the dissipative zone on crack edges. The ΔK value depends on the shape of the transformation zone and on the structure of the tensor of strains within the zone [5]. In the approximation of conditions of a plane strain and an opening-mode fracture, as well as of a complete polymorphic transformation of all particles within a stationary 'follow-up' dissipative zone (Figure 1.45), an asymptotic (at high $\Delta l/h$ values – h is the zone

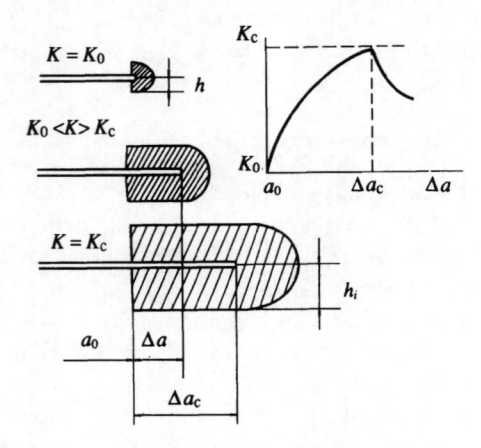

Figure 1.45 Evolution of polymorphic transformation zone in course of subcritical crack growth.

Structural materials

width) increment of the crack resistance, ΔK_c, resulting from the screening action of the dissipative zone is [5]

$$\Delta K_c = cE_c V \varepsilon_{ij}^T h/(1-v), \qquad (1.23)$$

where E_c is the composite's elastic modulus, V the volume fraction of the particles undergoing the polymorphic transformation, ε the dilatometric effect of the transformation, h the dissipative zone width, and v the Poisson's ratio; the factor c is equal to 0.22 when the transformation is initiated by a hydrostatic field of stresses and to 0.38 when deformations in shear bands at the crack tip play the dominating part.

The increment in the fracture toughness, ΔG_c, corresponding to the integral dissipation of energy within the follow-up zone of the process is estimated in terms of the energetic concept. The estimates derived in terms of the force and energetic concepts are qualitatively identical [5].

Since the service behaviour of materials is determined not only by the K_c value, but also by the fracture energy, parameters of the R-curve of resistance to the subcritical crack propagation are important.

According to generally adopted concepts [5, 9], rising R-curves are due to the evolution of the geometry of the dissipative zone and, possibly, to an increase in energy dissipation density in the zone with increasing crack length. The relative variation of the width of the zone of the process, h_i/h, at a transformation toughening as a function of the relative crack length increase (h is the width of a stationary zone of the process at $\Delta K \to \Delta K_c$) is shown in Figure 1.46. A real structure of the zone of the process depends, of course, on conditions of initiation of polymorphic transformation of the disperse phase.

The analytical description of R-curves, which, as will be shown below, is applicable not only for the case of polymorphic transformations, but

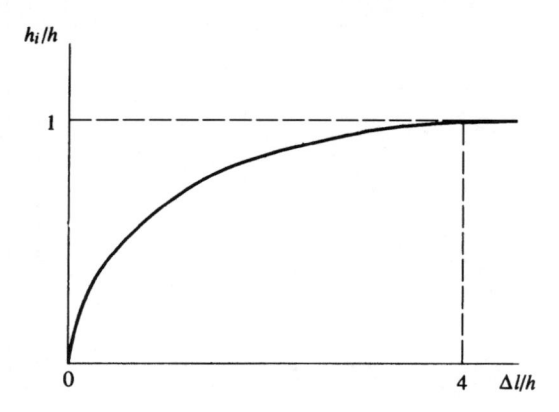

Figure 1.46 Variation of relative width h_i/h of transformation zone of process with relative crack increase $\Delta l/h$.

also for other dissipative processes, can be based on the energetic approach initially developed in [10] for the case of a sub-critical growth of a fatigue crack in metallic materials. The energy balance at a crack length increase by dl is analysed:

$$dW = dW_e + dW_a, \qquad (1.24)$$

where d$W = Gdl$ is the change in the work of external forces, G the deformation energy release rate, d$W_e = [K(1 - \nu)E_c]dl$ the stored elastic energy, E_c the composite's elastic modulus, and d$W_{d\alpha}$ the change of the irreversible work in the dissipative process. For the case of a frontal 'cylindrical' zone of the process (transformation toughening, microcracking [5]),

$$dW_d = w_d r_d dr_{d\alpha}, \qquad (1.25)$$

and for the case of a linear follow-up zone of the process (formation of connecting bridges between crack edges),

$$dW_d = w_d dr_{d\alpha} \qquad (1.26)$$

where w, is the energy dissipation density and r_d the linear dimension of the zone of the process. An infinitesmal increment in the crack length is determined by the increment in the stress intensity factor, dK, and its instantaneous value, and depends on the crack propagation history. From (1.24) and (1.25) it follows that the change in the fracture resistance ΔG at the stage of the sub-critical crack growth (the R-curve) is

$$\Delta G = K^2(1 - \nu^2)/E + dW_d/dl. \qquad (1.27)$$

Setting w_ρ in (1.25) or (1.26) and integrating (1.27) yields an analytical description of R-curves for various dissipative processes in the crack tip region.

For the case of the transformational zone of the process, according to [8], $r = h$,

$$w_\rho = 2V\sigma_c h \varepsilon_{ij}^T, \qquad (1.28)$$

and

$$h = \frac{3(1 + \nu)^2}{12\pi}\left(\frac{K}{\bar{\sigma}_c}\right), \qquad (1.29)$$

where $\bar{\sigma}_c$ is the mean stress causing the polymorphic transformation. Transformation of (1.27) taking into account expressions (1.25), (1.28), (1.29) and integration yields the relation

$$\Delta l = \beta_1 \frac{K^2 - K_0^2}{K_c^2} + \ln \frac{K_c^2 - K^2}{K_c^2 - K_0^2}, \qquad (1.30)$$

where β_1 is a factor depending on the parameters appearing in (1.28) and (1.29). Expanding the expression in brackets in equation (1.30) into a series on condition that $\Delta K/K_c < 1$, substituting $\beta_1 = \beta_2(K_c/K_0)^2$, and shifting the origin of coordinates to the point with $K = K_0$ results in an approximate relation

$$\Delta l = \beta_2[(K_0 + \Delta K)/K_c]^2. \quad (1.31)$$

Thus, the R-curve of resistance to the crack propagation in composite ceramic materials containing particles of a second phase which undergoes a polymorphic transformation is described by a parabolic function. The β_2 value characterizes the 'extent' of the R-curve and can be considered as a criterion, which substantially supplements the criterion of critical stress intensity [11, 12].

(b) *Bridging between crack edges*
This effect occurs in ceramic materials toughened with plastic metallic particles or reinforced with whiskers of fibres.

According to [13], the asymptotic increment in the crack resistance, ΔK_c, resulting from formation of a zone of connecting bridges by plastic metallic particles, which are deformed without hardening, at the crack tip (Figure 1.47) can be estimated as

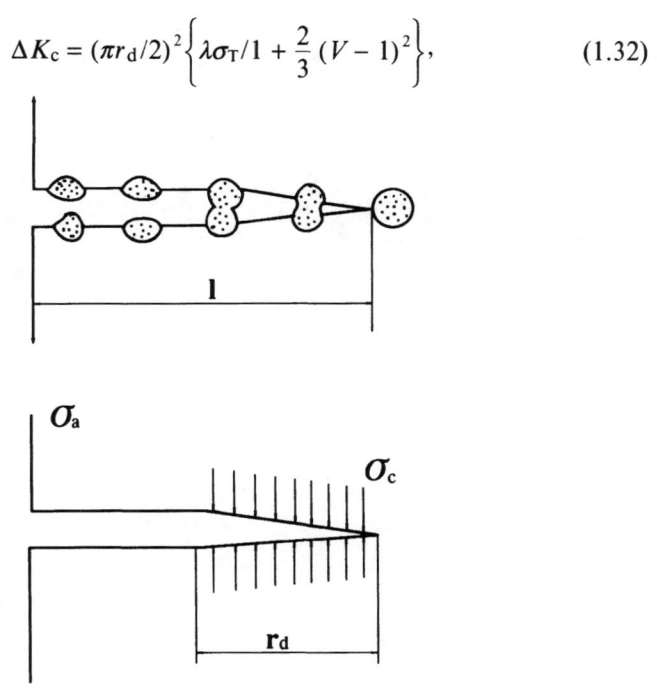

$$\Delta K_c = (\pi r_d/2)^2 \left\{ \lambda \sigma_T/1 + \frac{2}{3}(V-1)^2 \right\}, \quad (1.32)$$

Figure 1.47 Zone of connecting bridges between crack edges, formed by plastic particles.

where r_d is the length of the zone of connecting bridges (the distance from the crack tip to the first fractured particle), σ_T the flow stress of metallic particles, V their volume fraction and λ a factor whose value is $\simeq 3$.

R-curves of resistance to the crack propagation can be described proceeding from an assumption of evolution of the follow-up zone of the process. The expression for the energy dissipation density in this case is

$$w_p = V\sigma_T \varepsilon \qquad (1.33)$$

where ε is the limiting strain up to fracture of particles that are deformed without hardening. The length of the zone of connecting bridges can be estimated on the basis of the Dugdale model [14] as

$$r_d = (\pi/16)(K/\sigma_T)^2. \qquad (1.34)$$

Integration of (1.27) taking account of (1.26), (1.33), and (1.34) yields

$$\Delta l = \beta_1 \ln[(1 - K^2/K_c^2)/(1 - K_0^2/K_c^2)], \qquad (1.35)$$

where β_1 is a factor depending on the parameters appearing in expressions (1.33) and (1.34). Simplifying transformations of (1.35) result in a parabolic dependence

$$\Delta l = \beta_1[(K/K_c)^2 - (K_0/K_c)^2] \qquad (1.36)$$

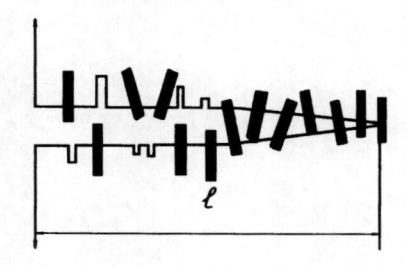

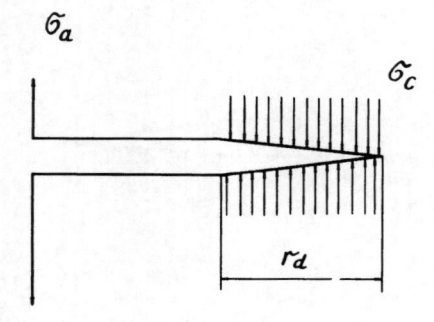

Figure 1.48 Zone of connecting bridges between crack edges, formed by whiskers or discrete fibres.

or, when the origin of coordinates has been shifted to a point with $K = K_0$ and $l = l_0$, in expression (1.31).

A similar approach can be used when the bridging zone is formed by whiskers or discrete fibres (Figure 1.48). The energy dissipation density in the bridging zone is the sum of the specific work of elastic deformation of fibres and the work done to overcome the forces resisting the pull-out of a fibre from the matrix. The expression for w_ρ in this case is [14]:

$$w_\rho = 4\pi N \tau^2 l_{db}^3 / 3E_f, \qquad (1.37)$$

where $N = V/\pi r^2$ is the number of fibres per unit area of the crack surface; r and V are respectively the radius of cross-section and the volume fraction of fibres in the composite; l_{db} is the bond disruption length; and τ is the shear strength of the fibre–matrix interface.

Results of the micromechanical modelling of fracture of whisker-reinforced ceramic composites show the limiting increment in the crack resistance, resulting from formation of the connecting bridge zone, to be

$$\Delta K_c = \sigma_f \{[Vr/6(1 - v^2)] (E_c/E_f)(\gamma_m/\gamma_i)\}^{1/2}, \qquad (1.38)$$

where σ_f is the strength of whiskers; E_c and E_f are elastic moduli of the composite and whiskers; γ_m and γ_i are specific works of fracture of the matrix and matrix–whisker interface.

The variation of the crack propagation resistance with the crack growth is described by relation (1.36), where the parameter β_1 is [15]

$$\beta_1 = [\pi^2 N l_{db}^3 / 6(1 - v^2)](\tau/\sigma_m)^2 (E_c/E_m). \qquad (1.39)$$

In its physical meaning, the β_1 value corresponds to the maximum size of the bridging zone (at $K \rightarrow K_c$).

(c) *Microcracking and crack branching*

Microcracking is an effective process of improving the crack resistance of composites containing both disperse particles and fibres in ceramic matrices. It can occur in structure regions with local tensile stresses or with a weakened bond between structural components of the composite. Propagation of a main crack in a material liable to microcracking involves an additional dissipation of the work of external forces as well as a decrease of the elastic modulus in some zone of the process and a shielding of the crack.

The dilatational effect of crack resistance improvement as a result of microcracking is described in the same manner as is the effect of the transformational improvement of the crack resistance [5]:

$$\Delta K_c = 0.32 \, E \varepsilon_c \sqrt{h}, \qquad (1.40)$$

where ε_c is the strain due to the microcrack formation and h the process zone width. The ε_c value for the regions of the structure, acted upon by tensile stresses, is [5]:

$$\varepsilon_c = (16/3)(1 - v^2)\rho_0\sigma_c/E, \qquad (1.41)$$

where ρ_0 is the volume fraction of the structure regions that undergo microcracking and σ_c the critical stress.

The elastic modulus change effect, which depends on the shape of the process zone, can be expressed as follows [16]:

$$\Delta K_c/K_c = \left[K_1 - \frac{5}{8}(G/\bar{G} - 1) + \left(K_2 + \frac{3}{4}\right)(\bar{v}G/\bar{G} - v)\right]/(1 - v), \qquad (1.42)$$

where K_1 and K_2 depend on microcracking criteria; G and $\bar{G}$ are shear moduli of the material in the initial state and after it underwent microcracking; v and $\bar{v}$ are Poisson's ratios for the material in the initial state and after the microcracking.

In the general case, the effects of the dilatation in the cracking zone and the elastic modulus change on the crack resistance are not additive [5].

An analytical description of R-curves can be derived within the scope of concepts of a frontal zone of the process, extending in the course of sub-critical crack propagation [17], and on the assumption that the forming microcracks do not interact with one another and their density is described by a statistical function of the following form [18]:

$$\rho/\rho_0 = 1 - \exp[-(K/\sigma_0 r^{1/2})^m], \qquad (1.43)$$

where ρ is the density of microcracks; σ_0 and m are distribution parameters. According to the analysis [17], the functional relation between the crack length increment, Δl, and the relative change in the stress intensity factor has the same form as that for the case of transformation-toughened ceramic composites. The factor β_1, characterizing the length of the sub-critical crack growth stage, depends in this case not only on σ and r_c (critical linear dimension of the process zone), but also on statistical distribution parameters from (1.43).

For layered-structure ceramic composites, experiments disclosed a significant increase in the specific work of fracture, associating with processes of reorientation and branching (bifurcation) of the main crack [19, 20]. The following approach can be used to estimate the effect of the crack branching process on the specific work of fracture [21]. The dW_d value in energy balance equation (1.24) is proportional to the number of branching points n_i, to the work G_i spent per unit length of the increment of a secondary crack and depending on the orientation of the latter, and to the secondary crack increment dl_i:

According to [22],
$$dW_d = \alpha_2 n_i G_i dl_i. \tag{1.44}$$

$$G_i = K^2 f(\phi_i)/E_c, \tag{1.45}$$

where $f(\phi_i)$ is a function of the angle of propagation of the secondary crack, whose values are given in [22]. From (1.24) and (1.45) it follows that

$$G = \left[\alpha_1 + \alpha_2 f(\phi_i) n_i \frac{dl_i}{dl}\right] K^2 (1 - v^2)/E_c. \tag{1.46}$$

Thus, the fracture toughness of layered composites is substantially dependent on the thickness of the layers, which determine the number n of potential crack branching points, as well as on the orientation of the layers with respect to the direction of propagation of the main crack, which affects the value of the function $f(\phi)$.

(d) *Reorientation of main crack*
Main crack reorientation processes occur in the case where the structure of a ceramic composite contains regions of action of local residual stresses or regions with weakened interfaces between components. Model representations of the fracture process under such conditions, allowing a quantitative representation of the effect of reorientation on the fracture toughness, have been developed in [22, 23]. They assume that propagation of a crack along a nonplanar path can be described with the use of a linear combination of local stress intensity factors k_1, k_2 and k_3, which characterize the stress field near the crack tip and correspond to the opening-mode, the sliding-mode and the tearing-mode fracture respectively. Here, with the use of some functions of crack tilt and twist angles, the values of the local intensity factors can be related to the intensity of the stresses applied to the material. Having introduced the concept of the probability of crack interaction with a structural component of the composite, it is possible to estimate the influence of the volume fraction and shape of the reinforcing phase on crack resistance improvement. For example, when a ceramic is reinforced by spherical particles, the fracture toughness increase is [22, 23]:

$$\Delta G_c = (1 + 0.87V)G_m, \tag{1.47}$$

where G_m is the fracture toughness of the matrix and V the volume fraction of spherical particles. Reinforcement of ceramic materials by disc-shaped particles was shown [22, 23] to be the most effective, the fracture toughness increase effect rising linearly with increasing disc diameter–thickness ratio.

(e) *Processes of fracture of ceramic composites reinforced with continuous fibres*
The fracture of continuous fibre-reinforced composites involves a combination of various processes that develop successively in the course of

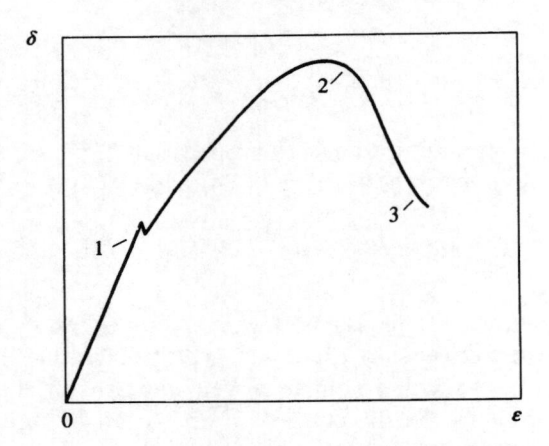

Figure 1.49 Stress–strain diagram for continuous fibre-reinforced ceramic composite: (1) matrix cracking; (2) fracture of fibres; (3) pull-out of fibres from matrix.

loading (Figure 1.49). Manifestation of stages of matrix cracking, fracture of fibres and their pull-out of the matrix, and hence the possibility of attaining a high fracture toughness are to a considerable extent dependent on the state of the fibre–matrix interface in the composite, which is an important factor for achieving high mechanical properties also for metal–matrix fibrous composites [24]. Other factors governing mechanical properties of continuous fibre-reinforced ceramic composites are the volume fraction V of fibres, shear resistance τ in the pull-out of a fibre, residual stresses at phase boundaries, elastic and strength characteristics of constituents and crack resistance of the matrix [5, 25, 26]. According to [27], the following conditions should be met to attain a high level of mechanical properties of fibrous ceramic composites: $\gamma_i/\gamma_f < 1/4$; $\tau = 2$–40 MPa; and $\varepsilon \leqslant 3 \times 10^{-3}$, where γ_i is the interface fracture energy; γ_f the fibre fracture energy; τ the shear stress at the fibre–matrix interface after disruption of the interface bond; ε the mismatch between fibre and matrix strains.

The matrix cracking (first stage of fracture of a composite) stress σ_c is determined by the relation [5, 26]:

$$\sigma_c = \sigma^* - \sigma_r E/E_m, \qquad (1.48)$$

where σ_r is the axial residual stress in the matrix,

$$\sigma^* = [6\tau\gamma_m V^2 E_f/(1-V)E_c E_m^2 R]^{1/3}, \qquad (1.49)$$

γ_m is the matrix fracture energy; E_f, E_m and E_c are elastic moduli of the fibre, matrix and composite respectively; R is the fibre radius.

The matrix cracking process (at absence of fracture of fibres) can reach saturation at a definite distance between microcracks.

Structural materials

The ultimate (breaking) strength σ_b of a composite is associated with fracture of fibres. Estimates with the use of the statistical theory of strength, based on the weak-link hypothesis, result in the relation

$$\sigma_b = V\hat{\sigma}\exp\left\{-\frac{1-(1-\tau d/R\hat{\sigma})^{m+1}}{(m+1)[1-(1-d/R)^m]}\right\}, \qquad (1.50)$$

where the magnitude of the stress $\hat{\sigma}$ is expressed through statistical distribution parameters; d is the distance between stationary microcracks at the 'saturation' of the matrix cracking process; R is the fibre cross-section radius.

The level of increase of the crack resistance of continuous fibre-reinforced ceramic composites depends on operation of the following energy dissipation processes at fracture: microcracking of the matrix; fracture on the fibre–matrix interface; overcoming of sliding friction forces at the pull-out of fibres from the matrix; and dissipation of the elastic energy of deformation of fibres after their fracture. Simplest estimates (within the scope of the energetic conception of fracture) result in the relation [5]:

$$\Delta G_c = Vd[\sigma_f^2/E_c - E_c\varepsilon^2 + 4\gamma_i/R(1-V)] + 2\tau V l_p^2/R, \qquad (1.51)$$

where l_p is the length of fibre pull-out of the matrix. The first term in the right-hand part of equation (1.51) represents the contribution of the elastic energy stored in reinforcing fibres by the moment of fracture; the second term, the decrease in the elastic energy because of cracking of the matrix and fracture of the fibre–matrix interface; the third term, the work spent for formation of new interfaces at fracture on the interface between components of the composite; and the fourth term, the energy dissipation in the process of fibre pull-out of the matrix. As noted in [5], the greatest contribution to the increase in the work of fracture of a composite results from overcoming the friction forces in the pull-out of fibres. It follows that a correct organization of the fibre–matrix interface is of crucial importance in developing the technology of continuous fibre-reinforced ceramic materials.

1.1.5.3 Mechanical properties.–some experimental data

(a) *Transformation toughening and microcracking in ceramics with disperse second-phase particles*
The effect of transformation toughening, discovered in partly stabilized zirconia [28], has been successfully used in ceramic composites with various matrices [29, 30], but the widest practical application has been found by alumina-matrix composites because of specific features of the chemical and thermomechanical compatibility of alumina and zirconia. The latest achievements in development of transformation-toughened

ceramic composites of the Al_2O_3–ZrO_2 system have been reviewed in [31]. The addition of disperse particles of tetragonal zirconia increases the strength σ_b and the crack resistance K_{Ic} of alumina ceramics from typical values of 500 MPa and 4 MPa m$^{1/2}$ to $\simeq$ 1000 MPa and 8–9 MPa m$^{1/2}$ respectively for a composite. The strength and crack resistance increase, however, is accompanied by a steady decline of the elastic modulus and hardness (from 390 to $\simeq$ 279 GPa and from 17.5 to 14 GPa respectively at the addition of 50% vol. ZrO_2 stabilized with 2% mol. yttria). The transformation toughening effect is temperature-dependent. An increase of the crack resistance of Al_2O_3–ZrO_2 composites to 12 MPa m$^{1/2}$ at 78 K has been found. The upper temperature limit of the favourable effect of addition of metastable tetragonal zirconia to ceramic matrices is restricted by the temperature of polymorphic martensitic transformation in zirconia, which is usually not over 700 °C (depending on the material structure). The pattern of R-curves of resistance to the crack propagation in transformation-toughened ceramics depends on the evolution of the process zone shape. The study [32] demonstrated the possibility of an extremum in R-curves. On the whole, mechanical properties of transformation-toughened ceramic composites depend on the relative content of particles of the metastable phase that undergoes the polymorphic transformation in the stress field at the crack tip; size of particles of the phase; type of stabilizing additive and doping of the solid solution (e.g., of zirconia with hafnia). Mechanical properties are significantly affected by the distribution of second-phase particles, in particular by the degree of their agglomeration [32].

The formation of a microcracking zone at the crack tip in ceramic composites containing disperse second-phase particles is less effective for upgrading the crack resistance than are polymorphic transformations. A favourable effect of microcracking has been experimentally ascertained, e.g., in such systems as Al_2O_3 (matrix) – disperse monoclinic zirconia particles or SiC (matrix) – disperse titanium diboride particles. Thus, an increase of the crack resistance of a SiC–TiB_2 system composite by K_{Ic} = 6 MPa m$^{1/2}$ over that of the matrix has been found in [5]. It seems that maximum attainable K_{Ic} values in materials of such a type cannot exceed 10 MPa m$^{1/2}$ [5]. The main drawback of the materials is the low level of their strength characteristics.

(b) *Materials with disperse metallic particles and interpenetrating skeletons of ceramic and metallic constituents*
The state of the phase interface is of crucial importance in the improvement of mechanical properties of brittle matrices at addition of disperse plastic metallic particles. Prerequisites for an effective interaction of a crack with particles of the tough phase are a strong bond between a particle and the matrix, and the absence in the matrix material of tensile

stresses because of which the crack can go round the particles. The connection between the particles and matrix should be formed so as to avoid formation of brittle interlayers with a lower fracture resistance. For example, when disperse nickel particles are added to an oxide matrix, a nickel oxide (NiO) layer with a laminated structure is formed at phase boundaries. On the contrary, the use of pre-oxidized aluminium powders provides a high strength of the connection and, accordingly, a high fracture toughness [13]: a sevenfold increase of K_{Ic} and $\simeq$ 60-fold increase of G_c were attained for materials produced by hot pressing of silica-based glass and oxidized aluminium powder at 20% vol. of the latter.

In materials with interpenetrating skeletons of the brittle ceramic and tough phases, there is provided an effective stabilization of the crack propagation process, which results in increasing first of all the values of energy criteria of crack resistance. Thus, impregnation of a porous zirconia ceramic with polymers increases the specific work of fracture from 30–40 to 100–300 J/m^2 [33]. The specific work of fracture of ceramic-metallic materials depends on the level of residual stresses: creation of periodic fields of residual stresses, effectively interacting with a crack, in ZrO_2–Zr materials by a thermal treatment increased the specific work of fracture from 38 to 65 J/m^2 [34]. The effect of organization of the structure of phase interfaces on mechanical properties was illustrated by $LaCrO_3$–Cr system cermets of a 70:30 phase ratio [9, 35]. The organization of the structure of interfaces at a high-speed pressing, which activates the subsequent sintering, makes it possible to attain a cermet crack resistance $K_{Ic} > 5$ MPa m$^{1/2}$ and a specific work of fracture $\gamma_F > 50$ J/m^2 along with a more than twofold increase in the thermal crack resistance over that of material manufactured by the traditional ceramic technology.

(c) *Layered-structure ceramic materials*
Implementation of processes of crack reorientation at its propagation in an alumina ceramic with a layered-granular structure increased the specific work of fracture to $\gamma_F = 300$ J/m^2 and the γ_F/γ_1 ratio ($\gamma_1 = K_{Ic}/2E$) to 15 as against $\gamma_F = 10$ J/m^2 and $\gamma_F/\gamma_1 = 0.8$–1 for alumina ceramics with the traditional disperse structure [36]. The analysis of kinetic crack resistance diagrams obtained by the double twisting method indicated the existence of stages of periodic acceleration and deceleration of a crack with a corresponding change of coefficients of the kinetic equation

$$\mathrm{d}l/\mathrm{d}t = AK^n, \qquad (1.52)$$

where t is the time, A and n are constants. A considerable (up to 1000 J/m) increase of the specific work of fracture of ceramic-metallic layered-granular materials of the alumina–chromium system was also ascertained [20, 36].

(d) Whisker- and continuous fibre-reinforced materials

Figure 1.50 shows the dependence of the strength σ_b at a four-point bend and of the crack resistance K_{Ic} of an Al_2O_3 (matrix)–SiC (whiskers) composite on the volume fraction of whiskers [37]. The materials were produced by hot pressing in a 10 mmHg vacuum at temperatures of

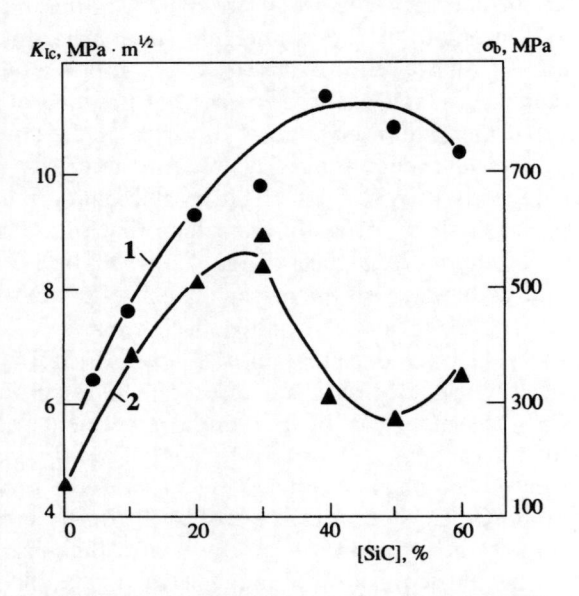

Figure 1.50 Variation of (1) strength and (2) crack resistance of Al_2O_3–SiC composite with volume fraction of SiC whiskers.

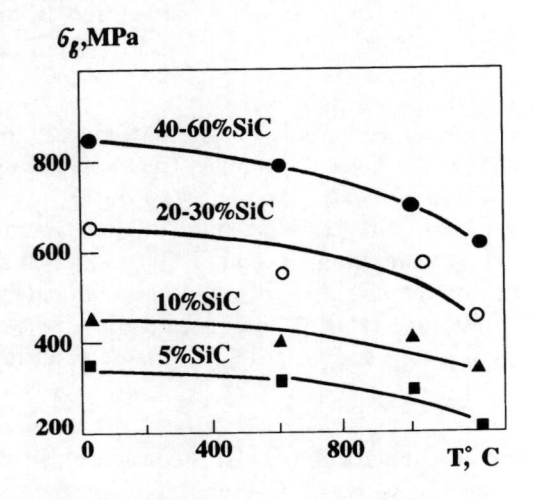

Figure 1.51 Temperature dependence of four-point bend strength of Al_2O_3–SiC composites with various contents of SiC whiskers.

1500–1800 °C after a wet mixing of constituents and drying. As can be seen, the strength and crack resistance of the composite rise considerably over those of the matrix, but at a SiC content over 30% vol. the crack resistance declines. As reported in [37], the effect of strengthening and crack resistance increase remains at temperatures up to 1200 °C. Moreover, reinforcement with SiC whiskers greatly enhances the Al_2O_3 resistance to a slow sub-critical crack propagation and to a high-temperature creep.

Figure 1.51 shows the temperature dependence of strength at a four-point bend of Al_2O_3–SiC whiskers composites (whisker diameter, 0.8 μm; length, 20–30 μm) with various volume fractions of whiskers, produced by a vacuum pressing [38]. A rapid strength decrease begins at temperatures over 1000 °C, which is associated with a nonelastic deformation of the matrix, while the crack resistance K_{Ic} increases as seen in Table 1.17.

Table 1.17 Temperature dependence of crack resistance K_{Ic} at varying compositions of Al_2O_3–SiC

Temperature (°C)	Crack resistance K_{Ic} (MPa m$^{1/2}$)		
	22	1000	1200
Al_2O_3–(20% vol. SiC)	7.4	7.0	–
Al_2O_3–(40% vol. SiC)	6.2	6.4	8.7

A further improvement of mechanical properties of Al_2O_3–SiC system composites is attained by the use of a composite matrix, based on Al_2O_3, with a disperse phase of tetragonal zirconia [37, 39]. The authors of [37] demonstrated the possibility of increasing K_{Ic} from 8.2 to 10 MPa m$^{1/2}$ by adding 20% vol. ZrO_2 to the matrix of the composite. A strength of 1200 MPa of an Al_2O_3–(15% vol. SiC) composite was attained, but its crack resistance was not over 5 MPa m$^{1/2}$ [39]. Authors of [40], however, believe that the effects of crack resistance increase as a result of the polymorphic transformation of disperse particles and of the reinforcement with whiskers are additive (the effect of K_{Ic} increase to 13.2 MPa m$^{1/2}$ was ascertained).

The crack resistance of Al_2O_3–SiC composites is substantially dependent on the state of the matrix–whiskers interface. When whiskers with a low oxygen content are used, the strength of bond at the interface is lower, and with a high oxygen content, greater. In the former case the composite crack resistance is as high as 8.3 MPa m$^{1/2}$, while in the latter case it is as low as 4.2 MPa m$^{1/2}$ [14].

The whisker reinforcement can also significantly improve other properties of ceramics. As shown in [41], the rate of erosion of Al_2O_3–SiC composites under action of abrasive Al_2O_3 or SiC particles in an airflow (300 m/s; abrasive powder feed rate, 2–4 g/min) is about four times as low as that of a nonreinforced matrix. Reinforcement of cordierite materials with SiC whiskers considerably increases their thermal shock resistance [42].

Reinforcement of silicon carbide materials with continuous SiC fibres yields the highest crack resistance characteristics for ceramic materials: $G_c \simeq 10 \text{ kJ/m}^2$ and $K_{Ic} = 39\text{--}41$ MPa m$^{1/2}$ [43]. A favourable effect of a carbon barrier coating of silicon carbide fibres on mechanical properties of SiC–SiC composites was shown in [44], where a composite strength of 1000 MPa was attained. Under cyclic loading of such material, hysteresis was observed and a fall in elastic modulus was noted. In the process of fracture, there occurred a fibre pull-out from the matrix for a length of up to 300 µm. Mechanical properties of ceramic composites with a spatial reinforcement with three-dimensional reinforcing elements or with a combination of two-dimensional ones are presented in [45, 46]. A high stability of mechanical properties of the materials at high-temperature service was noted.

Table 1.18 presents some typical mechanical properties of ceramic composites with various types of reinforcing elements.

Table 1.18 Mechanical properties of ceramic composites

Structure type	Dissipation mechanisms	Crack resistance K_{Ic} (MPa m$^{1/2}$)	Specific work of fracture γ_F (J/m^2)	Strength σ_b (MPa)
Transformation-toughened (Al_2O_3–ZrO_2)	Polymorphic transformation	Up to 15	–	800–1200
With disperse metallic phases (Al_2O_3–Al, SiO_2–Al)	Crack bridging, plastic deformation	Up to 25	Up to 700	–
Layered structure (Al_2O_3–Al_2O_3, Al_2O_3–Cr)	Crack reorientation, crack branching	Up to 6	Up to 1000	up to 500
Whisker-reinforced (Al_2O_3–Zro_2–SiC wisker)	Crack bridging, crack path deflection	Up to 12	–	Up to 1200
Continuous fibre-reinforced (SiC–SiC, lithium aluminosilicate–SiC)	Microcracking, fracture and pull-out of fibres	30–40	Up to 20,000	Up to 1000

Thus, the use of composite structures provides a means for substantially upgrading the mechanical properties of ceramic materials. The attained effect depends on the reinforcement type, content of the reinforcing phase, relation between properties of the matrix and of the reinforcing phase, state of interfaces and residual stresses. Materials of such a class are very promising for application in modern engineering.

1.1.6 Engineering
A. G. Romashin, A. D. Burovov and A. A. Postnikov

Nomenclature

P_v	fracture probability
σ_s	equivalent reduced stress in volume element dV according to one of theories of strength for brittle materials
σ_0	characteristic strength
m	Weibull modulus
σ_m	minimum strength
V	total part (specimen) volume
V_{ef}	effective part (specimen) volume
V_0	volume of specimen in tensile test
$(\sigma_0)_{v=1}$	parameter of strength of unit volume
b, h	specimen cross-section dimensions
L, l	distances between outer and between inner supports
σ^{max}	maximum stress in the part
σ_{ch}	stress at check tests
σ_{op}	stress in operation

1.1.6.1 General principles of designing

The ever increasing attention paid to structural ceramics and their practical applications is due to the feasibility of attaining radically new performance and economic characteristics. Ceramics as a structural material, however, has hardly been studied. The designing and practical fabrication of metal structures have been dealt with in extensive and profound studies and in hundreds of thousands of publications, a vast experience in their commercial applications has been gained, but for load-bearing ceramic structures the work is only beginning. At present a designer has no practical experience in designing ceramic load-bearing elements, and basic data for a reliable analysis of the serviceability of ceramic materials are often lacking. The reason is that ceramics have always been used as a refractory or building material and studied only in this respect.

The design should result in documents for an economically expedient manufacture of parts performing their technical functions in the operation of a machine. The service lives of parts and the probability of failures should be known. The practice often relies on a trial-and-error method: a ceramic part is designed on an empirical basis, is then manufactured and put into operation; if it fails, it is changed until positive results are attained. This method has been used rather successfully and widely, but in view of modern requirements and capabilities it has become obsolescent. Repeated failures of ceramic parts may bring about unjustified losses because of breakdown of a machine as a whole.

A distinguishing feature of ceramics as a structural material is that parts from them have to be made at once, as a whole. So far it is practically impossible to made them by welding or by the mechanical working of a blank; they cannot be assembled by bolts or screws and cannot be adhesively bonded together as is usual for metallic structures.

This determines a rigid relation between the design of a product made from a ceramic material and its manufacturing technology, a strong dependence of actual properties of the material in a product itself on the design and technology.

Thus, the actual strength of a structure may be strongly impaired by residual stresses and defects, arising in the moulding and sintering of products because of their incorrect design that failed to take into account the specifics of ceramic materials.

It follows that the work on developing critical ceramic products and elements should, from the very beginning, rely on a complex approach to the solution of the problem, whose essence is that the development of the design, the development of the material and the development of the technology have to be continuously interrelated, i.e., the design has to be selected so as to suit the technology and material and, conversely, the material and technology have to suit the design and the service conditions of the product.

More than two decades of experience of developing ceramic and glass structures for operation under high thermal stresses has convincingly demonstrated that only such an approach will provide a real possibility of developing load-bearing structures from brittle materials. Designers, production engineers, test engineers and materials researchers have to work concurrently on a common task, under a common guidance and according to one and the same philosophy.

Ceramic structures call for radically different design approaches. For example, it is impossible simply to replace a metallic blade in a gas turbine engine (GTE) with a ceramic one; both the blade and the whole assembly have to be designed quite differently, with allowance for brittleness of ceramics and thermal action conditions. A certain experience in designing and developing load-bearing ceramic structures for operation under high thermal stresses has been acquired by now. Theoretical and practical studies have resulted in elaboration of basic aspects of a scientific designing of parts and elements from brittle materials and formulation of mandatory principles of their designing. Some of these are discussed below.

The first is the principle of minimum stress concentrations. Its essence is that a ceramic part should have no stress concentrators, especially in a strongly loaded zone. The importance of this principle is convincingly

Structural materials

illustrated by ceramic products containing mechanical bolted joints through holes in the ceramic. Cyclic thermal tests result in the appearance of cracks emanating from the holes. The holes are the only crack nucleation source and hence the cause of fracture of the product.

With ceramic materials, the existence of stress concentrations, characteristic for bolted joints in general, is aggravated by the absence of their redistribution through plastic deformation of the material and unavoidable damage of the material (appearance of microcracks) as the holes wear out. As convincingly indicated by numerous tests, the serviceability of such joints under load fails to be provided even when deformable bushings, spacers, springs, etc., are used.

Any grooves, sharp shoulders and thickness changes, which are also stress concentrators, are extremely undesirable in parts made of brittle materials. High stress concentrations arise at ceramic–metal contacts. Because of this, surfaces to be mated should be fitted to each other with a particular thoroughness so as to avoid point contacts. It is expedient to use elastic spacers. A material with an elastic modulus as low as possible should be selected for the part to be mated with a ceramic one [1].

The second principle of designing is the principle of free thermal deformation of a ceramic element. It requires that in the course of service the thermal expansion of all elements mated with a ceramic element should be close to the deformation of the latter. The principle of free thermal deformation can be implemented in structures either through an appropriate selection of linear thermal expansion coefficients of the ceramic and metal, or through the use of elastic compensating elements. For nonstationary conditions, an unfavourable role is played by the fact that the heat capacity of ceramics is twice that of a metal. This means that under otherwise equal conditions, a metallic element mated with a ceramic one will be heated twice as fast as will the ceramic, so that their thermal deformations will differ, with the result that thermal stresses will arise.

The third principle is that of thermal uniformity of a ceramic element. Its essence is that the temperature field of a ceramic part should be maximally uniform throughout the part volume. Violation of this principle gives rise to additional thermal stresses, which impair the serviceability of the ceramic part. Figure 1.52 shows examples of designs where the principle of thermal uniformity is observed (a) and violated (b, c). In Figure 1.52(c), a metallic lining thermally insulates a part of a cylinder, which gives rise to thermal stresses. A similar case occurs in heating a plate or rod with an abrupt thickness change (Figure 1.52b). Fracture of cylindrical parts at places of an abrupt wall thickness change is often encountered in practice; such fractures can occur without application of any external loads.

The fourth principle is that of a purposeful creation of stresses in a ceramic element. Its essence is to ensure such a character of application and

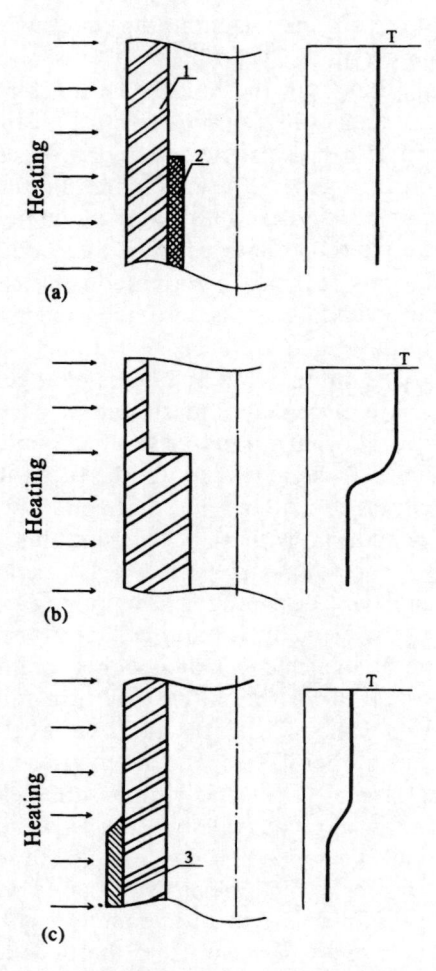

Figure 1.52 Examples of design (a) observing and (b, c) violating thermal uniformity principle: (1) ceramic element (cylinder); (2) functional sublayer; (3) metallic lining; T represents temperature profile over height of element.

distribution of loads on a ceramic element, which produces predominantly compressive stresses. The importance of this principle is well illustrated by the following example, taken from practical experience.

In developing a protective glass enclosure for a deep-water luminaire, a design in the form of a hollow cylinder was selected. It was clamped at its ends by flanges tied together by bolts. Thus, it seemed that the glass cylinder was always acted upon by compressive stresses since it was compressed at its ends by flanges and the external pressure also compressed it on the cylindrical surface. It turned out, however, that, at a considerable thickness of the enclosure, non-uniform contact stresses arose in its end cross-sections which resulted in fracture of the enclosure. To reduce the effect of the stresses, the enclosure ends were made semicircular across the thickness instead of flat ones and the mating surfaces of flanges were fitted accordingly. As a result of this modification, the fracturing pressure rose from 240 atm with flat ends to 880 atm with rounded ones. The theoretically optimal shape of the end-face is, in this case, a curve close to an ellipse.

The fifth principle is the principle of dispersion of edge stress concentrations. Its essence consists in a geometric dispersion of stress concentrations produced by ceramic part fastening elements and by edge stress concentrators, in particular at ends of a ceramic element, as well as by zones of local damages of the ceramic material at its machining.

This principle has clearly demonstrated its importance in the case of aircraft transparencies made from silicate glasses. Two aircraft transparency types were tested. In the first case, a three-layer transparency was supported on a carrying frame, simulating the aircraft cockpit canopy, strictly along the perimeter (Figure 1.53a), while a 5 mm overhang of the transparency beyond the supporting frame was made in the second case (Figure 1.53b). The transparency was a flat panel of 690 × 370 mm, consisting of two 5 m quenched glasses adhesively bonded together by a polymeric film. Tests were conducted by creating a pressure difference across the transparency, which simulated aerodynamic loads in a flight. Three transparencies of each version were tested. Transparencies with an overhang fractured at a pressure difference 2.2 times as high as for those without the overhang.

The sixth principle is the principle of absence of stresses in a ceramic element at its inoperative state or long-term storage. This means that neither macroscopic or microscopic residual stresses nor permanently acting stresses arising in the course of assembling or manufacture of an

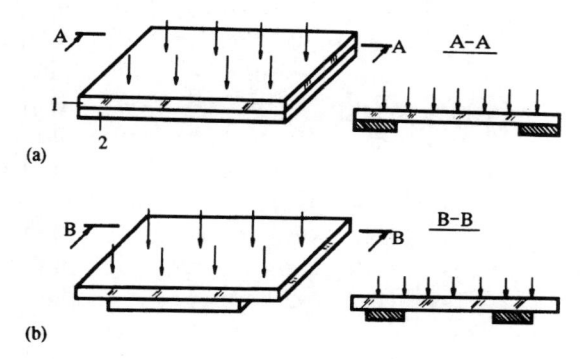

Figure 1.53 Fracture of silicate glass transparency: (a) glass (1) and supporting frame (2) of identical perimeters; (b) transparency with overhang beyond supporting frame.

assembly as a whole should remain in a ceramic part. The point is that in this case ceramics are subjected to a prolonged action of stresses while the long-term strength of ceramics is incommensurably less than the short-term one [1].

Production-induced residual stresses arise, for example, when a ceramic is joined with metal by a shrink fit or when it is soldered or lined with a melt at elevated temperatures. Residual stresses at storage stem in these cases from different thermal deformations of the ceramic and metal in cooling down to the ambient temperature.

The above principles hold for practically any structure from glass, ceramics, composites based on them, or any other brittle material, when a structure is acted upon by a force. The essence of the principles consists in creating the most favourable conditions of loading a brittle material, first of all allowing for absence of stress relaxation. The importance of fulfilling each of the principles will vary with specific requirements placed on a product and its specific service conditions, but the analysis of every structure for compliance with all the principles has to be conducted without fail, and the more thoroughly and in the greater detail, the more stressed the structure is.

The principles of design do not at once yield a specific design for a ceramic element in an assembly unit of a machine, but the analysis of specific designs for compliance with the principles allows wrong design features to be precluded and will give warning to the designer in doubtful cases.

The compressive strength of ceramics, glass and other brittle materials is many times higher than their tensile strength. This should be taken into consideration in the development of structures, but developing a structure subjected in service solely to compressive stresses is practically impossible. Even in the simplest cases, as a rule, bending stresses arise. When a region of a ceramic element is compressed, tangential stresses that may result in fracture of the structure arise at the boundary with an unloaded region.

Technical progress is, to a considerable extent, due to the development of new structural materials. On the other hand, just the progress itself generates a need for developing new materials with higher performance characteristics for implementation of new design features. Modern engineering trends towards ever higher thermal and mechanical loads, requirements for a higher stability in corrosive media, etc., set the task of developing structural materials for various purposes, offering, for example, high strength and crack resistance. Furthermore, the materials should often exhibit such a set of properties which has not been offered by earlier developed materials. The solution to this problem involves a deliberately planned narrow specialization of materials in terms of their application.

The prospects for advance in mechanical engineering are at present being primarily associated with development and extensive application of new composite materials. Development of ceramic composites with predetermined properties and a controlled anisotropy can provide serviceability of ceramic elements in highly loaded assemblies of structures. In this case, it is expedient to begin the designing of a product by designing the required structure of the material and selecting an appropriate technology for the manufacture of the ceramic elements. This makes it reasonable to introduce the principle of compositeness and optimization of the structure of ceramic materials along with the above-listed principles [2].

Also included in these important principles must be the principle of manufacturability. Manufacturing properties of structural mechanical engineering materials should provide the minimum labour intensity in manufacture of parts and structures. The manufacturability is the ability of a material to acquire a predetermined shape under the action of various factors (temperature, pressure, etc.), to lend itself to machining, to be joined by various methods (welding, adhesive bonding), etc. The manufacturability, as well as the cost of a material is of particular importance in mass production processes [3, 4, 5].

1.1.6.2 Probabilistic approach to analysis of strength and fracture of materials and structures

The principal requirement placed on ceramic materials employed for manufacture of thermally stressed parts of structures is the retention of their high strength and reliability at working temperatures. In their level of mechanical properties, silicon nitride-based ceramic materials mostly meet this requirement. At the same time, the fracture of structural elements manufactured on their basis is, in most cases, of a brittle character. A higher sensitivity of ceramic materials to the presence of macro- and microdefects calls for the improvement of the established principles of calculation and design of structures on the basis of using statistical models. The analysis of the strength reliability of ceramics with the aid of the Weibull's weakest-link model [6] is very fruitful. The model relies on the hypothesis that the strength of a brittle body is fully determined by the strength of its weakest element, the fracture of which results in the fracture of the body as a whole. According to this model, the probability of fracture of a part, as a measure of the strength reliability, is determined from the formula [6]

$$P_V(\sigma_0) = 1 - \exp\left\{-\frac{1}{V_0}\int_{\sigma_s > \sigma_m}\left(\frac{\sigma_s - \sigma_m}{\sigma_0 - \sigma_m}\right)^m dV\right\}, \qquad (1.53)$$

where σ_s is the equivalent reduced stress in a volume element dV; σ_0, σ_m, the characteristic and the minimum strength respectively, and m, the

Weibull modulus, determined from results of tests of specimens of the volume V_0.

While the approach may seem simplistic, the use of the weakest-link model in the form of equation (1.53) requires that a number of experiments be conducted and problems be solved beforehand [7], to give the information below.

1. Experimental determination of the material parameters σ_0 and m with speciments of the volume V_0 in their tensile test.
2. Experimental determination of strength criteria for the ceramic material in question.
3. Calculation of the stressed–strained state (SSS) of the part being designed, caused by action of force and thermal loads, throughout the volume of the part under all operating conditions.
4. Experimental check of the applicability of the weakest-link model for estimating the strength reliability of the ceramic material in question in real structures.

Fulfilling the above-listed tasks is the prerequisite for a successful application of this model. We will now present some basic results obtained with the use of the weakest-link model, and the difficulties to be overcome in solving such problems.

The material parameters obtained in a tensile test of V_0 volume specimens have to be used in formula (1.53). Such tests are generally not conducted because of technical difficulties involved in making special specimens and, above all, of great errors in results of such tests. Particular difficulties and substantial errors arise in high-temperature tests. Due to this, ceramic materials are most often subjected to the three- or four-point bend test. Difficulties are also involved in determining the volume V_0, which depends not only on the geometric dimensions of the specimen, but also on the Weibull modulus. For most investigated silicon nitride-based ceramic materials it turned out that the parameter $\sigma_m \to 0$ and a simpler two-parameter Weibull distribution [6, 8], i.e., $\sigma_m = 0$, is convenient to use for describing the probabilistic properties.

To exclude the dependence of strength parameters of a ceramic on dimensions of samples being tested and on the loading type, the parameter of strength of a unit volume, $(\sigma_0)_{v=1}$, is introduced, which is found from the following formula [8]:

$$(\sigma_0)_{v=1} = \sigma_0 \left\{ \int_v \left(\frac{\sigma_s}{\sigma_{\max}} \right)^m dV \right\}^{1/m}. \tag{1.54}$$

For a prismatic beam tested for a three- or a four-point bend, the following analytic expression can be used:

$$(\sigma_0)_{v=1} = \sigma_0 \left(\frac{b \times h}{2}\right)^{1/m} \left(\frac{L + l \times m}{(m + 1)^2}\right)^{1/m}, \qquad (1.55)$$

where b and h are specimen cross-section dimensions and L and l are distances between the outer, and between the inner supports respectively (at a three-point bend, $l = 0$). For other types of specimens, including those for the tensile test, the integration in the formula (1.54) is to be carried out numerically.

If the area S of the working surface of the specimen is assumed as a measure characterizing the fracture, then, retaining the approach, the formula for calculation of S_0 can be derived. In this case the value of the characteristic strength for the same test scheme is reduced to a unit area [8]:

$$(\sigma_0)_{S=1} = \sigma_0 \left(\frac{2L(b + h)(m + 2)}{4(m + 1)^2}\right)^{1/m}.$$

In testing fibrous materials it may happen that their strength depends not so much on their volume and surface area as on the length of fibres. Then it is natural to substitute in formula (1.53) the integration over the volume with the integration over the length [6].

The material parameters σ_0 and m can be estimated from specimen test results in various ways, but the conclusion on reliability of the obtained data has to be drawn very carefully. In [7] the parameters were estimated with the use of the method of least squares in linear and nonlinear modifications and the method of maximum verisimilitude. Domains of applicability of data-processing algorithms were determined through a numerical statistical experiment based on the Monte Carlo method. Comparison was carried out for different sampling sizes, taken within 10 and 100. Basic conclusions from the experiment are that the method of least squares provides a good convergence to the true values of the distribution parameters at relatively small sampling sizes, whereas the method of maximum verisimilitude is better for providing a smaller dispersion. This gives grounds to use the latter at a sampling size $n > 30$.

To test the validity of this approach, ceramic specimens of reaction-bonded silicon nitride were subjected to the three-point bend tests [8, 9]. Two batches of 50 specimens each were tested; specimens in one batch were $7 \times 7 \times 70$ mm, and in the other, $5 \times 5 \times 50$ mm; the distance between supports was 60 and 40 mm respectively. Parameters of the Weibull distribution (1.53) were estimated through minimizing the standard deviation of the theoretical distribution from the experimental one in linear and nonlinear statements. Results of the data processing are shown in Figure 1.54 in logarithmic coordinates. Numerals 1 and 2 designate results for $7 \times 7 \times 70$ mm and $5 \times 5 \times 50$ mm samples respectively. The Weibull modulus value defines the slope of straight lines. The characteristic

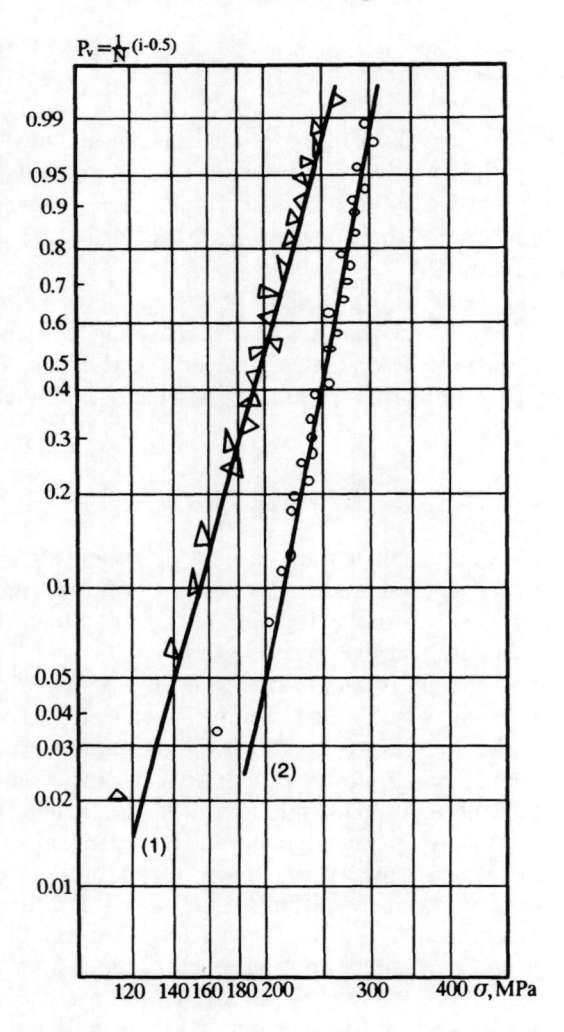

Figure 1.54 Weibull distribution for three-point bend test of reaction-bonded silicon nitride specimens of different sizes; (1) 7 × 7 × 70 mm and (2) 5 × 5 × 50 mm.

strength value corresponds to a fracture probability $P = 0.632$. Results of the tests are summarized in Table 1.19.

Table 1.19 Weibull modulus and strength results for the different sized batches

Specimen size (mm)	Weibull modulus	Strength (MPa)		
		characteristic	unit volume	unit area
7 × 7 × 70	9.44	247	325	369
5 × 5 × 50	9.39	266	313	368

Tensile stresses are, as a rule, responsible for the fracture of ceramic parts; this predetermines the necessity to determine the ultimate strength of the material from tensile tests [7]. For brittle ceramic materials, there are no standards for tensile test specimens. At the same time the need exists to determine strength properties of structural ceramics in tensile tests. These include the test for the long-term strength and especially for creep. This need is brought about by conditions of such tests, since the specimen should have a part of a certain length, where a constant and uniform stress will act. An analysis of various design versions of such tensile specimens indicated their imperfection. The drawback of these specimens in tests of brittle materials is a low measurement accuracy because of a large spread of material strength characteristics in the course of a test, which is due to the presence of a stress concentrator at the transition part and a high sensitivity of brittle materials to such concentrators. Since brittle ceramic materials exhibit a full absence of the relaxability with respect to action of high local stresses, this results in a premature fracture of a specimen beyond its working part. To eliminate these drawbacks, a new tensile specimen has been developed [10]. Its main feature is that the surface of the working part from the heads to the midpoint of the specimen has a profile similar to that of a stream of an ideal liquid flowing out under gravity through a hole in the bottom of a vessel. The stress concentration, typical for a specimen where the transition part has the shape of a circular arc, is fully excluded on the surface of such a profile. Comparison of the proposed specimen with a standard one demonstrated the accuracy of determining the strength parameters to increase 1.7-fold [7].

It should be pointed out that the accuracy of determining the material parameters, especially the Weibull modulus, is significantly affected by the conditions of tests, whatever their type. Thus, it was found that in the bend test, the existence of friction in supports may change the spread of results nearly twofold.

A transitional stage to the analysis of the strength reliability of ceramic parts is the calculation of their stressed state under the action of thermal and force loads. This problem is solved with the use of the most advanced numerical method: the finite element method in two-and three-dimensional versions. The distribution of principal stresses throughout the volume of the part under investigation, found for separate moments of time of a nonstationary process, is used to calculate the probability of fracture.

To analyse the stressed state and work out the requirements for the strength of ceramics, the concept of 'effective volume' is introduced, the value of which is found from the following formula [8]:

$$V_{\text{ef}} = \int_{\sigma_s > 0} \left(\frac{\sigma_s}{\sigma^{\max}} \right)^m dV,$$

where $\sigma^{\max}$ is the maximum stress in the part and σ_s the equivalent reduced stress in a volume element dV according to one of the theories of strength for brittle materials.

The V_{ef} value does not depend on the absolute values of loads, but characterizes the type of the stressed state for a specific calculation scheme. Then the probability of fracture of a specific part for any load and adopted calculation scheme is determined without integration from the following formula [8]:

$$P_v(\sigma_s) = 1 - \exp\left\{-\frac{V_{ef}}{V}\left(\frac{\sigma^{\max}}{\sigma_0}\right)^m\right\}.$$

To estimate the degree of loading now of a whole class of parts regardless of the absolute value of load and dimensions, it is expedient to introduce a dimensionless quantity called the 'load factor':

$$K_N = \frac{V_{ef}}{V}. \tag{1.56}$$

The use of the concept of an 'effective' volume (area) and of the load factor made it possible, on the one hand, to determine the probability of fracture of a part from the characteristic strength of specimens and, on the other hand, to formulate requirements for the strength of material from results of tests of specimens and the predetermined probability of their fracture:

$$(\sigma_0)_{v=1} = \sigma^{\max}\left\{-\frac{K_N V}{\ln(1 - P_v)}\right\}^{1/m}. \tag{1.57}$$

At the analysis of a delayed brittle fracture of a part, to calculate the time until fracture, τ (a service-life characteristic of a material), an additional factor $(\tau/\tau_s)^b$, allowing for the material strength decrease, should be included in formula (1.53). In this case the quantity σ_s in formula (1.53) will acquire the meaning of the fracture stress at the time base of the test, τ_s, and the strength decrease degree is characterized by the parameter b [6]. A programmed implementation of the model with account of the temperature dependence of σ_0 and m is provided to solve such problems.

To illustrate the efficiency of the approach, results of the analysis of applicability of silicon nitride-based ceramic materials for manufacturing thermally stressed elements of a gas turbine engine are presented. The blade of a nozzle vane was investigated under conditions of a nonstationary heating, corresponding to the engine start-up. The heat exchange conditions were given as the gas temperature and convective heat transfer coefficient at the inner and outer sides of the vane. Thermal and stressed-strained states of the part were determined by solving, by the finite element method, the thermoelasticity problem in a quasi-spatial statement.

Table 1.19 Properties of some grades of silicon nitride-based materials

Properties	Grade of reaction-bonded silicon nitride-based materials		
	OTM-904	OTM-907	OTM-908
	Si_3N_4	Si_3N_4, Y_2O_3, MgO, Fe_2O_3	Si_3N_4, BN
Density (kg/m^3 × 10^3)	2.5–2.7	3.1–3.3	2.4–2.5
Bending strength (average) (MPa):			
$T = 293$ K	180–300	500–600	160
$T = 1693$ K	200–250	450–500 (1493 K)	
Elastic modulus (GPa)	113	220	100
K_{Ic} (MPa m$^{1/2}$)	2.1–2.3	6.0	2.0
Linear thermal expansion coefficient ($\alpha \times 10^6$ K^{-1}):			
$T = 293$–573 K	2.07–2.35	2.26–2.40	2.00
$T = 293$–773 K	2.50–2.62	2.60–2.90	2.40
$T = 293$–1173 K	2.50–2.90	2.90–3.40	2.80
Poisson's ratio	0.20–0.23	0.25	0.19
Thermal conductivity (W/m K):			
$T = 373$ K	6.03	13.60	9.40
$T = 573$ K	6.04	11.80	9.30
$T = 973$ K	6.67	15.80	10.40
$T = 1173$ K	7.44	23.70	11.40
Heat capacity (kJ/m K):			
$T = 373$–1173 K	1.38–2.09	0.90–1.50	0.80–1.40

Table 1.20 Attributes of the solid nozzle vane under nonstationary conditions

Property	Materials		
	OTM-904	OTM-907	OTM-908
Maximum principal tensile stress σ_1^{max} (MPa)	53.2	99.2	42.0
Load factor $K_N = V_{ef}/V$	0.0698	0.0652	0.0667
Probability of fracture $P_v(\sigma_1^{max}) \times 10^5$	0.54	0.14	0.45

Consider that materials for the vane were reaction-bonded, chemically compacted, and modified silicon nitride of grades OTM-904, OTM-907, and OTM-908 respectively [9], whose properties are presented in Table 1.19. Values of maximum principal tensile stresses, load factor and probability of fracture of the vane are summarized in Table 1.20. Variation of K_N and σ^{max} in the course of heating of the vane are shown in Figure 1.55 and 1.56, where curves 1, 2 and 3 correspond to materials

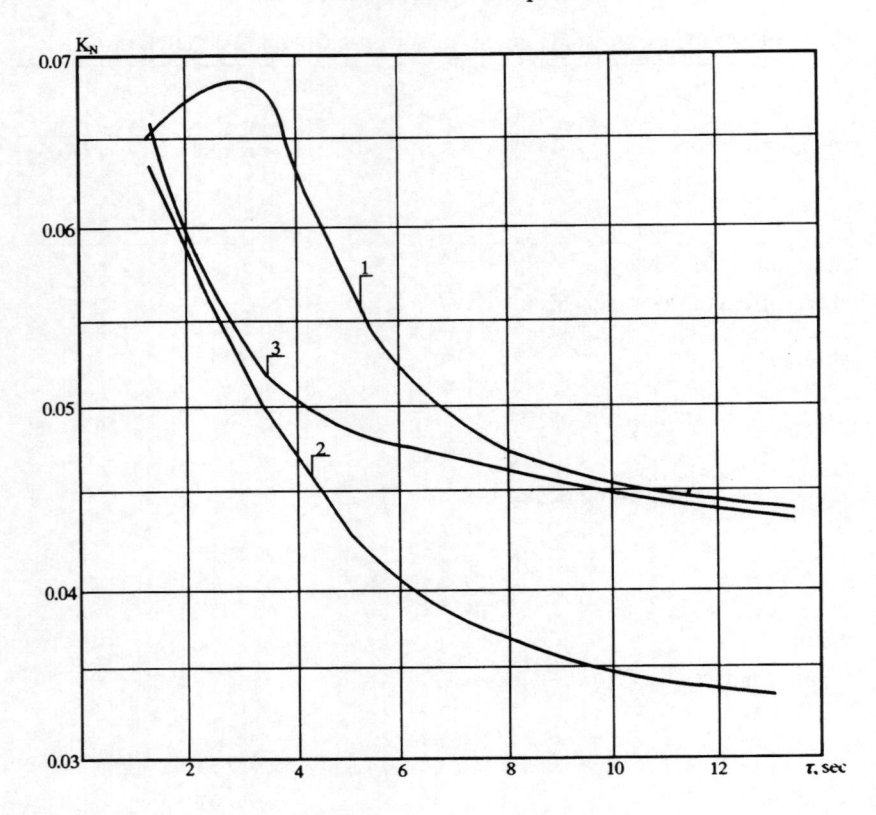

Figure 1.55 Variation of load factor in course of heating of nozzle vane made from (1) OTM-904; (2) OTM-907; and (3) OTM-908.

OTM-904, OTM-907 and OTM-908 respectively. Their analysis shows that, in spite of high absolute values for the vane made from OTM-907 (curves 2), the vane loading with tensile stresses, responsible for fracture, is less, and hence the probability of its fracture is less as well.

Based on the data of Table 1.20, the characteristic material strength to be obtained in the three-point bend test of standard specimens, which ensures the probability of fracture of the future part not over some predetermined value, can be determined from the known volume of the part and the Weibull modulus from formula (1.57) or, conversely, limiting loads at known properties of the material and geometry of the part at the predetermined probability of its fracture can be found. The latter problem arises in specifying the level of loads for check or screening tests of ceramic parts; it can be solved only by calculations.

Selecting the type of check tests of parts from brittle ceramic materials and specifying the check load value are a complex task. The pro-

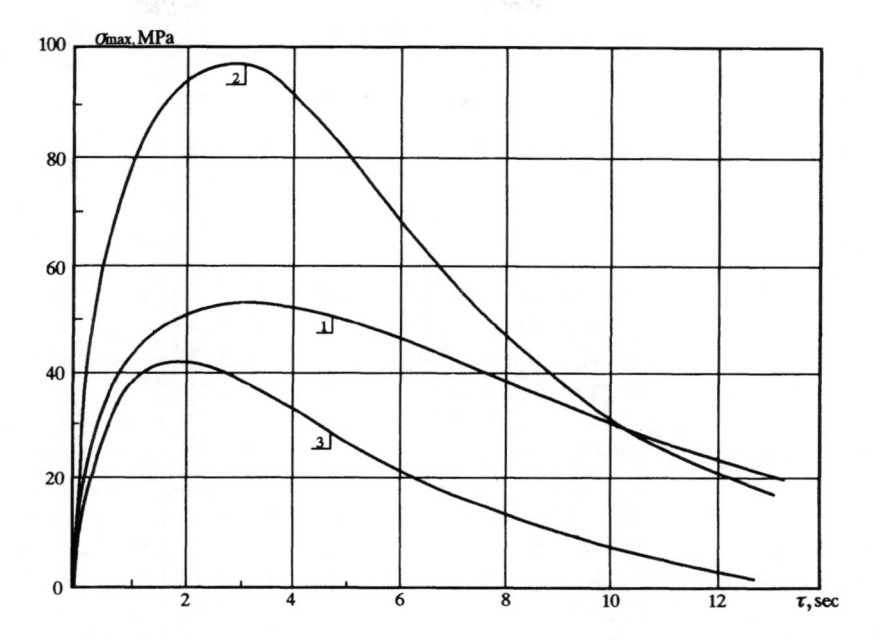

Figure 1.56 Variation of principal maximum tensile stresses in course of heating of nozzle vane made from (1) OTM-904; (2) OTM-907; and (3) OTM-908.

posed approach to specifying the check tests of ceramic parts relies on the principle of equality of the probabilities of fracture of parts from the action of the service and check loads. This principle differs from the traditional method of tests based on the equality of maximum stresses.

The scale factor inherent in ceramics shows up in that the value of limiting fracture stresses is affected both by the volume and by the character of distribution of tensile stresses in a structure [6, 7]. Thus, two fully identical structures of the same volume, made from the same material, and having the same strength and Weibull modulus, but subjected to different loading schemes, will have different ultimate fracture stresses. This indicates that the method of specifying the check load, used for materials with determinate properties, is unsuitable for ceramic structures. The check load for the latter should correspond to the level that produces such maximum limiting stresses (σ_{ch}^{max}) at check tests which are equal to maximum operating stresses (σ_{op}^{max}). The character of distribution of tensile stresses in a structure at check tests can differ from that in actual service, and therefore, taking into account the specifications of a ceramic material, it is proposed to select such a check load, for which the probability of its fracture equals the probability of fracture from the service load. Thus, having determined the 'effective volumes' for selected loading schemes either by calculation or experimentally, and knowing the value of the maximum operating stress, we find the maximum stress at the check load:

$$\sigma_{ch}^{max} = \sigma_{op}^{max} \left\{ \frac{\int_v (\sigma_{op}(x, y, z)/\sigma_{op}^{max})^m dV}{\int_v (\sigma_{ch}(x, y, z)/\sigma_{ch}^{max})^m dV} \right\}^{1/m}. \quad (1.58)$$

Now the absolute value of the check load is determined from the condition of creation of stresses equal to $\sigma_{ch_0}^{max}$ in the structure.

The proposed approach to specifying the check load and determining its magnitude simplifies requirements for the check test type and imposes no special constraints on the latter. Any test type, most convenient for a given class of products, easily and accurately performed in practice, may be choosen. Thus, the check load for ceramic parts can be a distributed pressure, tensile force or thermal action. If the load is a single-parameter one, such as a tensile force, then the V value is independent of the magnitude of the load, is a constant for a specific product and can be calculated beforehand. In this case, the V value uniquely determines the magnitude of the tensile force in the check tests. At the same time the service load can also not be a single-parameter one, such as a combination of thermal and force loads, but with an exactly known distribution of stresses in the structure under the action of the loads. When the service load and hence also the probability of fracture of the product vary with time, $P_v = f(\sigma_s, \tau)$, then to ensure the check test it suffices to vary only a single value with time, using formula (1.58) for every moment of time.

A part having passed such tests without fracture is regarded as serviceable for the operating loads. The probability of fracture of the part after the test declines substantially, but is not fully excluded, and can be found from

$$P_v^{res}(\sigma) = \begin{cases} 0 & \text{if } \sigma_{ch}(x, y, z) \geq \sigma_{op}(x, y, z); \\ 1 \times \exp\left\{ -\frac{1}{V_0} \int_{\sigma_{op} > \sigma_{ch}} \left(\frac{\sigma_{op}(x, y, z)}{\sigma_0} \right)^m dV \right\}. \end{cases} \quad (1.59)$$

if $\sigma_{op} > \sigma_{ch}$

The analysis of formula (1.59) leads to an important conclusion: to ensure a zero probability of fracture after a check or screening test, it must be given provided that stresses

$$\sigma_{ch}(x, y, z) \geq \sigma_{op}(x, y, z) > 0 \quad (1.60)$$

take place throughout the volume of the part in the course of the tests. To attain this condition, several types of successive tests can be carried out. For example, cooling and heating a part make it possible to create

Structural materials

in separate zones of the part, various stresses, whose magnitude can be determined by calculations or experimentally.

Thus, the use of the probabilistic approach at the designing stage allows a more substantiated selection of design versions on the basis of the minimum probability of their fracture, elaboration of requirements for ceramic materials at given loading conditions or, conversely, finding of limiting service loads as well as loads for the check (screening) tests at known properties of the material, geometry of the part and at a given probability of its fracture.

1.1.6.3 Mechanical properties and applications of ceramic composites in modern engineering

The application of ceramic materials in diverse fields of modern engineering is determined by the set of properties attained at their manufacture. Ceramic composites make possible the programming of their properties through selection of starting constituents, their fractions, geometric shapes, arrangement and dimensions [5]. Improvement of the technology and methods for designing products from ceramic structural materials (CSM) and their extensive introduction in mechanical engineering are urgently needed, and seem promising for the tasks faced by advanced technology developers.

Of a broad class of materials included into the advanced or promising structural-purpose ceramics, only a limited scope of compounds – silicon carbide and nitride, sialons, boron nitride and zirconia – are used. Efforts on developing and improving products from these compounds are continuously under way in industrial countries [3, 4, 9, 11, 12]. Principal lines of research include synthesis of powders and whiskers; development of ceramic-matrix composites; development of modern consolidation methods; development of coatings and ceramic-to-metal joints; methodology of product design.

Table 1.21 Market of promising structural ceramics in the USA (all figures are US$m)

Fields of application	Years		
	1987	1995	2000
Automotive/heat engines	29	310	820
Cutting tools	32	246	500
Wear-resistant parts and other industrial equipment	75	320	720
Heat exchangers	7	50	100
Materials for space and military purposes	20	200	445
Bioceramics	8	34	60
Total:	171	1160	2645

Ceramic-matrix composites

Extensive research and development efforts on ceramic parts for automotive and gas turbine engines (GTE) are under way in view of the possibility of an enormous economic gain from the use of structural ceramic materials in the engine and power industries, etc. Highly effective ceramic materials for GTE of various purposes are being developed, which, together with the heat recovery and perfect aerodynamics, will allow fuel consumption to be reduced by 50% by 2000.

An analysis of the structure and trends of the world production of ceramics show the maximum growth rate occurring at present for structural-purpose ceramics [3,13]. Data on the advanced ceramics market in the USA is presented in Table 1.21 [14]. Modern developments of ceramic materials are conducted with a close interrelation of the 'composition–technology–structure–property' and a wide use of the principle of compositeness and optimization of the structure [15]. Most of the ceramic

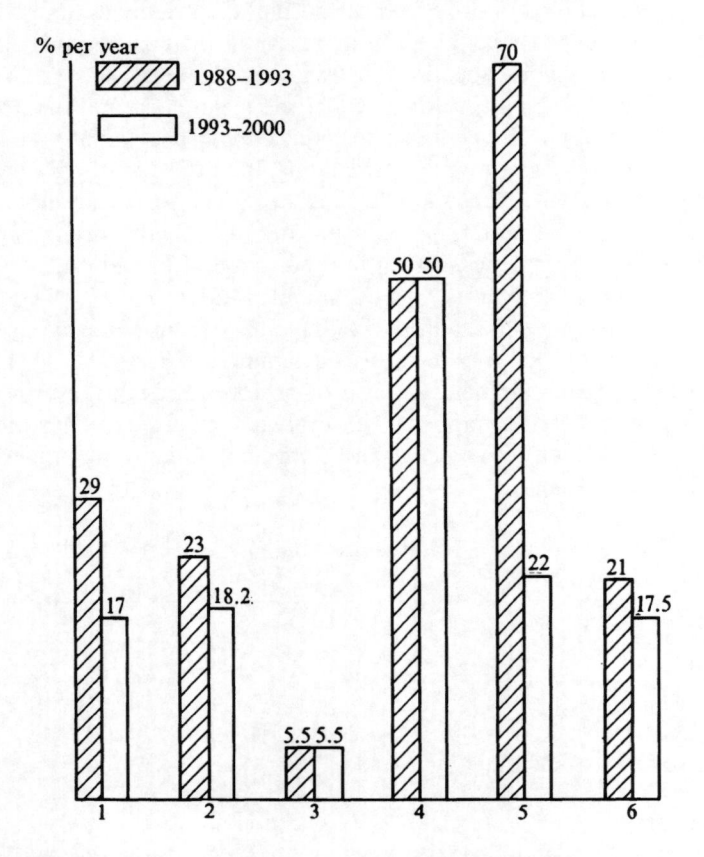

Figure 1.57 Annual growth of ceramic composite market 1 = cutting tools; 2 = wear-resistant parts; 3 = parts for space and military purposes; 4 = engines; 5 = power plants; 6 = total amount of production.

Structural materials

structural materials being developed are composites. It is ceramic composites whose development will occur at the fastest rate (Figure 1.57) [16].

The main field of application of structural ceramics is the engine industry: gas-turbine engines (GTE) and internal combustion engines (ICE) of various types. Power plants for aircraft, automobiles, tractors, ships, for gas-pumping through trunk pipelines, etc., are being developed in many countries. The use of structural ceramic materials allows the efficiency of heat engines (piston and turbine ones) to be substantially increased through increasing the working medium temperature. New engine designs call for materials capable of operation at temperatures up

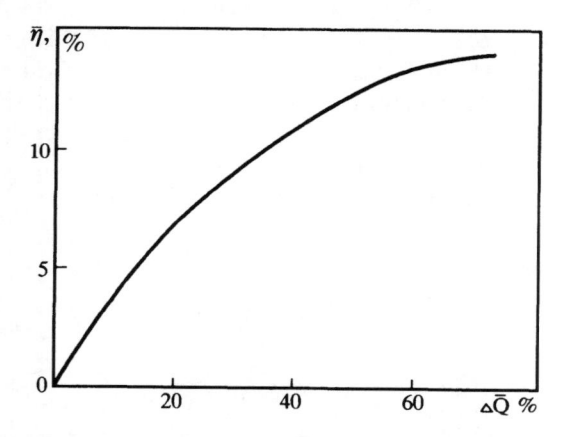

Figure 1.58 Relative specific fuel consumption of a diesel engine as a function of relative loss through engine structure.

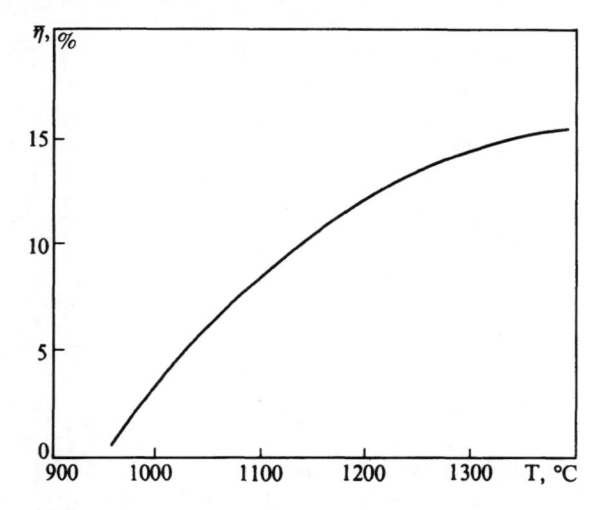

Figure 1.59 Relative specific fuel consumption of a gas turbine engine as a function of gas temperature at inlet to turbine.

to 1200 °C for piston engines, and up to 1700–2100 °C for gas turbine ones.

Figure 1.58 shows a relative decrease in the specific fuel consumption by a diesel engine as a function of a relative reduction of the heat loss through the engine structure by thermally insulating the combustion chamber with a ceramic material [9]. Figure 1.59 shows the effect of the temperature rise at the inlet to the CTE turbine on a relative decrease in the specific fuel consumption [9].

Along with upgrading the fuel economy of engines, the use of ceramics makes it possible to increase the specific power output (by 15–20%) and to reduce the mass and dimensions of engines (by 30–50%), to simplify the design, and to reduce the inertia of rotating parts, which on the whole is exceedingly important for engines of aircraft and of other transportation facilities. As an example, Table 1.22 compares charateristics of three automotive GTE: (1) a modern one, made of metal; (2) a GTE being developed, where the turbine will be made of uncooled metallic parts; (3) a future one, with a ceramic gas turbine [9].

Table 1.22 Characteristics of automotive gas turbine engine

Engine version	Temperature (°C) at inlet		Engine efficiency	Engine mass (kg)
	to turbine	to heat exchanger		
1	1010	704	0.26	270
2	1038	982	0.33	165
3	1370	1093	0.46	131

The development of new generation engines involves first of all increasing the gas temperature in the working space in adiabatic turbocompound or gas turbine engines for the ground, air and sea transport.

The use in engines of uncooled ceramic elements and assemblies, whose density is 2–2.5 times lower than that of metals, allows the gas temperature at the inlet to the turbine to be raised up to 1400–1700 °C without cooling, which in turn will increase the efficiency by 10–15% and the specific power output by 20–30%, reduce both the fuel consumption and the consumption of scarce chromium–nickel alloys, etc.

A stoichiometric combustion of the fuel in the engine will reduce environmental pollution with toxic gases. The elimination of the cooling system, e.g., in gas turbine and diesel engines, will upgrade a failure-free performance, manoeuvrability and pick-up of the machines, reduce the overall dimensions, and reduce the noise and smoke of the engine [9].

Many interesting ceramic composites based on a ceramic matrix with fillers in the form of fibres, whiskers, particles of refractory compounds and metals have now been developed. The strength and crack resistance characteristics of ceramic composites are several times higher than those of monolithic ceramics. The mechanisms of toughening of ceramic composites have, to date, been developed only in a general outline, although a complex enough set of equations of the mechanics of ceramic composites has already been derived.

A number of requirements on ceramic composites that offer superior strength and crack resistance to those of dense matrix materials can be formulated [17].

1. A ceramic matrix should have a high strength, should be oxidation-resistant, and should provide a uniform load distribution.
2. Ceramic fibres (e.g., SiC, Al_2O_3 or Si_3N_4) should exhibit a long-term thermal stability.
3. The porosity of the matrix should be the minimum possible to prevent oxygen transport into the matrix and a subsequent oxidation of materials.

An analysis of mechanisms of fracture and toughening of ceramic composites indicates that developments of the following materials having a high crack resistance are promising.

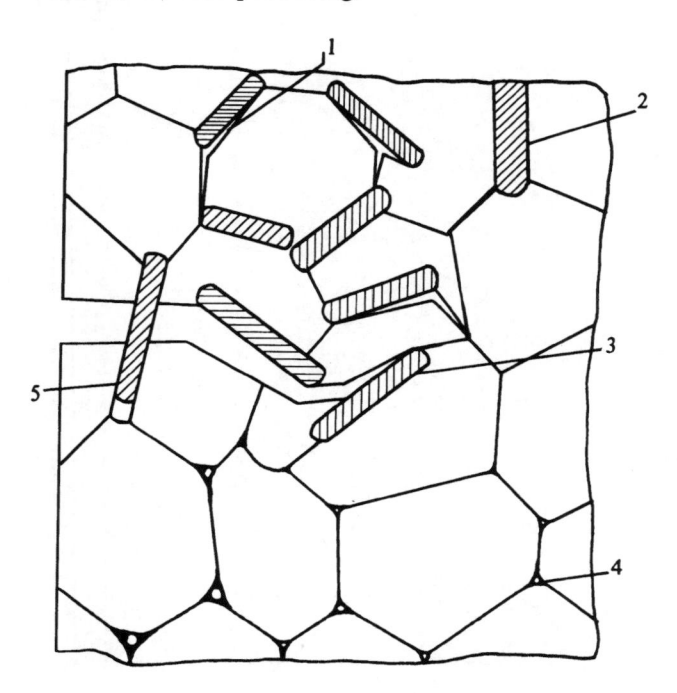

Figure 1.60 Schematic of real structures of ceramic composites with higher toughness, high-temperature strength, and creep resistance: (1) microcrack; (2) dispersions of particles or whiskers; (3) crack deflection; (4) crack nucleation in glass phase residue; (5) bridge connection, pull-out mechanism.

1. ZrO_2, HfO_2 (utilization of the mechanism of modification transformations).
2. Al_2O_3–ZrO_2, Si_3N_4–SiC, SiC–TiO_2 (microcracking).
3. Al_2O_3–Al, ZrB_2–Zr, Al_2O_3–Ni, WC–Co (addition of metallic particles).
4. Si_3N_4–SiC, Al_2O_3–SiC (addition of whiskers).
5. Glass ceramics: SiC, Al_2O_3, SiC–SiC [18].

Typical structures are shown in Figure 1.60 [19].

Using a combination of several toughening mechanisms in developing a composite can greatly increase the toughening degree. Thus, combination of two toughening mechanisms for a mullite-based composite, namely its reinforcement by SiC whiskers and fine ZrO_2 particles, yields a much higher fracture toughness than that attainable with the use of either of the two mechanisms separately (Figure 1.61) [20].

(a) *Silicon nitride-based materials*
Among a broad range of refractory compounds, several carbides, nitrides and oxides can be distinguished, which, on the one hand, have been

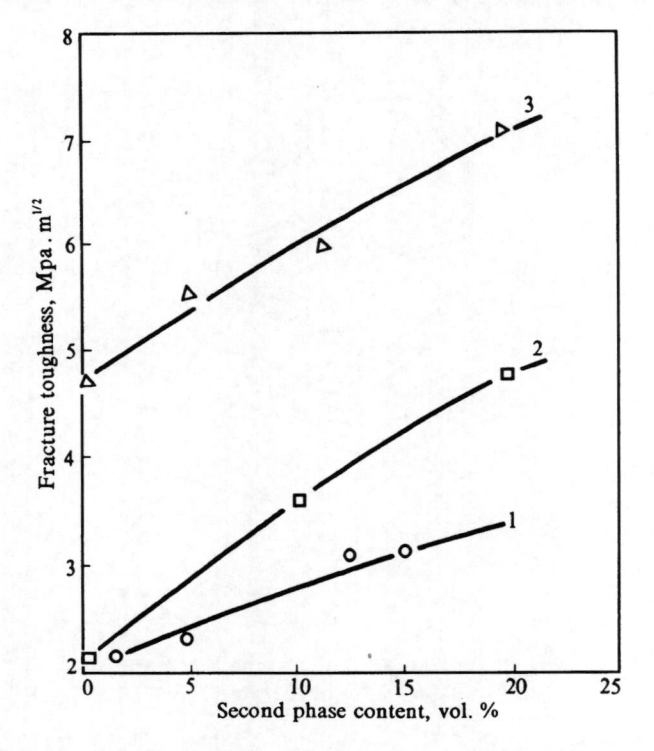

Figure 1.61 Increase of critical fracture toughness of mullite-based composite by various toughening mechanisms: (1) addition of ZrO_2 particles; (2) addition of whickers; (3) addition of 20% vol. SiC whiskers and ZrO_2 particles.

relatively more studied and, on the other hand, are interesting for their variety and wideness of use. Silicon nitride, owing to a low linear expansion coefficient, high scaling resistance and hardness, semiconductor and dielectric characteristics, and a low density, shares the lead in this group of compounds. An easy availability of the starting stock is also an important factor, opening broad possibilities for developing silicon nitride and composites on its base.

Silicon nitride-based materials are attractive for mechanical engineering owing to their strength characteristics, very low linear thermal expansion coefficient, high hardness, wear resistance in combination with a low density, inertness in many corrosive media, and ready availability of the starting stock [21].

Silicon nitride can be used in such fields as elements of gas turbine and piston engines, cutting tools and abrasives, products with high wear and corrosion resistance, needed for the chemical industry, metallurgy, mechanical engineering, etc.

Although these materials have been very actively developed and studied (especially in the last 10–15 years), the problem of improving their strength characteristics remains to be solved so far.

Silicon nitride composites are typical brittle materials, exhibiting no macroscopic plasticity at fracture over a broad temperature range. In this respect, Si_3N_4 behaves similarly to many other refractory compounds: carbides, borides, nitrides, oxides. The level of properties of such materials is much dependent on the strength of bond at boundaries of structure elements, which can be controlled by addition of various fillers, variation of characteristics of starting powders and of the product manufacturing technologies. The strength of materials of such a type is greatly affected by the content and character of macro- and microdefects, stressed state type and strain rate. Their fracture is governed by features of initiation and propagation of brittle-fracture cracks.

Figure 1.6 shows various schemes of structural transformations in production of silicon nitride-based composites [19]. With respect to the structure, ceramic materials are divided into monolithic (Figure 1.62a) and composite ones (Figure 1.67b). In the former case, the materials are produced from powder blends. As a result of thermal treatment, directional crystalline elements which effectively toughen the composite can form in the material. In the latter case, reinforcing elements in the form of whiskers or fibres (directional or specially oriented) are included in the composition of the material. The former type of materials includes numerous composites which have gained a wide use: reaction-bonded, sintered, and hot-pressed silicon nitride-based materials.

The strength and crack resistance of silicon nitride-based materials have been improved within certain limits through the use of doping additives. Numerous publications in the published literature deal with

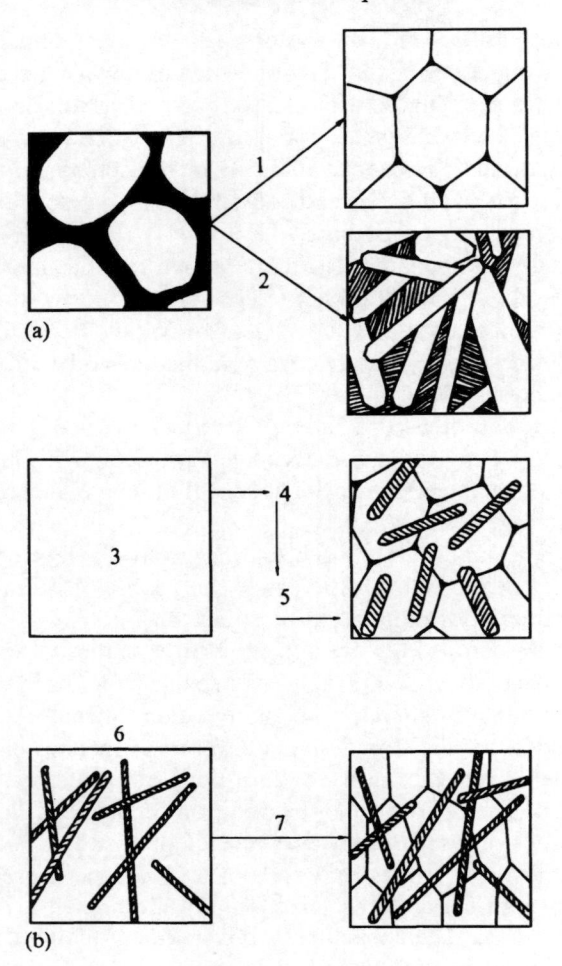

Figure 1.62 Various microstructures of composites produced by various methods [19]: (a) of monolithis ceramics and (b) composite ceramics: (1) intermediate liquid-phase sintering (hot-pressed β'-Si_3N_4); (2) liquid-phase sintering (sintered hot-pressed Si_3N_4); (3) liquid eutectic; (4) glass phase; (5) crystalline phase (Si_2N_2O–YAG); (6) whiskers; (7) impregnation, sintering (composite: β-SiC in Si_3N_4).

sialons, solid solutions of silicon nitride and alumina. The term 'sialon' is at present used for a family of materials whose structural units are tetrahedrons of a composition (Si, Al) (O, N)$_4$ or (Si, M)(O, N)$_4$, where M is a metal. Sialons are in their structure similar to silicates and can be based on α- and β-Si_3N_4, silicon oxynitride and carbide, aluminium nitride, spinel, etc. [4].

A potential advantage of sialons over a pure silicon nitride is that they can be sintered to a high-density state in an inert atmosphere at a temperature of 1600 °C. It is in sialons that all the achievements of the 'old' approach to increasing the material toughness are concentrated in the most complete form [3, 21]. Their potential applications are cutting tools, welding set elements, extruder dies, seals, bearings, drilling tools and other wear-resistant parts.

The material OTM-907 (Si_3N_4–MgO–Y_2O_3–Al_2O_3) has been developed for manufacture of parts subjected to high mechanical loads [22]. After the reaction sintering at 1300 °C the ceramic was subjected to an additional heat treatment at temperatures up to 1750–1800 °C for the formation of a dense microstructure, consisting of elongated β-Si_3N_4 grains (length, 16 μm; diameter, 0.5–0.7 μm). Just these thermal treatment conditions resulted in a material of the optimal structure with a high level of mechanical properties and with a crack resistance as high as 6.7 MPa m$^{1/2}$.

The idea of development of ceramic composites through the self-reinforcement of ceramics in the course of silicon nitride dissolution and recrystallization was implemented in a material of the Si–Y_2O_3–MgO system. The material features a number of unique properties. It retains the bending strength at a temperature of 1300 °C at a level of 530 MPa and has K_{Ic} values as high as 8.5–10.2 MPa m$^{1/2}$ over a range from room temperature to 1300 °C. The material has a specific structure with elongated β-Si_3N_4 grains of a regular hexagonal cut, having a cross-sectional size of 0.5 μm and an elongation degree over 10. Silicon nitride grains intergrow from silicon grains and, orienting in different directions, reinforce the ceramic, increasing its strength and crack resistance. Fracture of the ceramic has a transcrystalline character and occurs through the mechanism of pulling the grains out of the matrix. At the same time, the silicon grains, uniformly distributed throughout the material, serve as dispersions which retard the crack propagation. X-ray diffraction studies found compressive stresses over 100 MPa to arise in the Si_3N_4–Si ceramic composite.

In a number of cases, composite ceramic products have to serve as thermal insulators. Silicon nitride-based materials with the heat conductivity lowered through addition of the second phase have been developed for this purpose. Selected as the second-phase additives are components that would not significantly reduce the strength and thermal shock resistance. Table 1.23 presents properties of reaction-bonded silicon nitride-based materials with additions of porous Al_2O_3 and Si_3N_4 microspheres as well as of finely dispersed Al_2TiO_5 powder. The material with the addition of Al_2TiO_5 is preferable from the standpoint of the structural strength and thermal shock resistanse, whereas the Si_3N_4–Al_2O_3 material may be preferred when higher heat-insulating properties at a moderate strength are needed.

The material with an Al_2TiO_5 addition (grade OTM-911) [23] retains the structural strength also at 1500 °C, but then a vitreous film starts forming on the ceramic surface in the air. A diesel engine cylinder linear insert, made of this material, withstood 385 thermal cycles at 1050–1100 °C without fracture with a holding time of 5 min when cooled in air.

Table 1.23 Characteristics of silicon nitride-based heat-insulating materials

Additive	Additive content (% mass)	Apparent density (g/cm^3)	Porosity (%)	Bending strength (MPa)		Thermal shock resistance (heating-water) (°C)	Thermal conductivity (W/m K)
				$T = 20$ °C	$T = 1400$ °C		
Al$_2$O$_3$ microspheres	5	2.50	18.0	190	–	850	
	15	2.53	20.0	140	130	900	
	30	2.55	18.8	100	100	970	0.5–1.0
Si$_3$N$_4$ microspheres	5	2.55	16.6	180	–	800	
	15	2.53	14.0	170	–	870	
	30	2.54	13.5	150	150	915	4–5
Al$_2$TiO$_5$	10	2.45	20.0	180	230	980	
	20	2.49	18.7	180	210	1100	
	30	2.53	20.5	120	130	1200	4–5

A Si$_3$N$_4$–BN material for the use in above-rotor seals was produced by reaction sintering [23]. The BN content of up to 30% mass in it provided the required tribologic characteristics; its microhardness varied between 20 and 2.5 GPa depending on the boron nitride content and presence or absence of SiO$_2$-based protective film. The bending strength of the material (grade OTM-908) was of 156 MPa at 20 °C and of 175 MPa at 1400 °C [23].

Despite a high level of properties of monolithic ceramics, the principal course of developments in the structural materials science involves the creation of reinforced ceramic composites. The amount of research on developing new composite parametric materials grows abreast of the expansion of the market of refractory fibres and whiskers. Reviews [24, 25] note that today's achievements in the field of continous fibre-reinforced ceramic composites are modest enough because of thermal instability of fibres. Moreover, the surface of fibres usually needs special coatings to protect the fibres and control their interaction with the matrix material. However, Si$_3$N$_4$–SiC fibre composites with high mechanical properties (σ_{bend} = 500–600 MPa; K_{Ic} = 13 MPa m$^{1/2}$) have been developed [26]. Mechanical characteristics of some Si$_3$N$_4$-based composites are presented in Table 1.24.

A disappointing feature of reinforced materials is the loss of high mechanical properties under the action of prolonged thermal and vibrational loads.

Structural materials

Table 1.24 Properties of reinforced silicon carbide-based composites

Filler (% vol.)	Mechanical properties	Manufacturing method	Source
SiC fibres (30)	σ_{bend} = 465 MPa; K_{Ic} = 8.5 MPa m$^{1/2}$ (initial: 930 and 7.2 respectively)	Hot pressing 1750 °C, 27 MPa	[27]
C fibres (30)	σ_{bend} = 454 MPa; K_{Ic} = 15.6 MPa m$^{1/2}$; fracture energy 4770 J/m^2 (initial: 473, 3.7 and 19.3 respectively)	Hot pressing of Si_3N_4; additive: ZrO_2; 1450 °C	[28]
C fibres (up to 20)	$\sigma_{bend}^{20\,°C}$ = 400–700 MPa; $\sigma_{bend}^{1400\,°C}$ = 500–800 MPa; K_{Ic} = 3.3–3.7 MPa m$^{1/2}$	Hot isostatic pressing with additives: B_4C, BP, BN, AlN	[29]
W fibres (15–42)	$\sigma_{bend}^{20\,°C}$ = 668 MPa; $\sigma_{bend}^{1400\,°C}$ = 280 MPa (initial: 506 and 617 respectively)	Hot pressing	[30]
SiC whiskers (30)	σ_{bend} = 700 MPa	Hot pressing 1800 °C	[28]
SiC whiskers	σ_{bend} = 500 MPa; K_{Ic} = 10.5 MPa m$^{1/2}$	Hot pressing 1850 °C	[28]
SiC whiskers (10)	σ_{bend} = 950 MPa; K_{Ic} = 9 MPa m$^{1/2}$	Pre-sintering and hot isostatic pressing	
SiC whiskers (30)	$\sigma_{bend}^{20\,°C}$ = 8.6 MPa; $\sigma_{bend}^{1250\,°C}$ = 5.0 MPa (initial: 6.2 and 5.0 respectively)		[31]

The striving to develop ceramic composites should be reasonably justified. A spectacular increase of the strength and crack resistance of a composite at room temperature may change to their decrease in operation under the action of high temperatures, gaseous atmosphere, etc. It is also not always rational to employ, e.g., sintering additives for silicon nitride. The use of high-purity starting stock and of modern isostatic equipment yields silicon nitride without additives with a stable strength over 500 MPa up to a temperature of 1400 °C [32].

(b) *Silicon carbide-based materials*
Silicon carbide has, to date, gained exceedingly wide industrial application as a refractory material for electric heaters, varistors, abrasive tools, and sealing elements for operation in corrosive media. A higher heat resistance than that of silicon nitride, a high thermal conductivity, and a relatively simple manufacturing technology all allow silicon carbide

to be considered as a structural material for engine elements and assemblies operating at temperatures above 1600 °C. Silicon carbide is almost the sole refractory compound capable of retaining its properties in an oxidizing atmosphere at such high temperatures and not fracturing under cyclic thermal actions. Such refractory compounds as nitrides and carbides of titanium, tungsten, tantalum, niobium and hafnium start oxidizing appreciably at temperatures over 1000 °C and feature a low heat resistance because of high linear thermal expansion coefficients. The high oxidation resistance of dense silicon carbide stems from appearance on its surface of a SiO_2 film which retards diffusion of oxygen or of fuel combustion products to the surface of the material. As compared with silicon nitride, silicon carbide operates better at lower heating rates.

Comparing the silicon nitride and carbide properties shows that nitride ceramics exhibit higher strength and heat resistance, while carbide ceramics feature high thermal conductivity and elastic moduli. These properties predetermine their basic applications. Thus, a low thermal conductivity in combination with an elevated thermal shock resistance are needed for combustion chambers, impeller discs, nozzle vanes and impeller blades of GTE and piston caps of adiabatic ICE. These requirements are met by Si_3N_4 and sialon-based materials. High thermal conductivity, heat resistance and thermal stability are needed for various fuel nozzles of combustion chambers and heat-exchange devices. A high thermal conductivity, a low thermal expansion coefficient and a high elastic modulus are required for power optics elements of lasers, since these characteristics ensure a dimensional stability of mirrors under the action of powerful energy fluxes. These requirements are met by high thermal conductivity silicon carbide ceramics, featuring a lower density than those of most alloys used in the metal optics [9].

A silicon carbide-based structural ceramic material for turbocompressors, gas turbines, etc., has been developed [33]. It contains additions of silicon nitride, sialon, niobium nitride, and less than 10% mass activating oxide additives. Products are manufactured by pressing and subsequent sintering (2300 °C) or by hot pressing (2050 °C). The resulting ceramic material has $K_{Ic} = 10\text{--}17$ MPa m$^{1/2}$ and a bending strength of 400–700 MPa, which little changes over a temperature range of 20–1050 °C. After oxidation for 2000 h at 1200 °C and 1500 °C the strength of the ceramic remains practically unchanged. A rotor made from this material operated without apparent changes at a gas temperature of 1300–1500 °C and a rotation speed of 30 000 rpm for 1000 h.

SiC fillers in the form of dispersions, fibres or whiskers, in spite of their sensitivity to oxidation, are employed in development of most high-temperature ceramic composites. The maximum permissible service temperature for SiC fibres was taken in [34] at 1350 °C. Problems of increasing

Structural materials

the chemical and thermal stabilities of SiC fillers and of composites containing them have been dealt with in numerous studies.

C–SiC and SiC–SiC system composites, capable of a prolonged operation in an oxidizing atmosphere at 1000–2000 °C, have been developed for use in engines and space technology [3, 35]. The application of SERCAPBINOX (C–SiC) and SERASEP (SiC–SiC) composites for chambers and nozzles of turbojet engines has been reported [35]. The service time of products from these materials is up to 10 h at 1600 °C without visible erosion traces after tests. Numerous other examples of SiC-based composites with a set of unique properties can be presented. This material is constantly under the scritiny of advanced technology developers.

(c) *Boron nitride-based materials*
Boron nitride belongs to the class of especially heat-resistant high-temperature compounds. The most important fields of application, where boron nitride-based materials already at present have no competitors, are listed below.

1. Chemical and metallurgy industry, which have an acute need for reusable crucibles for evaporation of metal, melting of glasses and alloys, vacuum deposition of films.
2. Radio engineering and electronics, which need high-frequency insulating materials, stable over wide frequency and temperature ranges.
3. The development of new types of engines, where refractory materials with a high thermal shock resistance, capable of operation at 1200–1500 °C, have to be used in above-rotor seal elements.
4. Elements of SHF devices.

There are many more promising applications of boron nitride ceramics than the practically implemented ones.

Crystallographic features of graphite-like boron nitride determine advantages of this material: high thermal conductivity; low thermal expansion; high electrical resistance; stability of dielectric characteristics over a range from radio to superhigh frequencies; high-temperature strength; and chemical inertness [36].

Physical properties of hexagonal boron nitride exhibit a pronounced anisotropy stemming from the nature of the bonds between elements of this compound. Manufacture of products from boron nitride both by pyrolytic deposition and by powder metallurgy methods involves, as a rule, formation of texture. Layered elements of boron nitride are arranged with a predominant orientation. In terms of the uniform compaction model the texture of the material can be described by the function [37]:

$$I(k, \varphi) = \frac{1}{\pi}\left[1 + \frac{1-k^2}{1+k^2}\cos 2\varphi\right], \tag{1.61}$$

where k is the orientation parameter, determined by the X-ray method and φ the angle of observation with respect to the axis of compaction. The case of an isotropic material (absence of texture) corresponds to $k = 1$, whereas $k = 0$ corresponds to a material textured to the limit as a result of a uniform deformation. To a real hot pressed material, there corresponds $k \approx 0.5$; sintered materials have $k = 0.9$–0.95.

When parameters of anisotropy of a single crystal and the texture function of a polycrystalline material are known, the expression to describe anisotropy of the material can be derived:

$$\alpha^p(\theta) = \tfrac{1}{2}(\alpha_a + \alpha_c) + \frac{1}{4} \times \frac{1-k^2}{1+k^2}(\alpha_a - \alpha_c)\cos 2\theta, \qquad (1.62)$$

where α^p is the parameter of some physical property described by a second-rank tensor (mechanical strength, thermal expansion, thermal conductivity, permittivity); θ is the angle of observation with respect to a plane perpendicular to the pressing axis; α_a and α_c are single crystal parameters along the a- and the c-axis.

Anisotropy of boron nitride-based materials is determined not only by the texture. An important role is played by the structure of the void space, additions of doping materials, as well as effects associated with micro-inhomogeneous displacements of structural elements at temperature changes.

The materials science concerning boron nitride of a graphite-like structure has an almost three-decade history; three basic trends in the technology of manufacture of ceramics on its basis have formed over this time: pyrolytic vapour deposition; hot pressing; and sintering of boron nitride powders with sintering additives [36].

Mechanical properties of pyrolytic boron nitride exhibit a remarkable property of retention (and even rise) with increasing temperature (Figure 1.63). This feature invariably draws the attention of developers of high-temperature facilities to pyrolytic boron nitride and composites based on it. Elements of high-speed aircraft, spacecraft and launching units, operating at exceedingly high temperatures, are developed from these materials. A method for manufacturing parts from pyrolytic composites, which yields dense shells with a predetermined orientation of various fibres through a gas-phase compaction of pre-assembled skeletons from high-strength fibres, is very effective [3].

As a rule, boron nitride-based pyrolytic materials feature a strong anisotropy. A specific lamination under load is typical for products with a wall thickness of 5 mm and more. Manufacture of products is very costly and involves the use of toxic substances.

The problem of sintering of boron nitride to an adequate structural strength has been solved when powders of a turbostratic structure, active in sintering, had been produced [38]. However, the application of sintered

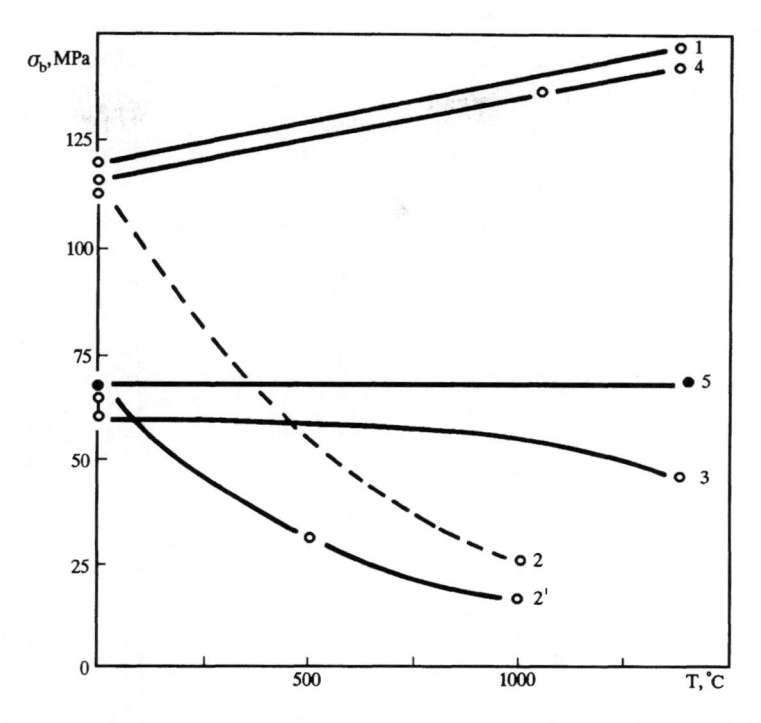

Figure 1.63 Temperature dependence of mechanical strength in lateral bending of BN-based materials: (1) pyrolytic BN_P; (2 and 2′) hot-pressed BN (made by ESK company) perpendicular (2) and parallel (2′) to pressing direction; (3) reaction-sintered BN (rBN); (4) rBN toughened with OSC pyrolysis products; (5) – BN_{hex}-based material, produced by thermal molecular cross-linking method.

and reaction-sintered boron nitride-based materials in engineering is restricted by their relatively low strength (Figure 1.63). In the last decade a number of sintered boron nitride-based composites have been successfully produced with the use of pyrolysis of organo-element compounds (OEC). This nontraditional approach opens broad possibilities and prospects for the development of new materials and modification of existing ones. It provides obtaining quite a number of ceramic products: fibres, powders, coatings, hollow spheres and composites of various compositions.

The basic idea of this method consists in using active products of the OEC pyrolysis for engineering of inorganic compounds of predetermined composition, structure and properties at the molecular level, which makes it possible to produce a number of inorganic compounds, including oxides, nitrides, carbides, borides and their combinations. The method opens up prospects for developing a simple and cheap technology for production of materials from a number of hard-to-sinter compounds [36].

146   *Ceramic-matrix composites*

To toughen boron nitride-based ceramics, it is most efficient to use organosilicon compounds (OSC). The use of reaction-sintered boron nitride ceramic as a porous, chemically inert matrix for its impregnation with OSC, followed by pyrolysis, increased two- to three-fold the strength of the starting material and yielded an isotropic material with a strength close to that of pyrolytic boron nitride.

The development of strong enough synthetic mullite whiskers in the 1970s made it possible to create a number of ceramic composites.

A set of such properties such as high thermal shock resistance, high thermal conductivity, low thermal expansion coefficient, high enough mechanical strength over a broad temperature range, and a relatively high oxidation resistance allows boron nitride-based materials to be used

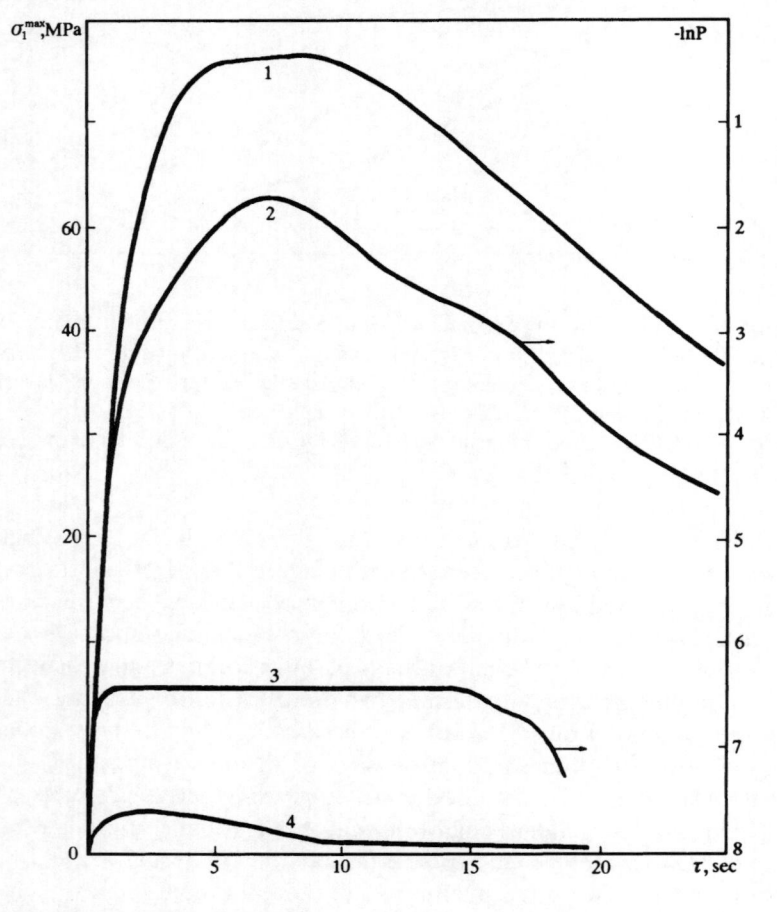

Figure 1.64 Variation of maximum tensile stresses (1, 4) and of fracture probability logarithm (2, 3) in the course of heating of combustion stabilizers made from OTM-909 (3, 4) and OTM-904 (1, 2).

for manufacturing structural elements of the hot zone of ICE and GTE. Thus, a material based on boron nitride and oligomethylhydridesilazane (grade OTM-909) was successfully tried for a GTE combustion stabilizer.

The combustion stabilizer functions directly in the flame and controls the flow, ignition and combustion of the fuel–air mixture. Its functions are to increase the flow turbulence, to stabilize the flame and to increase the fuel combustion rate.

High-temperature alloys employed to manufacture the stabilizers provide their long-term serviceability only up to 1270–1320 K; increasing the working temperatures calls for replacement of metals by more heat-resistant materials.

A batch of products made of OTM-909 was successfully tested in a combustion chamber in a high-temperature (1800 K) flow of products of kerosene combustion in air. The total test time was of 20 h; no visible changes in the stabilizers were found.

The thermally stressed state of the combustion stabilizer made of OTM-909 was analysed in terms of the magnitude of maximum principal tensile stresses (σ_1^{max}) (Figure 1.64, curve 4) for the engine start-up conditions, where gas temperature differences are the greatest. The calculation indicated that the maximum principal stress of 3.2 MPa occurs at 2.45 s and the zone of its action is at the centre of the base of the part. The pattern of stress fields is shown in Figure 1.65. The calculation of the probability of fracture (P) of the element in the course of heating (Figure 1.64, curve 3), based on the Weibull distribution's statistical parameters $m = 6.5$ and $\sigma_0 = 138.4$ MPa, demonstrated it to be negligible over the whole engine start-up period.

It is of interest to compare the above with results of analysis of the thermally stressed state of a combustion stabilizer made of a reaction-bonded silicon nitride-based material (grade OTM-904) (Figure 1.64, curves 1, 2). The comparison shows that the maximum tensile stresses in this stabilizer are greater and, despite high strength properties of OTM-904, the probability of its fracture in operation is much higher than those for the part made of OTM-909. The crucial role is played here not by the level itself of mechanical characteristics, but by the set of strength and thermophysical properties, which is more favourable for the boron nitride-based material.

Boron nitride-based materials can also be used in other thermally stressed elements of the GTE hot path. Unique tribological properties of boron nitride (low friction coefficient, excellent lubricity due to a specific softness and graphite-like structure) allow materials based on it to be used in sealing elements of GTE nozzle sets in the form of above-rotor inserts, coatings on the casing inside, etc., where an easy breaking-in of moving turbine parts to counterbodies is needed for an economic and dependable engine operation.

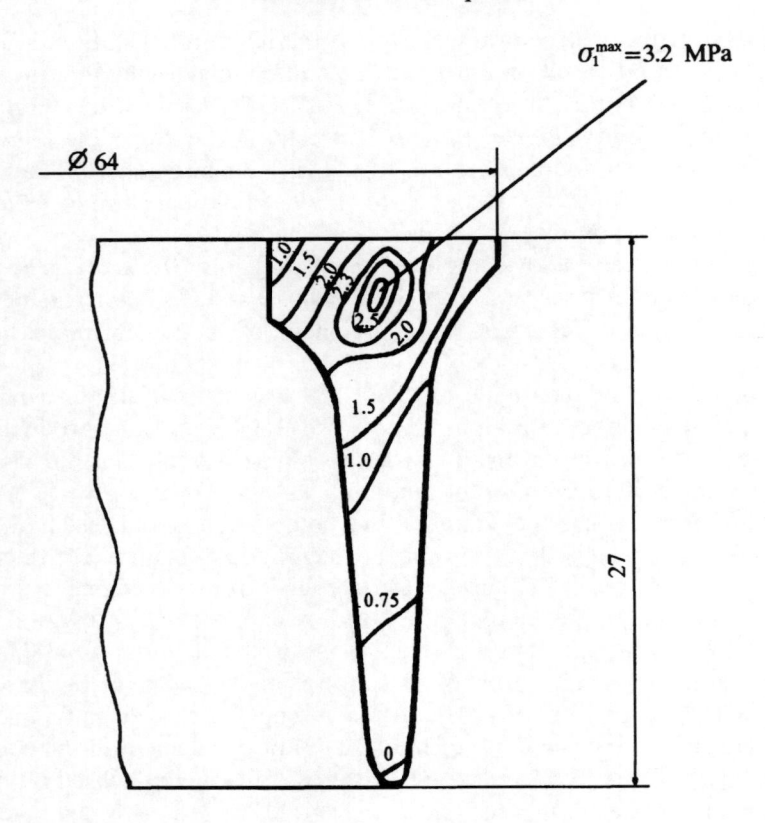

Figure 1.65 Isolines of principal stresses σ_1 in cross-section of combustion stabilizer made from OTM-909 at 2.45 s after engine start-up.

As shown by tests, above-rotor seals made of BN–SiO$_2$ material, produced in the BN–OSC system, ensure the cutting-in of steel disc's knife-edges in a few seconds without visible wear of the surface of the knife-edges and ensure GTE serviceability, being superior to a number of known materials of the BN–Si$_3$N$_4$ system [23] and at the same time cheaper (two to six times).

It is important that tribological characteristics of boron nitride-based materials are stable over a broad temperature range; this property in combination with an adequate wear resistance under abrasive loads make such materials attractive for high-temperature sliding bearings and provide an effective enough serviceability of the latter without lubrication.

Boron nitride-based BN–SiO$_2$ composites (of the BN–OSC system) differ from a pure sintered boron nitride by a higher structural strength and can thus be recommended for a number of elements in aerospace technology.

For example, an electrically insulating (in the form of a washer) aircraft engine vibration pick-up from a BN–SiO$_2$ material has been developed. It operates over a temperature range of 210–1020 K at the following vibration load parameters: acceleration, up to 980 m/s^2; overload, up to 100 g; frequency range, 5–5000 Hz.

Excellent stable dielectric properties of BN–SiO$_2$ materials over broad temperature and frequency ranges allow their use for radiotransparent ports of aircraft, microwave oscillators, chemical microwave and high-frequency heating plants.

Studies of numerous materials ascertained that addition of boron nitride to the composition of a ceramic directionally changes its electrophysical properties, upgrades the thermal shock resistance, improves the machinability and tribological characteristics, and reduces the mass of finished structures.

The effect of boron nitride on mechanical properties of β'-sialon was described in [39]. The material was produced by hot pressing of a blend of silicon nitride, aluminium, boron and alumina powders in a nitrogen atmosphere. The presence of hexagonal boron nitride in the structure of the composite made a nonelastic deformation of the material possible, through displacement of boron nitride layers relative to one another under the action of load. The nonelasticity degree varied from $\chi = 0.7$ at 20 °C to $\chi = 0.29$ at 1400 °C; the bending strength was 170–210 MPa at 20 °C and declined by 10–15% at 1200 °C. The significant nonelasticity of the material promoted the stress relaxation in it, reduction of the fracture process intensity, with the result that the homogeneity factor (Weibull modulus) increased considerably: to 14.6 and 10.6 at 20 °C and 1200 °C respectively.

(d) *Zirconia-based materials*

A special place among ceramic composites toughened with disperse particles is held by the group of transformation-toughened materials [42, 43]. A classical example of this type of ceramics is partially stabilized zirconia (PSZ). It is a coarse-grained (50–100 μm) polycrystalline material based on cubic zirconia, in whose grains fine (< 0.1 μm) inclusions of metastable tetragonal phase are dispersed. A material of such a structure is produced by sintering at a temperature over 1700 °C of ZrO$_2$ powders with a stabilizing additive (MgO, CaO, Y$_2$O$_3$), whose amount is insufficient for a complete stabilization of the cubic modification. After the sintering, for growing of tetragonal inclusions in the cubic matrix, the material is fired at an intermediate temperature of 1000–1200 °C. PSZ is one of the most strong and crack-resistant ceramic materials. Its bending strength is 600–1000 MPa and the critical stress intensity factor is $K_{Ic} = 8$–15 MPa m$^{1/2}$ [41].

Owing to its high thermomechanical properties, PSZ is gaining an ever wider application as a structural material in various fields of engineering. It is used to manufacture wear- and corrosion-resistant nozzles, plungers in the steelmaking industry, tools to extrude chromium–nickel alloy tubes and drawing dies. The combination of a low thermal conductivity and a linear thermal expansion coefficient close to that for cast iron makes this material attractive for use in thermally loaded engine parts.

Metastable tetragonal zirconia particles are employed for toughening other ceramic materials. The best results have been attained in development of $Al_2O_3-ZrO_2$ system composites.

The Al_2O_3 (maxtrix)–ZrO_2 (inclusions) system is a typical example of a ceramic composite with a higher fracture toughness. As reported in [42], at 10–15% vol. ZrO_2 the critical stress intensity factor rises from 5.2 to 9–9.5 MPa m$^{1/2}$. The composite strength was in this case somewhat less than the matrix strength and amounted to about 480 MPa. A similar fracture toughness increase mechanism is characteristic for a material consisting of a mullite matrix and a ZrO_2 filler. This material is a good candidate for manufacture of die inserts of press tools for metals, bearings for operation at elevated temperatures and in corrosive environments, erosion-resistant nozzles and cutting tools.

A substantial improvement of mechanical characteristics has also been obtained by adding tetragonal zirconia particles to other ceramic matrices: alumomagnesian spinel, aluminium titanate, silicon nitride and zirconium boride. The resulting increase in the strength and fracture toughness of these materials through transformation toughening allows

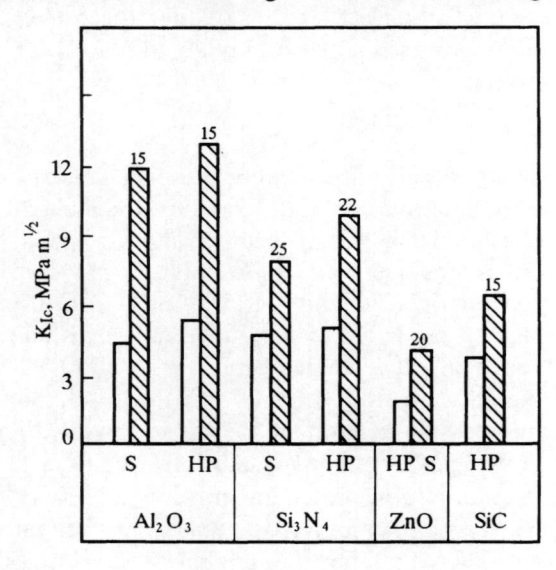

Figure 1.66 Fracture toughness increase at zirconia particles addition into ceramic matrix. Numerals over cross-hatched bars: volume fraction of added zirconia. Adjacent unhatched bar: fracture toughness of matrix material. S – sintering; HP – hot pressing.

their use for critical products: cutting tools, drawing dies, sealing elements, and other products where a combination of wear and corrosion-resistance is needed.

Figure 1.66 shows the increase in the toughness of a number of ceramic materials at addition of zirconia particles to a ceramic matrix [41].

The greatest effect in developing high-impact strength ceramic composites of such a type can be attained through implementation of several toughening mechanisms. For example, a simultaneous addition of ZrO_2 (monoclinic or tetragonal) particles and SiC whiskers raises the fracture toughness more efficiently than does the addition of any one of them. Results of studying a dense hot-pressed mullite with fillers, uniformly dispersed ZrO_2 particles (of about 0.6 μm in diameter) and SiC whiskers, in the mullite matrix of particles sizing 1–2 μm are reported in [43]. Because of a mullite matrix, no interaction between SiC and ZrO_2 did occur. (Such a reaction could result in SiO_2 and ZrC, which is quite probable for an alumina matrix.) Studies of the fracture toughness at 800 °C (Figure 1.67) indicate that a thermal treatment (1130 → 800 °C) of samples of mullite reinforced with SiC whiskers and tetragonal ZrO_2 particles results in the greatest increase of the fracture toughness ($K_{Ic} > 10$ MPa m$^{1/2}$) over that for the monoclinic form ($K_{Ic} = 7$–7.5 MPa m$^{1/2}$).

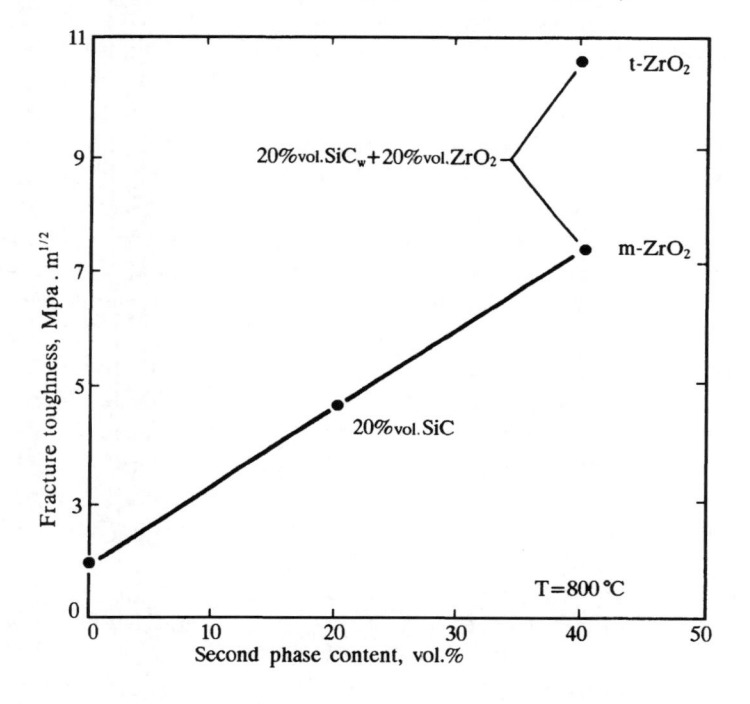

Figure 1.67 Fracture toughness of mullite reinforced with SiC whiskers and ZrO_2 particles [43]

Tetragonal ZrO_2-based materials reinforced with SiC whiskers are promising to attain a high level of high-temperature properties. Whiskers are regarded as preferable to fibres, having a higher Young's modulus. Moreover, it is believed that they can better withstand the processing conditions, which reduce mechanical and thermal properties of fibres [41].

Addition of up to 30% vol. whiskers to tetragonal ZrO_2 considerably changed its mechanical properties. The fracture toughness at room temperature doubled (up to 6–12 MPa $m^{1/2}$), while the bending strength decreased by half (from 1200 to 600 MPa). However, the addition of SiC whiskers doubled the strength of the material at 1000 °C. A decline in mechanical properties because of oxidation of SiC whiskers was noted at a prolonged service of such materials in the air.

Some possibilities for improving mechanical properties of zirconia-based composites can be used in moulding the preforms of products. Thus, employing the principle of forming composites with a layered-granular structure, the authors of [44] succeeded in upgrading the density, thermal shock resistance, and strength of ZrO_2 (85% mass)–CeO_2 (12% mass)–Y_2O_3 (3% mass) system materials used for composite electrodes. Preforms were pressed from layered cubic-shaped granules consisting of 150–200 μm thick layers of different densities. As compared with grain-structure materials, those with the layered-granular structure exhibited a much greater thermal shock resistance (2.5–4 times higher, up to 45 thermal cycles of 1200–20 °C), a lower porosity (14–18% as against 22–24%), and a higher tensile strength (17–22 MPa as against 10–15 MPa).

The manufacture of fibre-toughened and layered ceramic materials through a directional crystallization of ceramic eutectics has been advanced in recent years. Production of ceramic materials of eutectic composition yields by itself the effect of a three-dimensional reinforcement, so that a special addition of reinforcing elements becomes unneeded. A characteristic feature of the process of formation of such composites is the fact that fibres and the matrix phase are in a thermodynamic equilibrium. High-strength zirconia-based materials of the Al_2O_3–ZrO_2, $CaZrO_2$–ZrO_2, ZrO_2–ZrC–ZrB_2 and other systems were produced by this technique. Methods for producing such materials were adopted from the technology of growing single crystals of refractory compounds. The main advantage of ceramic materials produced by the directional crystallization method is a high thermal stability of the structure, i.e., the ability to retain the composite structure right up to a temperature equal to 0.98 of the melting temperature. Due to this, such materials retain their strength and other structure-sensitive properties at a high-temperature service. Thus, a directionally crystallized Al_2O_3–ZrO_2 eutectic retains a tensile strength of 520 MPa to a temperature over 1650 °C. Because of this, these materials are considered as substitutes of superalloys for uncooled GTE turbine blades.

The combination of high strength, toughness and chemical stability provides for zirconia-based materials a wide application in various structures operating at high loads and in corrosive environments. Various cutting tools of these materials are popular. Products of the materials are employed not only under industrial conditions for cutting metals, plastics, paper, etc., but also for domestic (kitchen knives, hair-cutting and shaving blades) and medical purposes (surgical scalpels).

The work on application of zirconia in engines is carried out very intensively. The combination of a high strength, a low thermal conductivity and a thermal expansion that matches that of metals makes zirconia a potential candidate for use in such engine elements as the piston cap, supporting plates and cylinder liners. The application of ZrO_2-based materials requires practically no significant changes in the engine design. Engines incorporate some elements whose service life is determined by the wear resistance of materials, such as cams, tappets, eccentrics and exhaust valves.

A partially stabilized ZrO_2-based material with an apparent density of 5.8 g/cm^3, bending strength of 480–670 MPa at 20 °C and of 150–180 MPa at 1300 °C, and an impact strength of 7–8 MPa m$^{1/2}$ was tried for manufacture of some ICE parts. Tests demonstrated the wear of ceramic tappets to be less than for metallic ones, the wear of steel cams decreasing at the same time.

Shafts (diameter 9 mm, length 105 mm) of a pump for corrosive media, made by 'Serfilco', operated for more than 600 h without failure. Compressor piston caps (diameter 130 mm, height 5–6 mm) operated for 1 h at a pressure of 20 MPa and a temperature of 1000 °C [45].

(e) *Quartz ceramics-based materials*
Materials produced from quartz glass by the ceramic technology are called quartz ceramics. These offer many valuable properties of quartz glass: low thermal expansion coefficient, high thermal shock resistance, good electrophysical characteristics, high chemical and thermal stability. Both high-density quartz ceramics (with a porosity of 1–2%) and foam ceramics (with a porosity of 90% and more) find application in engineering.

Among other advantages of quartz ceramics, its high-temperature strength is notable [46]. Carbide and nitride ceramics feature an insignificant bending strength change up to temperatures of 1200–1300 °C, and main types of oxide ceramics, a considerable drop of this characteristic, while SiO_2, Al_2O_3–SiO_2 ceramics exhibit a growth of strength characteristics, which is due to a specific mechanism of fracture of these materials.

Quartz ceramics are chiefly used in new or specialized fields, where materials with an excellent thermal shock resistance, constant electrophysical properties, high heat-insulating properties, and the ability to retain their dimensions at high temperatures are needed. Quartz ceramics find application as both general- and special-purpose refractories, for space-rocket technology, and in a number of other fields [46]. They are attractive for manufacture of numerous components of rocket and space facilities, such as leading edges, head parts (nose cones) of rockets, aerial fairings, radiotransparent ports, nozzles and bushings of rocket engines.

In view of the extension of fields of application of quartz ceramics, a need has arisen to improve their characteristics. While retaining the uniquely high thermal stability of quartz ceramics, it is necessary to improve the strength, thermophysical, dielectric and other properties, to upgrade the operational reliability of products. One efficient technique for this purpose is doping. Used as dopants are oxides of rare earth elements and of transition metals, oxygen-free compounds such as SiC, BN, Si_3N_4, Si_3B_4. The doping of quartz ceramics is effected by several methods, described in [47]. Addition of chromium (up to 5% mass) to quartz ceramics increases 2–3 times the integral degree of blackness, upgrades the stability in high-temperature gas flows, and reduces by half the attenuation of a radio signal at high temperatures [47]. As reported in [47], an effective amount of dopant can be as little as 1.5% mass Cr_2O_3. The resulting material has a bending strength of 30–80 MPa, an elastic modulus of $(2.4–4.5) \times 10^4$ MPa; its thermal expansion, heat capacity and dielectric characteristics over a temperature range of 20–1000 °C are close to those of ordinary quartz ceramics.

A material with a negative temperature coefficient of permittivity is obtained by adding 15–25% mass TiO_2 to quartz ceramics. The closeness of casting properties of aqueous suspensions of these materials makes it possible to produce, by a layer-by-layer moulding, layered composites and products with various combinations of doped ceramics: products with vacuum-tight and porous layers; products whose surface or certain zone is made of a material having a higher emissivity or reflectivity, a higher permittivity or conductivity, hardness or impact strength. Long (over 1 m) blanks in the form of bodies of revolution with a high dimensional accuracy are successfully manufactured. Combined products based on quartz ceramics and quartz glass, quartz and boron nitride ceramics, quartz ceramics reinforced with a metal wire, etc. can be made. This extends the design capabilities of products and upgrades their performance characteristics. Ceramic plates with optical light guides from a transparent quartz glass, built into the plates at predetermined places, have been produced, making possible an optical monitoring of processes proceeding behind the ceramic screen.

When a material is doped through impregnation with salt solutions, a refractory dopant penetrates into the intergrain space and reduces a viscous flow of the material at quartz glass grain boundaries. Thus, a quartz ceramic doped by impregnation with an $Al(NO_3)_3 \cdot 9H_2O$ solution retains its high-temperature strength up to 1300 °C, whereas the initial material retains it only to 1200 °C [47]. A need to dope only surface layers or separate zones of a quartz ceramic product arises in some cases. The surface treatment of large products is carried out by dipping the whole product or its part into an appropriate solution or by deposition [47].

In cases where a material is used at elevated temperatures and high loads for a short time, it is expedient to toughen quartz ceramics by impregnating them with polymers. Such an impregnation results in 'healing' of microcracks and scratches. When hardening, the polymer shrinks, additionally toughening the material through redistribution of stresses in it. The polymer for impregnation in the ceramic is selected depending on the specific use of the product taking into account the following characteristics: thermal stability, adhesion to the ceramic material, mechanical strength, contact angle of wetting, solubility in organic solvents, dielectric properties, thermal conductivity and thermal expansion. In particular, the manufacture of radiotransparent products operating at high temperatures calls for the use of thermally stable polymers, offering a good adhesion to ceramics, etc. This set of requirements is met by polyorganosiloxanes, polyorganosilazanes, carboranes, organosilicon–titanium compounds, which belong to the class of high-molecular organoelement compounds.

Polymer-toughened quartz ceramic composites differ from initial ceramics by a higher mechanical strength (increase by 200% and more, the dispersion of this characteristic declining substantially), a higher (by up to 80%) impact strength, and by high moisture resistance characteristics. On the whole, the composites feature higher physicomechanical properties while retaining such parameters as dielectric properties, thermal expansion and elastic modulus practically unchanged.

An effective composite based on a quartz foam ceramic impregnated with phenolic resins has been developed. When heated at the entry to dense atmosphere layers, the impregnated foam ceramic returns a considerable part of the heat acquired by it back into the space by radiation. The filler absorbs heat due to pyrolysis. Gaseous decomposition products cool the ceramic as they arrive from the pyrolysis zone to the surface.

The development of reusable aerospace facilities necessitated the creation of radically new heat-protection and heat-insulation materials. The technical feasibility and economic expedience of development of the *Shuttle* and *Buran* space vehicles depended on the attainable set of properties of the heat-insulation material.

A comprehensive analysis of all the possible thermal protection types indicated that the best one is a ceramic i nsulation from ultrathin quartz fibres. The technology for manufacturing reliable and highly effective heat-insulating blocks was developed. More than 30 000 heat-insulating tiles were installed on the *Buran* vehicle, of which as few as six failed. Successful missions of the *Buran* and *Shuttle* confirmed the feasibility of utilization of ceramic materials, despite their brittleness and relatively low strength, in most critical structures. The point is, how strictly and reliably the whole technology of their manufacture and inspection is implemented and how the design and manufacturing process corresponds to operating conditions and material characteristics.

The developed technology makes it possible to vary the density, strength and thermal conductivity of the material over broad ranges to suit one or other operating conditions of aerospace facilities or other objects.

Several modifications of the thermal-protection material are commercially produced at present (Table 1.25).

Table 1.25 Properties of some commercially available thermal-protection materials

Grade	Density (g/cm^3)	Tensile strength (MPa)	
		guaranteed	average
TZMK-10	0.144	0.25	0.37
TZMK-12	0.120	0.20	0.27
TZMK-20	0.200	0.40	0.65
TZMK-25	0.250	0.45	1.00

Dimensions of tiles can be varied up to 350 mm at a thickness up to 150 mm. The thermal conductivity of the material under ordinary conditions is not over 0.1 W/(m K); in particular, for materials TZMK-10 and TZMK-25 it amounts to 0.05 and 0.07 W/(m K) respectively.

The application temperature range varies from 120 to 1500 K depending on the duration and cyclicity.

The anisotropy of mechanical properties of the thermal-protection materials was studied by acoustic techniques. It was ascertained that the materials feature a pronounced transversal anisotropy of acoustic, elastic and strength properties. Coefficients of anisotropy of the tensile strength, compressive strength and elastic moduli at a static tension and compression are respectively of 1.6–3.7; 1.0–2.1; 2.2–5.0; and 2.0–5.0.

Owing to a unique set of properties, the TZMK thermal-protection materials have found application, apart from the aerospace technology, in various engineering and industrial objects, in particular, as the heat insulation of various thermal equipment, in metallurgy, in fire-fighting facilities, in the gas and oil industry, in electrically insulating and dielectric systems, etc.

1.1.6.4 *Problems and prospects of development of ceramic composites*

In spite of optimistic predictions, commercial production of ceramic materials over the world proceeds at an exceedingly slow rate. Engines with ceramic elements (friction pairs, tappets, exhaust ducts, etc.), cutting tools, filters, measuring tools of silicon nitride and carbide, zirconia, boron nitride, and other ceramics are gradually entering the market. At the same time the advance of technical progress calls continuously for newer and newer materials meeting high requirements in operating temperature, strength and stability in oxidizing and corrosive environments during a long service life of machines.

Ceramic materials, while being superior to metals and alloys in quite a number of properties, suffer, nevertheless, from one serious drawback: a low fracture toughness. As stated by leading ceramic experts in the USA, Japan, Germany, USSR, and the UK, for application in critical elements of automotive and aircraft engines, gas turbines, and spacecraft, ceramic materials should have a bending strength of at least 700 MPa at 1400 °C and a fracture toughness of 30 MPa m$^{1/2}$ [3]. At present, there are no materials with such a fracture toughness.

In the field of engineering ceramics, much attention will be, as before, given to materials based on silicon carbide and nitride, sialons and zirconia; the potentialities of monolithic ceramics having been exhausted, attention will be focused on ceramic composites, including those toughened with ceramic fibres and whiskers. Development of a number of new attractive AlN–Al composites is contemplated; the interest in using such materials as TiB_2, ZrB_2, BN, TiC in ceramic composites is growing.

In the field of development of ceramic materials with predetermined properties, a successful advance has been gained by the concept of development based on conceptions of engineering of composite structures. This is due, on the one hand, to a considerable demand for new materials, without which technical advance is impossible, and, on the other hand, to marked advances in the technology of the structural (non-reinforced) ceramics, the base for development of ceramic composites. Researchers at present see no other ways for improving the fracture toughness of ceramic materials except the development of heterogeneous composite structures, where energy-intensive processes of dissipation of the work of external forces could be implemented.

The demand of the engineering of new materials spurred the advance in manufacturing technology for reinforcing elements: continuous and discrete fibres and whiskers of refractory compounds. The extension of the scales of production reduced their cost, which, along with the relative cheapness of matrix materials, makes real an extensive commercial application of ceramic composites [11].

The attained level of thermomechanical properties and high permissible operating temperatures render ceramic composites indispensable for a number of engineering fields.

Many problems involved in the development of such materials, however, remain unsolved so far. These include problems of optimization of properties of the fibre–matrix interface, of a homogenous distribution of discrete reinforcing elements in the matrix, and of production of high-density defect-free matrices. Numerous studies on the surface treatment of reinforcing elements for their protection in oxidizing and corrosive environments will have to be conducted. A successful solution of the problems calls for specialized production equipment with automated monitoring and control systems.

1.2 FUNCTIONAL MATERIALS

1.2.1 Ceramic composites for electronics
V. V. Skorokhod and V. A. Dubok

Functional ceramic composites for electronics, or electronic ceramic composites (ECC), are ceramic materials where the combination of constituents is employed to control the electronic-ionic processes which determine the function of a material.

The ECC constitute a large class of materials with disordered and organized structures, produced by ceramic technology, which are used to make electronic devices and circuits, ceramic sensors, active and passive elements in optoelectronics, acoustoelectronics, magneto-optics, etc., and also as electromagnetically transparent and absorbing, and other special materials.

Structures of the ECC feature a wide range of sizes of constituents added to a material to attain its functional properties: from randomly disposed inclusions that are either associates of a few atoms or nanometre-thick layers, to ordered macroscopic-scale elements.

The ECC designing principles are based first of all on calculations of characteristics of polyphase composites with various combinations of electrophysical – scalar, vector and tensor – properties with allowance for the specifics inherent in the ceramic technology: existence of pores, segregation of impurities and other phases at grain boundaries, dissimilar morphologies of different phases, etc.

The required electrophysical properties of ECC constituents are attained on the basis of the insight into the nature of corresponding physical processes in them and ensuing methods for controlling individual properties, such as various components of electrical conduction – electron, hole, anionic, cationic, protonic, as well as the dielectric permittivity and loss, ferro-, piezo-, pyroelectric and other properties, conditions of phase transitions, which determine the properties, with allowance for environmental parameters.

The ECC manufacturing technology should yield the required combination of properties of the constituents and their spatial–structural organization, which, taken together, provide realization of the physical principles that form the basis of functioning of a given ECC.

The list of various ECC types and designing principles, presented below, which is far from being complete, outlines in a most general way the scope and importance of this problem. Due to the limited size of this section, it includes elements of the theory of physical properties of polyphase ceramic materials as well as brief descriptions of the physical concepts of various types of electrical conduction, dielectric polarization and loss, employed in the ECC. Attention is focused on mechanisms controlling the electrophysical properties; in particular, the effect of departure from the stoichiometry and of doping on ionic and electronic components of the electrical conduction, electrochemical doping and intercalation of oxides are discussed by way of example of oxides, which are most typical ECC constituents.

The usefulness of the presented concepts is illustrated by their application to specific EEC and devices based on them: bulk resistors and heaters, ceramic capacitors, various sensors, posistors, varistors and other devices.

1.2.1.1 Theory of physical properties of polyphase ceramic materials

Inherently, ceramic composites are structurally inhomogeneous, consisting of two or more phases. A phase in this case implies a collection of structural elements characterized by a chemical and a crystallographic homogeneity. Phases in a composite may not be in thermodynamic equilibrium.

Polyphase systems also include most real ceramic materials containing a glass phase and residual porosity. The main task of the macroscopic theory of physical properties of polyphase materials is to derive methods for calculation of effective values of one or other property at given values of properties of individual phases, their volume fractions, microstructural parameters of the system, etc. In the simplest case of so-called scalar properties (which include density, heat capacity, penetrating radiation absorption coefficient) the task becomes trivial: the effective value of a property is calculated by the rule of additivity and does not depend on structural parameters of a polyphase system [1]:

$$c_{\text{ef}} = \sum_i c_i \theta_i, \qquad (1.63)$$

where c_{ef} is the effective value of a scalar property, c_i the corresponding value of the property for the ith phase, and θ_i the volume fraction of the latter.

For a much wider class of properties, determining the relation between various vector- and tensor-nature fields outside and within a polyphase material, however, the task becomes by no means trivial and acquires a great practical importance. For example, depending on microstructural features of a dielectric–conductor mixture, its permittivity can vary by several orders of magnitude and the electrical resistance can vary from that of an insulator to that of a conductor at one and the same volume concentrations of phases. In other words, varying the contents and microstructures of phases provides, in a number of cases, a means to control the physical properties of heterophase materials no less successfully than by varying the properties themselves of the phases that constitute the material.

Further discussion calls for classification of properties both from a formal mathematical and a thermodynamic standpoint. It is also expedient to systematize polyphase materials on the basis of the determining structural parameters, bearing in mind first of all the geometry of interphase surfaces.

(a) *Classification of physical properties*
Physical properties of solids are commonly defined as some factors of proportionality, which relate parameters of a physical field applied to a body, to parameters of the field induced thereby in it. Both applied and induced fields are of a vector or a tensor nature, and therefore the set of factors defining a linear dependence of components of one tensor on components of another tensor is in itself a tensor, whose rank is the sum of ranks of the two tensors (a vector is considered as a tensor of first rank, and a scalar as a tensor of zero rank). For example, electrical conductivity factors relate components of the vector of the current density in a body, j_i, to components of the electrical field vector E_k and are therefore a second-rank tensor λ_{ik}:

$$j_i = \lambda_{ik} E_k \qquad (1.64)$$

(addition over repeating subscripts is implied here and hereinafter [2]).

The permittivity ε_{ik}, defining a linear relation between components of electrostatic induction vector D_i and components of electric field vector E_k, is also a second-rank tensor.

Elastic constants define a linear relation between components of the stress tensor σ_{ik} and of strain tensor e_{lm} and are therefore a fourth-rank tensor c_{iklm} [3]:

$$\sigma_{ik} = c_{iklm} e_{lm}. \qquad (1.65)$$

Relation (1.65) expresses Hooke's law in the general form.

The applied and induced physical field can be of different types and described by tensors of different ranks. Physical properties corresponding to such relations can also be described by odd-rank tensors. Thus, electromechanical force factors, defining the magnitude and sign of the piezoelectric effect, are a third-rank tensor (linear relations between components of the electrostatic induction vector and of the mechanical stress tensor):

$$D_i = \chi_{ikl}\sigma_{kl} \tag{1.66}$$

Vector- and tensor-type fields can result from a scalar field. Thus, a change in the intensity of a uniform temperature (scalar) field brings about a change in the elastic strain field because of the thermal expansion or in the electric field (for pyroelectrics). In the former case, linear thermal expansion coefficients are components of a second-rank tensor α_{ik}, while in the latter case, pyroelectricity coefficients, as components of the electric field intensity vector, constitute a first-rank tensor. The classification of physical properties is summarized in Table 1.26. Physical properties which are mathematical analogues (dielectric and magnetic permittivities, on the one hand, and electrical conductivity, thermal conductivity and diffusion coefficient, on the other) may belong to two basically different types of physical phenomena.

Table 1.26 Classification of physical properties

Property	Tensor rank	Thermodynamic characteristic
Electromagnetic		
Dielectric permittivity	Second	Equilibrium
Electrical conductivity	Second	Nonequilibrium
Magnetic permittivity	Second	Equilibrium
Magnetic susceptibility	Zero	Equilibrium
Thermal		
Heat capacity	Zero	Equilibrium
Thermal conductivity	Second	Nonequilibrium
Thermal expansion coefficient	Second	Equilibrium
Mechanical		
Elastic constants	Fourth	Equilibrium
Viscosity coefficients	Fourth	Nonequilibrium
Elastic compressibility	Zero	Equilibrium

Physical properties of one type involve a set (sequence) of equilibrium states arising in a solid as external conditions change (elastic deformation, polarization, orientation of magnetic dipoles, etc.). Properties of this type can be conventionally called equilibrium ones. Physical properties of the other type characterize the intensity of fluxes arising in a field under the action of the thermodynamic forces, i.e., a substantially nonequilibrium state [3]; these are so-called nonequilibrium or kinetic properties. The former case involves accumulation of an energy in a body, which, as the external action has been removed, is returned to the medium external with respect to the body, whereas in the latter case, there occurs a continuous dissipation of the energy, arriving from the environment, i.e., its conversion into heat, which is dissipated in the environment. It is to be borne in mind that in the general case the variational principles, governing reversible and irreversible processes, are different [2], and therefore the analogy between equilibrium and kinetic properties is not complete.

(b) *Classification of heterophase systems*
Polyphase systems, which include ceramic composites, can be subdivided into two large groups:
   (I) regular (ordered) systems;
   (II) irregular (disordered) systems.
Matrix (M) and skeleton (S) structures, differing in the topology of the phase interface, can exist within either of the two groups. The classification of heterophase systems is summarized in Table 1.27.

Table 1.27 Structural classification of heterophase systems

Geometric characteristic	Anisotropy degree	Dimensionality
(1) Regular structures		
Parallel layers	Strong anisotropy	One-dimensional (perpendicular to layers)
Parallel fibres in matrix	Strong anisotropy	Two-dimensional (perpendicular to fibres)
Spherical inclusions in matrix	Weak anisotropy	Three-dimensional
Interpenetrating skeletons	Weak anisotropy (orthotropy)	Three-dimensional
(2) Irregular structures		
Randomly oriented fibres in matrix	Isotropy	Three-dimensional
Randomly disposed spherical inclusions in matrix	Isotropy	Three-dimensional
Statistical mixture of isomeric polyhedrons	Isotropy	Three-dimensional

Regular systems exhibit a certain symmetry and periodicity of structure. They generally have a translational symmetry. A representative element, whose translation in space constructs the whole system, can always be distinguished in such systems. Regular systems include three-dimensional and quasi-two-dimensional ones. The latter comprise layered structures and structures reinforced with parallel-laid continuous fibres [4]. Three-dimensional regular structures can be matrix or skeleton ones.

The former (in the case of two-phase systems) comprise inclusions of one phase, which are of a geometrically regular shape, of identical linear dimensions, identically oriented in space, and regularly arranged in the other phase, matrix. Here one of the phases (matrix) is continuous, while the other (inclusions) consists of structural elements not connected with one another. The phase interface is multiple-connected.

The latter, i.e., skeleton structures, are such that one of the phases is in the form of a continuous spatial skeleton, or framework, featuring some regular symmetry. The second phase can here be considered as the matrix one, but the interface between the phases is generally of a complex topology; it is mainly convex on the side of the skeleton and concave on the side of the matrix.

In the simplest case of parallelepipedal elements of the structure, the two phases can be treated as two interpenetrating frameworks. Such a structure is symmetrical with respect of the volume content of phases, whereas in matrix structures the phases are structurally unequal at any of their volume contents.

Irregular matrix and skeleton structures are distinguished in that characteristic geometric parameters of structural elements become random variables obeying some distribution. Such parameters can be the size of inclusions, their spacing, angle of orientation of nonisomeric inclusions in the space – for matrix structures. For skeleton structures, random values can be taken by the cross-sectional area of the skeleton, local orientations of its elements, spacing of skeleton junctions.

Regular structures are in the general case anisotropic. Structures having a cubic symmetry are orthotropic; they exhibit isotropy for physical properties described by tensors of not more than second rank.

Irregular structures can be either anisotropic or isotropic. The latter are of the most general interest, since the overwhelming majority of heterophase ceramic materials, not subjected to a special treatment, are isotropic.

A special class among irregular heterophase structures is formed by so-called statistical isotropic mixtures, where all the phases are represented by identical structural elements, polyhedrons, which fill the whole space of the system. A statistical character of such systems is determined so that, at the place of any selected structural element, there can be the ith phase with a probability

$$P_i = \theta_i; \quad \sum_i P_i = 1,$$

where θ_i is the volume fraction of the ith phase in the mix [5]. Here all the phases are structurally equal. A distinguishing feature of statistical mixtures is that they change their character depending on volume concentrations of phases. This can be illustrated in more detail by the simplest two-dimensional model of a two-phase statistical mixture, where structural elements are represented by regular hexagons, the coordination

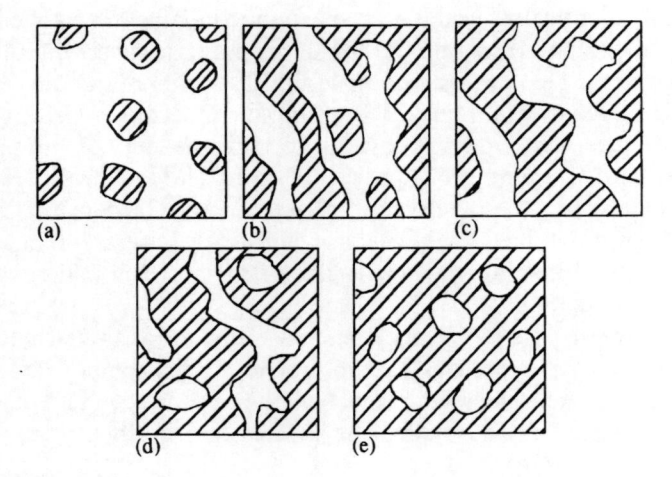

Figure 1.68 Variation of character of two-dimensional statistical two-phase mixture at various phase concentration ratios: (a) 0.1:0.9 (b) 0.3:0.7 (c) 0.5:0.5 (d) 0.7:0.3 (e) 0.9:0.1

number being six. If a fixed element of the first phase is selected and the cluster formed by it and its closest neighbours (seven elements in all) is examined, it becomes obvious that the average concentration of the first phase will be equal to the ratio between the number of first-phase elements and the total number of elements in the cluster. At an average concentration $\theta_1 \leq 1/7$, the probability that the fixed element of the first phase will be the only one in the cluster of seven elements is close to one. This means that at $\theta_1 \leq 1/7$, the statistic mixture will be of a matrix character (first phase has the form of separate inclusions; second phase is a physically continuous matrix). At $\theta_1 \geq 3/7$ the system will, with a probability close to unity, acquire the character of two interpenetrating skeletons. At $1/7 < \theta_1 < 3/7$, the mixture has an intermediate character. The above is, of course, symmetrical with respect to interchange of the phases. The character of variation of the relative arrangement of phases in a two-dimensional statistical mixture with their concentrations is shown in Figure 1.68

(c) *General principles of macroscopic theory of physical properties of polyphase materials*

There exist two fundamentally different approaches to the solution of the general problem of calculating the physical properties of a heterophase material from given values of properties of individual phases and their volume concentrations.

The first approach relies on an adequate modelling of a real heterogeneous system by geometrically defined elements of the microstructure.

The selection of the model is unambiguous only in some simplest cases (layered or unidirectionally reinforced structures, spherical inclusions at their low volume concentration). In the general case such a selection unavoidably involves a number of simplifying assumptions, substantially affecting the rigour of the solution and reliability of its results.

As noted above, scalar properties can be calculated by the rule of additivity; either a property itself (heat capacity, specific volume) or its inverse quantity (density) can be additive. All the non-scalar properties, however, regardless of the rank of the tensor describing them, obey the following important theorem: the true value of a physical property of a heterophase system is always confined in the interval between its average value and the inverse of the average value of its inverse, i.e.,

$$\bar{\lambda} > \lambda_{\text{ef}} > \left(\overline{\frac{1}{\lambda}}\right)^{-1},$$

where λ is a physical property (conductivity, permittivity),

$$\bar{\lambda} = \sum_i \lambda_i \theta_i \text{ (averaging by Voight)};$$

$$\overline{\frac{1}{\lambda}} = \sum_i \frac{\theta_i}{\lambda_i} \text{ (averaging by Reiss)}. \qquad (1.67)$$

λ_i is the value of the property of the i-th phase and θ_i is the volume fraction or concentration of the phase. The limiting values of a physical property of a polyphase mixture have been named the Voight–Reiss interval [4, 6].

The geometric interpretation of the Voight and Reiss averagings corresponds to the effective value of a physical property of a polyphase system, consisting of plane-parallel layers, along the direction of the layers and perpendicular to the direction. Determining the Voight–Reiss interval is an effective method for estimating the expected value of a physical property of a system modelled by geometrically regular elements arranged with a certain symmetry in space [1].

In other cases, especially for random-structure systems (both matrix and statistical ones), use is made of the self-consistent field method, which will be discussed below.

The second approach consists in searching for such an averaging method which would at once yield the effective value of a physical property of a polyphase system regardless of its geometric model. This can be attained rigorously enough only in some cases at substantial constraints. Thus, for polyphase systems where the maximum difference in the values characterizing a generalized conductivity of individual phases corresponds in the order of magnitude to its average value, i.e.,

$$(\Delta\lambda)_{\max} \approx \bar{\lambda}, \qquad (1.69)$$

the cube root of the value of the conductivity,

$$\lambda_{\text{ef}}^{1/3} = \sum_i \lambda_i^{1/3} \theta_i, \qquad (1.70)$$

is additive with an adequate degree of accuracy [2]. The generalized conductivity implies that here all the physical properties are described by a second-rank tensor: permittivity, electrical conductivity, thermal conductivity, etc. [1, 3]. An exact solution for a two-dimensional random-inhomogeneous two-phase system was derived for the case of equal volume concentrations of phases [7]. In this case

$$\lambda_{\text{ef}}^{(2)} = \sqrt{\lambda_1 \lambda_2}, \qquad (1.71)$$

i.e., the conductivity of the mixture is equal to the geometric mean of the phase conductivities. The whole concentration range can be calculated with the aid of relation (1.71) by calculating the conductivity successively for concentration ratios 1:3, 1:7, 1:15, etc. Relation (1.71), however, becomes unsuitable if λ_1 or λ_2 goes to zero. It is also unclear how this result is to be generalized to a three-dimensional case. It can only be argued that the conductivity of a two-dimensional system will always be less than that of a similar three-dimensional system (at the same phase concentrations), and therefore for two-dimensional systems the interval of possible values can be narrowed:

$$\bar{\lambda} > \lambda_{\text{ef}} > (\lambda_1 \lambda_2)^{1/2}, \qquad (1.72)$$

since $\lambda_{\text{ef}}^{(2)}$ is greater than Reiss-averaged λ.

The application of the formulas for the generalized conductivity to calculation of the permittivity and electrical conductivity proper of two-phase system has its specific features.

In the former case it should be remembered that the minimum value of the permittivity (for void) is 1, and therefore the Reiss averaging as well as the use of formula (1.71) is always possible. For most ordinary ceramic dielectrics the ε value amounts to several units or several tens. Only for ferroelectrics it can be as high as hundreds and thousands. Because of this, in calculation of the permittivity of a ceramic composite the ratio of values of properties of individual phases cannot be very great and in any event never tends to infinity. The method of self-consistent field, described below, is quite acceptable for such systems, while calculation with formula (1.70) or (1.71) yields adequate results at a phase permittivity ratio less than 10.

In contrast, the electrical conductivity proper of phases in a ceramic composite can vary over a very broad range, from values on the order of 10^{-20} ohm^{-1} cm^{-1}, characteristic for good insulators, to those of 10^3–10^4 ohm^{-1} cm^{-1}, i.e., the ratio of characteristic values of phase conductivities can be as high as 10^{23}–10^{24}. In this case quasi-continual approaches can yield basically incorrect results; the effective conductivity of a polyphase system becomes strongly structure-sensitive. It is useful to employ here the principles of the percolation theory, which are discussed below.

(d) *Calculation of conductivity of two-phase composites with account of their microstructure*

The theoretical calculation of conductivity-type properties of two-phase systems can be effected with the use of the self-consistent method and solutions of the problem of polarization of simple-shape bodies in a uniform external field.

The microstructure of two-phase materials is here taken into account by selecting both the most suitable configuration of phase elements (sphere, ellipsoid, cylinder, infinite plate) and the pattern of their relative disposition (statistical and matrix two-phase mixtures).

The degree of 'matricity' of ceramic materials is determined by conditions of their sintering (solid- or liquid-phase one).

The derived relations make it also possible to use the measurement of conductivity properties for checking the number of phases in a material or for evaluating its structure at a given composition and a substantial difference in the phase conductivities.

The self-consistent method for calculating the conductivity of two-phase bodies relies on computing the average values of the electric field in elements of each phase on the assumption that the external polarizing field is a given quasi-homogeneous field in the two-phase body [1, 5]. The equation for the sought-for value of the macroscopic conductivity is an expression describing a macroscopically uniform flux (electric or magnetic induction, electric current, heat flux, etc.):

$$\bar{j} \equiv \lambda \bar{E} = \sum_i \lambda_i \bar{E}_i \theta_i, \qquad (1.73)$$

where i is the phase index, j the macroscopic flux, $\bar{E}$ the macroscopically homogeneous field, λ the sought-for value of the conductivity of the two-phase body (electrical conductivity, dielectric or magnetic permittivity, thermal conductivity), λ_i the conductivity of the ith phase, $\bar{E}_i$ the average value of the field in the latter, and θ_i the volume content of the ith phase.

If the phases are structurally equal, their mixture will be of a statistical character. In this case, the configuration of equiaxial isomeric polyhedrons, which in turn can be substituted with equivalent spheres, can be assigned to elements of phases with an adequate approximation within some range of volume concentrations.

Using a known expression for the field within a sphere [2] at given homogeneous external field and conductivities of substances of the sphere and surrounding medium,

$$E_i = \frac{3\lambda}{\lambda_i + 2\lambda} E \qquad (1.74)$$

(E_i is the field within the sphere, E the external polarizing field, λ_i the conductivity of the substance of the sphere, and λ the conductivity of the external medium, in our case the sought-for conductivity of the two-phase body), we obtain the equation for λ in the form

$$\sum_i \frac{3\lambda_i \theta_i}{\lambda_i + 2\lambda} = 1 \qquad (1.75)$$

or, using the identity

$$\sum_i \frac{\lambda_i + 2\lambda}{\lambda_i + 2\lambda} \theta_i \equiv 1,$$

we transform equation (1.75) to the form proposed by Odelevskij [1]:

$$\sum_i \frac{\lambda_i - \lambda}{\lambda_i + 2\lambda} \theta_i = 0. \qquad (1.76)$$

In most cases, however, phases are structurally unequal. Most often one phase forms the matrix into which the second-phase inclusions are embedded. A system of such a type is called a matrix. When calculating their conductivity, we may assign one or other geometrically regular configuration to the inclusions. If a sphere, cylinder, ellipsoid, or infinite plate is selected as the geometric image of the inclusions, the field within the latter will be homogeneous. So as not to calculate the average value of the inhomogeneous field in the matrix, the following procedure can be resorted to [2, 5].

Consider the identity

$$\overline{j - \lambda_0 E} \equiv j - \lambda_0 E, \qquad (1.77)$$

where j is the current density, λ_0 the matrix phase conductivity, E the electric field, and the bar denotes the averaging over volume V:

$$\overline{j - \lambda_0 E} = \frac{1}{V} \int_V (j - \lambda_0 E) \mathrm{d}V. \qquad (1.78)$$

The expression under the integral sign in (1.78) becomes zero in the matrix phase and hence the integral will be proportional to the volume of inclusions. Since the field within inclusions is homogeneous, then

$$\bar{j} - \lambda_0 \bar{E} = \theta_1(\lambda_1 - \lambda_0)E_1, \qquad (1.79)$$

where E_1 is the field within inclusions and θ_1 the volume fraction of the latter. Taking into account relation (1.78), we obtain the equation for the conductivity of a two-phase matrix-type mixture:

$$\lambda = \lambda_0 + \theta_1(\lambda_1 - \lambda_0)K_1, \qquad (1.80)$$

where $K_1 = E_1/E$ is the coefficient of polarization of inclusions. If inclusions are spherical, we can use the expression for the field within a sphere having a conductivity λ_1, surrounded by a spherical interlayer with a conductivity λ_0 and immersed into a medium with the sought-for conductivity λ [2, 6]:

$$K_1 = \frac{9\lambda\lambda_0}{(\lambda_0 + 2\lambda)(2\lambda_0 + \lambda_1) - 2(\lambda_0 - \lambda)(\lambda_0 - \lambda_1)(r/R)^3}, \qquad (1.81)$$

where r and R are respectively the inside and the outside radius of the spherical interlayer. Just the condition of existence of an interlayer with a conductivity λ_0 between an inclusion and a quasi-homogeneous two-phase mixture is the condition for a matrix character of the mixture. Having substituted (1.81) into (1.80) and taken into account the obvious relation $(r/R)^3 = \theta_1$, we obtain the equation for λ:

$$\lambda = \lambda_0\left[1 + \frac{9\lambda(\lambda_1 - \lambda_0)\theta_1}{(\lambda_0 + 2\lambda)(2\lambda_0 + \lambda_1) - 2(\lambda_0 - \lambda)(\lambda_0 - \lambda_1)\theta_1}\right]. \qquad (1.82)$$

This is a quadratic equation, where only its positive root

$$\lambda = \lambda_0 \frac{\lambda_1(1 + 2\theta_1) + 2\lambda_0(1 - \theta_1)}{\lambda_1(1 - \theta_1) + (2 + \theta_1)\lambda_0} \qquad (1.83)$$

has a physical meaning [6]. Just such an expression for λ was earlier derived by Odelevskij [1] by the interpolation method.

In the above-described model of a matrix-type mixture we admitted a substantial inaccuracy: the matrix-phase interlayer was assumed to have a constant thickness, which, of course, is not the case for a spherical shape of inclusions. On the other hand, there often occur such two-phase structures where inclusions of one phase surround polyhedral equiaxial grains of the second phase by a uniform layer. Structures of such a type are also matrix ones, but they are better described by a model of randomly disposed infinite plates penetrating the body of the main phase. Such a model is, of course, acceptable if the volume fraction of the 'plate-like' phase is much less than unity.

In this case, we consider the field in a plate randomly oriented in an external homogeneous field. Let φ be the angle between the direction of the polarizing field and the normal to the planes defining the plate. Components of the polarizing field, normal and parallel to the plate, are given respectively by the formulas

$$E_n = E \cos \varphi; \qquad E_p = E \sin \varphi. \tag{1.84}$$

The component E_p in the plate remains unchanged regardless of the ratio of conductivities of the plate and main phase. The normal component, from the condition of conservation of the normal derivative of the current density at the interface [2], will vary inversely with the ratio of conductivities of the plate and external medium, i.e.,

$$E_n^{(0)} = E_n \frac{\lambda}{\lambda_0}. \tag{1.85}$$

Averaging the components of the field in the plate over all possible values of the angle φ from 0 to $\pi/2$ yields

$$E_0 = \frac{1}{2}\left(1 + \frac{\lambda}{\lambda_0}\right) E. \tag{1.86}$$

Here λ_0 denotes the conductivity of the substance of the plates, since it forms a physically continuous phase-matrix. Using relation (1.80), we obtain the first-degree equation for λ:

$$\lambda = \lambda_1 + \frac{\theta_0(\lambda_0 - \lambda_1)(\lambda_0 + \lambda)}{2\lambda_0}, \tag{1.87}$$

from which it follows that

$$\lambda = \lambda_0 \frac{2\lambda_1 + \theta_1(\lambda_0 - \lambda_1)}{2\lambda_0 - \theta_0(\lambda_0 - \lambda_1)}, \tag{1.88}$$

where λ_1 is the conductivity of the main phase, λ_0 the conductivity and θ_0 the volume fraction of the matrix phase. The greatest difference between conductivities of statistical and matrix-type two-phase mixtures with identical volume fractions of phases takes place when $\lambda_0 \ll \lambda_1$. At an inverse relation, $\lambda_0 \ll \lambda_1$, substantial differences between the conductivities occur only at $\theta_0 \ll 1$.

The structures of real sintered two-phase materials can, to various degrees depart from a statistical character, acquiring one or other degree of 'matricity'. Tendencies toward a matrix character of a mixture show up at a solid-phase sintering and in cases where a two-phase powder mixture is composed of particles greatly differing in size [6]. The trend toward 'matricity', however, manifests itself most pronouncedly in materials produced by sintering in the presence of a liquid phase. In the case

Functional materials

where the matrix phase has a much lower conductivity or permittivity, properties of a sintered material depend substantially on the production process. On the other hand, the measurement of a physical property can be a sort of check of the structure, allowing a quantitative evaluation of the degree of 'matricity' of a two-phase mixture.

The fraction of 'matricity' in an alloy can be estimated on the assumption that regions of a statistical and of a matrix structure are randomly mixed throughout the volume of the alloy. As shown above, the law of additivity of the cube root of the phase conductivity is valid when phase conductivities do not differ too much (see formula 1.70).

Applying this relation to the case of conductivity of a mixed-type structure, we obtain the following expression for the 'matricity' fraction ω:

$$\omega = \frac{\lambda_{st}^{1/3} - \lambda^{-1/3}}{\lambda_{st}^{1/3} - \lambda_{m}^{1/3}}, \qquad (1.89)$$

where λ_{st} and λ_m are respectively the conductivities of statistical and matrix mixtures with a given volume content of phases and λ is the conductivity of an intermediate-character mixture. The 'matricity' fraction

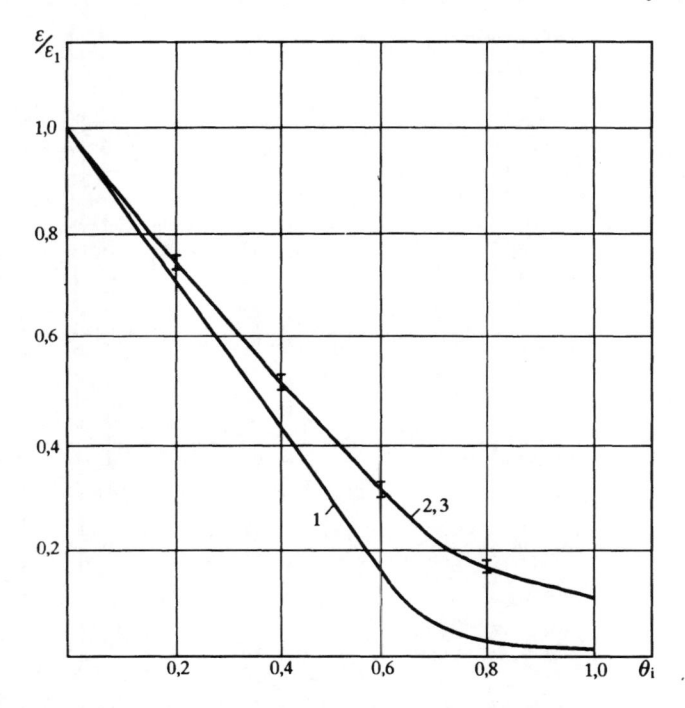

Figure 1.69 Concentration dependence of permittivity of two-phase dielectric (in arbitrary units): (1) $\varepsilon_1/\varepsilon_2 = 10^3$; (2) $\varepsilon_1/\varepsilon_2 = 10$ (by formula 1.76); (3) – $\varepsilon_1/\varepsilon_2 = 10$ (by formula 1.70).

calculated from formula (1.89) tends to decline in a liquid-phase sintering with increasing content of the liquid phase.

As indicated in section 1.2.1.1(c) formula (1.76) is acceptable for calculating the permittivity of a ceramic composite with structurally equal phases without practically any restrictions. Examples of such calculations are presented in Figure 1.69.

It is of interest to determine the concentration dependence of the conductivity of a two-phase ceramic (cermet), containing a conducting and a nonconducting phase, at various microstructure types. In this case, the conductivity of one of the phases can be equated to zero. For a statistical mixture,

$$\lambda_{ef} = \lambda_{cond}\left(1 - \frac{3}{2}\theta_{ins}\right), \qquad (1.90)$$

where λ_{cond} is the conductivity of the conducting phase and θ_{ins} the volume fraction of the nonconducting phase. For a matrix-type mixture, according to formula (1.83),

$$\lambda_{ef} = \lambda_{cond}(1 - \theta_{ins}) / \left(1 + \frac{1}{2}\theta_{ins}\right). \qquad (1.91)$$

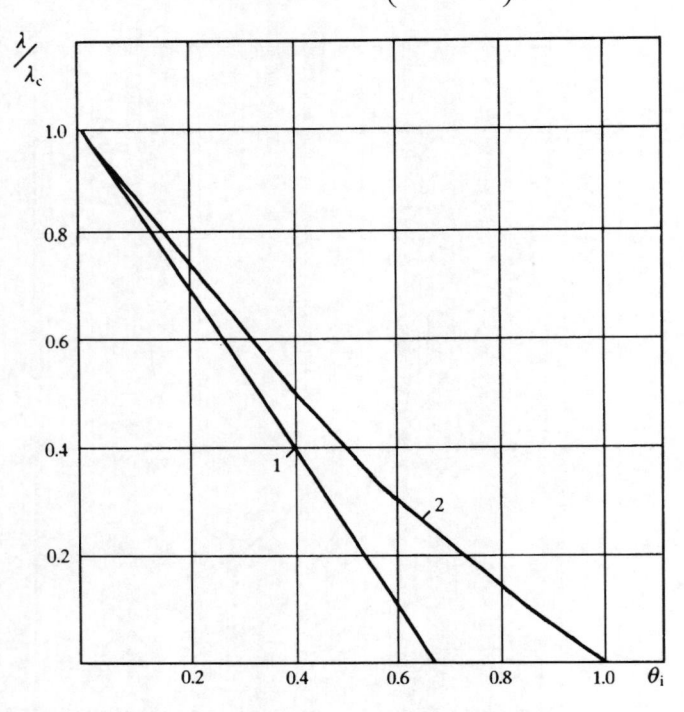

Figure 1.70 Conduction of conductor–insulator composite as a function of volume concentration of nonconducting phase (1) for statistical mixture; and (2) matrix-type mixture.

Functional materials

The two dependences are compared in Figure 1.70. The tendency of the conductivity to zero by formula (1.90) at $\theta_{ins} \to 2/3$ is in actual fact a consequence of the approximation of the self-consistent method. The problem of the conductivity of a system of such a type is more rigorously solved by percolation theory methods, which will be described in the next paragraph.

Using formula (1.88), one can evaluate the dependence of the permittivity of a ferroelectric ceramic containing some amount of a glass phase on the volume concentration of the latter (on the assumption that the glass phase envelopes ferroelectric phase crystals by a uniform layer). Substituting the conductivity λ in formula (1.88) with the permittivity ε and setting $\varepsilon_1 \gg \varepsilon_0$ and $\theta_0 \ll 1$ yields

$$\varepsilon_{ef} = \varepsilon_1 \frac{2}{2 + \frac{\varepsilon_1}{\varepsilon_0}\theta_0}. \tag{1.92}$$

Figure 1.71 shows the dependence over a range of $0 < \theta_0 \leq 0.05$ at $\varepsilon_1/\varepsilon_0 \approx 10^3$, which indicates that for matrix-type structures even fractions of a volume per cent glass phase can reduce several-fold the permittivity of ferroelectric ceramics.

(e) *Interpolation methods for calculation of conductivity of two-phase systems*

A rigorous calculation of the conductivity or permittivity of two-phase systems is far from always possible. At great differences of conductivities of phases and their irregular, random geometry, one has to be content with determining the possible interval of effective conductivity values or to resort to various interpolation methods [1, 8].

The so-called Voight–Reiss interval has been presented above. In a generalized form it can be written as

$$\lambda_{ef}^n = \theta_1 \lambda_1^n + (1 - \theta_1)\lambda_2^n. \tag{1.94}$$

At $n = 1$ there is the Voight averaging; at $n = -1$, the Reiss one. If $-1 < n < 1$, then formula (1.94) describes various intermediate cases (as indicated above, at $n = 1/3$ the relation (1.94) describes the conductivity of a mixture with close values of phase conductivities). At $n = 0$, relation (1.94) converts into a logarithmic relation [2, 9]

$$\ln \lambda_{ef} = \theta_1 \ln \lambda_1 + (1 - \theta_1)\ln \lambda_2, \tag{1.95}$$

which can be rewritten as

$$\lambda_{ef} = \lambda_1^{\theta_1} \lambda_2^{\theta_2}. \tag{1.96}$$

At $\theta_1 = \theta_2 = 0.5$ this formula coincides with relation (1.71)[8].

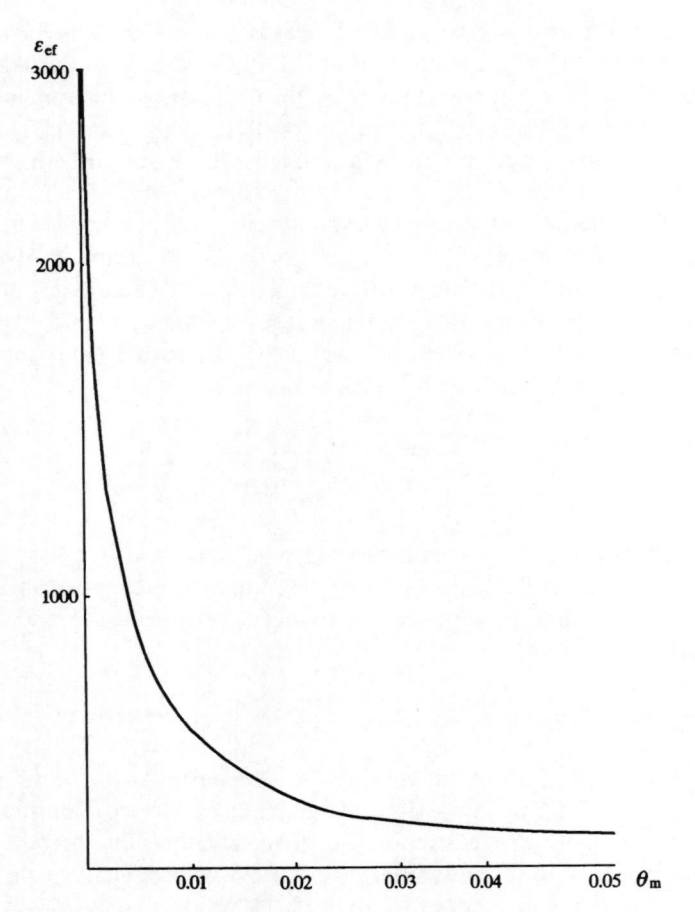

Figure 1.71 Decrease of permittivity of two-phase matrix-type mixture with increasing volume concentration of matrix (glass phase) at $\varepsilon_1/\varepsilon_0 = 10^3$.

The geometric approach to evaluation of conductivity properties of two-phase mixtures, summarized in the review [8], considers as the basic criterion of a mixture the relative arrangement of bonds between particles of each phase, the so-called connectivity. Particles of each of the phases can be isolated (such a state is designated by the index 0), can form linear chains (index 1), flat layers (index 2), or three-dimensional structures (index 3). In all there are ten various connectivity types: 0–0, 0–1, 0–2, 0–3, 1–1, 1–2, 1–3, 2–2, 2–3 and 3–3. Permutations of indices are symmetrical with respect to interchange of particles of the phases. This indexing system makes it possible to designate regular two-phase systems whose classification was presented in section 1.2.1.1(b). Thus, a 0–3 system is a matrix structure with first-phase inclusions uniformly distributed throughout the matrix; a 2–2 system is one with alternating layers of the phases; and a 3–3 system is a system of interpenetrating skeletons. Most of the systems in a pure state exhibit a considerable anisotropy, but real two-phase materials in macroscopic volumes are most often isotropic, although they may be locally anisotropic in microvolumes. Limiting values of the conductivity of random two-phase mixtures of such a type, calculated by the variational method in [8], are presented by the relations (at $\lambda_1 > \lambda_2$)

$$\lambda_{max} = \lambda_1 + \frac{\theta_2}{\left(\dfrac{1}{\lambda_2 - \lambda_1} + \dfrac{\theta_1}{3\lambda_1}\right)}; \qquad (1.97)$$

$$\lambda_{min} = \lambda_2 + \frac{\theta_1}{\left(\dfrac{1}{\lambda_1 - \lambda_2} + \dfrac{\theta_2}{3\lambda_2}\right)}. \qquad (1.98)$$

Comparing relations (1.97) and (1.98) with formula (1.83) shows their complete identity. This means that limiting values of the conductivity of a random two-phase system are conductivities of matrix-type systems with the same volume concentrations of phases, in one of which the phase with a higher conductivity forms the matrix, and in the other, inclusions (the matrix being formed by the lower-conductivity phase). This interval is considerably narrower than the Voight–Reiss interval (1.67).

An important place among ceramic composites is held by sintered polycrystalline dielectrics with inclusions of a conducting phase. The conductivity of such systems is strongly dependent on the volume concentration of the conductor; when the volume fraction of the conducting phase reaches some critical value, called the passage threshold, there occurs a dielectric-conductor microstructural phase transition. The theory of conductivity of such systems has been named the percolation theory [9, 10, 11].

The essence of the theory consists in calculating the probable sizes of clusters formed by conducting phase particles at their random disposition in space.

The passage threshold corresponds to an infinitely large cluster, when the current from one electrode to the other can pass along continuous pathways, chains of conductor particles. For a homogeneous mixture of conducting and nonconducting spherical particles of identical diameters, packed regularly or randomly, the passage threshold is $\theta_c = 1/6$. Determining the passage threshold for a statistical mixture analytically is impossible. Calculating experiments on computers yield for three-dimensional random two-phase systems a value of $\theta_c = 0.15$–0.19 [10, 11]. When conducting particles are smaller than dielectric ones, a statistical mixture will exhibit a tendency towards matricity for the conductor, and the passage threshold will decline. Thus, at $R_i/R_c = 30$ (R_i is the radius of nonconducting and R_c of conducting spheres), $\theta_c = 0.03$ [10, 11]. At an inverse ratio the passage threshold can rise, tending for an ideal matrix-type mixture to 1.

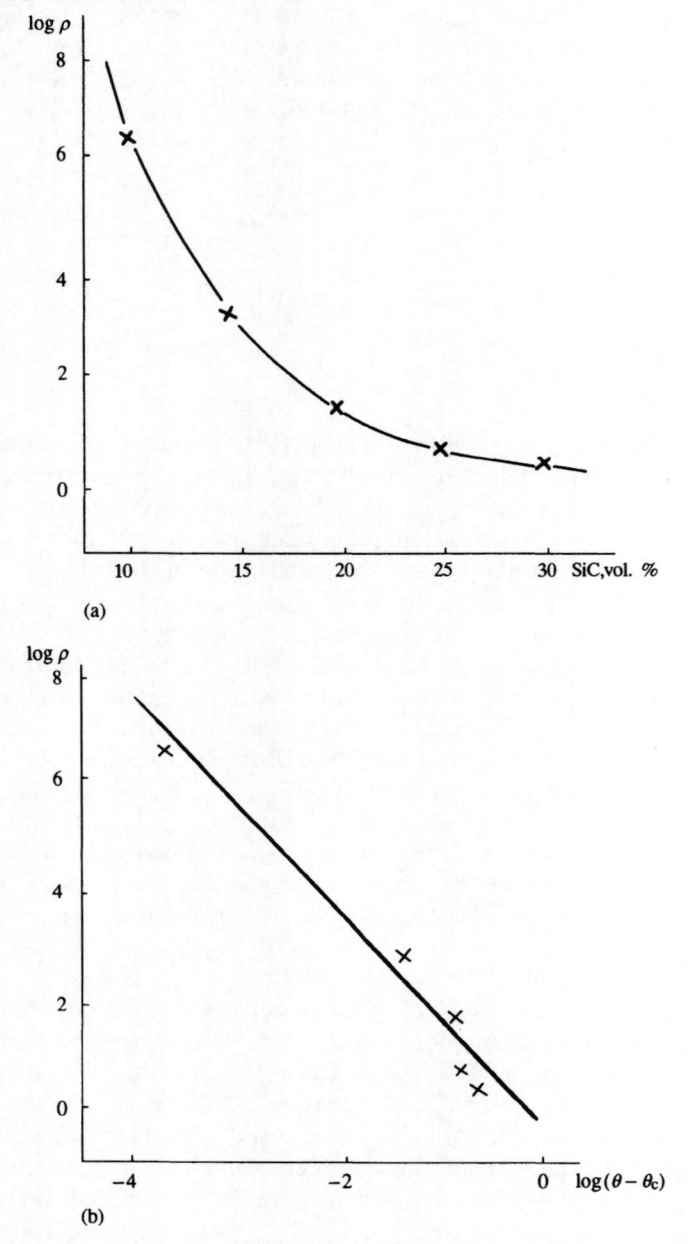

Figure 1.72 (a) Logarithm of the variation of resistivity of Si_3N_4–SiC composite with volume content of conductor (SiC); and (b) rectification of experimental data by formula (2.36) in coordinates $\log\rho$–$\log(\theta - \theta_c)$.

Above the passage threshold the dependence of the conductivity on the volume fraction of the conducting phase, θ, is described by the interpolation formula

$$\lambda_{ef} = \lambda_0(\theta - \theta_c)^t, \tag{1.99}$$

where t is the so-called critical index, which for a three-dimensional space is equal to 1.6–1.95 [10].

Figure 1.72 shows the dependence of the electrical resistance of ceramic composites of the Si_3N_4–SiC system on the volume concentration of the conductor (SiC) as well as the same dependence processed in accordance with the formula

$$\log \rho_{ef} = \log \rho_0 - t \log (\theta - \theta_c), \tag{1.100}$$

which follows from (1.99). In this case the passage threshold is $\theta_c = 0.09$–0.1 and the critical index is $t \approx 2$ [12].

That passage threshold values are somewhat lower than the theoretically predicted value of 0.15–0.19 are evidence that in these ceramics the size of conducting SiC particles is somewhat smaller than that of dielectric Si_3N_4 grains. The t value in this case turns out to exceed somewhat the theoretical one.

To conclude, consider the impact of anisotropy of reinforcing elements on the conductivity and permittivity of composites. Most typical cases are reinforcements with elongated (needle-like) or oblate (plate-like) inclusions.

A rigorous solution of such a problem is possible only in the case of a low volume concentration of inclusions having the configuration of regular (oblong or oblate) ellipsoids of revolution [2, 13]. In such an ellipsoid, placed into a homogeneous dielectric medium, the polarization field in the presence of an external polarizing field E will be:

along the rotational x-axis:

$$E_x = \frac{\varepsilon}{\varepsilon - n_x(\varepsilon - \varepsilon_i)} E; \tag{1.101}$$

and across the rotational axis (along the y-axis):

$$E_y = \frac{\varepsilon}{\varepsilon - n_y(\varepsilon - \varepsilon_i)} E, \tag{1.102}$$

where n_x and n_y are so-called depolarization coefficients, depending on the eccentricity of the ellipsoid; $n_x + 2n_y = 1$ [2]. For spheres, $n_x = n_y = 1/3$. For very elongated particles (fibres), $n_x \to 0$, $n_y \to 1/2$. Conversely, for infinitely thin discs, $n_x \to 1$, $n_y \to 0$. At a random orientation in space the average field in nonisometric particles is

$$\bar{E} = \tfrac{1}{2}(E_x + E_y). \tag{1.103}$$

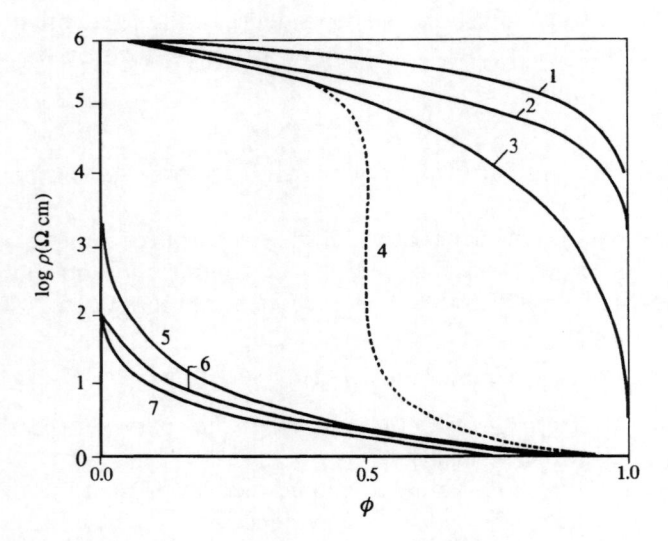

Figure 1.73 Calculated composition dependences of resistivity for two-phase mixture consisting of high-resistance ($\rho_h = 10^6$ ohm cm) and low-resistance ($\rho_l = 1$ ohm cm) phases with particles of different shapes: (1) series-connected phases; (2) upper limit of calculation of anisotropic mixtures by variational method; (3, 4, 5) calculations by methods of generalized theory of effective media for various phase parameters; (6) lower limit of calculation of anisotropic mixtures by variational method; and (7) parallel-connected phases.

Similar formulas for conducting systems can be obtained by substituting ε_i by λ_i [13].

At arbitrary concentrations of phases whose particles are of arbitrary configurations, one has to content himself with interpolation relations of the following type:

$$\frac{\lambda_1^t - \lambda_{\text{ef}}^t}{\lambda_1^t + A\lambda_{\text{ef}}^t}\theta_1 + \frac{\lambda_2^t - \lambda_{\text{ef}}^t}{\lambda_2^t + A\lambda_{\text{ef}}^t}(1 - \theta_1) = 0, \qquad (1.104)$$

where A and t are experimentally determined parameters characterizing the anisotropy of particles of the two phases [8].

The anisotropy and orientation of particles also strongly affect the critical conductivity indices in the percolation theory.

Figure 1.73 shows various dependences of the electrical conductivity of ceramic composites on the phase concentrations (the electrical conductivities of the phases differing by six orders of magnitude), calculated from formula (1.104) for various parameters t and A as well as from formulas (1.97) and (1.98).

It is obvious that varying the phase concentrations and the structures of constituent phases is a powerful means for controlling the physical

Functional materials

properties of functional ceramic composites, such as insulators, capacitor materials, dielectric substrates, resistors, etc.

1.2.1.2 Electrical conductivity of ceramics

(a) *Electrical conductivity values and types*
Substances widely varying in the nature of the chemical bond and structure of electronic shells are used as constituents of electronic ceramic composites (ECC), so that the range of their electrophysical properties is also very broad. Thus, substances such as boron nitride, alumina, magnesia and silica are the best known high-temperature inorganic dielectrics: their electrical conductivity at room temperature being as little as 10^{-17}–10^{-21} ohm^{-1} cm^{-1}, a low enough electrical conductivity being retained up to high temperatures, of the order of 1500 °C and sometimes up to 2000 °C. At the same time, such materials as titanium nitride, zirconium carbide, some other nitrides, carbides as well as oxides, borides, silicides, and other compounds in their electrophysical properties belong to semimetals; their electrical conductivity is about 10^4 ohm^{-1} cm^{-1} [14].

A relatively small number of intermetallic compounds, sulphides and other compounds exhibit superconductivity at temperatures up to 20 K, while some oxides exhibit a high-temperature superconductivity, retained up to temperatures on the order of 100 K.

The electrical conductivity interval between dielectrics and semimetals, amounting to about 20 orders of magnitude, is filled with substances classed with semiconductors and including most inorganic compounds. The basic feature of semiconductors is their ability to change considerably their electrical conductivity under the effect of most diverse external actions (temperature, pressure, radiations, etc.) as well as of the composition and structural factors (impurities, departure from stoichiometry, crystal imperfections, etc.). It is these changes that underlie the functions performed by various functional ECC.

The above-mentioned is commonly related to electron–hole processes in a crystal, studied by solid-state electronics. The solid-state ionics, studying ion transport phenomena, stemming from a higher mobility of one of crystal lattice components as well as of some impurity ions, protons, or hydronium ions, has also been actively advanced recently. Ionic processes underlie operation of many ECC employed in sensors, capacitors, storage cells, energy converters, etc.

The range of ionic conductivities of ceramics is also very broad: from values on the order 10^{-20} ohm^{-1} cm^{-1} (only such a value of the ionic and electronic conductivity can illustrate the existense of electrets) to 10^3 ohm^{-1} cm^{-1}, which is close to those of salt solutions and melts.

It must be pointed out that electronic and ionic processes in solids are usually considered separately, while such an approach is a very simplified and rough approximation of reality. In actual fact, the two phenomena are inseparably related to each other, and just their interrelation is the basis for the insight into such phenomena as electric ageing, catalysis, some types of dielectric polarization, etc.

(b) *Physical nature of electron conduction*

Current carrier concentration The electrical conductivity of a substance (σ) is determined by the concentration of mobile current carriers (n), their mobility (u) and charge (q):

$$\sigma = qnu. \qquad (1.105)$$

The carrier concentration for the electron conduction of solids varies over the greatest range, i.e., it is essentially the main factor determining the electrical conductivity.

This stems from the fact that current carriers in an ideal crystal of a dielectric or semiconductor should be absent, as all the electrons there are incorporated in chemical bonds and therefore do not respond to a weak, as compared with the chemical bond energy, action of an external electric field and cannot participate in the electric conduction. The discrete energy spectrum of electrons in an atom is in a crystal transformed into quasi-continuous energy bands, of greatest interest among which from the standpoint of the electrical conduction are the highest-energy band, formed by bonding orbitals (valence band), and the lowest-energy band of antibonding orbitals (conduction band).

For defect-free crystals of dielectrics and semiconductors, the valence band is completely filled with electrons, while the conduction band is completely empty; hence, electrons of such a crystal cannot increase their energy at the expense of an external electric field, and a directional motion of electrons, an electric current, is impossible. Only those electrons whose energies are close to unfilled energy levels, i.e., electrons in the conduction band or in a partly filled valence band, can participate in charge transport. Such a state is attained as a result of transfer of electrons from the valence to the conduction band due to fluctuation of the thermal motion or other pulsed actions whose energy exceeds the forbidden gap width (light quanta, fast particles, etc.). After such transfers, the concentrations of mobile electrons (n_i) and free sites in the valence bands, holes, (p_i) are always equal, and therefore such a semiconductor is called an intrinsic semiconductor, and its conduction, an intrinsic conduction.

The concentration of thermal equilibrium carriers in an intrinsic semiconductor (n_i) is an exponential function of the forbidden gap width (ΔE):

$$n_i = p_i = 2\left(\frac{2\pi kT}{h^2}\right)^{3/2}\left(\frac{m_e^*}{m_h^*}\right)^{3/4}\exp\left(-\frac{\Delta E}{2kT}\right), \qquad (1.106)$$

where k and h are respectively the Boltzmann and Planck constants, m_e^* and m_h^* are effective masses of an electron and a hole [15].

The forbidden gap width, the most important quantity, characterizing the intrinsic electronic conduction of solids, for the best electrical insulators (BN, Al_2O_3, MgO) amounts to 6–10 eV; for typical semiconductors, to less than 3 eV; and in metals, is absent since not all bonding orbitals are filled with electrons. The effect of the forbidden gap width on the carrier concentration can be illustrated as follows. Assume that m_e^* and m_h^* are equal to the electron mass; then the calculated concentrations of conduction electrons at 300 K for a semiconductor with $\Delta E = 1$ eV will amount to about 10^{10} cm^{-3}, and for $\Delta E = 2$ eV, to about 10^2 cm^{-3}. At 1000 K for semiconductors with $\Delta E = 1, 2$ and 5 eV the calculated carrier concentrations will be respectively of about 10^{17}, 10^{15} and 10^7 cm^{-3}.

Impurity current carriers The conduction resulting from the transfer of electrons across the forbidden gap is called the intrinsic one. Electron and hole concentrations in an intrinsic semiconductor are always equal. When a crystal contains impurities or defects, the picture gets complicated, as energy levels of electrons in an impurity atom differ from those in the main atoms of the crystal. This results in distortion of the energy pattern of the crystal: appearance of free and filled energy levels which are disposed within the forbidden gap and can either accept electrons transferred from the valence band (acceptor levels) or give off electrons to the conduction band (donor levels). Such a conduction is called the impurity conduction. Electron and hole concentrations in an impurity semiconductor are dissimilar: the electron conduction prevails in a crystal with donor impurities, and the hole conduction, with acceptor impurities.

The temperature dependence of the carrier concentration in an impurity semiconductor is expressed by the formula

$$n = N_1 \exp\left(\frac{-\Delta E}{2kT}\right) + N_2 \exp\left(-\frac{\Delta E_{im}}{kT}\right), \qquad (1.107)$$

where N_1 and N_2 are constants representing the densities of electronic states associated with main atoms of the crystal and the impurity and ΔE_{im} is the impurity ionization energy.

Since $\Delta E_{im} \ll \Delta E$, a temperature range can exist where all impurities have already been ionized, but the contribution of the transfers across the forbidden gap to conduction is insignificant. In this temperature

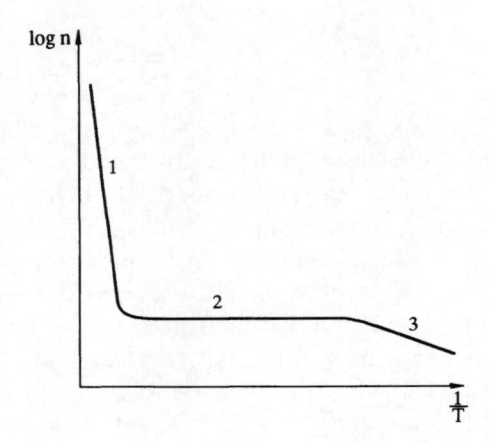

Figure 1.74 Temperature dependence of current carrier concentration in doped semiconductor: (1) transitions across forbidden gap; (2) impurity depletion region; (3) transitions from impurity level.

range the carrier concentration does not depend on the temperature (impurity depletion region) (Figure 1.74). The energy of ionization of impurities in crystals varies over a broad enough range. In typical covalent semiconductors, such as germanium and silicon, the energy of ionization of most impurities is not over 0.1 eV, and therefore such impurities are fully ionized at room temperature. At the same time in crystals of typical electrical insulators, such as Al_2O_3, MgO, BN, SiO_2, Si_3N_4, the energy of ionization of any impurities exceeds several eV, i.e., their ionization is unlikely even at very high temperatures. Just this property of electrical insulators accounts for the existence in the nature and in the technology of numerous electrically insulating materials; such materials are often manufactured from commercial grade substances containing various impurities.

It should also be noted that, due to recombination, the following relation is valid for any impurity semiconductor:

$$np = n_i^2. \tag{1.108}$$

Mobility The mobility of current carriers (velocity of an electric charge in a unit electric field) is substantially different from that of, e.g., free electrons in vacuum. Forces of the chemical bond of electrons to atoms in a crystal greatly exceed forces exerted by an external electric field, but, due to wave properties of electrons, the energy barrier between adjacent atoms is not an obstacle for electrons. The periodicity of a crystal results in equivalence of electron energies at different lattice sites, and therefore at the existence of free energy levels close in energy, electrons can raise their energy: get accelerated by the electric field, although their effective mass differs from the mass of free electrons, is different for electrons and holes, depends on the structure of energy bands, and can be both smaller and greater than the free electron mass m_0 and even negative. A negative effective mass means that electrons are slowed down rather than accelerated by the external electric field. Such a behaviour for a free particle is hard to imagine, but just because electrons near the upper boundary ('top') of the valence band have a negative effective mass, holes, i.e., electron vacancies, behave as positively charged particles with a positive effective mass. In contrast to a scalar mass of a free electron, the effective mass is a third-rank tensor.

Because of the complex structure of energy bands, overlap of various bands, carriers with different effective masses and different mobilities can coexist in a crystal, such as 'light' and 'heavy' holes in germanium and silicon, whose effective masses, e.g., in germanium are of $0.044\,m_0$ and $0.28\,m_0$, and mobilities, of 15 000 cm^2/V s and 1900 cm^2/V s respectively (at 300 K). The temperature dependence of mobility is determined by the character of irregularities where electron waves get scattered, i.e., by carrier scattering mechanisms. The mobility generally declines with increasing temperature. If, for example, the scattering on thermal vibrations of the crystal lattice predominates, then the mobility variation with temperature is described by the formula

$$u = u_0 T^{-3/2}. \tag{1.109}$$

Electron and hole mobilities (as also mobilities of ions and of any charges being in the thermal equilibrium with the crystal) are related to the diffusion coefficients by the Nernst–Einstein relation

$$\frac{D}{u} = \frac{kT}{q}. \tag{1.110}$$

Many substances used as constituents of ceramics, in particular oxides, feature a considerable fraction of an ionic component of the chemical bond. In such substances, a free electron or hole brings about polarization of the surrounding crystal lattice and its motion occurs simultaneously with movement of the state of polarization. Such a state of a carrier is called 'polaron'. The effective mass of a polaron can be tens or hundreds of times as great as the free electron mass, and therefore carrier mobilities in ionic crystals are generally much lower than in covalent ones.

In oxides of transition d and f elements the s and p electrons of oxygen form hybrid bonding and antibonding sp^3 orbitals, which in the band pattern of an oxide correspond to broad energy bands σ and σ^*, one of which is fully filled with electrons while the other is empty. The energy states corresponding to d and f electrons are split by the crystalline field into several groups of levels, but, due to a greater localization of d and f orbitals and their lower interaction with one another, the splitting of the energy levels, i.e., the width of the energy bands formed by them, is much smaller than that for s and p electrons. These bands, filled or free, are usually disposed in the energy gap between the σ and σ^* bands. Some of the d and f states are practically not split by the crystalline field, i.e., form a system of local energy levels.

Dimensions of polarons formed by carriers that are in such narrow bands or at local levels are comparable with those of an elementary cell of a crystal, due to which these polarons are called small-radius polarons. Depending on the temperature and the width of the d band (or f band), two different mechanisms of motion of small-radius polarons in such crystals are possible.

At low temperatures the wavelength of electrons is great, they are not localized and move in narrow polaronic zones, their mobility decreasing with increasing temperature. When the temperature becomes high enough, the uncertainity of the energy of an electron exceeds the width of the permitted band where it is, and the concept 'polaronic zone' loses its sense. Over-barrier jumps of a polaron from a site to a site of the crystal lattice, occurring through the thermal motion energy, start playing the principal role. The carrier mobility in this case exponentially increases with temperature, although its magnitude remains low enough: 10^{-4}–10^{-5} cm^2/V s. Such a character of charge transport can be interpreted as exchange of charges between ions of one and the same element in different valence states. Thus, a Fe^{3+} ion in FeO is equivalent to a hole, and a Fe^{2+} ion in Fe_2O_3, to an electron [16].

(c) *Ionic conduction*

Ionic conduction types Apart from electrons and holes, various ions can participate in charge transport in ionic crystals: both intrinsic anions and cations of a crystal and ions of various impurities, protons, hydronium ions, etc. Motion of ions in an ideal close-packed crystal at a low temperature is impossible, but at the sites of crystal defects – intensive thermal vibrations, vacancies, interstitial atoms – the ionic conduction can be effected by ion-vacancy exchanges or through interstices.

The total value of the ionic conduction σ_i is the sum of contributions by various ions

$$\sigma_i = \sum_k q_k n_k u_k, \qquad (1.111)$$

where q_k, n_k and u_k are charges, concentrations and mobilities of various ions.

Contributions by ions of various types to the ionic conduction of a perfect crystal are primarily determined by geometric factors: structure of the crystal, cation-to-anion size ratio, size and arrangement of interstitial vacancies.

The electric transport of impurity ions through interstices is a widespread ionic conduction type characteristic not only for ionic, but also for covalent crystals with a loosely packed crystal structure, such as quartz and silicon nitride crystals. A cationic conduction is exhibited by most silicate and borosilicate glasses as well as by some oxides, sulphides and halides. Mobile ions in all these substances are ions of alkali metals, although conduction by alkaline earth metal ions is possible as well.

Anionic conductors are also known, where current is carried by fluorine, chlorine, oxygen and sulphur ions. Conduction by fluorine and other halogens is exhibited, e.g., by fluorides of alkaline earth elements with a large radius of cations – calcium, barium, strontium, as well as of some transition 3d elements and lanthanides. Typical examples of oxygen-ionic conductors are hafnia-, zirconia-, thoria- and ceria-based solid solutions with anionic vacancies.

The participation of protons and hydroxonium ions in ion transport depends on the combination of the crystal structure and physicochemical properties of atoms and cannot always be predicted. Such a type of conduction occurs in some phosphates as well as molybdates, tungstates and niobates (usually at 400–600 K). A higher temperature (at 900–1000 K) protonic conduction has recently been found in compounds of a $BaCeO_3$-, $SrCeO_3$-type and some other perovskite-structure compounds as well as in yttria [17] and other rare earth element oxides (REEO). Doping of REEO with calcia increases the protonic conduction by about an order of magnitude [18].

Motion of main crystal components in an electric field is possible only for crystals with a chemical bond of ionic character and is also determined by relative sizes of anions and cations, their charges, and crystal structure. In close-packed NaCl-type crystal structures (such as in magnesia crystals) the intrinsic ionic conductivity becomes appreciable at temperatures of the order of 0.7–0.8 of the melting temperature, when a great enough number of thermally equilibrium structural defects of a crystal has been formed.

In crystals with a lower density of packing of structural elements (such as those with a CaF_2-type structure, into which substances such as ZrO_2, HfO_2, CeO_2 crystallize) the energy of formation of structural defects is lower, and therefore the ionic conduction is already observed at temperatures of the order of $(0.4–0.5)T_{melt}$. It must be pointed out that mobilities of anions and cations as a rule significantly differ from each other. When anionic and cationic charges are equal, smaller-size ions are always most mobile; when they are different, the decisive factor for a higher mobility is a smaller charge.

As in the case of electron conduction, the ionic conduction can be changed by several orders of magnitude by doping a crystal with certain types of impurities. The ionic conduction is enhanced by addition of impurities that generate vacancies in the most mobile sublattice of a crystal. Thus, the ionic conduction of magnesia, where cations are most mobile, can be increased by doping it with scandia, which results in formation of cationic vacancies:

$$MgO + Sc_2O_3 \rightleftharpoons 2[Sc_{Mg}]^\bullet O + O_0 + V''_{Mg}. \quad (1.112)$$

In yttria the mobility of anions is higher, and therefore the ionic mobility is increased by its doping with calcia, which promotes formation of anionic vacancies:

$$Y_2O_3 + CaO \rightleftharpoons [Ca_Y]'_2O_3 + V_0^{\bullet\bullet}. \quad (1.113)$$

Doping can also reduce the ionic conduction. Thus, if the concentration of anionic vacancies in magnesia is increased, e.g., by dissolving lithia in it,

$$MgO + Li_2O \rightleftharpoons 2[Li_{Mg}]'O + V_0^{\bullet\bullet}, \quad (1.114)$$

then, due to recombination, the concentration of cationic vacancies and the ionic conduction of magnesia decrease:

$$[V''_{Mg}] = \frac{K_s}{V_0^{\bullet\bullet}}, \quad (1.115)$$

where K_s is the equilibrium constant in the reaction of generation of Shottky defects in undoped magnesia

$$Zero \rightleftharpoons V''_{Mg} + V_0^{\bullet\bullet} \quad (1.116)$$

The mobility and coefficient of diffusion of ions of any type, as those of electronic carriers, are always interrelated by the Nernst–Einstein formula, similar to (1.110). For example,

$$\frac{D_{Mg}}{u_{Mg}} = \frac{kT}{q_{Mg}}. \quad (1.117)$$

Many processes in ceramics, such as sintering, recrystallization, diffusion creep, involve the necessity of mass transport, i.e., a simultaneous movement of anions and cations. In the case of the so-called ambipolar diffusion (called also mass-transport diffusion) the total diffusion coefficient is determined by the limiting process, diffusion of the slowest component of a given crystal. Therefore, in order to accelerate the diffusion mass transport and to improve the sinterability of a ceramic, it is expedient to reduce the coefficient of diffusion of the most mobile particles (and the total ionic conduction of the crystal), thereby increasing the diffusion coefficient and electrical conductivity in the crystal sublattice of the inverse sign. Indeed, production of a nonporous ceramic or even of a ceramic sintered to an optically transparent state from yttria is attained by adding to it a few molar per cent of zirconia or hafnia; similar ceramics from magnesia are produced by adding lithia to them.

Ionic superconductors The above-discussed examples of controlling the ionic conduction rely on varying the concentrations of vacancies of various types, randomly disposed in a crystal. A special case is presented by crystals where the crystal lattice structure itself predetermines a complete or a partial disorder in the sub-lattice of atoms of one kind at an ordered structure of remaining atoms. In this case, the ionic conductivity of the crystal becomes comparable with the electrical conductivity of liquid electrolytes, solutions or melts of salts (10^{-1}–10^{-3} ohm^{-1} cm^{-1}). Such substances are called solid electrolytes or ionic superconductors (a less correct name, superionic conductors, is used as well). The essence of the ionic superconduction phenomenon consists in the existence in the crystal of one-dimensional channels or planes, within which there are no considerable potential barriers to motion of ions of any one kind.

A typical example of such a structure is the AgJ crystal lattice, where in an elementary cell for two silver ions there are 42 permitted positions for their location. Similar structure types include $RbAg_4J_5$, Ag_2S, Ag_3SJ, CuB_2, and $CuCl$.

A group of cationic superconductors of the β-Al_2O_3 (β-alumina) type has also been studied to a sufficient detail. This term denotes sodium polyaluminate ($Na_2O.11Al_2O_3$), which is usually synthesized with a significant departure from stoichiometry, so that its composition is more correctly expressed by the formula $Na_2O.nAl_2O_3$, where $n = 5.3$–8.5. The crystalline structure of β-alumina can be visualized as layers of Al_2O_3 molecules, interconnected by bridging oxygen ions. Sodium ions are interposed between the bridging ions; the spacing between layers (about 0.2 nm) suffices for sodium ions to move freely in this plane as well as for the arrangement there of additional sodium ions, which is indicated by a departure of the composition from the stoichiometric one (Figure 1.75). The number of bridging oxygen ions can be reduced and thereby the conduction of β-alumina can be enhanced by adding magnesium ions to it, which substitute aluminium ions and at the same time create vacancies in the oxygen sub-lattice.

Sodium ions in β-alumina can be substituted with other alkaline ions – lithium, potassium, rubidium – as well as with silver ions by means of the ionic exchange or by an electrochemical method. Attempts to produce such compounds by the chemical synthesis have failed so far.

The achievements in studying β-alumina spurred the search for crystalline structures with three-dimensionally arranged channels for the motion

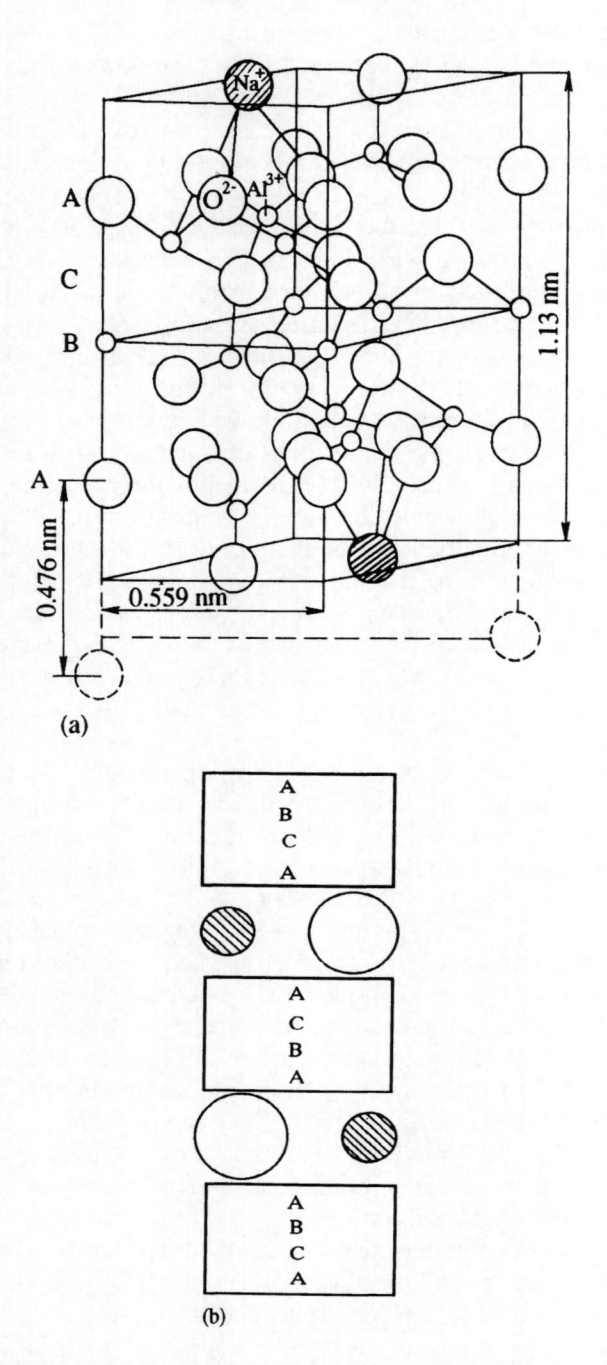

Figure 1.75 Arrangement of ions in (a) elementary cell of sodium β-alumina and (b) model of arrangement of layers in it.

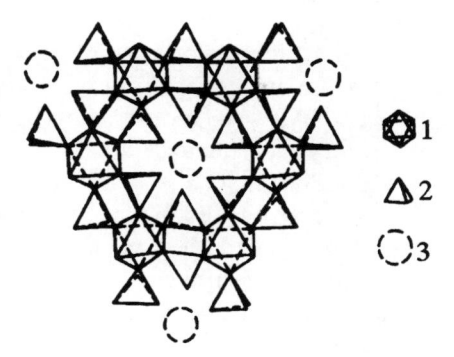

Figure 1.76 Schematic diagram of $Na_3Zr_2Si_2PO_{12}$ nasicon crystal: (1) ZnO_2 octahedron; (2) SiO_2 or OP_4 tetrahedron; (3) positions that can be occupied by Na ions.

of ions, which would assist in overcoming the main drawback of aluminas, namely, the anisotropy of electrical conductivity. The efforts resulted in synthesis of cationic solid electrolytes of the nasicon (*Na*trium *Super*ionic *Con*ductors) type (Figure 1.76). The composition of a typical

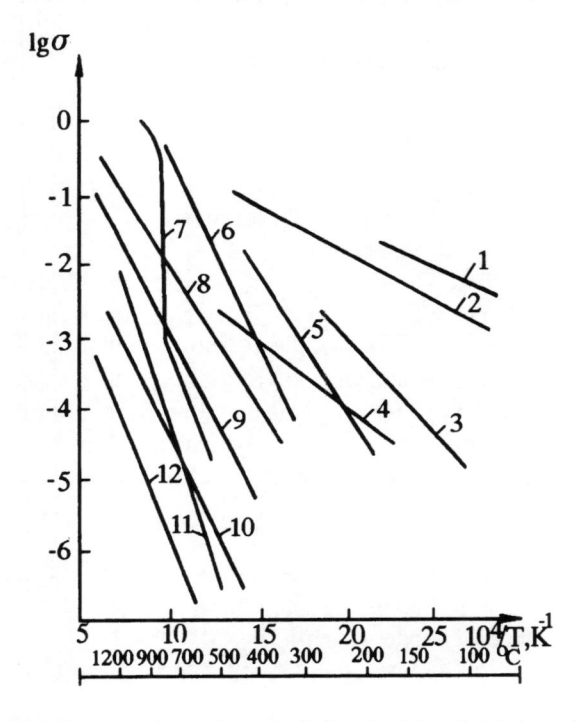

Figure 1.77 Temperature dependence of electrical conductivity of some anion-conducting solid electrolytes.

nasicon is $Na_{1+x}Zr_2Si_xP_{3-x}O_{12}$ ($0 \leq x \leq 3$). The conductivity of nasicons exceeds that of β-alumina in a number of cases.

Anionic superconductors, featuring a considerable oxygen-ionic conduction, are solid solutions of $MO_2-R_2O_3$ and MO_2-BO types, where M is Zr, Hf or Th; R is a rare earth element; and B is an alkaline earth element. The anionic conduction in such substances is due to a significant (8–15% mol.) concentration of a dissolved additive which stimulates emergence of an appropriate number of vacancies in the anionic sublattice. Similar properties are also exhibited by solid solutions based on some oxides of rare earth elements with oxides of alkaline earth elements, but in this case the maximum solubility of an alkaline earth oxide is not over 1% mol. and therefore the conductivity is less.

The highest oxygen-ionic conductivity among all superconductors is exhibited by $Bi_2O_3-Y_2O_3$ solid solutions, which, in contrast to bismuth oxide, undergo no phase transformation into a nonconducting phase (Figure 1.77). Temperature and partial oxygen pressure ranges where the oxygen conduction of bismuth-containing anionic conductors is retained, however, are smaller than for doped zirconia and especially for hafnia and thoria [17].

Both cationic and anionic conductors find numerous technical applications as selective electrodes, sensors of the chemical compositions of gases and liquids, and also in galvanic cells and storage cells.

1.2.1.3 Dielectric polarization and loss

(a) *Polarizability, equivalent circuit of dielectric, dielectric loss*
While the electrical conductivity of dielectrics characterizes a through transport of electric charges across a substance, the dielectric polarization and loss characterize displacement of electric charges under the action of an external field within a dielectric, which gives rise to an electric dipole moment in the volume of the dielectric, i.e., to polarization. The measure of polarization, polarizability P of a dielectric, is equal to the electric dipole moment of unit volume of the dielectric in unit electric field or, which is the same, to the surface charge arising on unit surface area of the dielectric as a result of polarization:

$$P = (\varepsilon - 1)\varepsilon_0 E, \qquad (1.118)$$

where ε_0 is the dielectric constant, equal to 8.854×10^{-12} F/m and ε is the relative permittivity of the substance, which can be defined as the ratio of the capacitance of a capacitor with the dielectric (C) to the capacitance of the same capacitor without the dielectric in vacuum (C_0):

$$C = \varepsilon C_0. \qquad (1.119)$$

The product $\varepsilon\varepsilon_0$ is also called the absolute permittivity of a substance.

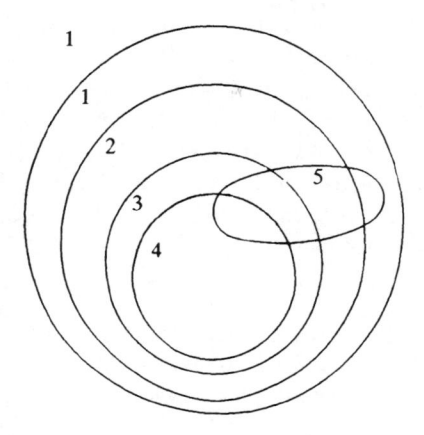

Figure 1.78 Interrelations between various types of dielectrics.

The permittivity of crystals is a tensor quantity (third-rank tensor).

It is convenient also to present the permittivity as a complex quantity

$$\varepsilon = \varepsilon' - j\varepsilon'', \qquad (1.120)$$

where the real part ε' characterizes the polarizability, i.e., the reactive component of the current through the dielectric, I_r, while the imaginary part ε'' characterizes the active component of the current through the dielectric, I_a.

$$\frac{\varepsilon''}{\varepsilon'} = \frac{I_a}{I_r} = \tan \delta, \qquad (1.121)$$

where δ is the dielectric loss angle.

If the dielectric loss is due only to the electric conductivity of the dielectric, then

$$\tan \delta = \frac{1.8 \times 10^{12} \sigma}{f \varepsilon}, \qquad (1.122)$$

where σ is the electrical conductivity (ohm^{-1}cm^{-1}) and f is the frequency (Hz).

An equivalent circuit of a dielectric with a loss can be represented as the dielectric without a loss and an equivalent active resistance, parallel- or series-connected with a capacitance (Figure 1.78); tan δ is expressed in terms of parameters of the equivalent circuit as follows:

$$\tan \delta = \frac{1}{R_p C_p \omega} = \omega C_s R_s. \qquad (1.123)$$

(b) *Physical mechanisms of dielectric polarization*

The polarization of a dielectric involves, as a rule, several processes, differing in the physical nature and the settling time.

Thus, in all substances, there occurs an elastic electronic polarization: a small displacement of electrons of outer electronic shells with respect to nuclei in an external electric field. The settling time of this polarization is of the order of the period of natural vibrations of valence electrons of an atom, of ~10^{-15} s, which corresponds to the UV region of the spectrum. The magnitude of the elastic electronic polarization of typical ceramics amounts to 2.5–4.5 and is related to the optical refractive index of crystals, n, by the formula

$$\varepsilon = n^2. \tag{1.124}$$

Like the refractive index, the elastic electronic polarization is slightly temperature-dependent; it slightly declines with increasing temperature because of the thermal expansion of a crystal and decrease of the number of atoms in unit volume.

The elastic ionic displacement polarization arises in an ionic crystal as a result of displacement of all positive and negative ions with respect to their equilibrium positions under the action of an external electric field. The permittivity value at such a polarization type is usually within 7 and 40 (up to 90 for TiO_2) and the polarization settling time is of the order of the period of natural vibration of ions, 10^{-13}–10^{-14} s, which corresponds to the IR region of spectrum. The magnitude of the elastic ionic polarization increases with temperature, since the increase of interionic distances because of the thermal expansion weakens the bonding forces between ions and increases their displacement in the electric field, i.e., raises the ionic polarization. This greatly exceeds the effect of decrease of the number of particles being polarized in the volume because of the thermal expansion. The temperature coefficient of variation of the permittivity, $TC = \dfrac{1}{\varepsilon}\dfrac{\partial \varepsilon}{\partial T}$, resulting from this process, is for, e.g., a NaCl crystal of 3.4×10^{-4} K^{-1}.

Thus, elastic types of polarization result from a direct action of the electric field on electrons or ions of a crystal. In contrast to this, so-called nonelastic or relaxation types of polarization – ionic-relaxational or electronic-relaxational – result from a relatively small disturbance, introduced by the electric field into the random thermal motion of some crystal particles that can be displaced for distances exceeding the size of an elementary cell of the crystal. Such a capability is exhibited by electrons or ions, disposed near defects of the crystal: point defects, their associations, as well as dislocations and their aggregates, grain boundaries, etc. Although only a small share of all charged particles of a crystal participate in such a polarization, the displacement length and the resulting dipole moment can nevertheless be very considerable, so that the attainable permittivity value amounts to 10^2–10^3. The settling time of relaxational polarization types, 10^{-7}–10^{-2} s, depends on the nature of relaxation centres. A characteristic feature of relaxational polarization types is an exponential decline in the polarization settling time with increasing temperature, which displays the nature of thermal fluctuations and diffusion, controlling the process [18].

In inhomogeneous dielectrics, consisting of phases with different electrical conductivities, the motion (migration) of electrons in an external electric field gives rise to electric charges at phase interfaces, which make a considerable contribution to the polarization. Such a polarization, called a migrational or an interlayer one, cannot be related to every point within a crystal, and therefore only the effective value of the permittivity can be considered here. Such a phenomenon is utilized, e.g., for developing so-called capacitors with barrier layers (BL-capacitors), where nonconducting layers at grain boundaries are created in strontium or barium titanate ceramics, doped to increase the electrical conductivity. The effective permittivity of such ceramics is of about 60 000, but the dielectric loss tangent is great: $\tan \delta = 0.51–0.25$ [19].

Still another process, often encountered in inorganic dielectrics, is a near-electrode polarization, where the charge carrier (electron or ion) build-up place is the energy barrier between a dielectric and an electrode. Such a situation occurs, e.g., in ionic crystals at elevated temperatures or in superionic conductors when the voltage across electrodes is less than the decomposition voltage, i.e., the discharge of ions (electrolysis) cannot occur. Utilizing this phenomena makes it possible to obtain capacitors with a record capacitance, up to 10^6 F/cm^3, but the maximum possible working voltage of such a capacitor is low, of the order of 0.7 V. The settling time of the interlayer and near-electrode polarization, as of the relaxational one, exponentially decreases with increasing temperature and, like the electrical conductivity of dielectrics, varies within many orders of magnitude (approximately within 10^{-5}–10^5 s). These polarization types can therefore cause an instability with time of the capacitance of capacitors which are under a constant voltage, but can be used effectively in capacitors operating at sonic frequencies. For the described polarization types, due to symmetry of a crystal, the electric field within it is isotropic and therefore does not directly participate in the emergence of the dipole moment in the crystal volume. In less symmetrical crystalline lattices – in 20 crystal classes of 32 possible ones – the ionic displacement polarization is always accompanied by deformation of a crystal: the application of an electric field results in a change of crystal dimensions in a definite direction and, conversely, a mechanical deformation results in appearance of charges at certain crystal faces. Such crystals are called piezoelectrics.

In piezoelectric crystals of ten least symmetric classes, having a special polar axis, centres of gravity of '+' and '−' charges in an elementary cell of a crystal do not coincide, and therefore every elementary cell has a dipole moment, which arises at the moment of crystallization. Such a polarization is called spontaneous. A considerable potential difference across faces of such a crystal is usually counterbalanced by the space charge of mobile electrons or holes as well as by charged particles adsorbed from the environment. However, as the spontaneous polarization is temperature-dependent, the electric potential of some crystal faces varies when the crystal is heated or cooled. This phenomenon is called the pyroelectric effect. Typical pyroelectrics – turmaline and potassium tartrate – develop such a high e.m.f. at their faces that they are broken down by the electric field. Thus, a 1 mm thick turmaline plate, cut out perpendicular to the pyroelectric axis, when heated by 10 °C, acquires an electric charge of about 5×10^9 C/cm^2 and the potential difference across its faces increases by 1200 V.

The direction of the spontaneous polarization in most pyroelectrics (linear pyroelectrics) cannot be changed by an external electric field at any temperatures right up to melting, but in some pyroelectrics an external field of 10^2–10^3 V/cm can reorient the spontaneous polarization axis. Such crystals are called ferroelectrics. Adjacent regions in ferroelectrics, featuring the spontaneous polarization (domains), are usually oriented oppositely and cancel the electric moments of each other. A predominant orientation of domains, nonelastic domain polarization, occurs in an electric field. A strong enough field can orient the polarization of all the domains in one direction, converting a ferroelectric crystal into a single-domain one, which accounts for the nonlinear properties typical of such substances: a strong field dependence of ε and a hysteresis loop in the polarization versus electric field coordinates. The crystalline structure at which a substance exhibits ferroelectric properties is usually retained over a limited temperature range; the temperature of transition from a ferroelectric phase to a paraelectric one (generally to a cubic-system phase) is called the Curie point. The permittivity of a ferroelectric increases according the Curie–Weiss law

$$\varepsilon = \frac{K}{T - T_0} \quad (1.125)$$

(K is the Curie constant and T_0 a temperature differing by 5–10 °C from the Curie point) as the Curie point is approached from either higher or lower temperatures. At temperatures near the phase transition on the permittivity of ferroelectrics is maximum, reaching 20 000–100 000 [20].

A special type of dielectric polarization is exhibited by electrets, dielectrics capable of retaining the state of polarization and creating in the surrounding space an electric field for a long time. Any dielectric with a very low electrical conductivity can be, in principle, transformed into an electret by creating in it for a time an electrical conduction by any method and placing it for this time in a strong electric field. Depending on the method of creation of the electrical conduction, there are distinguished photoelectrets (such as sulphur, CdS), thermoelectrets, coronaelectrets (where the conduction is created by a corona discharge), etc. The lifetime of an electret can be as long as several years (for this the conductivity of a dielectric should be less than $10^{-20}\,\text{ohm}^{-1}\,\text{cm}^{-1}$), but declines rapidly with increasing temperature, moisture of the ambient air, in presence of dust, etc. Interrelations between various types of dielectrics are shown in Figure. 1.78.

(c) *Physical nature of dielectric loss*
The dielectric loss is the electric field energy which converts into heat in the volume of a dielectric. In a constant field, a dielectric is heated due to the electrical conduction; the dielectric loss power in this case is

$$P = E^2 \sigma. \quad (1.126)$$

The energy spent for the establishment of all types of the dielectric polarization – electronic- and ionic-relaxational (relaxation loss), transport of charges to internal boundaries between different phases in an inhomogeneous dielectric (migration loss), etc. – also converts into heat. The power of the dielectric loss, spent for the establishment of the polarization, is

$$P = E^2 \omega \varepsilon \varepsilon_0 \tan \delta. \quad (1.127)$$

The dependence of $\tan \delta$ on the frequency of the electric field, temperature of the dielectric and other parameters is determined by the nature of polarization processes in the dielectric. If charged particles are coupled to the electric field by quasi-elastic forces (elastic-displacement electronic and ionic polarization), then a considerable energy dissipation occurs only when the frequency of the electric field coincides with the natural frequency of vibration of particles. This so-called resonant dielectric loss is not temperature-dependent and occurs in a narrow frequency band in the ultraviolet (electronic polarization) or infrared (ionic polarization) spectral region.

Relaxation and migration dielectric losses also have a maximum at a frequency corresponding to $\omega \tau = 1$, where τ is the polarization settling time. Broad maxima of the relaxation dielectric loss are usually observed in the region of sonic or radio frequencies; they span a frequency variation of two to four orders of magnitude. These frequency maxima broaden still more if relaxation processes with close relaxation times are possible in the dielectric, which is characteristic, in particular, of non-single-phase ceramics and crystals with a large number of defects.

When a dielectric is heated, the maximum of the relaxation dielectric loss shifts towards higher frequencies because of the τ decrease, and

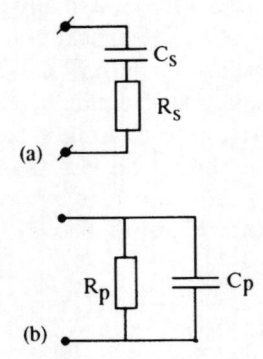

Figure 1.79 Equivalent circuit of dielectric with loss: (a) parallel and (b) series.

therefore the temperature dependences of tan δ are in this case similar to its frequency dependence.

The least tan δ value (10^{-5}–10^{-4}) is observed in dielectrics with elastic mechanisms of polarization: crystals with the maximum forbidden gap width, minimum amount of structural defects and impurities. Examples of materials with the minimum dielectric loss are pyrolytic boron nitride, sapphire and magnesia single crystals. In dielectrics with a relaxational and an interlayer polarization, the tan δ value in the region of the maximum of the dielectric loss varies from 10^{-2} to several units depending on the concentration of relaxation centres. For ferroelectrics the tan δ at the maximum of the dielectric loss is from 10^{-3} to 10^{-2}, the maximum being close to the Curie point. Thus, a typical dielectric loss spectrum, i.e., the tan δ variation with the electric field frequency, comprises usually several maxima of different widths, corresponding to those frequencies at which the lower-frequency polarization mechanisms are inoperative (Figure 1.79).

1.2.1.4 Effect of nonstoichiometry and doping on electrical conductivity of oxides

(a) *Heterovalent doping and nonstoichiometry of oxides*
Doping is commonly exemplified by formation of substitutional solid solutions in the cationic sub-lattice. Higher-valence cations, which substitute main cations of an oxide (donor impurity), are local positively charged centres, whose charge, due to the crystal's electroneutrality condition, should be balanced out by an equal negative charge; the latter can be provided by conduction electrons or ionic defects of the crystal [21].

The type of the forming solid solution is determined by relative energies of formation of electronic and ionic defects: those defects whose formation energy is lower appear in a crystal. Counterbalancing the donor impurity charge by electrons is more typical for oxides with a mixed ionic–covalent type of chemical bond, a relatively narrow forbidden gap, and readily polarizable cations. Such a solid solution type results, e.g., from zinc oxide doping with bismuth oxide:

$$ZnO + Bi_2O_3 \rightleftharpoons 2Bi(Zn)^+O + 2e + {}^1/_2 O_2. \qquad (1.128)$$

In oxides with predominating ionic type of chemical bond and a considerable forbidden gap width, the donor charge is usually cancelled by negatively charged ionic defects, such as cationic vacancies (V'_c). Such a solid solution is exemplified by alumina-doped magnesia

$$MgO + Al_2O_3 \rightleftharpoons 2Al(Mg)^+O + V''_{Mg}. \qquad (1.129)$$

The maximum concentration of conduction electrons or ionic defects, attainable at doping, is determined by solubility of the admixture, i.e., by the structure of the phase diagram of the interacting oxides. The solubility of admixtures generally declines with lowering temperature, which physically shows up as separation of admixture (impurity) atom associations, clusters, from the solution. Clusters can be treated as the smallest-size component of a composite, actively affecting the electrical conductivity of ECC at temperature oscillations.

The regularities at a heterovalent substitution of a main cation by a smaller-charge cation (acceptor impurity) are similar.

An acceptor centre is negatively charged with respect to an undoped crystal and its charge is counterbalanced by those positively charged defects whose formation energy is the lowest: holes or anionic vacancies. Thus, the hole conductivity of nickel (II) oxide rises substantially at dissolution of lithia:

$$NiO + Li_2O + {}^1/_2 O_2 \rightleftharpoons Li(Ni)'O + P. \qquad (1.130)$$

At the same time, the dissolution of yttria in ZrO_2 appreciably increases the anionic component of the electrical conductivity of zirconia, i.e., the energy of formation of anionic vacancies is in this case much less than the energy of excitation of an electron from a filled valence band of the oxide (consisting of 2p states of oxygen) to the acceptor impurity level, unfilled 4d orbital of an yttrium ion.

From formulas (1.128)–(1.130) as well as (1.112)–(1.114) it follows that doping of oxides by heterovalent ions with counterbalancing their charge by ionic defects involves no oxygen exchange with the environment. In contrast, doping of oxides by donors, which increases the electron conduction, is always accompanied by release of oxygen into the atmosphere (as in reduction of an oxide), whereas doping by acceptors with counterbalancing their charge by holes always involves absorption of additional oxygen (as in oxidation).

Formations of ionic and electronic defects that cancel the impurity charge are independent processes; in other words, these defects are always formed simultaneously. Since the concentration is an exponential function of the energy of formation of defects, defects of one kind predominate usually, but if energies of defects are comparable, then contributions by different defect types to the counterbalancing is to be taken into account.

For example, the mobility of electrons and holes greatly exceeds that of charged ionic defects (the ratio of the mobilities can be as high as 10^6 and more), and therefore an appreciable increase in the electron conduction of an oxide can occur even when the concentration of the ionic defects which counterbalance the impurity charge considerably exceeds the concentration of electrons or holes. The same mechanism accounts for differences in the electrical conductivity of different oxides at identical concentrations of a dissolved impurity: this occurs not only because of different mobilities of electrons (or holes) in them, but also because different fractions of the dissolved impurity are counterbalanced by generation of electrons (or holes) [2]. It is also obvious that the contribution of the electron conduction in such a donor-doped oxide will grow with decreasing oxygen content in an equilibrium gaseous phase, while the contribution of the hole conduction (at doping by acceptors) will rise with increasing oxygen content.

Important features of all oxides (as also of other compounds) are departures from a stoichiometric composition and their impact on the electrical conductivity.

From the thermodynamic condition of equality of chemical potentials of oxygen in an oxide and in a gas which is in equilibrium with the oxide, it follows that at a given temperature there exists only one value of the partial pressure of oxygen (and at a given pressure, only one temperature) where the oxide will be strictly stoichiometric. At a departure from this pressure (or temperature) to any side the equalization of chemical potentials is attained through a departure from the stoichiometry of the oxide: excess or lack of oxygen as compared with the stoichiometric formula.

The charge of nonstoichiometric defects is counterbalanced by electrons at a lack, and by holes, at an excess of oxygen in accordance with the equations

$$ZnO \rightleftharpoons {}^{3/2} O_2 + V_O'' + 2e; \qquad (1.131)$$

$$ {}^{3/2} O_2 \rightleftharpoons Y_2O_3 + 2V_Y''' + 6. \qquad (1.132)$$

Anionic and cationic vacancies are not a sole possible type of ionic defects accompanying a departure from the stoichiometry; those defects arise whose energy is the lowermost. The concentration of nonstoichiometric defects as well depends solely on their energy: the higher the energy of origination of a pair of nonstoichiometric defects (an ionic defect jointly with an electron or a hole), the lower their concentration at which the change in the chemical potential of oxygen, needed for the equilibrium, occurs. Thus, the difference between berthollides and daltonides consists only in the energy of formation of nonstoichiometric defects [22].

To calculate the dependence of the concentration of nonstoichiometric defects and electrons (or holes) on the partial pressure of oxygen, the law of mass action is applied to equations of the type of (1.131) and (1.132), and the resulting equation is solved jointly with the electroneutrality equation [23]. For example, for (1.131) the law of mass action gives

$$P_{O_2}^{1/2} n^2 [V_O''] = K_n, \quad (1.133)$$

where K_n is the equilibrium constant of this reaction; $[V_O'']$ and n are concentrations of oxygen vacancies and electrons. The electroneutrality equation, in the simplest case where impurities and other defects are absent, is

$$n = 2[V_O'']. \quad (1.134)$$

Substituting (1.134) into (1.133) and solving the resulting equation for n yields

$$n = (2K_n)^{1/3} P_{O_2}^{-1/6}. \quad (1.135)$$

Similarly, for the case of (1.132):

$$P_{O_2}^{3/2} = K_p [V_Y''']^2 \cdot p^6; \quad (1.136)$$

the electroneutrality equation is

$$3[V_Y'''] = p; \quad (1.137)$$

and the dependence of the hole concentration on P_{O_2} is

$$p = \left(\frac{3}{K_n}\right)^{1/8} P_{O_2}^{3/16}. \quad (1.138)$$

The above case ignored intrinsic defects of a crystal, such as Schottky defects. Allowing for the defects, it is not difficult to plot the theoretical dependence of concentrations of various defects on P_{O_2} at a fixed temperature, e.g., for a R_2O_3-type crystal (Figure 1.80).

The doping of an oxide changes the character of dependences of electronic and ionic defects on P_{O_2}. Calculated dependences of concentrations of various defects on P_{O_2} for a R_2O_3-type crystal doped by a donor and an acceptor impurity for the case where Schottky defects prevail in a stoichiometric crystal, similar to those of Figure 1.80, are shown in

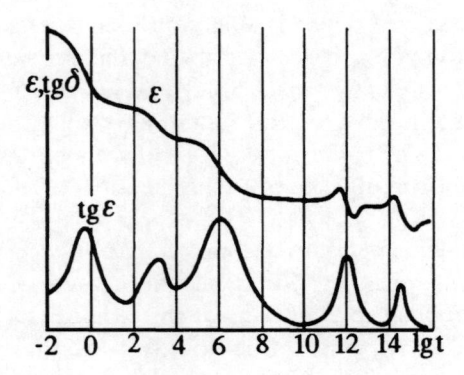

Figure 1.80 Variation of permittivity (ε) and loss-angle tangent (tan δ) of dielectric over broad frequency band.

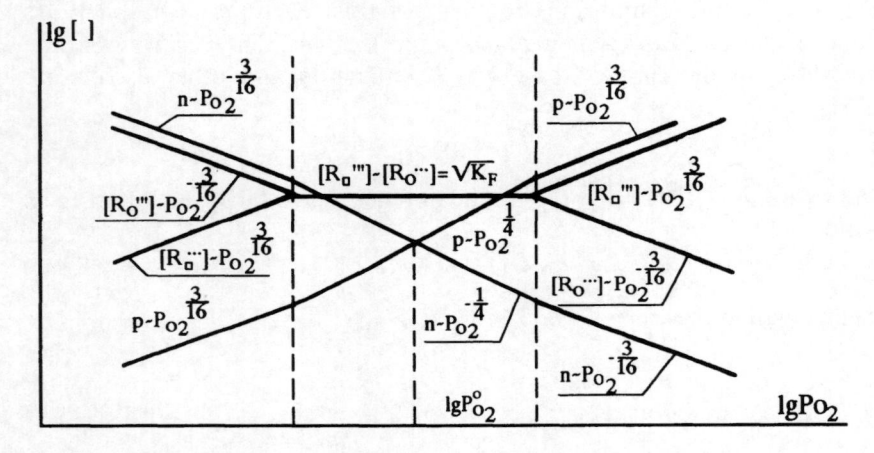

Figure 1.81 Defect concentration dependence on partial oxygen pressure at fixed temperature for undoped R_2O_3 crystal with predominating Schottky disorder.

Figure 1.81 and Figure 1.82. Since various charge states of defects and predominance of various intrinsic defect types (Schottky or Frenkel defects, electron–hole disorder) are possible, dependences like those presented in Figure 1.80–1.82 turn out to be diverse enough [22], but they are just the principal ones in selecting methods to control the electrical conductivity of oxides.

A temperature rise brings about not only increase in absolute concentrations of intrinsic defects, but also a shift of the partial pressure of oxygen, corresponding to stoichiometry of the oxide, towards higher P_{O_2}. In other words, increasing the temperature at a constant P_{O_2} promotes reduction of an oxide. This can be illustrated by copper oxides: copper (II) oxide (CuO) is stable in air at low temperatures; copper (I)

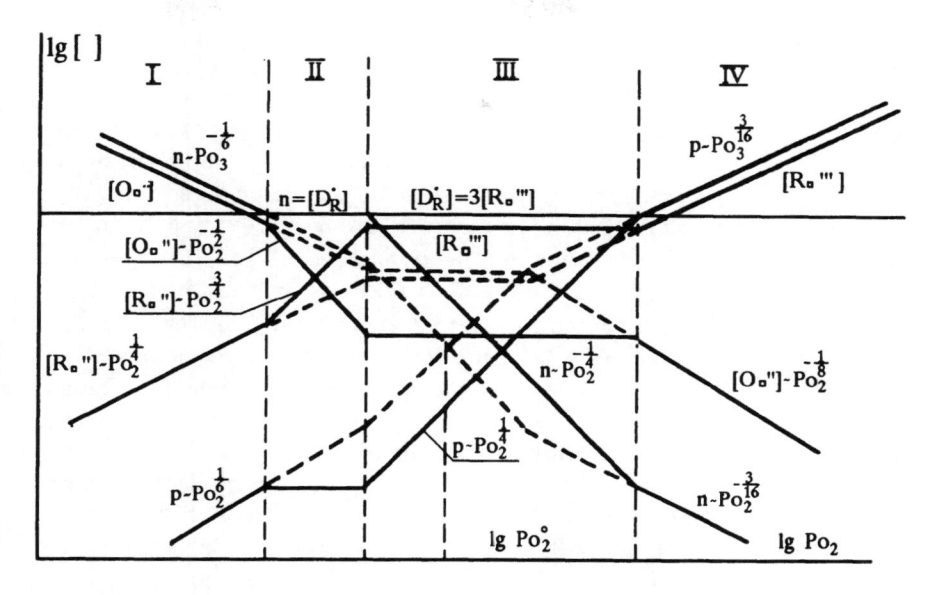

Figure 1.82 As in Figure 1.81, but for crystal doped with donor (D) impurity.

oxide Cu_2O, at 800–1100 °C; a further heating reduces the oxide Cu_2O to metal, and therefore, when copper is melted in the air, an oxide film on its surface is absent.

An important conclusion from the presented diagrams of nonstoichiometric disorder is also the following: if, e.g., an acceptor impurity creates primarily anionic vacancies in an oxide, then the hole concentration in the doped oxide will be much higher than in the pure one and will rise with increasing P_{O_2}, which, with account of the ratio of mobilities, will affect both the electrical conductivity being measured and its dependence on P_{O_2}. In other words, the excess charge of an impurity ion can be neutralized, with some probability, both by ionic and by electronic charges, i.e., there exists no distinct boundary between solid solution types described, e.g., by equations (1.128) and (1.129): they progressively turn into each other as energies of formation of electronic and ionic defects of the same sign approach each other or (which is the same) always coexist in one and the same oxide at one and the same time, the probability of the existence of either of them depending on the energetics of defects [24].

In view of specific features of electronic states in the valence and conduction bands of oxides and formation of polarons with various energies and sizes, the electron-to-hole mobility ratio in most oxides differs considerably from unity. Practically, it results in that only one type of electronic current carrier can be observed in many oxides under any conditions, for example, an electronic conduction in zinc and tin oxides, a hole conduction in nickel and cobalt oxides and lantanum chromite. The hole conduction is more typical for oxides [25].

Temperature dependences of the electrical conductivity σ of both pure and doped oxides are usually described by the Arrhenius law $\sigma = \sigma_0 \exp(-B/kT)$, where B is the activation energy and k the Boltzmann constant. The doping of an oxide can either increase or reduce not only the value of electrical conductivity, but also its activation energy, which can be understood on the basis of the presented concepts. A general regularity of temperature dependences of the electrical conductivity of oxides, as of other substances, is a decline in the impact of doping on the electrical conductivity with increasing temperature, although for different oxides and different impurity types the regularity shows up dissimilarly. Thus, in doping of many oxides of rare earth elements and their compounds with calcia, the electrical conductivity rises by a factor of 10^2–10^3, while the conduction activation energy declines, the activation energy at high temperatures being lower than at low ones. At the same time, doping the same oxides with hafnia or zirconia reduces the electrical conductivity by the same factor of 10^2–10^3; the activation energy in this case remains the same as for an undoped oxide, but grows with increasing temperature (Figure 1.83) [27]. Physical causes of such regularities stem both from a much lower mobility of electrons than that of holes in REE oxides and from the type of defects forming at dissolution of impurities and their interaction processes. Thus, a calcia impurity increases the electrical conductivity of REE oxides through increasing the number of holes,

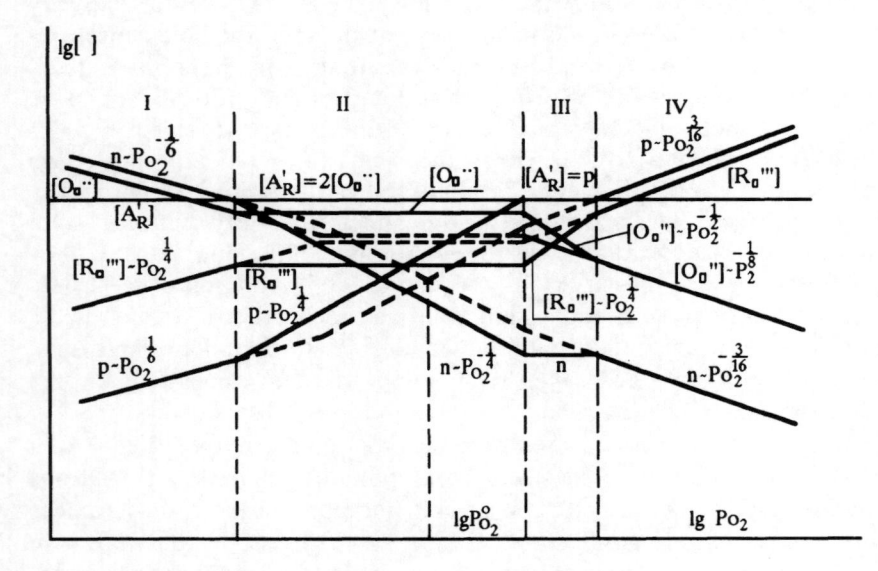

Figure 1.83 As in Figure 1.81, but for crystal doped with acceptor (A) impurity.

predominating both in the mobility and in the concentration in a pure oxide. The following phenomena, depending on the temperature according to a law close to exponential and affecting the conduction activation energy, can thus be distinguished for the electrical conductivity of an oxide doped in this manner: (1) increase in the hole mobility with temperature, which is characteristic of oxides of 3d and 4f elements (activation energy E_1); (2) increase in the hole concentration with temperature because of ionization of hole-impurity ion centres (E_2); (3) for impurity concentrations exceeding the solubility limit (about 0.5% mol. for calcia), increase in the impurity concentration because of growth of the solubility limit with temperature (E_3). The total observed activation energy is $E = E_1 + E_2 + E_3$. As the temperature increases, E_2 goes to zero at a complete ionization of the impurity, while E_3 becomes zero when the solubility of the impurity exceeds its concentration in a given sample. E_1 is generally not temperature-dependent. With increasing impurity concentration, E_1 and E_2 as a rule decrease, whereas E_3 remains unchanged. The above-described qualitative regularities are in a good accord with experimental data (Figure 1.83).

When most REE oxides are doped with donor impurities, such as zirconia and hafnia, the electron concentration in them rises and the hole concentration declines due to the hole–electron recombination in accordance with the equation $np = K_i$, the total electrical conductivity decreasing because of a low electron mobility. The main contribution to the electrical conductivity being in this case, as before, made by holes, the conduction activation energy, in compliance with the above-presented regularities, coincides with the activation energy for an undoped oxide. More energy-consuming processes of carrier generation become conspicuous on the background of the total low electrical conductivity only at high enough temperatures (about 1400 °C in Figure 1.83), with the result that the total electrical conductivity and its activation energy increase.

(b) *Electrochemical doping and intercalation*
In contrast to formation of solid solutions of oxides, effected by means of a solid-phase diffusion or fusion, in the electrochemical doping of substances with a prevailing cationic conduction, an impurity is introduced by means of an ionic exchange at the solid electrolyte–electrode interface and a progressive substitution by the impurity of ions of the main substance in compliance with electrolysis laws. For example, an electrochemical substitution of sodium by silver in β-alumina occurs at a temperature of about 500 °C in the following cell

$$+\text{Pt}/\text{Ag}/\text{AgJ}/\text{Na-}\beta\text{-Al}_2\text{O}_3/\text{K}_2\text{0.7Fe}_2\text{O}_3/\text{Pt} - . \qquad (1.139)$$

An external electric field of the indicated polarity creates a flow of Ag^+ ions, which get embedded in the alumina structure while an equivalent amount of Na^+ ions pass into the cathode material [27]. The intermediate AgJ layer serves as an ionic membrane. Its incorporation into the galvanic circuit drastically increases the rate of implantation of silver ions owing to reduction of polarization of the anodic boundary of $Na\text{-}\beta\text{-}Al_2O_3$. The cathode material, potassium polyferrite, has a broad enough region of homogeneity in cations and, receiving excess sodium ions, retains a single-phase structure; it also has a combined electron–ion conduction with a high level of the ionic component and does not restrict the exchange rate. An external potential difference at which the ionic exchange process proceeds at a high rate is of about 0.6 V. The quantity of electricity passed through the circuit serves as an exact measure of the ionic exchange.

The electrochemical doping process is applicable for transport not only of alkali, but also of alkaline earth and some other cations. Electrochemical doping is one of the basic mechanisms accompanying the passage of current through substances with a combined ion–electron conduction, and therefore it is to be taken into consideration in the development of many electronic ceramic composites.

The above-described scheme of electrochemical doping is also suitable for carrying out the intercalation: implantation of ions, atoms or molecules into a layered crystal between weakly bonded atomic layers. Intercalated compounds were first produced on the basis of crystals of transition metal dichalcogenides MeX_2. The crystals consist of layers, each of which is a sandwich of two chalcogen atom layers with a metal atom layer in between. Atoms of metals and hydrogen, ammonia and even large organic molecules can be implanted between the chalcogen atom layers, which are bonded to each other by weak van der Waals forces. In the latter case (large organic molecules) the layers move apart to a spacing of about 50 Å, whereas a normal spacing is close to 3 Å. The intercalation is also one of the probable results of charge transport across a boundary between two bodies, although this process is encountered less often than the electrochemical doping.

1.2.1.5 Some ECC-based devices

(a) Composite resistors

High-power bulk resistors as well as thick-film resistors, widely used in modern electronics, are in essence a sintered two-phase composite of conducting particles and an insulating phase. Commonly selected as the conducting phase are powders of metals (tungsten, tantalum, ruthenium, palladium–silver alloy, etc.) as well as powders of refractory compounds that have either a metallic or a semimetallic conduction (nitrides, carbides, borides, silicides, oxides, etc.). The selection of conducting phase is governed by the aspiration to reduce the temperature coefficient of resistance (TCR) and to extend the range of resistor ratings. Thus, a high stability is featured by resistors where nonstoichiometric borides of rare earth elements serve as the conducting phase.

The requirements placed on the insulating phase include a good adhesion to conducting particles (to guarantee the strength of the composite) at not very high fluidity and wetting of conducting particles by the insulator melt. Silicate or borosilicate glasses of various compositions as well as glass ceramics are used for the insulating phase.

Most powerful bulk resistors are moulded from a blend of powders of electrically conducting and nonconducting refractory compounds (e.g., Si_3N_4 and TiC, TiN and AlN, $CrSi_2$ and BN, etc.), which are then sintered or hot pressed at 1500–1900 °C. The above-listed requirements for the conducting phase are in this case added to by that of oxidation resistance, which determines both the working temperature range and the maximum power dissipated by a resistor. Some refractory compounds are greatly superior to metals in the oxidation resistance.

(b) *Varistors*

Varistors are one version of bulk resistors, featuring a nonlinear current-voltage characteristic, which is usually approximated by the formula

$$I = I_0 e^{\beta(U - U_0)}, \qquad (1.140)$$

where I is the varistor current; U the voltage; β the nonlinearity factor, varying from 5 to 50; and I_0, U_0 are constants.

Varistors with the maximum nonlinearity at some specified voltage U_0 increase the current passed through them by more than 10^6 times at a voltage change of 5–10%. The range of the specified (or, as they are called, classification) voltage values of varistors varies from a few volts to hundreds of kV, and the currents passed by them, from milliamperes to kA/cm^2.

In spite of such a broad range of parameters, most varistors are based on zinc oxide ceramics, whose nonlinear I–U characteristics were first studied in 1957 [28]. Varistors of this type are now among the most widely used semiconductor devices (their manufacture in Japan alone approaches 400 billion units annually); they provide unified components for the overvoltage protection of electronic and electric devices, from integrated circuits to solid-state arresters for high-voltage power transmission lines. The basic element of varistors is sintered zinc oxide ceramic doped with one of the following combined additives: (1) TiO_2 + glass phase; (2) CoO, MnO, Bi_2O_3, Sb_2O_3, Cr_2O_3; (3) Co_2O_3, MnO_2, Bi_2O_3, Cr_2O_3, NiO, Sb_2O_3, Al_2O_3. Each of the combined additives includes admixtures affecting the nonlinearity of the I–U

characteristic, microstructure and bulk electrical conductivity of the ceramic (Table 1.28).

Table 1.28 Types and application of varistors

Additive type	I–U characteristic	Application
(1)	Weakly nonlinear with nonlinearity factor $\beta \approx 5$	Varistors VNKS-25
(2)	Tunnel-type with $\beta \geq 40$	Varistors SN2-1, ZNR, GE-MOV
(3)	Tunnel-type with $\beta \geq 50$ and improved nonlinearity at heavy currents through doping of crystals	Varistors for high-power high-voltage arresters

The nature of nonlinearity of the I–U characteristic is associated with formation at interfaces within the ceramic of an excess concentration of doping atoms (e.g., Co^{3+}, Bi^{3+}), which create at the interface an inversion of electrical conductivity, i.e., a 0.6–0.8 eV high potential barrier for current carriers. With a small enough thickness of the interface and a sufficient potential difference applied across it, the probability of tunnelling of electrons across the interface rises exponentially with the applied voltage. The energy diagram of such a tunnel-transparent interface is shown in Figure 1.84.

According to existing concepts, the number of tunnel-transparent interfaces in ceramics is of about 30% of the total number of interfaces. As indicated by electron microscopy and Auger spectroscopy, their thickness is within 50 Å. These are as a rule low-angle boundaries, such

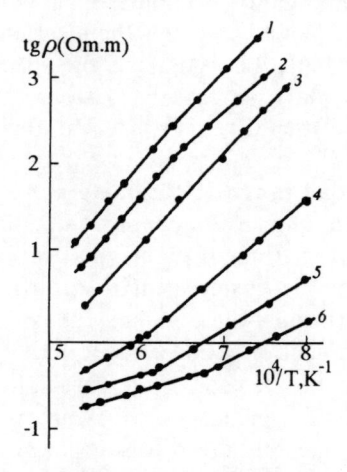

Figure 1.84 Temperature dependence of electrical conductivity for $GdScO_3$ ceramics doped with hafnia and calcia: (1) 1% mol. HfO_2; (2) 5% mol. HfO_2; (3) undoped; (4) 0.1% mol. CaO; (5) 1% mol. CaO; (6) 10% mol. CaO.

Functional materials

as twinning ones. A specific feature of zinc oxide ceramics is an oriented grain substructure, at which the six-fold axis has the same direction for all blocks of a given crystallite. A unit block has an acicular shape with a lateral dimension of a hexagon of 1–2 µm. A slight misalignment of axes of adjacent blocks forms between them low-angle boundaries, just which, when the required admixtures are present at them, serve as elementary microvaristors.

(c) *Composite ceramic heaters*
The oxidation resistance confers on refractory-compound ceramics stable performance characteristics even at fairly high temperatures in the air. This property underlies the development of silicon carbide- (SiC) and molybdenum disilicide- ($MoSi_2$) based high-temperature heaters. To reduce the electrical conductivity, silicates or silica are added to the blend; a film of the silica is also formed on the heater surface due to oxidation. The maximum working temperature is determined by the fluidity of the film (which decreases, e.g., as the amount of alkali metal admixtures is reduced). For the majority of commercially produced heaters, this temperature is of 1550 °C, and for low-porosity high-purity ones, of up to 1700 °C. Molybdenum oxide additions increase the silica melt viscosity, owing to which the best types of molybdenum disilicide heaters can be operated in the air at temperatures up to 1850 °C. In reducing atmospheres, especially in hydrogen, the working temperature of such heaters is 300–500 °C lower because of silane formation.

Oxide ceramics, in particular those based on REE chromites and zirconia, are also used as high-temperature heaters for oxidizing atmospheres.

The possibility of a considerable increase of the hole conduction of REE chromites by doping them with oxides of alkaline earth elements was first discovered by Tresvjatskij [31]. Oxides exhibit an absolute oxidation resistance since they are thermodynamically in equilibrium in oxidizing atmospheres. Melting temperatures of REE chromites exceed 2160 °C, but their maximum working temperatures are limited by chromium oxide sublimation, which intensifies markedly above 1700 °C. To reduce this effect, a composite technology is resorted to: heaters are manufactured from molten chromite grains, sintered with addition of a disperse phase of chromite of another composition.

Zirconia-based heaters are manufactured by a similar technology, but here the technique of phase doping and creation of controlled nonuniformities of the structure pursues a different goal: improving the thermal shock resistance and reducing recrystallization of the ceramic. These heaters can operate at up to 2200 °C, but optimum conditions for their operation are in continuous-heating furnaces (because of their inadequate thermal shock resistance). Zirconia is a dielectric at room temperature, and therefore such heaters require a starting heating-up to about

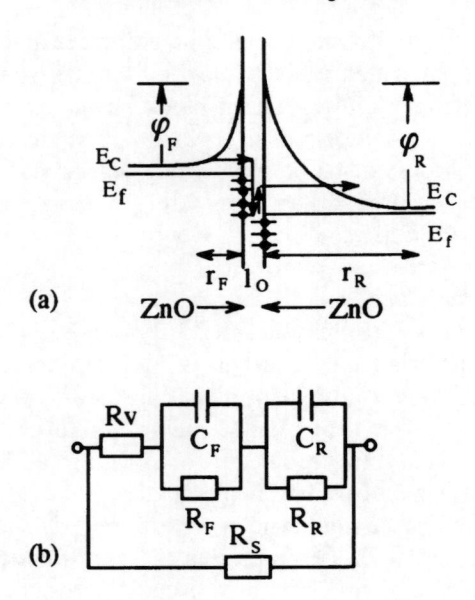

Figure 1.85 (a) Energy diagram of intercrystallite boundary and (b) scheme of its substitution. φ_F, φ_R and r_F, r_R are heights and widths of potential barrier at grain boundary; E_f is the Fermi level; E_c is the conduction band bottom; l_0 is the intercrystallite layer thickness. Arrows indicate tunnellings of electrons with participation of surface states.

700 °C, and current leads to them are made from platinum or REE chromites [34].

(d) Ceramic capacitors
Ceramic capacitors are most often manufactured from oxide ceramics having ferroelectric or antiferroelectric properties – barium, strontium, and calcium titanates ($BaTiO_3$, $SrTiO_3$, $CaTiO_3$), lead zirconate ($PbZrO_3$), as well as niobates, tantalates, stannates and other compounds. A drawback of all capacitors of such a type consists in a drastic (by tens of times) increase in the permittivity of the material near the Curie point. To smooth out this dependence and to provide the maximum capacitance and at the same time stability of capacitors, use is made of multicomponent ceramics whose composition is selected so that Curie points of individual components are uniformly spaced over the whole working temperature range of a capacitor. This technique manages to retain the relative permittivity of a ceramic at a level of thousands over a temperature range of more than 100 °C (Figure 1.85). This is possible, however, only when the ceramic is not an ideal solid solution (for which only one fixed Curie point would exist), but a compound with inhomogeneous chemical and phase compositions, resulting from an incomplete

solid-phase reaction. Every element of the compound makes its contribution to the permittivity of the ceramic, which can be calculated by the above-described effective medium theory methods. As the temperature changes, dielectric characteristics of composite constituents change as well, and polarization effects, whose contribution to the dielectric characteristics is also significant, arise at places of inhomogeneities.

Based on these effects, the technology of ceramic multilayered capacitors with a range of ratings of up to 100 μF at a working voltage of up to 250 V was developed; their dimensions and cost are less than those of electrolytic capacitors.

(e) *Optically transparent ceramic materials*
One of greatest achievements of oxide ceramic sintering technology is the production of optically transparent sintered and hot pressed ceramic materials. The first example of such ceramics was an alumina-based transparent (more exactly, translucent) ceramic, synthesized by Coble in the early 1960s [33]. This material essentially accomplished a revolution in lighting engineering by making it possible to manufacture high-pressure sodium lamps, which are among the most efficient compared with any other light sources. At present, there are tens of recipes for synthesizing such materials, optical characteristics of the best of them little differing from those of high-quality single crystals. Transparent ceramics have gained numerous applications as IR-transparent materials, optical filters, effective luminophores, including converters of X-radiation into a visible one, etc.

The technology of synthesis of practically all materials of this class relies on the principle of a composite material: addition to a ceramic of special additives which concentrate at boundaries of growing grains to form layers that inhibit the growth of grains, and at the same time promote an accelerated diffusion along grain boundaries and a complete elimination of all pores in sintering. Since the thickness of these layers is much less than the visible light wavelength and their refractive index is not too different from that of the main substance, such layers exert practically no effect on the transparency of the ceramics.

It can also be noted that additives exist, such as LiF and MgO, which almost completely evaporate at the last stages of sintering of transparent ceramics. This is in fact a new type of composite material, whose structure changes with time [34, 35].

(f) *Ceramic sensors*
Notwithstanding that the above-presented examples cover a very broad scope of various ceramics, there is every reason to assert that the principal ECC application are ceramic sensors. Sensors are devices converting information on nonelectrical quantities into an electric signal.

Of all the sensor types – optical, electromagnetic, etc. – ceramic-based sensors turned out to be the most diverse in the principles employed and information obtained, the most miniature, reliable, cheap and therefore the most numerous. They are at present used in all spheres of human activity: control of production processes, robotics, environmental monitoring, military engineering, etc. The amount of information received with the aid of ceramic sensors is usually so great that it can be processed only by computers. Thus, computers and ceramic sensors excellently supplement each other and make it possible to develop qualitatively new technologies, modern machinery and devices. In general, it can be stated that ceramic sensors have become an integral part of the modern society and have striking prospects for the future [36].

Practically all ceramic sensors are composites, since only an appropriate combination of several dissimilar phases can provide the conversion of a nonelectrical signal into an electrical one and its further transmission to a data processing system.

It is expedient to classify ceramic sensors proceeding from the physical principles underlying their action as well as according to their purposes.

Thermistors These are temperature-sensitive resistors with a negative temperature coefficient of resistance (TCR) and are employed for temperature measurement and control. The action of thermistors relies on an exponential decrease of their electrical resistance R with increasing temperature T, which is characteristic of semiconductors, which also include many oxides,

$$R = R_0 \exp\left(\frac{B}{T}\right), \qquad (1.141)$$

where B is the activation energy and R_0 a constant.

The TCR of a thermistor is

$$\alpha = \frac{1}{R}\frac{dR}{dt} = -\frac{B}{T^2}. \qquad (1.142)$$

The maximum TCR is exhibited by ceramic thermistors made from a blend of several transition metal oxides (iron, nickel, cobalt, manganese, bismuth and some other oxides) doped with oxides of magnesium, strontium, barium as well as of indium, gallium and of some other elements. The doping of oxides is aimed at increasing the overall electrical conductivity and its temperature coefficient as well as reducing the impact of the partial oxygen pressure on the electrical conductivity. Phase and microstructural studies evidence that ceramics in thermistors are usually inhomogeneous and consist both of grains of individual oxides and of products of their interaction. Contacts between doped oxide grains of different compositions are like heterojunctions in classical semiconductors and promote an increase of the TCR. As to the design, thermistors have the form of ceramic pellets, beads, plates, etc., a few millimetres in size, with metallic electrodes applied by the burning-in, chemical, electrochemical, and other methods. An interesting version are thermistors whose main component is doped oxide ferrite having a great coercive force, which allows such thermistors to be attached to steels and other ferromagnetic materials by means of magnetic interaction.

Posistors These are thermistors with a positive TCR showing up as a rule over a small temperature range. Typical examples of posistors are ceramic elements from barium titanate doped by oxides of rare earth elements (REE) or oxides of stibium, niobium and of some other elements. Trivalent ions of REE substitute barium ion in $BaTiO_3$, stibium and niobium ions substitute titanium ion, i.e., all these elements serve as donor impurities and increase the electrical conductivity of $BaTiO_3$.

Electrical properties of such a doped ceramic, sintered in the air or in oxygen atmosphere, are distinguished in that when it is heated to some temperature, of about 120 °C, its electrical resistance increases greatly, by 3–4 orders of magnitude. A positive temperature coefficient of resistance remains over a small temperature range, as little as 20–30 °C, and at a further heating the resistance starts somewhat decreasing once again (Figure 1.86). The initial temperature of this so-called posistor effect exactly enough coincides with the temperature of the barium titanate phase transformation from the low-temperature tetragonal modification (where barium titanate exhibits ferroelectric properties) to the high-temperature cubic one. It is of interest that this effect is fully

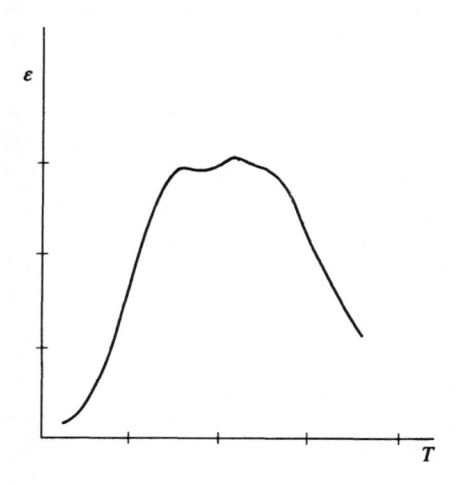

Figure 1.86 Temperature dependence of permittivity of polyphase composite ferroelectric.

absent in single crystals of the same composition as well as in ceramics fired in a reducing atmosphere.

Modern concepts relate the nature of this effect to the composite structure of a doped barium titanate ceramic. Due to segregation, a barium deficiency already exists at grain boundaries in this ceramic, which at firing in an oxidizing atmosphere creates acceptor levels at grain surfaces. Their interaction with donors results in formation at grain surfaces of an insulating layer, a potential barrier for current carriers, the so-called Schottky barrier, whose height φ is

$$\varphi = \frac{eN_D l^2}{2\varepsilon\varepsilon_0}, \qquad (1.143)$$

where N_D is the donor concentration; $2l$ the barrier width; and ε the relative permittivity.

At heating to the phase transformation temperature the permittivity declines so that the potential barrier height grows, which greatly increases the resistance [37].

The phase transformation temperature and the positor effect can be varied from room temperature to about 300 °C by partly substituting barium with strontium or lead, and titanium, with zirconium or tin. Owing to this, such ceramics find diverse applications as heaters with a preset heating temperature, in temperature stabilization devices, etc. [21].

Gas composition sensors These are a rapidly growing group of miniature ceramic sensors, whose use obviates the need for complex and cumbersome gas analysis instruments. Operating principles of the sensors are widely varying and rely on utilization of various electrophysical properties of ceramics and ceramics–environment interaction processes specific for the properties. Some examples of such sensors are described below.

Resistive sensors serve to measure the content or, more exactly, the partial pressure of oxygen in a gas, using for this purpose the dependence of electrical resistance of an oxide, which is in a thermodynamic equilibrium with the surrounding gaseous phase, on the partial pressure of oxygen. Such a dependence, in accordance with the regularities, described in section 1.2.1.4, is

$$R = R_0 P_{O_2}^n \qquad (1.144)$$

where the exponent n usually varies within $-1/4$ and $+1/6$ depending on the character of the electrical conduction of the oxide and the type of nonstoichiometric defects in it. The oxygen sensitivity of the sensors reaches 10^{-10} Pa [38].

As to the design, such a sensor is a miniature ceramic element made of an oxide of titanium, tin, iron, niobium, etc., doped in accordance with the regularities presented in section 1.2.1.4. Each of the oxides is

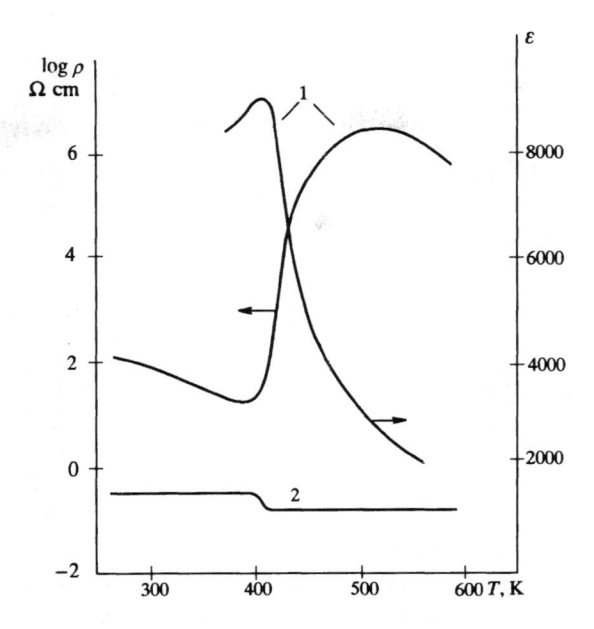

Figure 1.87 Variation of resistivity and permittivity with temperature of BaTiO$_3$ (99.5% mol.)–Sm$_2$O$_3$ (0.5% mol.) solid solution: (1) sintered ceramic; (2) single crystal.

serviceable over certain ranges of temperatures and partial pressures of oxygen. For a rapid enough (in fractions of a minute) establishment of equilibrium with a gaseous phase the ceramic element temperature should be not less than 600–700 °C, and therefore a ceramic or metallic heater as well as a thermistor which accurately controls the heating temperature are incorporated in the sensor. The temperature control is needed because the electrical resistance of the gas composition-sensitive element decreases exponentially with temperature. All these components as well as current leads to them and separating insulators can be integrated into a miniature ceramic package (Figure 1.88).

Adsorption-resistive sensors, in contrast to the preceding ones, exhibit a high (down to 10^{-5}%) sensitivity and a relatively high selectivity to various gases: CO, CO$_2$, CH$_4$, H$_2$S, H$_2$, vapours of alcohols and of some other organic substances. They have the form of small (on the order of a few millimetres) elements consisting of sintered (usually porous) three- to five-component blend of oxides of such elements as tin, titanium, niobium, zinc, indium and molybdenum with additions of oxides of bismuth, vanadium, stibium and some others. In one popular design of such sensors, the oxide mixture is sintered to the outside of a glass bead, into which platinum current leads are soldered; a platinum heater providing under working conditions the sensor heating to 200–400 °C, is built into the bead.

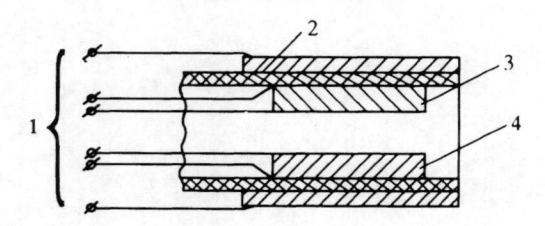

Figure 1.88 Design of a resistive oxygen content sensor: (1) leads; (2) quartz tube with heater; (3) titanium dioxide sensing element; (4) thermistor element.

Despite the external simplicity of such sensors, their operating principle is fairly complex and involves such processes as a catalytic selective adsorption of gases by active sites on the sensor surface, formation of acceptor bonds between adsorbed gases and the surface layer of oxides, transition of electrons from the oxide semiconductor to the bonds, and the resulting change in the stoichiometry and electrical conductivity of the oxide. Thus, a sensor comprises several functionally close related elements integrated in the composition of the composite ceramics. The selectivity to definite gases is attained by an appropriate selection of the type of active catalytic centres and of the sensor temperature. The temperature dependence of sensitivity of a sensor has the form of a curve with a maximum; the temperature corresponding to the maximum differs with the gas because molecule desorption energies for different gases are dissimilar (Figure 1.89).

The sensitivity drop with increasing temperature is accounted for by a rapid desorption, i.e., by a short time of residence of a molecule in an adsorbed state, when it affects the electrical conductivity of an oxide.

Drawbacks of such sensors include unsatisfactory selectivity, poor stability and reproducibility of properties at a mass production, and sensitivity loss resulting from surface contamination.

Solid-electrolyte oxygen sensors These are essentially a membrane of a ceramic where current carriers are only oxygen ions, fitted with inert gas-permeable electrodes that do not block the ionic current. The required type of electrical conduction is offered by doped oxides with high-mobility oxygen vacancies. Zirconia-based ceramics are commonly used, but oxides of thorium, hafnium, bismuth and some others are also employed. The sensor signal E under conditions of a thermodynamic equilibrium of membrane surfaces with various gaseous mixtures is described by the Nernst formula

$$E = \frac{RT}{4F} \ln \frac{P'_{O_2}}{P''_{O_2}}, \qquad (1.145)$$

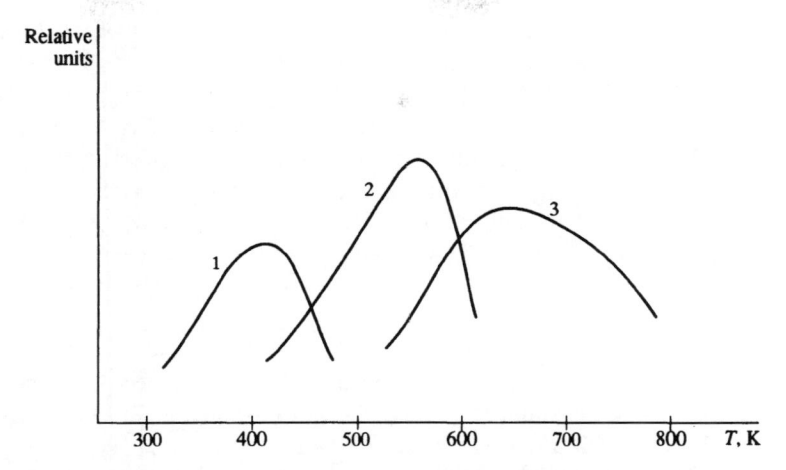

Figure 1.89 Temperature dependence of signal of adsorption-resistive gas composition sensor for various gases: (1) CH_4; (2) H_2S; and (3) CO_2.

where R and F are the Boltzmann and Faraday constants, P'_{O_2} and P''_{O_2} the partial oxygen pressures near different membrane surfaces. Such a sensor usually has the form of a test-tube or pellet from a solid-electrolyte ceramic. In operation, it is heated to a temperature where the oxygen-ion conduction becomes considerable (e.g., about 700 °C for zirconia). Electrodes are applied to opposite end-faces of a pellet or sides of a tube; one of them is flown over by the gas being measured, and the other, by a gas with a known oxygen content, such as air (Figure 1.90). Miniature designs where sintered electrically-conducting complex oxides (cobaltites, manganites, cuprites) are used as electrodes and an oxide-metal blend (also sintered) as the reference electrode have been successfully developed recently. Such a sensor is a typical ceramic composite with both a geometrically and functionally organized structure.

Zirconia-based solid-electrolyte sensors represent the main type of sensors used at present in motor vehicles, at thermal power stations, on industrial boilers to control the air–fuel ratio. A great signal from such

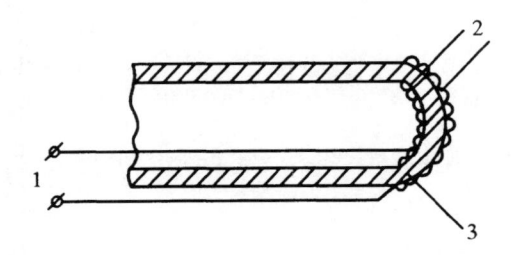

Figure 1.90 Schematic design of solid-electrolyte sensor of partial oxygen pressure: (1) leads; (2) electrodes; (3) solid electrolyte.

sensors (about 0.7 V) and a steep change in it at an air–fuel ratio departure from the stoichiometric can allow this ratio to be maintained accurately to within ±0.5%, which practically cannot be attained by means of other air and fuel feed-monitoring instruments. In this case, the optimum fuel-burning conditions and, with the aid of a catalyst, the most complete neutralization of main toxic components of exhaust gases – nitrogen oxides, carbon monoxide and methane – are attained. A system including an above-described sensor and a neutralizer is therefore called a triple-action system [17].

(g) *Conclusion*
The above-presented examples far from exhaust the list of electronic ceramic composites and their applications as sensors and various electronic components. This list also includes ceramic moisture sensors, oxygen, hydrogen, sulphur and nitrogen concentration sensors capable of operation both in a gas atmosphere and in high-temperature metallic and nonmetallic melts, as well as numerous examples of piezo- and ferroelectric ceramics and transducers, thermo-tribo- and pyroelectric transducers and sensors, ceramic switches with N- and S-shaped current–voltage characteristics, electro-, cathodo- and X-ray luminescent ceramic materials, thermo-, photo- and other emission materials, various types of magnetic ceramics, electromagnetically absorbing and transparent ceramics, and many other rapidly advanced ECC types. In general, it is to be concluded that EEC, as no other type of advanced ceramics, use the most modern ideas of solid-state physics, physical chemistry, electrochemistry and other fundamental science fields, combining them with ideas of materials science and ceramic technology and with them, produce actual products and functional devices.

On the other hand, the evolution of the ECC repeats, albeit in more diverse forms, the evolution of solid-state microelectronics and is aimed at implementation of ideas, utilized initially in cumbersome and complicated-enough devices, in the form of miniature ceramic units with functionally organized components. Advantages of such a course has been convincingly confirmed by integrated circuit technology.

From the above list, it also becomes clear why the cost of various types of electronic-purpose ceramics, whose major part falls at the EEC, accounts for about 85% of the cost of all ceramic materials and why this relation will hardly change in the foreseeable future.

1.2.2 Ferroelectrics, piezoelectrics and high-temperature superconductors
M. D. Glinchuk

An important place in modern technology is held by devices whose functioning relies on utilization of the set of properties of materials of the oxygen-octahedral type. Such materials include many ferro- and piezoelectrics as well as recently discovered high-temperature superconductors. Most piezo- and ferroelectrics are dielectrics featuring a low dielectric loss in combination with the maximum dielectric strength. Special dielectric, electromechanical, optical and other properties of the materials are essential for their applications in electronic devices. The dielectric materials differ advantageously from their semiconductor prototypes by a high radiation resistance, a low dependence of properties on variation of environmental parameters, a long-term stability, etc.

The advance in work on oxygen-octahedral materials started with the discovery in 1944 of ferroelectric properties of $BaTiO_3$. Numerous materials belonging to this family have been discovered by now: $SrTiO_3$, $KTaO_3$, $LiNbO_3$, $NaNbO_3$, $PbTiO_3$, $PbZrO_3$, etc., as well as more complex structures, such as $PbZr_{1-x}Ti_xO_3$ (PZT), $KTa_{1-x}Nb_xO_3$ (KTN), $PbMg_{1/3}Nb_{2/3}O_3$ (PMN), etc.

All these materials exhibit a set of properties, providing for their successful use in various electronic devices and opening up wide prospects for their applications in the piezoelectric facilities, acoustoelectronics, optoelectronics, nonlinear optics, etc. [1, 2, 3]. Thus, piezoelectrics, which convert mechanical energy into electrical energy, and vice versa, find application in piezoelectric transformers, piezoelectric motors, piezofilters, piezoelectric oscillators, ultrasonic radiators and other devices. Pyroelectrics, which convert thermal energy into electrical energy, and vice versa, are proving promising as electric power sources and find application in sensitive radiation detectors (from the SHF to the UV band), including infrared imagers, monitoring system sensors, including medical diagnostics ones, etc. Unique physical properties of ferroelectric materials make them attractive candidates for application in data processing and storage devices [4].

The broad scope of potential applications of superconductors, especially high-temperature ones, is well known [5]: cryopower engineering, cryoelectronics, magnetic-field engineering, including development of high-efficiency electric motors and generators, creation of strong magnetic fields needed for operation of magnetohydrodynamic generators, development of controlled thermonuclear fusion plants. NMR tomographs, magnetic-suspension vehicles, new fast-acting computer facility elements, high-sensitivity instruments to detect the electromagnetic radiation of electric and magnetic fields, etc.

This list of practical applications of oxygen-octahedral-type materials, which is far from complete, is to a considerable extent determined by anomalies of properties, associated with the phase transitions (structural, ferroelectric, electronic) which occur in these substances. Emergence of a structural instability in definite temperature ranges is a characteristic feature of the materials under consideration. Thus, in all high-temperature superconductors known at present – $MeBa_2Cu_3O_{6+\delta}$ (Me denotes all rare earth ions except Ce and Pr), $Bi_2(Sr_{1-x}Ca_x)Cu_2O_{8+y}$, $Tl_2BaCa_2Cu_3O_{10}$, $Ba_{1-x}K_xBiO_3$, etc. – there occur not only the phase transition to a superconducting state, but also structural transitions, including possibly also ferroelectric phase transitions [6]. On the other hand, the existence of superconductivity in materials with a perovskite structure, characteristic for ferroelectrics, such as $BaPb_{1-x}Bi_xO_3$ ($x = 0.25$; $T_c \approx 12$ K), was known earlier [7].

A common feature of all the above-listed materials of the oxygen-octahedral type is a perovskite-like structure, for which the presence of an oxygen octahedron encompassing the ions of one of the metallic sub-lattices is characteristic. Figure 1.91 shows, as an example, the structure of perovskite (general formula, ABO_3), characteristic of many ferroelectrics, and the structure of one of the most studied high-temperature superconductors of type 123. All these materials feature the existence of oxygen vacancies and the associated change in the state of charge of some ions of the metallic sub-lattice. The resulting crystal lattice defects significantly affect the properties and can play the governing part in mechanisms

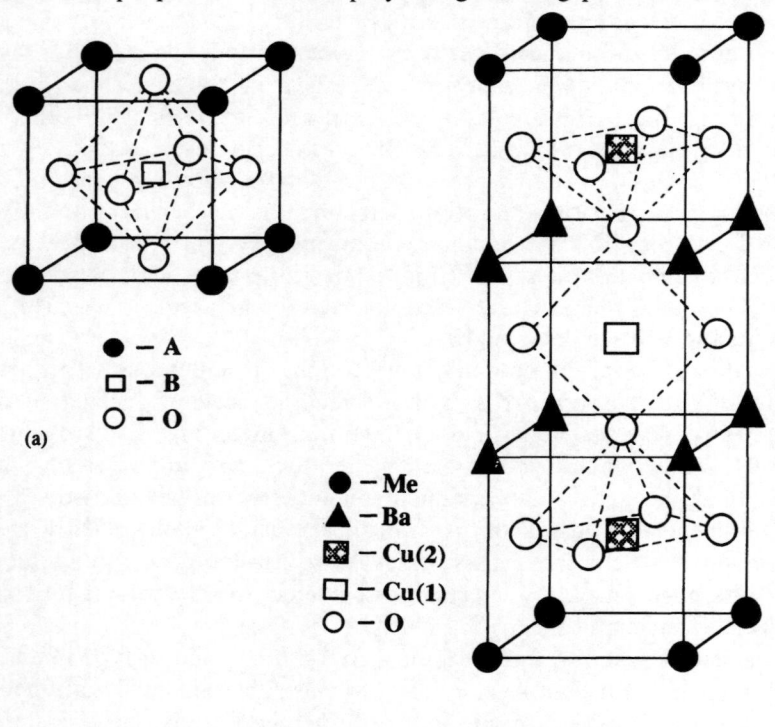

Figure 1.91 Structure of materials of oxygen-octahedral type: (a) perovskite ABO_3; (b) high-temperature superconductor $MeBa_2Cu_3O_7$ (Me is Y, Eu, etc.).

of structural and superconducting phase transitions [8, 9]. Attention is to be given to a high polarizability of oxygen, promoting phase transitions in the systems under consideration. The existence of phase transitions results in anomalously strong impacts on properties not only of external fields, but also of internal fields created by intrinsic defects or impurities. This fact is widely used in technology to create a set of properties useful for applications and to control them. Thus, one of the main materials science trends in developing new materials for surface acoustic wave devices is the attempt to control the sound velocity and its temperature behaviour by a programmed doping of the matrix material. The modification of the matrix (incorporation of a small amount, less than 10%, of impurities and additives) as a production technique is employed in manufacturing the most widely used piezoceramic, PZT.

Furthermore, a metered addition of impurities is used to shift the phase transition temperature to a region convenient for operation of an instrument, to smoothen steep temperature dependences of working parameters of a material, etc. Although the addition of impurities as a production technique has long been known [10], the physical causes of the strong impact of impurities and structural defects on properties of materials which exhibit phase transitions have become clear only in recent years.

1.2.2.1 Ferro- and piezoelectrics of oxygen-octahedral type

The effect of defects on physical properties of ferro- and piezoelectric materials has received great attention in the world scientific publications in recent years (see, e.g., [11]). The impact of defects on the properties is especially strong near the phase transition temperature T_c. Near T_c the structure of a substance becomes exceedingly soft, yielding to any, even weak, actions, both external and internal, produced by defects. Due to this, local defect-induced distortions of the matrix, which in ordinary crystals extend only to the nearest atoms adjacent to a defect, can in crystals exhibiting structural phase transitions cover large regions, whose size (on the order of the correlation radius r_c) greatly exceeds the lattice constant; at $T \to T_c$, $r_c \approx \varepsilon^{1/2} \to \infty$. Thus, the specifics of the impact of defects on properties of crystals with structural phase transitions is primarily associated with the appearance of long-range actions by defects on the crystal matrix. Because of the long range of the actions, even at a low concentration of defects, their impact starts showing up not only as a change in local (i.e., near a defect) properties of a crystal, but also as a substantial change in the properties of the material as a whole.

It was theoretically shown [11, 12] that long-range fields extending over regions of r_c in size are, in ferroelectrics, created by defects of practically all types: point, linear and planar. As a result, contributions of defects to anomalies of various physical properties near T_c turn out to be temperature-dependent, which is not the case for ordinary dielectrics. The effect of electric point dipoles, linearly interacting with the order parameter (electrical polarization of the ferroelectric), turned out to be particularly strong. Such defects determine, in particular, the width of the central peak [13], the outcome of anomalies of dielectric, thermal, and elastic properties of the crystal matrix, in temperature- and impurity concentration-dependent features in the light and sound scattering. Indeed, in the case under consideration the scattering cross-section can be large, as it is proportional to $Nr_c^2 \to \infty$ at $T \to T_c$ (N is the impurity concentration).

The successes achieved in recent years in elucidating the mechanisms of the strong impact of impurities on properties of ferroelectrics have been associated with studying crystals with a metered amount of impurities of a certain type by research techniques providing for the study of not only macroscopic, but also local properties.

As known, one of the most informative techniques for identification of defects and of lattice changes associated with them is the electron paramagnetic resonance (EPR) method [14]. The defect structure of many dielectric materials has been examined in detail by this method (see [15]). In ferroelectrics, however, due to the existence of long-range fields, the appearance of features in spectra, in concentration and temperature dependences of the width and shape of EPR lines could be expected. The experimental observation of such anomalies made it possible, as will be shown below, to ascertain not only characteristics of defects, their concentration and distribution throughout the crystal, but also specific features of the electric and elastic fields created by them. The information was extracted by comparing experimental data with results of a theoretical analysis of expected anomalies in EPR spectra of ferroelectrics. One of the principal issues calling for an experimental substantiation is the proof of existence of long-range fields of defects in ferroelectric materials. To solve this problem, virtual ferroelectrics with a metered amount of defects of a certain type, exerting an especially strong influence on properties, were selected, since impurities in such materials can induce phase transitions, near which practically all, rather than individual, properties are anomalous [16, 17]. It is on this most important aspect of the impact of impurities on properties that attention will be focused below.

As indicated by examination of the world scientific literature, the main experimental results on the influence of impurities have been obtained in the last ten years in investigation of the virtual ferroelectric $KTaO_3$ with impurities of Li, Nb, Na. As discussed in [18], virtual ferroelectrics are crystals that undergo no phase transition right down to 0 K, but whose dielectric constant increases with decreasing temperature. However, rel-

Functional materials

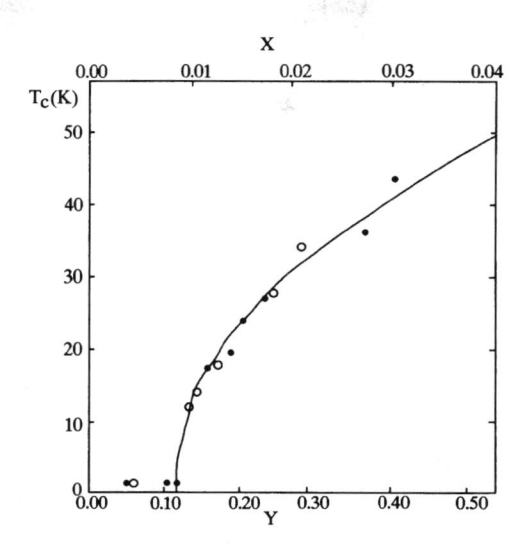

Figure 1.92 Variation of phase transition temperatures with concentration of Nb (X) and Na (Y) in $KTaO_3$.

atively small external actions (addition of impurities or pressure) can induce a ferroelectric phase transition. It turned out [19] that a few per cent Li, Na or Nb, added to $KTaO_3$, render it a real ferroelectric with $T_c \approx 100$ K, the T_c depending nonlinearly on the impurity concentration (Figure 1.92). It is clear that such materials are an example of systems with a controlled phase transition temperature and hence with a controlled temperature range of anomalies of physical properties.

The mechanism of impurity-induced phase transitions was shown by calculations [20] to be associated with the long-range character of fields created by impurity electric dipoles in a ferroelectric. The appearance of such dipoles in $K_{1-x}Li_xTaO_3$ is because a Li ion, substituting 9 K, does not come to the position at a site, but gets shifted to an amount on the order of 1 Å in one of six equivalent directions of the (100) type, i.e., is an off-centre one and can thus be treated as an electric dipole. It was ascertained that it can become reoriented between equivalent positions at a frequency $v = v_0 e^{-u/T}$; $u \approx 1000$ K, $v_0 \approx 2.6 \times 10^{12}$ Hz (see, e.g., [21]). Such reoriented electric dipoles at concentrations over $n > n_{cr}$, ($n_{cr} \geqslant 3-5\%$) lead to a ferroelectric ordering, while at $n < n_{cr}$, a dipole glass phase is possible, as in ordinary $K_{1-x}Li_xCl$-type dielectrics, where long-range fields of lithium dipoles are absent and $\sqrt{P}$ rather than P is the order parameter.

A direct experimental proof of the existence of long-range fields was obtained by the EPR method in $K_{1-x}Li_xTaO_3$, where an axial centre Fe^{3+} was the paramagnetic probe. It turned out that at different

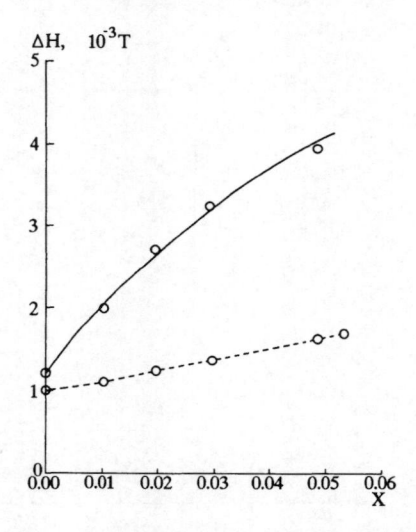

Figure 1.93 Concentration dependences of EPR linewidth at $T = 77$ K for $\theta = 90°$ (broken line) and $\theta = 30°$ in 8 mm band. Solid line: theoretical prediction [23] allowing for contribution of long-range fields.

orientations of the paramagnetic centre axis in the external magnetic field, characterized by the angle θ, the EPR line gets differently widened by the Li impurity (Figure 1.93). An ordinary linear concentration dependence occurs at $\theta = 90°$, when the paramagnetic probe does not interact with electric fields. The nonlinear concentration dependence of the EPR linewidth stems from the fact that lithium dipoles create electric fields which, apart from ordinary $(d/3)$-type terms, include additional contributions of the $(d/rr_c^2) e^{-r/r_c}$ type (d is the dipole moment and r the distance from an impurity to the observation point) [23]. It can be seen that, as the impurity concentration increases, when the mean distance becomes comparable with r_c ($nr_c^3 \geqslant 1$), the fields vary with the distance as $1/r$ rather than as $1/r^3$. At such a field variation law, the EPR line-shape should change from a Lorentzian to a Gaussian one, which was also observed experimentally [22].

The reorientational motion of dipoles results in a 'motor' narrowing of EPR spectra and anomalies of the spin-lattice relaxation. The observation of these features allowed the determination of frequencies of reorientation not only of electric, but also of elastic dipole moments, inherent in off-centre impurities [21, 24, 25]. It was found that while for Li^+ in $KTaO_3$ the frequencies of reorientation of electric and elastic dipole moments are identical, for Nb^{5+}, substituting Ta^{5+} in $KTaO_3$, the frequencies differ substantially, so that double dynamics, a high-frequency ($\simeq 10^{12} s^{-1}$) dielectric and a low-frequency ($\simeq 10^8 s^{-1}$) elastic one, take place. It is to be noted that specific features in the interaction of impurities which are elastic dipoles, i.e., linearly interacting with strain fields, are possible in highly polarizable lattices.

As shown by calculations [26], the strain fields created by such impurities in highly polarizable lattices also contain long-range contributions. As a result, such impurities at high enough concentrations can induce the ferroelectric phase transition in the lattice, and a quadrupolar glass phase is possible at lower concentrations [27].

In the above-discussed examples, impurities isovalently substituted ions of the basic lattice and increased the phase transition temperature. Note that another type of isovalent impurities (e.g., nonreorienting impurities) [12], which lower T_c, exists. Thus, introduction of Sr^{2+} or Ca^{2+} ions, substituting Pb in PZT ceramics, lowers the Curie point by about 10 °C per every introduced atomic per cent, with the result that the permittivity at room temperature increases.

The impact of impurities whose charge differs from that of the ion being substituted with them is known [28] to depend on the sign (positive or negative) of the excess charge brought by the impurity into the lattice. The excess charge is usually compensated through vacancies in cationic or anionic sub-lattices and sometimes through a simultaneous introduction of several additives which compensate one another. From the above it is clear that the state of charge of the impurities and hence also their effect on the properties of ceramics depend on many factors and that their investigation is one of the most important problems in scientific and applied aspects. The possibilities of studying the state of charge of impurities in PZT piezoceramics by the EPR method will be described below.

PZT-based ceramics exhibit excellent piezoelectric, pyroelectric, and electro-optical properties and are therefore widely used in practice [1, 28, 29]. Their properties are greatly dependent on the Ti:Zr ratio and on the metal oxides added (La_2O_3, Nb_2O_5, MnO_2, Fe_2O_3, etc). The temperature and time stability of piezoelectric parameters is particularly strongly affected by MnO_2. The mechanism of this effect depends on the state of charge of Mn and calls for detailed investigation, especially for a simultaneous addition of several oxides.

EPR spectra were measured in two frequency intervals, 9 GHz and 37 GHz, at room temperature. Two series of samples were used: (I) $Pb(Zr_{0.6}Ti_{0.4})O_3$ with additions of MnO_2 and La_2O_3; (II) $Pb(Zr_{0.52}Ti_{0.48})O_3$ with additions of MnO_2 and Nb_2O_5. Figure 1.94 shows, as an example, EPR spectra of samples of series I, containing no La_2O_3, for several molar concentrations x of MnO_2 [30]. As can be seen, right up to $x \simeq 0.05$, manganese is mainly in the state Mn^{4+}. The absence of correlation between x and the integral intensity of the EPR line of Mn^{4+} may result from the presence of Mn^{3+} ions as well as from build-up of the undissolved MnO_2 phase in the intergrain space, the EPR signal from

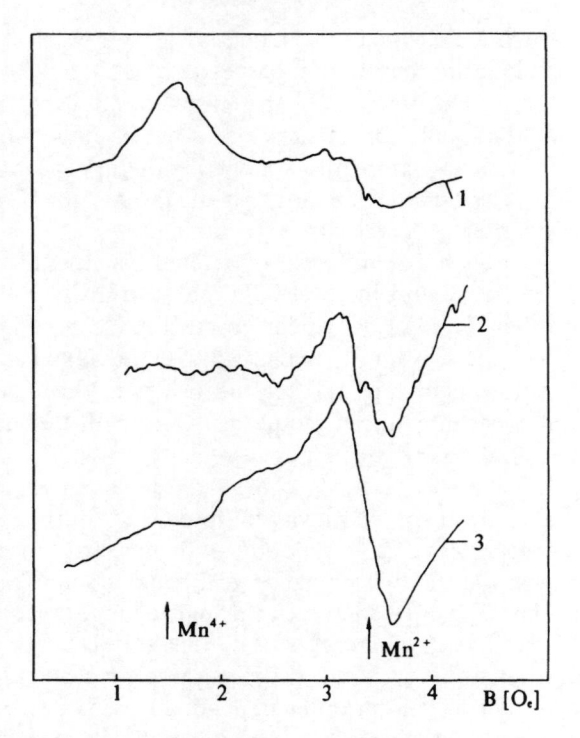

Figure 1.94 EPR spectrum of manganese in $Pb(Zr_{0.6}Ti_{0.4})O_3$ with admixture of MnO_2: (1) 0.006 mole; (2) 0.020 mole; (3) 0.050 mole.

which is not observed. Direct evidence of the predominating manganese distribution at grain boundaries at not too high concentrations was obtained for samples of series II for $x = 0.02$ with a low content y of Nb_2O_5 ($y = 0.007$) by the X-ray microprobe analysis. The Mn concentration at grain boundaries was found [31] to be 30 times that in the bulk of grains. As the Nb concentration increased, the picture changed substantially: Mn diffusion into the bulk of grains occurred, so that for $y > x$ the Mn distribution throughout the sample was practically uniform. Measurements of EPR spectra [31] of samples of series II demonstrated the EPR line of Mn^{4+} to occur only for samples with a low Nb content ($y \leq 0.0175$) and the line of Mn^{2+} to appear as the Nb content increased. Figure 1.95 shows variations of relative intensities of EPR signals of Mn^{4+} and Mn^{2+} with the Nb_2O_5 concentration. The minimum intensity of signals of the two ions in an interval $0.0175 \leq y < 0.0185$ is notable; its occurrence was ascribed to appearance of the maximum amount of Mn^{3+} ions. The Mn^{2+} content rises quickly in the region $0.03 < y < 0.07$. Comparing results of the EPR and X-ray microprobe analysis shows that Mn^{4+}, which requires no charge compensation, does

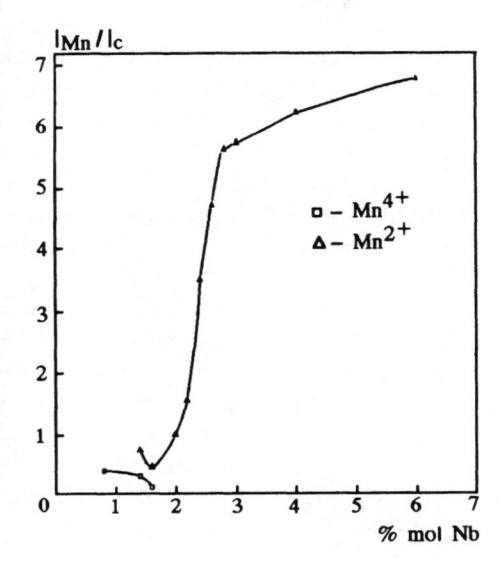

Figure 1.95 Relative intensities of EPR lines of Mn^{2+} and Mn^{4+} (I_c is intensity of EPR line of polycrystalline boron).

not dissolve in grains, but gets arranged mainly at their boundaries. As the concentration of Nb_2O_5, which brings an excess positive charge into the lattice, increases, the need for the charge compensation results in recharging: $Mn^{4+} \to Mn^{3+} \to Mn^{2+}$, accompanied by Mn diffusion into

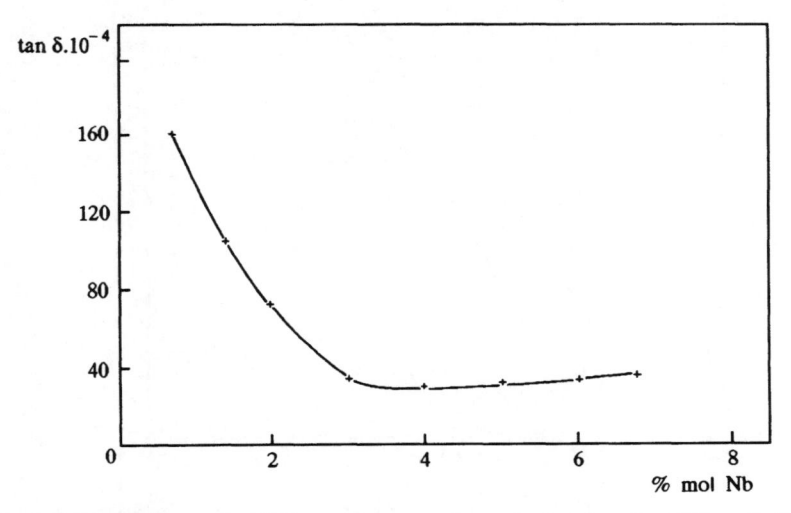

Figure 1.96 Effect of Nb content on loss tangent of solid solutions $PbMn_{0.02}(Zr_{0.52}Ti_{0.48})_{0.98-x}Nb_x$.

grains and growth of grains. At low Nb concentrations, its excess charge can be compensated by Pb vacancies, not changing the state of charge of Mn^{4+}. It is to be pointed out that, as well as the predominant existence of Mn^{4+} in PZT right up to $x \approx 0.05$, this explains the decline in the size of grains in PZT with a low Mn content ($0 \leq x \leq 0.04$), observed in [32].

The obtained results also account for the decline in the dielectric loss in samples of series II with the Nb concentration increasing right up to $y = 0.03$ (Figure 1.96), observed in [31]. From Figure 1.95 it is seen that in the region $0.01 \leq y \leq 0.03$ the Mn^{4+} concentration decreases while that of Mn^{2+} grows rapidly. At the same time, there occurs a growth of grains, and manganese leaves the intergrain space. It follows that the basic mechanism of the dielectric loss is associated with grain boundaries. This conclusion is in a good agreement with the results of [32], where the anomalously high dielectric loss in PZT with addition of Mn was shown to be due to the conductivity at grain boundaries.

A qualitatively similar behaviour was also observed [30] for samples of series I. Figure 1.97 shows the variation of the relative intensity of spectra of Mn^{4+} and Mn^{2+} with the La concentration. Characteristic

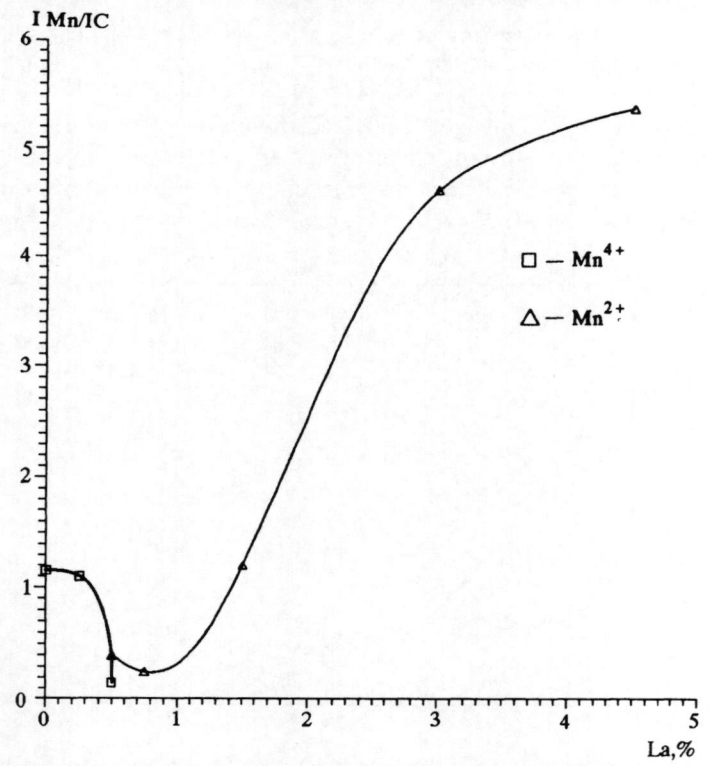

Figure 1.97 Relative intensity of EPR lines of Mn^{2+} and Mn^{4+} as a function of lantanum concentration in $Pb(Zr_{0.6}Ti_{0.4})O_3$.

features here – a minimum at an La concentration $y_0 = 0.0075$, a fast growth of the Mn^{2+} concentration at $y > y_0$, and asymmetry of the curve – evidence that the recharging $Mn^{4+} \to Mn^{3+} \to Mn^{2+}$ with increasing La content occurs in this case as well. It can be believed that this process involves the Mn passage from intergrain interlayers into grains, whose volume increases. The Nb and La concentrations y_0, corresponding to minima of curves in Figures 1.95 and 1.97 may be regarded as some characteristic concentrations, at which the Mn^{4+} content is the minimum, the Mn^{3+} content is the maximum, and a high-rate recharging $Mn^{3+} \to Mn^{2+}$ begins, accompanied by the Mn passage from the intergrain space into grains growing in their volume. As a result, a significant decline in the dielectric loss occurs in regions with $y \geqslant y_0$.

The above-discussed mixed systems $K_{1-x}Li_xTaO_3$, $K_{1-x}Na_xTaO_3$, $KTa_{1-x}Nb_xO_3$, $PbZr_{1-x}Ti_xO_3$ with various additions, where x can be varied over a broad range by varying the degree and character of ordering, may be regarded as materials prepared artificially with the aim of obtaining the properties needed for applications and elucidating the nature of phase transitions.

Of considerable interest for science and practice are also such natural mixed systems as $PbMg_{1/3}Nb_{2/3}O_3$ (PMN) and $PbZn_{1/3}Nb_{2/3}O_3$, belonging to the class of ferroelectrics with diffused phase transitions. Investigations into such strongly disordered (due to a random arrangement of two different ions in B-type sites) systems started long ago [33], but the nature of anomalous properties of these compounds remains unclear so far. At the same time, their characteristics such as a high permittivity and electrostriction over a broad temperature range make many of them attractive for various applications [34, 35].

PMN-type compounds have the structure of perovskite (Figure 1.91a), where Mg^{2+} and Nb^{5+} ions are disorderly disposed at centres of elementary cells. Since direct X-ray studies are difficult because of the X-ray amorphism of the material, no data on the character and parameters of the disorder of these systems have been available until recently. The first such information was obtained recently by the NMR method [36].

NMR spectra of ^{93}Nb nuclei were measured over a temperature range of 170–500 K in PMN single crystals and ceramics. Since the main conclusions in the two cases are identical, while processing of the obtained data for single crystals is simpler, the results of studying the latter will be presented. Figure 1.98 shows NMR spectra recorded for a number of temperatures at a magnetic field orientation along the (001) axis. At $T = 300$ K the spectrum consists of a narrow, 15 kHz wide, intense line and a broad, 50–70 kHz wide, pedestal which practically does not vary with temperature. At the same time the narrow line broadened considerably with decreasing temperature, reaching a width of 30–40 kHz at $T = 180$ K. As the temperature increased, the line narrowed, and a

$\nu = 49$ MHz

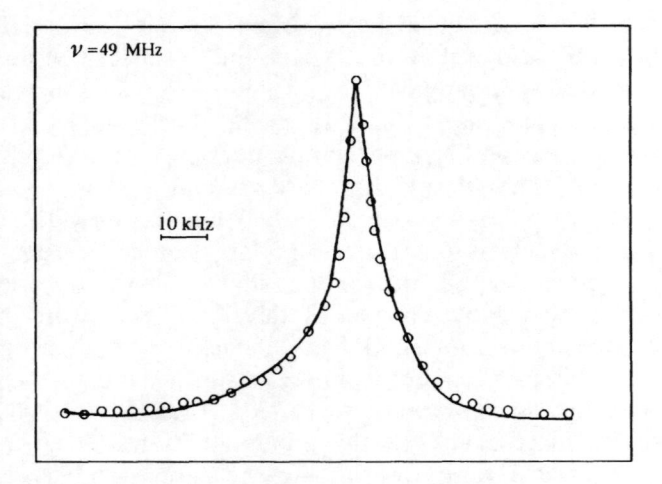

10 kHz

Figure 1.98 NMR spectrum of ^{93}Nb in PbMg$_{1/3}$Nb$_{2/3}$O$_3$ for $T = 450$ K. Solid line is experiment; points are calculation.

narrower line of $\Delta\nu \approx 3$ kHz in width separated out from it at $T \geqslant 400$ K. With increasing spectrometer frequency the NMR spectrum narrowed.

The ^{93}Nb nucleus has a great quadrupole moment, so that the NMR spectrum is highly sensitive to the existence of electric field gradients (EFG) and to their possible changes. As known, with a cubic symmetry, all the EFG components are zero, so that quadrupolar effects in NMR spectra make it possible to extract information not only on macroscopic, but also on local symmetry changes. The symmetry can change both at displacement of ions from their equilibrium positions and at various charge compensation methods. Just this allows the NMR method to be employed to extract information on the lattice disorder degree.

The calculation of EFG components for an ideal structure corresponding to the formula PbMg$_{1/3}$Nb$_{2/3}$O$_3$, where every elementary cell is charge-compensated, as well as of various degrees of departure from the ideality, associated with randomness of the Mg and Nb disposition, was conducted in [36]. The observed spectrum was described on the assumption that the width and shape of lines are determined by quadrupole effects and magnetic dipole–dipole interaction. It was ascertained that the appearance at $T \geqslant 400$ K of a narrow ($\Delta\nu \simeq 3$ kHz) line whose intensity grows with increasing temperature is due to the existence of ideal structure regions, whose volume, however, is as little as 10–20 % at $T \simeq 400$ K. It seems that the polar phase originates just in these regions. In this phase the Nb ions are shifted in (111)-type directions with respect to the oxygen octahedron, which results in the change of the width of the line with temperature.

The coincidence of NMR spectra for ceramics and monocrystals evidences small sizes of such polar regions. As shown by the calculation, the observed NMR spectrum can be described only on the assumption of a substantial spread of the direction and magnitude of polarization in these small regions, which allows an assertion of a dipole glass-type state in PMN.

1.2.2.2 High-temperature superconductors with a 123 structure

The role of variable-valence ions in the emergence of high-temperature superconductivity is today generally accepted [37]. The most studied to date is the ceramic $YBa_2Cu_3O_{6+\delta}$ (123 phase of YBCO), where, as δ changes from 0 to 1, copper ions can be in three charge states: Cu^+, Cu^{2+}, and Cu^{3+}. This in turn results in a change of the structure and magnetic properties and in the appearance of superconductivity. Compositions of $0.7 \leq \delta \leq 1$, corresponding to the superconducting state with a transition temperature over 77 K, are characterized by the appearance of an oxygen hole, i.e., O^- [38]. Radiospectroscopy methods turned out to be very informative in studying all these different-charge ions and their locations in the YBCO lattice (Figure 1.91b) (see, e.g., [39].)

Recent studies by the EPR method extracted unique information about the coexistence of superconducting and dielectric regions in YBCO and made it possible to suggest techniques for the inspection and diagnostics of superconducting ceramics. We will briefly dwell on the results below.

The investigation of high-temperature superconducting metal oxide ceramics of the YBCO composition by the EPR method is based on studying the low-field nonresonant microwave absorption (the so-called 'low-field line') and the EPR line proper of bivalent copper Cu^{2+} in the g-factor region of $g \approx 2$. The diverse information obtained in studying the low-field line concerns primarily the magnetic properties of the HTSC ceramics, in particular, the character of penetration of magnetic vortices into the superconducting material, features of the metal oxide ceramics as the Josephson medium, magnitude of the critical magnetic field, etc. Furthermore, the appearance of the low-field absorption at the superconducting transition temperature T_c makes it possible to utilize the low-field line as a highly sensitive means for detection of the superconducting transition and determination of the T_c value.

On the other hand, numerous experimental results of studying the EPR line of Cu^{2+} ions in the YBCO ceramics are contradictory enough. This is mainly due both to the absence until now of an exact understanding of the nature of the observed EPR line of Cu^{2+} from the standpoint of the paramagnetic centre model and to the absence of the answer to the question of whether ions responsible for the EPR signal belong to the superconducting 123 phase proper of the YBCO ceramics or the observed signal is related to copper ions that are in nonsuperconducting impurity phases included in the composition of the HTSC ceramic under investigation. The answer to this question is exceedingly important, since in the former case, studying the EPR line of Cu^{2+} makes it possible to extract important physical information on properties of the 123 phase, such as the character of the local environment of copper ions, genesis of their state of charge, degree of localization of electrons participating in the conduction, etc., while in the latter case, the EPR method can serve for checking the phase composition of HTSC YBCO ceramics.

Concurrent measurements of the EPR spectra and magnetic susceptibility for samples with different values of the oxygen index δ and for Y_2BaCuO_5 (green phase), which is often present as an impurity in 123 ceramics, turned out to be useful to elucidate these issues [40].

Figure 1.99 shows the temperature dependences of magnetic susceptibility of $YBa_2Cu_3O_{6+\delta}$ and Y_2BaCuO_5 samples over a temperature range of 4.2–300 K ($\delta = 0.21; 0.54; 0.88$ and 0.93).

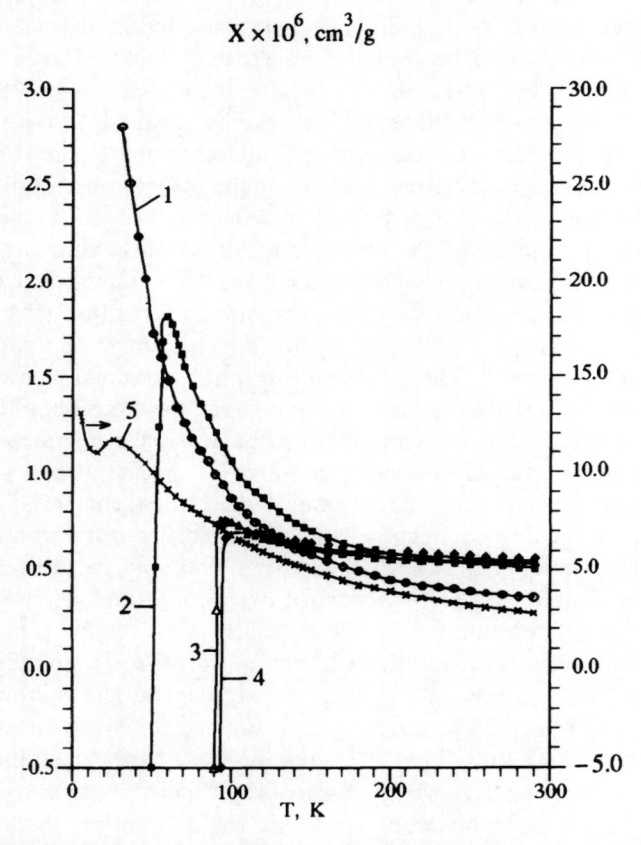

Figure 1.99 Temperature dependences of static magnetic susceptibility in samples of $YBa_2Cu_3O_{6+\delta}$; δ is (1) 0.21; (2) 0.54; (3) 0.88; and (4) 0.93; and (5) Y_2BaCuO_5.

As is known, semiconductor $YBa_2Cu_3O_{6+\delta}$ with the oxygen index values $0 \leq \delta \leq 0.5$ is an antiferromagnetic with a Néel temperature depending on the oxygen content ($T_N \approx 420$ K for $\delta = 0$). As seen from Figure 1.99, a considerable increase in the magnetic susceptibility of samples 1 and 2 with decreasing temperature presumes the existence in them of localized magnetic moments, contributing to the paramagnetic susceptibility of the substance. For samples 3 and 4 the contribution is much less. An abrupt transition of samples 2–4 to the diamagnetic state is associated with the beginning of the superconducting transition, so that T_c values (for the beginning of the transition) can be determined as 52 K, 90 K and 96 K respectively. For sample 1 the superconducting transition was not observed right down to $T \approx 4.2$ K. The presence of the 'low-field line', detected by the EPR spectrometer for samples 3 and 4 at $T \approx 77$ K, confirms the existence in them of the superconducting state at this temperature.

The temperature dependence of the magnetic susceptibility of Y_2BaCuO_5, exhibits a maximum at $T_N \approx 25$ K, corresponding to an antiferromagnetic ordering. A quantitative analysis of temperature dependences of the magnetic susceptibility demonstrated that over a broad temperature range $T > T_N$ the temperature dependence of magnetic susceptibility of Y_2BaCuO_5 obeys the Curie–Weiss law, $\chi \approx 1/(T - \theta)$, with a negative value of the Weiss temperature, $\theta \approx -35$ K. The obtained value is in good accord with the value $\theta = -38$ K, obtained earlier. As regards $YBa_2Cu_3O_{6+\delta}$ samples, the paramagnetic contribution to the magnetic susceptibility for samples 1 and 2 is dominating at $T < 150$ K and can be described by the Curie–Weiss law with $\theta \approx 0$ (Curie law). For superconducting samples 3 and 4 with an orthorhombic structure, this contribution is small and determines a weak temperature dependence of the magnetic susceptibility of the samples at $T < 200$ K. At high temperatures, $T > 200$ K, the principal contribution to χ appears to be made by the Pauli susceptibility and the susceptibility associated with antiferromagnetic fluctuations. Thus, different behaviours of paramagnetic susceptibilities of localized magnetic moments of Cu in $YBa_2Cu_3O_{6+\delta}$ and Y_2BaCuO_5 samples ($\theta \approx 0$ and $\theta \approx -35$ K respectively) indicate that the paramagnetic contribution to the magnetic susceptibility of $YBa_2Cu_3O_{6+\delta}$ is not associated with the 'green phase' impurity.

Consider now the results of EPR studies. An EPR signal typical for powder samples was observed for all the investigated $YBa_2Cu_3O_{6+\delta}$ samples (Figure 1.100). It is to be noted that shapes of EPR lines for a compact ceramic and a ceramic reduced to powder are different and depend on the degree of reduction. This can be due either to different angular distributions of granules in a compact and a powdery sample or to a change in the ratio between the skin layer depth and the particle size. The latter can result in different admixtures of the signal of dispersion to

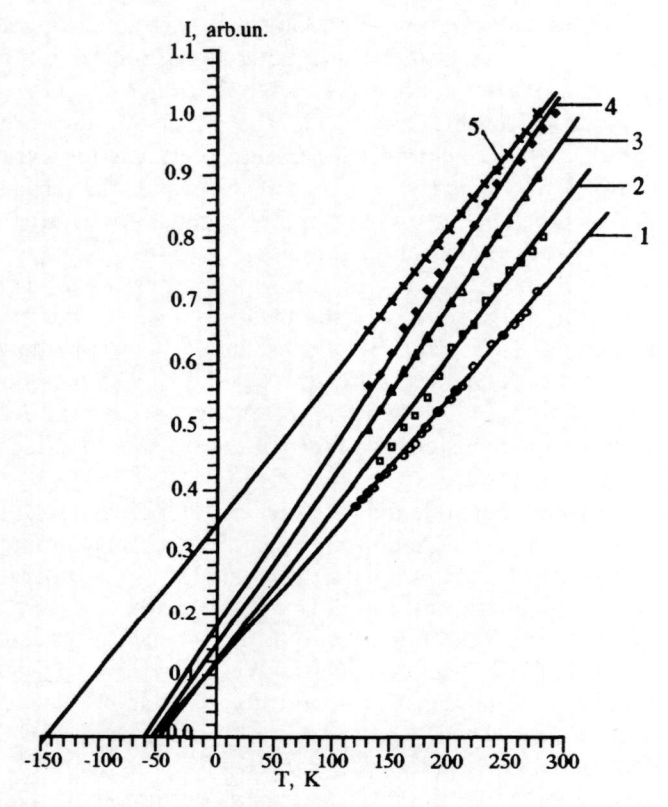

Figure 1.100 Temperature dependences of inverse intensity of EPR signal of Cu^{2+} in $YBa_2Cu_3O_{6+\delta}$ and Y_2BaCuO_5. Designations are the same as in Figure 1.99.

the EPR line being observed. The spectra shown in Figure 1.100 were obtained for powdery $YBa_2Cu_3O_{6+\delta}$ samples, comminuted to such a degree that a further comminution brought about no substantial change in the EPR line-shape being observed.

As the oxygen index increased, the signal intensity for $YBa_2Cu_3O_{6+\delta}$ samples declined with the linewidth practically unchanged. The width of the line (as well as its shape) did not change within the experiment error over the whole temperature range studied, 120 K < T < 300 K. The temperature dependence of peak-to-peak intensity of the EPR line, proportional to the inverse paramagnetic susceptibility, for $YBa_2Cu_3O_{6+\delta}$ samples is shown in Figure 1.101. As can be seen, these temperature dependences both for nonsuperconducting (tetraphase) and for superconducting orthorhombic samples are well described by the Curie–Weiss law with one and the same value of $\theta = -(55 + 5)$ K. The same figure shows also the corresponding temperature dependence of the inverse

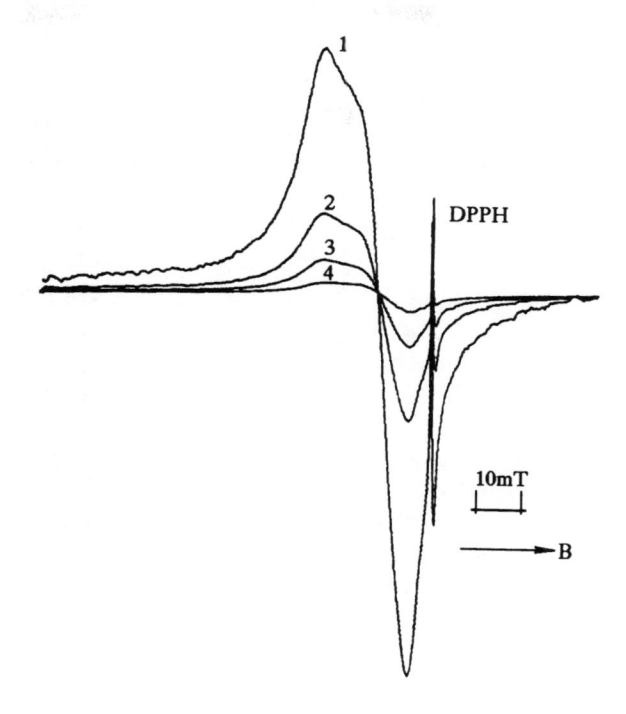

Figure 1.101 EPR spectra of Cu^{2+} in $YBa_2Cu_3O_{6+\delta}$ and Y_2BaCuO_5 at $T = 300$ K. Peak-to-peak intensity ratio: 1:3:7:20. (1)–(4) are as Figure 1.99.

intensity of the EPR signal for Y_2BaCuO_5, which obeys the Curie–Weiss law at a Weiss temperature $\theta \approx -140$ K.

Thus, temperature dependences of paramagnetic susceptibility of $YBa_2Cu_3O_{6+\delta}$ and Y_2BaCuO_5, measured by the EPR method, as in the case of magnetic measurements, are characterized by different Weiss temperatures. This indicates that the EPR signal of Cu^{2+}, observed for $YBa_2Cu_3O_{6+\delta}$, stems from the presence of localized magnetic moments in the 123 phase and is not related to the Y_2BaCuO_5 impurity phase. The accuracy of determining the green phase by the EPR method is within fractions of a per cent and hence greatly exceeds the sensitivity of X-ray methods.

Let us discuss the above-presented results. Magnetic properties of the $YBa_2Cu_3O_{6+\delta}$ system are determined by the magnetism of Cu ions which occupy in an elementary cell two nonequivalent positions: Cu (1) and Cu (2) (Figure 1.91b). The antiferromagnetism of semiconductor $YBa_2Cu_3O_{6+\delta}$ at $0 \leq \delta \leq 0.5$ is here primarily associated with an antiferromagnetic interaction of Cu (2) ions in two-dimensional layers $(CuO_2)_\infty$ through an indirect exchange via 2p orbitals of oxygen O_2. The behaviour of magnetic susceptibility near the Néel temperature is described within the scope of the model of the two-dimensional Geisenberg antiferromagnetic, which confirms the two-dimensional character of the magnetic system of Cu ions. With the composition $YBa_2Cu_3O_6$ the Cu (1) ions, having a double oxygen surrounding, are in a univalent state, Cu^+, and are not paramagnetic. Thus, the composition $YBa_2Cu_3O_6$ should not exhibit paramagnetic properties. As the oxygen content in $YBa_2Cu_3O_{6+\delta}$ increases, some Cu (1) ions change over to a bivalent state, Cu^{2+}, which gives rise to localized magnetic moments with spin $S = 1/2$ in positions of copper Cu (1). A weak magnetic interaction within a strongly diluted system of Cu (1) and between Cu (1)–Cu (2) ions (of a ferromagnetic character) predetermines the paramagnetism of Cu (1) ions. Because of this, the manifestation of a paramagnetic contribution in the static magnetic susceptibility, observed in a $YBa_2Cu_3O_{6.21}$ sample, as well as the presence of the EPR signal of Cu^{2+} ions in the sample does not appear to be surprising.

The evaluation of the number of paramagnetic Cu^{2+} ions from the integral intensity of the EPR signal has led to a conclusion that 5% of Cu (1) ions are in a bivalent state. Evaluations indicated the contribution of these ions to the paramagnetic susceptibility to be less than the observed one. This allowed an inference of the presence of an admixture of magnetic Cu^{3+} ions in the samples under consideration.

The $YBa_2Cu_3O_{6+\delta}$ transition to a metallic state at $\delta > 0.5$ is accompanied by destruction of the antiferromagnetic long-range order. At the same time, the EPR signal of Cu^{2+} in samples 3 and 4 drops steeply from $T_c \approx 90$ K (Figure 1.101) (the ratio of signals for the samples is of 1:20), while the paramagnetic contribution to the static magnetic susceptibility for the samples did not show up practically. The cause of the disappearance of localized magnetic moments of Cu in the orthophase of the HTSC YBCO ceramics is unclear so far. A model of the Andersson lattice with a high Kondo temperature, an exceedingly strong spin-lattice relaxation (10^{-13} s), etc., are being discussed. The observed weak EPR signal of Cu^{2+} in superconducting samples 3 and 4 also appears not to be connected with the orthorhombic phase. The invariability of the shape and spectroscopic characteristics of the EPR line for all the samples having been studied (1–4) and, first of all, the identity of temperature dependences of the intensities of EPR signals for these samples make it reasonable to conclude that the EPR signal observed for orthorhombic samples is due to the presence in them of regions of the tetraphase with a lower oxygen content. The volume of the regions declines and, accordingly, the temperature of transition to the superconducting state rises with increasing oxygen index.

1.2.2.3 Some applications of ferro- and piezoelectrics. Filtration elements of electronic microwave devices

The advance in development of ferro- and piezoelectric materials with a required combination of electrophysical and mechanical properties has created real prerequisites for developing a new generation of filtration elements for microelectronic microwave facilities, having a qualitatively new level of parameters and mass-and-size characteristics. Thus, improvement of the thermal stability, mechanical and dielectric quality factors of ceramic structures as well as achievements in the synthesis of materials with a considerably diffused phase transition (attained by incorporation of impurities and additives, as discussed in section 1.2.2.2) along with increasing their permittivity made it possible to attain the governing parameters and miniaturization degree for the basic series of microwave dielectronics, close to the limiting ones [41].

In particular, optimization of requirements for piezoceramic materials for piezotransformer elements (PTE) of filters, development on their basis of PTE designs meeting the requirements for multifrequency and matchability with other circuit elements, elaboration of methods for constructing wide-band filters based on PTE and dielectric resonators (DR), as well as development of piezoelectric actuators of precision displacements for controlling the DR position made it possible to realize two classes of microwave bandpass filters.

1. Tunable DR-based filters of the centimetre-wave band with a tuning range of 10–15% without worsening of the quality factor and with a loss in the pass band of 0.5 to 1.0 dB [42].
2. Small-size filters which effect selection or delay of a converted microwave signal at a fixed frequency over a range from 0.001 to 10 MHz with a pass (delay) band width of 0.01–10% and attenuation beyond the pass band of 40–60 dB, where the formation of characteristics and suppression of off-frequency resonances are attained also by ingenious design and circuit features [43]. Based on the same ferroceramic materials, monolithic low-pass filters (LPF) were developed, which allow plotting and integration with elements of microelectronic designs of microwave devices, with an attenuation of 20–30 dB in a filtration band from 0 to tens of megahertz.

Another line in the advance in the low-pass filtration involved the development of a novel type of devices [44] with an ultrawide filtration band, of 0.02 to 100 GHz, and a higher factor of attenuation of an alternating electromagnetic field (55–60 dB in the centimetre-wave band). The expansion of functional capabilities and optimization of characteristics of the proposed coaxial LPF are attained through the use of a design whose basic elements is a ferroelectric ceramic ($\varepsilon = 8000\text{--}10\,000$) capacitance and an active layer of a polymeric current-conducting microwave-absorbing mass, contacting with it. The basic design of this LPF has served as an origin for development of a technologically unified series of filtration elements of various versions, including a planar version and a design with a 100 dB attenuation level.

The commercial implementation of the two latter results made it possible to solve a complex design-circuit engineering problem, the electric sealing of microwave systems, through filtration of the alternating component in power supply circuits of individual devices and in complex radioelectronic microwave equipment as a whole.

(a) *Piezoelectric motors, positioners (actuators) and piezotransformers*
The basic material employed for manufacture of these devices is the PZT piezoceramics with various additions and admixtures [10] providing for an optimum set of useful properties and for their stability.

Piezoelectric motors and actuators are representatives of a wide class of devices whose operation relies on utilization of the inverse piezoeffect.

Piezoelectric motors have in a short time passed from invention to practical use as drives for various purposes [45–47]. In particular, piezomotors have been developed with an efficiency over 80%, a torque of up to 10 Nm, and a rotation speed from one to 2–3 thousands of revolutions per minute.

Micropositioners provide a unique possibility of attaining ultrasmall (from ångströms to a few millimeters) electrically controlled displacements, which is essential for applications in functional electronic devices. They make it possible to construct on their basis miniature relays with a component density of several elements per cm^2, printers with a response time less than 1 μs, gates with a force of up to 2 mN at a travel up to 0.1 mm and with a response time of 19 μs [48].

The advent of highly effective PZT-system piezoceramic materials spurred the efforts of development on the basis of piezotransformers (PT) for secondary power supplies (SPS). The use of PT reduces mass and size, and upgrades the efficiency of SPS and also, which is particularly advantageous, makes it possible to eliminate wound elements from them, i.e., to arrange their manufacture by a common microelectronic technology.

Two trends in development of piezotransformer-based SPS have now been advanced: high-voltage supplies based on so-called voltage piezotransformers and low-voltage ones based on current piezotransformers [49–51]. High-voltage power supplies of several types are being manufactured in the former Soviet Union and efforts on starting commercial production of low-voltage mains-power SPS are under way [51]. The power of a single-PT supply is limited to 3–5 W. Requirements for piezoceramic materials for PT stem primarily from their two features: the need for operation at high mechanical stresses and for transmission of a maximum power to the load. It follows that materials for PT should offer minimum electric and mechanical losses at electric field strengths of up to hundreds of volts per millimetre and frequencies of hundreds of kilohertz; the product of squared electromechanical coupling factor by the mechanical quality factor should be great enough and weakly dependent on temperature. Existing piezoceramic materials do not fully meet the requirements, but PT with a specific power more than 10 W/cm^3 and an efficiency over 90% have nevertheless now been developed.

High-voltage power units based on piezoceramic transformers can be used for the supply to anodes and focusing electrodes of cathode-ray tubes and other high-voltage devices.

Piezotransformers used in power units offer an operational dependability and advantages of a small mass and power saving over ordinary electromagnetic transformers.

(b) *Controllable actuators from electrostriction ferroelectric ceramics*
Ferroelectric ceramics with a diffused phase transition offer an optimal set of characteristics for development of controllable actuators and micromanipulators, capable of providing a displacement range up to several tens of micrometres with a resolution on the order of a few nanometres. Such micromanipulators and actuators are needed for joining optical fibres in fibre-optic communication lines; for elements for an optical correction of atmospheric distortion in astronomic telescopes and of a temperature drift of in-laser cavities; for moving the needle of a scanning tunnel microscope and of sets for nanolithography; for precision mechanical engineering; for tuning units of microwave waveguides and antenna array radiators; and for other systems.

Electrostriction actuators offer all the advantages of the above-discussed piezoceramic ones, while differing beneficially from the latter by a small electromechanical hysteresis (or even its absence), a zero residual deformation, and a high stability of properties with time.

A sole parameter in which electrostriction actuators are inferior to piezoceramic ones is the temperature stability. The optimum for their use is the temperature range near, and somewhat higher than, the mean Curie point of the diffused phase transition.

Developed multilayered electrostriction control actuators ensure a displacement from 8 to 60 μm at a high stability with time, great forces being generated, and a small hysteresis [52].

(c) *Prospects for the use of HTSC*
To conclude, it must be noted that prospects for application of high-temperature superconductors are intensely discussed in world literature (see, e.g., [5]).

The discovery of HTSC materials broadened the possibilities of application of essentially simple devices, whose cost was previously dominated by the cryogenic facilities needed for them. Thus, potential applications of SQUIDS, stroboscopic converters, voltage standards, simple analogue-to-digital converters extend drastically and new applications, such as a superconducting screening of electromagnetic fields or modulation of signals, become cost-effective.

Prototypes of all the devices have been made and tested, so that their extensive use in electronic devices could be expected in the not too distant future. As regards complex systems, towards which modern superconducting electronics is tending, in particular LSI and VLSI, the cost-effectiveness and prospects of using HTSC materials there are still being discussed. Possibly, it will turn out that such circuits are advantageously based on high-temperature superconductors, but operated at temperatures intermediate between the nitrogen and heluim ones (e.g., 20–30 K), at which the cost of cryocooling is already much less than at 4.2 K. A more precise definition of the prospects calls for a large scope of both fundamental and applied studies.

REFERENCES TO SECTION 1.1.1: THEORETICAL FUNDAMENTALS

1. Samsonov G. V. and Vinitsky I. M. (1976) *Refractory Compounds*, Metallurgija, Moscow.
2. Samsonov G. V., Upadhaja G. S. and Neshpor V. S. (1974) *Physical Metallurgy of Carbides*, Naukova Dumka, Kiev.
3. Andrievsky R. A. and Umansky Y. S. (1977). *Interstitial Phases*, Nauka, Moscow.
4. Andrievsky R. A., Lanin A. G. and Rymashevsky G. A. (1974) *Strength of Refractory Compounds*, Metallurgija, Moscow.
5. Grigorovich V. K. and Sheftel E. N. (1980) *Precipitation Hardening of Refractory Metals*, Nauka, Moscow.
6. Trefilov V. I., Milman Y. V. and Firstov S. A. (1975) *Physical Fundamentals of Strength of Refractory Metals*, Naukova Dumka, Kiev.
7. Trefilov V. I., Milman Y. V. and Gridneva I. V. (1984) Mechanical properties of covalent crystals. *Izv. Akad. Nauk SSSR, Neorg. Mater.*, **20** (6), 958–67.
8. Trefilov V. I., Milman Y. V. and Gridneva I. V. (1984) Characteristic deformation temperature of crystalline materials. *Crystal Res. Technol.*, **19** (3), 413–21.
9. Trefilov V. I., Milman Y. V. and Kazo I. F. (1984) On the temperature dependence of flow stress of crystalline materials. *DAN SSSR*, **276** (6), 1399–401.
10. Grigoriev O. N., Milman Y. V. and Trefilov V. I. (1978) Mechanism of plastic deformation and parameters of dislocation movement in diamond and boron nitride, in *Elementary Processes of Plastic Deformation*, Maukova Dumka, Kiev, pp. 144–65.
11. Samsonov G. V., Kuchma A. and Timofeeva I. I. (1970) Origin of homogeneity regions in transition metal compounds with carbon and nitrogen. *Poroshkovaja Metallurgija*, **9**, 69–74.
12. Samsonov G. V., Prjadko I. F. and Prjadko L. F. (1971) *Configuration Model of Substance*, Naukova Dumka, Kiev.
13. Khachaturjan A. G. (1974) *Phase Transformation Theory and Structure of Solid Solution*, Nauka, Moscow.

References

14. Belov N. V. (1947) *Structure of Ionic Crystals and Metallic Phases*, AN SSSR Press, Moscow.
15. Piljankevich A. N., Britun V. F., Olejnik G. S. and Kuzenkova M. A. (1991) On the effect of high pressure on the structure of polycrystalline AlN. *Poroshkovaja Metallurgija*, **3**, 38–44.
16. Russel G. J., Fellows A. T., Oktik S., Ture E. and Woods J. (1982) Mechanically induced phase transformations in CdS, CdSe and ZnS. *J. Mater. Sci. Lett.*, **1** (4), 176–8.
17. Kelly A. (1973) *Strong Solids*. Clarendon Press, Oxford.
18. Swain M. V. (1985) Inelastic deformation of Mg–PSZ and its significance for strength–toughness relationships of zirconia toughened ceramics. *Acta Met.*, **33** (11), 2083–91.
19. Vitek V. and Kroups F. (1966) Dislocation theory of sliding geometry and temperature dependence of strain–stress in b.c.c. metals. *Phys. Stat. Solidi*, **18**, 703–11.
20. Conrad G. (1963) Yielding and plastic flow of b.c.c. metals at low temperature, in: *The Relation Between the Structure and Mechanical Properties of Metals*. Proceedings of the Conference held at the National Physics Laboratory, Teddington, Middlesex, Jan. 1963, HMSO, London, pp. 225–54.
21. Masters B. and Christian I. (1963) Experimental evidence of Peierls–Nabarro force existence in niobium, vanadium, tantalum and iron. *Ibid.*, pp. 287–93.
22. Alefeld G., Chambers P. and Firle T. (1965) Peierls barrier influence on stress–strain dependence in presence of dislocations. *Phys. Rev. A*, **140**, 1771–83.
23. Borisenko V. A., Krashchenko V. P., Moiseev V. F. and Trefilov V. I. (1977) Thermoactivated plastic deformation in precipitation-hardened Mo alloys. *Ukr. Fiz. Zhurnal*, **22** (2), 43–8.
24. Puirier J. -P. (1985) *Creep of Crystals*, Cambridge University Press.
25. Startsev V. I., Il'ichev V. Y. and Pustovalov V. V. *Plasticity and Strength of Metals and Alloys at Low Temperatures*, Metallurgija, Moscow.
26. Krasovskij A. Y. (1980) *Brittleness of Metals at Low Temperatures*, Naukova Dumka, Kiev.
27. Fen E. K., Borisenko V. A., Koval'chenko M. S. *et al.* (1975) Temperature dependence of hardness of hot pressed transition metal oxides and metal-oxide compositions. *Gorjachee Pressovanie*, **2**, 158–64.
28. Yaroshevich V. D. and Rivkina D. G. (1970) On thermoactivated character of plastic deformation of metals. *Fiz. Tverd. Tela*, **12** (2), 68–77.
29. Yaroshevich V. D. (1971) On mechanism of thermoactivated plastic deformation of metals at low temperature. *Fizika Metallov i Metallovedenie*, **31** (4), 856–65.
30. Guyot P. and Dorn J. E. (1967) Critical review of Peierls mechanism of deformation. *Canad. Journ. Phys.*, **45**, 983–1004.
31. Ivens A. and Rollings P. (1969) Thermoactivated deformation of crystalline materials. *Phys. Stat. Solidi*, **34**, 9–26.
32. Engelke H. (1970) Theories of thermoactivated processes and their application to dislocation movement in crystals, in *Fundamental Aspects of Dislocation Theory*, (eds Simmons J. A., de Wit R. and Bullough R.) Nat. Bur. Stand. Spec. Publ., **2** (317), 1137–74.
33. Vitek V. (1966) Thermoactivated dislocations movement in b.c.c. metals. *Phys. Stat. Solidi*, **18**, 687–96.
34. Borisenko V. A. (1975) General regularities in temperature dependence of mechanical properties of heat-resistant metals. I. Temperature dependence of Mo strength. *Problemy Prochnosti*, **8**, 58–63.

35. Borisenko V. A. (1975) General regularities in temperature dependence of mechanical properties of heat-resistant metals. II. Temperature dependence of W strength. *Problemy Prochnosti*, **9**, 23–31.
36. Krasovskij A. Y. (1969) On temperature-rate dependence of the yield stress, in *Thermal Stress Resistance of Materials and Structural Elements*, vol. 5, Naukova Dumka, Kiev, pp. 143–50.
37. Basinski Z. S. (1957) Activation energy for creep of aluminium. *Acta Met.*, **5** (11), 684–6.
38. Conrad H. and Wiedersich H. (1960) Activation energy for deformation of metals at low temperatures. *Acta Met.*, **8** (1), 684–92.
39. Conrad H. (1958) An investigation of the rate controlling mechanism for plastic flow of copper crystals at 90 and 170 K. *Acta Met.*, **6** (5), 339–50.
40. Conrad H. (1961) On the mechanism of yielding and flow in iron. *J. Iron and Steel Inst.*, **198** (4), 364–75.
41. Conrad H. and Hayes W. (1963) Thermally activated deformation of the bcc metals at low temperatures. *Trans, ASM*, **56**, 249–62.
42. Conrad H. (1964) Thermally activated deformation of metals. *J. Metals*, **16** (7), 582–8.
43. Seeger A. and Shiller P. (1969) Dislocation kinks and their effect on intrinsic friction in crystals, in *Physical Acoustic*, (ed. W. P. Mason) vol. 3, Part A, Academic Press, New York and London, 1966. Russian translation, Mir, Moscow, pp. 428–562.
44. Petukhov B. V. and Pokrovskij V. L. (1972) Quantum and classical dislocations movement in Peierls relief. *Zhurn. Eksp. Teor. Fiz.*, **63** (8), 634–47.
45. Seeger A. (1981) The temperature and strain-rate dependence of the flow stress of body-centered cubic metals: A theory based on kink–kink interactions. *Z. Metallkunde*, **72** (6), 369–80.
46. Milman Y. V. (1985) The structural aspects of warm and cold plastic deformation of crystalline materials. *Fizika Metallov i Metallovedenie*, **6** (3), 2–8.
47. Gridneva I. V., Milman Y. V., Rymashevskij G. A. *et al.* (1976). On the effect of temperature on strength characteristics of zirconium carbide. *Poroshkovaja Metallurgija*, **8**, 73–9.
48. Grigoriev O. N., Trefilov V. I. and Shatokhin A. M. (1983) Temperature effect on brittle material failure under concentrated loading. *Poroshkovaja Metallurgija*, **12**, 75–81.
49. Evans A. G. and Davidge R. (1969) The strength and fracture of stoichiometric polycrystalline UO_2. *J. Nucl. Mater.*, **33** 249–54.
50. Gridneva I. V. Milman Y. V., Sinel'nikova V. S. *et al.* (1982) Temperature effect on mechanism of fracture and mechanical properties of single-crystalline NbC. *Metallofizika*, **4** (5) 74–9.
51. Aptekman A. A., Bakun O. V., Grigoriev O. N. *et al.* (1986) Diamond deformation and fracture regularities in bend tests in temperature range of 300–2100 K. *Dokl. Akad. Nauk SSSR*, **290** (4), 845–8.
52. Evans T. and Wild R. K. (1965) Plastic bending of diamond plates. *Philos. Mag.*, **12** (17), 479–89.
53. Evans T. and Sykes I. (1974) Indentation hardness of two types of diamond in the temperature range 1500 °C to 1850 °C. *Philos. Mag.*, **29** (1), 135–47.
54. Gnesin G. G., Gridneva I. V., Dyban Y. P. *et al.* (1978). Investigation of the temperature effect on character of self-bonded silicon carbide fracture. *Poroshkovaja Metallurgija*, **3** 76–80.
55. Bakun O. V., Grigoriev O. N. and Yaroshenko V. P. (1987) Temperature dependence of hardness and strength of sialon–TiN system materials. *Poroshkovaja Metallurgija*, **6**, 71–5.

56. Galanov B. A. and Grigoriev O. N. (1990) Fracture of elastically deformed solid heterophase materials with periodic microstructure and their fracture criteria. Preprint No. 9, IPM AN Ukr. SSR, Kiev.
57. Sanchez-Palencia E. (1980) *Non-Homogenous Media and Vibration Theory*. Lecture Notes in Physics, Springer-Verlag, Berlin.
58. Trefilov V. I., Milman Y. V., Ivashchenko R. K. *et al.* (1983) *Structure, Texture, and Mechanical Properties of Molybdenum Alloys in Deformed State*, Naukova Dumka, Kiev.
59. Kurdjumova G. G., Milman Y. V. and Trefilov V. I. (1979) On classification of micromechanisms of fracture. *Metallofizika*, **1** (2), 56–62.
60. Ioffe A. E. (1974) *Selected Works*, vol. 1, Nauka, Leningrad.
61. Evans A., Heuer A. H. and Porter D. Toughness of ceramics, in *Advances in Research on the Strength and Fracture of Materials*, (ed. M. R. Taplin), Pergamon, vol. 1, pp. 134–64.
62. Faber K. T. (1984) Toughening mechanisms for ceramics in automotive applications, in: *Ceramic Engineering and Science*. Proceeding of the American Ceramic Society, Columbus, vol. 5, pp. 408–39.
63. Claussen N. (1985) Strengthening strategies for ZrO_2–Al_2O_3 toughened ceramics at high temperatures. *Mat. Sci. and Eng.*, **71** (1), 23–38.
64. Hillig W. (1987) Strength and toughness of composites with ceramic matrix. *Ann. Rev. Mat. Sci.*, **17**, 341–83.
65. Bakun O. V., Grigoriev O. N., Kartuzov V. V. and Trefilov V. I. (1986) Fracture of dense modification of boron nitride heterophase polycrystals. *Dokl. Akad. Nauk SSSR*, **288** (6), 1351–3.
66. Cowley J. D. (1984) Overview of zirconia with respect to gas turbine applications. *NASA Techn. Paper 2'286*.
67. Evans A. G., and Cannon R. N. (1986) Toughening of brittle solids by transformation. *Acta Met.*, **34** (5), 761–800.
68. Wiederhorn S. M. (1984) Fracture mechanics and microstructural design. *Ann. Rev. Mater. Sci.*, **14**, 373–403.
69. Lange F. F. (1978) Fracture mechanics and microstructural design, in *Fracture Mechanics of Ceramics*, vol. 4, Plenum, New York, pp. 799–815.
70. Shvedkov E. L. (1987) Problems of developing tough ceramics. Preprint No. 7, IPM AN Ukr. SSR, Kiev.
71. Lange F. F. (1970) The interaction of a crack front with a second-phase dispersion. *Phil. Mag.*, **22** (179), 983–92.

REFERENCES TO SECTION 1.1.2: TECHNOLOGY

1. Shatt V. (ed.) (1985) *Powder Metallurgy. Sintered and Composite Materials*, Metallurgija, Moscow. 1985.
2. Evans A. G. and Langdon T. G. (1976) *Engineering Ceramics*. Plenum, New York.
3. Gnesin G. G. and Osipova I. I. (1981) Hot-pressed silicon nitride-based materials. *Poroshkovaja Metallurgija*, **4**, 33–45.
4. Billman E. R., Mehrotra P. K., Shuster A. F. and Beeghly C. W. (1988) Machining with Al_2O_3–Sic-whisker cutting tools. *Ceramic Bull.*, **67** (6), 1016–19.
5. Gnesin G. G., Gervits E. I., Shipilova L. A. *et al.* (1990) Resistive composition Si_3N_4–ZrC. 1. Concentration dependence of electrical conductivity. *Poroshkovaja Metallurgija*, **4**, 80–4.
6. Shklovskij B. I. and Efros A. L. (1975) Theory of flow and conductivity of strongly inhomogeneous media. *Uspekhi Fiz. Nauk*, **117** (3), 401–35.

7. Skorokhod V. V. (1982) *Powder Materials Based on Refractory Metals and Compounds*, Tekhnika, Kiev.
8. Gnesin G. G., Osipova I. I., Rontal G. D. et al. (1991) *Ceramic Tool Materials*, Tekhnika, Kiev.
9. Andrievskij R. A. and Spivak I. I. (1984) *Silicon Nitride and Materials Based on it*, Metallurgija, Moscow.
10. Emjashev A. V. (1988) *Gas-Phase Metallurgy of Refractory Compounds*, Metallurgija, Moscow.
11. Katzman H. A. (1990) Fiber coating for composite fabrication. *Mater. and Manuf. Process.*, **5** (1), 1–15.
12. Piller R. C., Fried S. J. and Denton I. E. (1989) Hiping of silicon carbide and silicon nitride. *Brit. Ceram. Process.*, **45**, 33–4.
13. Gogotsi Y. G., Grigoriev O. N. and Yaroshenko V. P. (1990) Mechanical properties of hot-pressed silicon nitride at high temperatures. *Silikattechnik*, **41**, 156–60.
14. Khejdemane G. M., Grabis Y. P. and Miller T. N. (1979) High-temperature synthesis of finely dispersed silicon nitride. *Izv. AN SSSR, Neorg. Mater.*, **4**, 595–8.
15. Merzhanov A. G. and Borovinskaja I. P. (1979) Self-propagating high-temperature synthesis in chemistry and technology of refractory compounds. *Zhurn. Vkho im. D. I. Mendeleeva*, 223–7.
16. Chen Lipong, Takashi G. and Hirai T. (1989) Preparation of silicon carbide powders by chemical vapour deposition of the SiH_4–CH_4–H_2 system. *J. Mater. Sci.*, **11**, 3824–32.
17. Johnson D. W. (1981) Nonconventional powder preparation techniques. *Amer. Ceram. Soc. Bull.*, **60** (2), 222–4.
18. 'Chemistry' seen as key to successful ceramics. *Mater. Eng.*, 1982, **96** (2), 46–7.
19. Roy R., Roy D. M. and O'Holleran T. P. (1984) Precision 'doping' of simple and complex ceramic powders by DMS technique. *Powd. Met. Intern.*, **16** (6), 274.
20. Segal D. L. and Woodhead J. L. New developments in gel processing, in *Novel Ceram. Fabr. Process and Appl. Meet.* Basic Sci. Sec. Inst. Ceram., Cambridge, 9–11 Apr. 1986. Stoke-on-Trent, 1986, pp. 245–50.
21. Cannon W. R., Danforth S. C. and Flint J. H. (1982) Sinterable ceramic powders from laser-driven reactions. *J. Amer. Ceram. Soc.*, **65** (7), 324–30.
22. Haggerty J. and Kennon U. R. (1984) Production of powders for sintering in laser-induced reactions in. *Laser-Induced Chemical Processes*, (ed.) J. J. Steinfeld, Plenum, New York and London, Russian translation, Mir, Moscow, pp. 183–206.
23. Mannetti C., Curcio F., Ghiglione G. and Musci M. Laser-driven synthesis of SiC ceramic powders in. *New Laser Technology and Applications*. Proceedings of the First International Conference, Olympia, 19–23 June 1983. Bologna, 1988, pp. 117–24.
24. Hoffmann B. (1988) Probleme der Feinstpulverherstellung für Mechanokeramik durch mechanische Aufbereitung. *Keram. Zeitschrift*, **2**, 90–6.
25. Dobrovolskij A. G. (1977) *Slip Casting*, Metallurgija, Moscow.
26. Bogojavlenskij K. N., Kuznetsov P. A. and Mertens K. K. et al. (1984) *High-speed Methods for Pressing Parts from Powder Materials*, Mashinostroenie, Leningrad.
27. Shelegov V. I. and Krot O. I. Use of hydrodynamic pressing method in technology of nitride powders, in *Research and Development of Theoretical Problems in Field of Powder Metallurgy and Protective Coatings*. Proc. of All-USSR Conf. Minsk, 1984, Part 2, pp. 134–6.

28. Maljushevskij P. P. and Tolstykh A. V. (1979) Use of electrohydraulic effect in powder metallurgy. *Poroshkovaja Metallurgija*, **5**, 22–6.
29. Gnesin G. G. (1987) *Nonoxide Ceramic Materials*, Tekhnika, Kiev.
30. Kovalchenko M. S. (1980) *Theoretical Fundamentals of Hot Pressure Working of Porous Materials*, Naukova Dumka, Kiev.
31. Larker H. T. (1986) On hot isostatic pressing of shaped ceramic parts, in. *Ceram. Compon. Engines.* Proceedings of the First International Symposium Hakone, 17–19 Oct. 1983. London, New York, pp. 304–10.

REFERENCES TO SECTION 1.1.3 MACHINING TECHNIQUES

1. Trent E. M. (1977) *Metal Cutting*, Butterworth, London.
2. Burke J. and Weiss V. (eds.) (1982) *Surface Treatment for Improved Performance and Properties*, Plenum, New York and London.
3. Rogov V. V. (1986) Diamond-abrasive machining of brittle nonmetallic materials, in *Synthetic Ultrahard Materials*, vol. 3. Application of ultrahard materials, (eds.) N.V.Novikov *et al.* Naukova Dumka, Kiev.
4. Maslov E. N. (1974) *Material Grinding Theory*, Mashinistroenie, Moscow.
5. Poduraev V. N. (1974) *Cutting of Hard-to-machine Materials*, Vysshaja Shkola, Moscow.
6. Rogov V. V. (1989) On problem of mechanism of diamond grinding of brittle nonmetallic materials. *Sverkhtverdye Materialy*, **5**, 57–61.
7. Bifano T. G., Blake P. N. Dow T. A. and Scattergood R. O. (1987) Precision finishing of ceramics. *Proc. Soc. Photo-Opt. Instrum. Eng.*, pp. 12–22.
8. Oisi Kendzi. (1988) Machining of fine ceramics. *Kikaj to Kogu (Tool Eng.)*, **32** (10), 72–6.
9. Nakajima T., Uno Y. and Fujiwara T. Mechanism of cutting of fine ceramics by single-crystalline diamond tool. *Prec. Eng.*, **11** (1), 19–25.
10. Rybitskij V. A. (1980) *Diamond Grinding of Hard Alloys*, Naukova Dumka, Kiev.
11. Zhed V. P., Borovskij G. V., Muzykant Y. A. and Ippolitov G. M. (1987) *Cutting Tools Fitted with Ultrahard and Ceramic Materials and their Application*, (reference book), Mashinostroenie, Moscow.
12. Lavrinenko V. I., Zlenko A. A. and Sytnik A. A. (1985) Working capacity of diamond wheels in grinding of cutting ceramic VOK-60. *Sverkhtverdye Materialy*, **4**, 45–7.
13. Melnikov V. A. (1990) Kazakov V. K. Structure and wear resistance of silicon carbide–aluminium nitride system ceramics. *Sverkhtverdye Materialy*, **2**, 28–31.
14. Kabaldin Y. G., Shepelev A. A. and Kovalev O. B. (1990) Improvement of working capacity and reliability of tools made of cutting ceramics. *Sverkhtverdye Materialy*, **1**, 48–53.
15. Iliemeroth J. (1987) Experimental study on various parameters of machining with special diamond and CBN tools for surface finish of silicon carbide, silicon nitride, and alumina ceramics, in *Intersociety Symposium on Machining of Advanced Ceramic Materials and Components*, pp. 121–30.
16. Zhang Bi, Hitoshi Tokura and Masanori Yoshikawa. (1988) Investigation into surface state in grinding fine ceramics. *Seimitsu Kogaku Kaisi (J. Japan Soc. Precis. Eng.)*, **54** (8), 1537–43.
17. Kurobe Tosidzi. (1988) Prospects for advance of fine ceramics grinding and polishing methods. *Kikaj to Kogu (Tool Eng.)*, **32** (11), 16–20.

REFERENCES TO SECTION 1.1.4 JOINING (BRAZING) OF CERAMIC MATERIALS

1. Helgesson C. J. (1968) *Ceramic to Metal Bonding*, Boston Technical Publishers, Cambridge, Massachusetts.
2. V. Houten G. R. (1959) A survey of ceramic-to-metal bonding. *Amer. Cer. Soc. Bull.*, **38** (6), 301–7.
3. Widman H. (1963) Keramik-Metal-Verbindungen. *Glas-Email-Keramik Technik*, **14** (6), 205–10.
4. Batygin V. A., Metelkin I. I., Reshetnikov A. M. (1973) *Vacuum Ceramics and their Brazed Joints with Metals*, Energija, Moscow.
5. Naidich Y. V. (1978) *Contact Phenomena in Metallic Melts*, Naukova Dumka, Kiev.
6. Naidich Y. V. (1981) The wettability of solids by liquid metals. *Progress in Surface and Membrane Science*, **14**, 353.
7. Naidich Y. V., Kolesnichenko G. A. and Kostjuk B. D. (1973) Adhesion of Metallic Melts to Molybdenum Films Deposited on Surface of Oxides. *Fizika i Khimija Obrabotki Materialov*, **6**, pp. 61–66.
8. Naidich Y. V., Kostjuk B. D. and Kolesnichenko G. A. (1975) Wettability in metallic melt – thin metallic film – nonmetallic substrate system. In *Physical Chemistry of Condensed Phases, Ultrahard Materials, and Their Interfaces*. Naukova Dumka, Kiev, pp. 15–27.
9. Pincus A. G. (1953) Metallographic Examination of Ceramic-to-Metal Seals. *J. Am. Ceram. Soc.*, **36**, (5), pp. 152–158.
   Pincus A. G. (1955) Mechanism of ceramic-to-metal adherence of molybdenum to alumina ceramics. Ceram. Age, 1954, **63**, (3), pp. 16–20, 30–32; *Ceram. Abstr.*, March, p. 53f.
10. Kirkjian C. and Kingery W. (1956) *J. Phys. Chem.*, **60** (7), 1961.
11. Naidich Y. V. (1965) Regularities of adhesion and wettability of nonmetallic bodies with liquid metals, in *Surface Phenomena in Melts and in Solid Phases Arising from Them*, Kabardin-Balkar, Nal chik.
12. Aksay J. A., Hoge G. E. and Pask J. A. (1974) Wetting under chemical equilibrium and nonequilibrium conditions. *J. Phys. Chem.*, **78** (12), 1178–83.
13. Enstathopoulos N., Chatain D. and Sangiorgi R. (1989) Work of adhesion and contact-angle isotherm of binary alloys on ionocovalent oxides. *J. Mater. Sci.*, **24**, 1100–6.
14. Naidich Y. V. and Zhuravlev V. S. (1974) Adhesion, wettability, and interaction of titanium-containing melts with refractory oxides. *Ogneupory*, **1**, 50–5.
15. Naidich Y. V. and Chuvashov Y. N. (1983) Wettability and contact interaction of gallium-containing melts with nonmetallic solids. *J. Mater. Sci.*, **18**, 2071.
16. Naidich Y. V., Krasovskij V. P. and Chuvashov Y. N. (1986) Wettability of zinc selenide and sulfide by metallic melts. *Adgesija Rasplavov i Pajka Materialov*, **17**, 40–4.
17. Naidich Y. V., Krasovskij V. P. and Chuvashov Y. N. (1986) Effect of sulfur and selenium on wettability of zinc sulfide and selenide by metallic melts. *Poroshkovaja Metallurgija*, **11**, 69–72.
18. Naidich Y. V., Chuvashov Y. N. and Krasovskij V. P. (1990) Wettability of magnesium, barium, and calcium fluorides by metallic melts. *Adgesija Rasplavov i Pajka Materialov*, **24**,
19. Naidich Y. V., Guravlev V. S. and Frumina N. I. Wetting of Rare-Earth Element Oxides by Metallic Melts. *J. Mater. Sci.*, 1990, **25**, pp. 1895–1901.

20. Naidich Y. V., Zhuravlev V. S., Frumina N. I. et al. (1988) Wetting of silicon nitride-based ceramics by metallic melts. *Poroshkovaja Metallurgija*, **11**, 58–61.
21. Naidich Y. V., Vojtovich R. P., Kolesnichenko G. A. and Kostjuk B. D. Wetting of inhomogeneous solid surfaces by metallic melts for systems with ordered arrangement of dissimilar regions. *Poverkhnost: Fizika, Khimija, Mekhanika*, **2**, 126–32.
22. Naidich Y. V. and Vojtovich R. P. (1991) Wetting of two-phase composites by metallic melts. *Poroshkovaja Metallurgija*.
23. Naidich Y. V., Gab I. I. and Zhuravlev V. S. (1983) Selection of optimum parameters of conditions for producing solid-phase metal-quartz joint through aluminium gasket. *Svarochnoe Proizvodstvo*, **6**, 18–19.
24. Naidich Y. V. and Gab I. I. (1989) Pressure welding of silicon nitride- and carbide-based materials. *Avtomaticheskaja Svarka*, **2**, 56–8.
25. Ishchuk N. F. and Naidich Y. V. (1989) Designing of telescopic brazed joints of glass materials with different thermal expansion coefficients. *Svarochnoe Proizvodstvo*, p. 23.
26. Naidich Y. V., Kostjuk B. D. and Poberezhnjuk V. L. (1990) Nonstressed brazed joint of ceramic and metallic parts. *Svarochnoe Proizvodstvo*, **6**.

REFERENCES TO SECTION 1.1.5: MECHANICAL PROPERTIES

1. Bolotin V. V. (1982) *Methods of Probability Theory and Reliability Theory in Calculation of Structures*, Strojizdat, Moscow.
2. Krasulin Y. L. and Barinov S. M. (1985) On reliability and safety margin of ceramic material. *Dokl. AN SSSR*, **285** (4), 883–6.
3. Shevchenko V. J. and Barinov S. M. Reliability criteria for engineering ceramics, in *Austceram '90*, Proceeding of the International Conference, (ed. by P. J. Darragh and P. J. Stead), Trans. Tech., Australia, pp. 344–50.
4. Barinov S. M. and Urjeva G. N. (1989) Ceramic-matrix composites, in *Science and Engineering News. New Materials, their Manufacturing and Working Technologies*, Izd-vo VINITI, Moscow, **9**, pp. 1–35.
5. Evans A. G. (1990) Prospective on the development of high-toughness ceramics. *J. Amer. Ceram. Soc.*, **73** (2), 187–206.
6. Bakun O. V., Grigoriev O. N., Kartuzov V. I. and Trefilov V. I. (1986) Fracture of heterophase polycrystals based on dense boron nitride modifications. *Dokl. AN SSSR*, **288** (6), 1351–4.
7. Galanov B. A. and Grigoriev O. N. (1990) Fracture of elastically deformable hard heterophase materials with periodic microstructure and their fracture criteria. Preprint No. 9. Institute for Materials Science Problems, Ac. Sci. of the Ukr. SSR, Kiev.
8. McMeeking R. M. and Evans A. G. (1982) Mechanism of transformation toughening in brittle materials. *J. Amer. Ceram. Soc.*, **65** (5), 242–7.
9. Krasulin Y. L., Barinov S. M. and Ivanov V. S. (1985) *Structure and Fracture of Materials from Refractory Compound Powders*, Nauka, Moscow.
10. Cherepanov G. P. (1974) *Mechanics of Brittle Fracture*, Nauka, Moscow.
11. Barinov S. M., Andriashvili P. I. and Tavadze F. N. (1989) Subcritical crack propagation in brittle materials. *Dokl. AN SSSR*, **304** (6), 1361–4.
12. Barinov S. M. (1988) R-curves of resistance to fracture of ceramic materials. *Zavodskaja Laboratorija*, **4**, 71–4.
13. Krstic V. V., Nicholson P. S. and Hoagland R. G. (1981) Toughening of glasses by metallic particles. *J. Amer. Ceram. Soc.*, **64** (9), 499–504.

14. Becher P. F., Hsueh C. -H., Angelini P. and Tiegs T. N. (1988) Toughening behavior in whisker-reinforced ceramic matrix composites. *J. Amer. Ceram. Soc.*, **71** (12), 1050–61.
15. Barinov S. M. and Shevchenko V. J. (1991) Crack bridging and R-curve behavior in whisker-reinforced ceramics. *J. Mater. Sci. Lett.*, **10** (6), 660–1.
16. Hutchinson J. W. (1987) Crack tip shielding by microcracking in brittle solids. *Acta Metallurgica*, **35** (7), 1605–19.
17. Barinov S. M. and Krasulin Y. L. (1982) Subcritical crack growth in brittle materials at microcracking. *Problemy Prochnosti*, **9**, 84–7.
18. Evans A. G., Heuer A. and Porter D. (1979) Crack resistance of ceramics, in *Fracture Mechanics*, Issue 17, Fracture of Materials, (ed. D. Taplin). Russian translation, Mir, Moscow, pp. 134–164.
19. Barinov S. M., Ivanov D. A., Kokhan S. V. and Fomina G. A. (1986) Structure designing and crack resistance of alumina ceramics. *Problemy Prochnosti*, **6**, 9–13.
20. Barinov S. M., Ivanov D. A. and Fomina G. A. (1991) Crack resistance of alumina–chromium system cerments. *Poroshkovaja Metallurgija*, **1**, 98–101.
21. Barinov S. M. and Andriashvili P. I. (1988) Crack propagation in TiAl compound. *Isv. AN SSSR. Metally*, **2**, 148–50.
22. Faber K. T., Evans A. G. (1983) Crack deflection process. 1. Theory. *Acta Metallurgica*, **31** (4), 565–76.
23. Faber K. T., Evans A. G. and Drory M. D. Crack deflection as a toughening mechanism, in *Fracture Mechanics of Ceramics*, (eds. R. C. Bradt, A. G. Evans, D. P. H. Hasselman and F. F. Lange), Plenum, New York and London, vol. 6, pp. 77–91.
24. Shorshorov M. K., Ustinov L. M., Gukasjan L. E. and Vinogradov L. V. (1989) *Physics of Strength of Fibrous Composites with Metallic Matrix*, Metallurgija, Moscow.
25. Kelly (1973) *Strong Solids*. Clarendon Press, Oxford. 1973.
26. Cao H. C., Bischoff E., Sbaizero O. *et al.* (1990) Effect of interfaces on the properties of fiber-reinforced ceramics. *J. Amer. Ceram. Soc.*, **73** (6), 1691–9.
27. Evans A. G. and Marshall D. B. (1989) The mechanical behavior of ceramic matrix composites. *Acta Metallurgica*, **37** (11), 2567–83.
28. Garvie R. C. Hannink R. H. J. and Pascoe R. T. (1975) Ceramic steel? *Nature*, **258** (12), 703–4.
29. Claussen N. (1982) Umwandlungsverstarkte Keramische Werkstoffe. *Ztschr. Werkstofftechn.*, **13**, 138–47.
30. Pilyankevich A. N. and Claussen N. (1978) Toughening of BN by stress-induced phase transformation. *Mater. Res. Bull.*, **13** (5), 413–15.
31. Wang J. and Stevens R. Review. Zirconia-toughened alumina (ZTA) ceramics. *J. Mater. Sci.*, **24** (11), 3421–40.
32. Stump D. M. and Budiansky B. (1989) Crack growth resistance in transformation-toughened ceramics. *Int. J. Solids Structures*, **25** (7), 635–49.
33. Timofeev V. N., Zabukas V. K., Machulis A. N. *et al.* (1980) Fracture characteristics of polymerceramics based on zirconia microspheres. *Fizika i Khimija Obrabotki Materialov*, **3**, 103–5.
34. Virkar A. B. and Johnson D. L. (1977) Fracture behavior of ZrO_2–Zr composites. *J. Amer. Ceram. Soc.*, **60** (11–12), 514–19.
35. Krassulin Y. L., Barinov S. M. and Ivanov V. S. (1986) Interparticle welding, mechanical properties and fracture behavior of $LaCrO_3$–Cr cermets. *J. Mater. Sci.*, **21** (11).

36. Barinov S. M., Ivanov D. A. and Fomina G. A. (1986) Crack resistance of ceramic-based layered composite. *Mekhanika Kompozitnykh Materialov*, **5**, 836–40.
37. Tiegs T. N. and Becher P. F. (1986) Whisker reinforced ceramic composites, in *Ceramic Materials and Components for Engines*. Proceedings of the Second International Symposium, Lubeck-Travemunde, Berichte DKG, pp. 193–200.
38. Becher P. F., Tiegs T. N., Ogle J. G. and Warwick W. H. (1986) Toughening of ceramics by whisker reinforcement, in *Fracture Mechanics of Ceramics*, (eds. R. C. Bradt, A. G. Evans, D. P. H. Hasselman and F. F. Lange, Plenum, New York and London, vol. 7, pp. 61–73.
39. Inoue S., Niihara K., Ushiyama T. and Hirai T. (1986) Al_2O_3/SiC ceramic composite. *Ceramic Material and Components for Engines*. Proceedings of the Second International Symposium, Lubeck-Travemunde, Berichte DKG, pp. 609–17.
40. Claussen N. and Swain M. V. (1988) Silicon carbide whisker reinforced and zirconia transformation toughened ceramics. *Mater. Forum*, **11**, 194–201.
41. Wada S. and Watanabe N. (1988) Solid particle erosion of brittle materials. *J. Ceram. Soc. Jap.*, **96** (2), 111–18.
42. Schneider G., Weisskopf K. -L., Greil P. and Petzow G. (1988) Thermal shock behavior of SiC-whisker reinforced composites. *Sci. Ceram. 14*. Proceedings of the 14th International Conference, Canterbury, 1988, pp. 819–24.
43. Gomina M., Chermant J. L., Osterstock F. *et al.* (1986) Applicability of fracture mechanics to fiber reinforced CVD ceramic composites, in *Fracture Mechanics of Ceramics*, (eds R. C. Bradt, A. G. Evans, D. P. H. Hasselman and F. F. Lange), Plenum, New York and London, vol. 7, pp. 17–32.
44. Grateau L., Lob N. and Parlier M. (1988) Microstructural studies of ceramic composition obtained by chemical vapor phase infiltration. *Sci. Ceram. 14*. Proceedings of the 14th International Conference, Canterbury, pp. 885–9.
45. NASA probes ceramic fibre reinforcement. *New Mater. Int.*, 1988, **3** (19), 5.
46. Yang J. -M. (1988) Processing of metal and ceramic matrix composites with three-dimensional braided reinforcement. *Adv. Mater. and Manu. Process.*, **3**, 233–45.

REFERENCES TO SECTION 1.1.6: ENGINEERING

1. Romashin A. G. (1981) Radiotransparent materials, in *Scientific Fundamentals of Materials Science*, Inst. of Solid State Physics, Ac. Sci. USSR, Nauka, Moscow, pp. 39–46.
2. Vasil'ev V. V., Protasov V. D., Bolotin V. V. *et al.* (1990) *Composites*, Mashinostroenie, Moscow.
3. Shvedkov E. L. and Judin A. G. (1990) *Ceramic Materials: State of the Art in Research and Development, Prospects for Advance. Analytical Review*. International Centre of Scientific and Technical Information, Moscow.
4. Pivinskij J. E. (1989) Structural ceramics and problems of its technology, in *Chemistry and Technology of Silicate and Refractory Nonmetallic Materials*, Nauka, Leningrad. pp. 109–25.
5. Pilipovskij J. E., Grudina T. V., Sapozhnikova A. B. *et al.* (1990) *Composites in Mechanical Engineering*, Tekhnika, Kiev.

6. Bolotin V. V. (1984) *Predicting Service Life of Machines and Structures*, Mashinostroenie, Moscow.
7. Postnikov A. A. and Chasovskoj E. N. (1990) Problems of weakest-link model applicability to estimation of fracture of ceramic materials, in *Mechanics and Physics of Fracture of Brittle Materials*, Izd. IPM AN Ukr. SSR, Kiev, pp. 111–15.
8. Postnikov A. A., Selunskij A. A. and Chasovskoj E. N. (1987) Analysis of strength and reliability of ceramic machine elements. *Mashinovedenie*, **3**, 36–41.
9. Romashin A. G., Vikulin V. V., Karpov J. S. and Osaulenko V. N. (1989) *Ceramic Materials in Designs of Aircraft Units*, Kharkov Aviation Institute, Kharkov.
10. USSR Inventor's Certificate No. 1.472,797. Specimen for Tensile Test of Brittle Materials. Granted to Chasovskoj E. N., Postnikov A. A. and Verevks V. G. Otkrytija, Izobretenija, (1988) No. 14, p. 184.
11. Barinov S. M. and Ur'eva G. N. (1989) Ceramic-matrix composites, in *Science and Engineering News. New Materials, their Manufacturing and Processing Technology*, VINITI, Moscow, **8**, 1–35.
12. Barinov S. M. (1988) Crack resistance of structural mechanical-engineering ceramics, in *Science and Engineering Results. Technology of Silicate and Refractory Nonmetallic Materials*, VINITI, Moscow, vol. 1, pp. 72–132.
13. Tretjakov J. D. and Metlin J. G. (1987) *Ceramics: Material for the Future*, Znanie, Moscow.
14. US market for advanced ceramics. *Ceramic Ind. J.*, 1989, Jan., p. 10.
15. Skorokhod V. V., Solonin J. M. and Uvarova I.V. (1990) *Chemical, Diffusion, and Rheologic Processes in Powder Technology*, Naukova Dumka, Kiev.
16. Strong growth predicted for several ceramic markets. *Amer. Cer. Soc. Bull.*, 1988, **67** (12), 1888–9.
17. Di Carlo James A. (1989) CMC's for the long run. *Adv. Mater. and Proces.*, **135** (6), 41–4.
18. Evans Anthony G. (1990) Perspective on the development of high toughness ceramics. *J. Amer. Cer. Soc.*, **73** (2), 187–206.
19. Lewis M. H. (1988) Microstructure of high temperature engineering ceramics in *New Materials and Their Applications*, Proceedings of the Institute of Physics Conference, Warwick, 22–25 Sept. 1987, Bristol, Philadelphia, pp. 41–51.
20. Wachtman J. (1988) Moving towards tougher ceramics. *Adv. Mater. and Proces. Metal Progress.* **133** (1), 45–6.
21. Andrievskij R. A. and Spivak I. I. (1984) *Silicon Nitride and Materials on its Base*, Metallurgija, Moscow.
22. Vikulin V. V., Dorozhkin A. I., Ogneva I. V. *et al.* (1989) Compacted chemically bonded silicon nitride and self-reinforced composite material. *Ogneupory*, **9**, 29–33.
23. Vikulin V. V. and Kurskaja I. N. (1990) Reaction-bonded silicon nitride-based structural ceramic materials. *Ogneupory*, **9**, 13–18.
24. Buljan S. -T., Pasto A. E. and Kim H. J. (1989) Ceramic whisker- and particulate composites: properties, reliability, and applications. *Amer. Ceram. Bull.*, **68** (2), 387–94.
25. Mecholsky J. (1989) Engineering research needs of advanced ceramics and ceramic-matrix composites. *Amer. Ceram. Bull.* **68** (2), 367–75.
26. Kadama H., Sacamoto H. and Miyoshi T. (1989) Silicon carbide monofilament-reinforced silicon nitride or silicon carbide matrix composites. *J. Amer. Cer. Soc.*, **72** (4), 551–8.

27. Shetty D. K., Pascucci M. R., Mutsudy B. C. and Wills R. R. (1985) SiC monofilament-reinforced Si₃N₄ matrix composite. *Ceram. Eng. and Sci. Proc.*, **6** (7-8), 632-45.
28. Tamari N. (1987) Production methods and properties of fiber-reinforced ceramic materials. *Ceramics (Japan)*, **22** (6), 502-7.
29. Japan Patent Application No. 63-107864. Tosihiko T. and Sigetaka V. Publ. 12 May 1988.
30. US Patent No. 3,914,500. Publ. 21 Oct. 1975.
31. Wereszcak A. A. and Rarvici-Majidi A. (1990) Effect of fracture temperature and relative crack propagation rate on the fracture behavior of whisker-reinforced ceramic matrix composites. *Ceram. Eng. and Sci. Proc.*, **11** (7-8), 721-33.
32. Isao T., Pezzotti G., Okamoto T. *et al.* (1989) Hot isostatic press sintering and properties of silicon nitride without additive. *J. Amer. Cer. Soc.*, **72** (9), 1656-60.
33. Japan Patent Application No. 60-33263. Myesi T., Takeda Y., Nagayama K. and Narisawa T. Publ. 20 Feb. 1985.
34. Jamet J. F. (1988) Ceramic-ceramic composites for use at high temperature. *New Materials and their Applications*. Proceedings of the Institute of Physics Conference, Warwick, 22-25 Sept. 1987, Bristol, Philadelphia, pp. 63-75.
35. Mathieu A., Monteuuis B. and Gount V. (1990) Ceramic matrix composite materials for a low thrust bipropellant rocket engine. *AIAA Paper*, **2054**, 1-7.
36. Rusanova L. N., Romashin A. G. Kulikova G. I. and Golubeva O. P. (1988) Problems and prospects of advance in boron nitride ceramics. *Poroshkovaja Metallurgija*, **1**, 23-31.
37. Buravov A. D. and Djakonov B. P. (1980) Texture of boron nitride-based materials. *Poroshkovaja Metallurgija*, **4**, 41-6.
38. Thomas J., Weston N. E. and O'Connor T. E. (1963) Turbostratic boron nitride, thermal transformation to ordered-layer-lattice boron nitride. *J. Amer. Chem. Soc.*, **84** (24), 4619-22.
39. Grushevskij J. L., Frolov V. I., Shabalin I. L. and Cheborjukov A. V. (1991) Mechanical behavior of boron nitride-containing ceramics. *Poroshkovaja Metallurgija*, **4**, 89-93.
40. Galakhov A. V. and Shevchenko V. J. (1990) Ceramic composites. *Ogneupory*, **6**, 53-8.
41. Stevens R. (1986) *Zirconia and Zirconia Ceramics*. Twickenham, UK.
42. Majdic A. (1981) Present trends in the field of high-temperature ceramics. *La Ceramica*, **31** (5), 24-30.
43. Becher P. F. and Tiegs T. N. (1987) Toughening behavior involving multiple mechanisms: whisker reinforcement and zirconia toughening. *J. Amer. Cer. Soc.*, **70** (9), 651-4.
44. Podkletnov E. E., Shepilov I. P. and Ivanova G. A. (1985) Thermal stability and mechanical properties of refractories based on zirconia and ceria with layered-granular structure. *Ogneupory*, **9**, 15-18.
45. Ivanova L. I., Romashin A. G., Buravova N. D. and Krjuchkov V. A. (1991) Production and properties of zirconia ceramics. *Ogneupory*, **2**, 6-8.
46. Pivinskij J. E. and Romashin A. G. (1974) *Quartz Ceramics*, Metallurgija, Moscow.
47. Borodaj F. J. (1990) Doping of quartz ceramics. *Steklo i Keramika*, **11**, 22-4.

FURTHER READING

Postnikov A. A. and Chasovskoj E. N. (1987) Probabilistic evaluation of strength of structurally inhomogeneous bodies. *Izv. Ac. Sci. of Azerb. SSR, Ser. Tekhn. Nauk*, **5**, pp. 21–26.

Ekström T. and Persson J. (1990) Hot hardness behavior of yttrium sialon ceramics. *J. Amer. Cer. Soc.*, **73**, (10), pp. 2834–2838.

REFERENCES TO SECTION 1.2.1: CERAMIC COMPOSITES FOR ELECTRONICS

1. Odelevskij V. I. (1951) Calculation of generalized conductivity of heterogeneous systems. *Zhurn. Tekhn. Fiz.*, **21** (6), 667–85.
2. Landau L. D. and Liftshitz E. M. (1982) *Electrodynamics of Continuous Media*, Nauka, Moscow.
3. Nye J. F. (1957) *Physical Properties of Crystals*, Clarendon Press, Oxford.
4. Dulnev G. N. and Novikov V. V. (1979) Conductivity of heterogeneous systems. *Inzh. Fiz. Zhurn.*, **36** (5), 900–9.
5. Skorokhod V. V. (1979) On electrical conductivity of conductor–nonconductor mixtures. *Inzh. Fiz. Zhurn.*, **11** (8), 51–8.
6. Skorokhod V. V. (1970) Methods for calculating physical properties of two-phase sintered alloys with account of their structure. *Proceedings of the Fourth International Conference on Powder Metallurgy*. Carlovy Vary, Czechoslovakia, pp. 29–41.
7. Ermakov G. A., Fokin A. G. and Shermergor T. D. (1974) Calculation of limits for effective permittivities of heterogeneous dielectrics. *Zhurn. Tekhn. Fiz.*, **44** (2), 249–54.
8. McLachlan D. S., Blaszkiewisz M. and Newnham R. E. (1990) Electrical resistivity of composites. *J. Am. Cer. Soc.*, **73** (8), 2187–203.
9. Kirkpatrik S. (1977) Percolation and conductivity, in *Solid-State Physics News*, **7**, Theory and Properties of Disordered Materials, (ed. Bonch-Bruevich V.L.), Mir, Moscow, pp. 249–92.
10. Dulnev G. N. and Novikov V. V. (1991) *Transport Processes in Heterogeneous Media*, Energoatomizdat, Leningrad.
11. Dulnev G. N. and Novikov V. V. (1983) Percolation theory and conductivity of heterogeneous media. *Inzh. Fiz. Zhurn.*, **45** (2), 136–41.
12. Kasjanenko A. A., Shipilova L. A. and Gervits E. I. (1992) Hot-pressed ceramic materials of Si_3N_4–SiC system, their structure and electrophysical properties. *Poroshkovaja Metallurgija*, in press.
13. Veinberg A. K. (1966) Magnetic susceptibility, electrical conductivity, permittivity, and thermal conductivity of medium containing spherical and ellipsoidal inclusions. *Dokl. AN SSSR*, **169** (3), 543–7.
14. Samsonov G. V. and Vinitskij I. M. (1976) *Refractory Compounds*, Metallurgija, Moscow.
15. Tsidilkovskij I. M. (1972) *Electrons and Holes in Semiconductors*, Nauka, Moscow.
16. Firsov J. A. (ed.) (1975) *Polarons*, Nauka, Moskow.
17. Norby T. and Kofstad P. (1976) Proton and native ion conductivities in Y_2O_3 at high temperatures. *Solid State Ionics*, **20**, 169–84.
18. Balakireva V. B. and Gorelov V. P. (1991) Proton and hole conduction of CaO-doped REE oxides in humid atmosphere. *Izv. AN SSSR, Ser. Neorg. Mater.*, **27** (1), 42–6.

19. Vecher A. A. and Vecher D. V. (1988) *Solid Electrolytes*, Univ. Izd., Minsk. 1988.
20. Tareev B. M. (1987) *Physics of Dielectric Materials*, Energoizdat, Moscow.
21. Maslennikova G. N., Mamaladze R. A., Midzuta S. and Koumoto K. (1991) *Ceramic Materials*, Strojizdat, Moscow.
22. Kolchin V. V., Barchukova N. I. and Balashova E. M. (1988) Ceramic dielectrics, semiconductors, and superconductors. *Steklo i Keramika*, **4**, 17–20.
23. Kroeger F. *Chemistry of Imperfect Crystals*.
24. Dubok V. A. and Lashneva V. V. (1982) Electrophysical properties of oxides of rare-earth elements and of solid solution based on them, in *New Materials from Oxides and Synthetic Fluorosilicates*, (ed. Tresvjatskij S. G.), Naukova Dumka, Kiev, pp. 37–76.
25. Kofstad P. (1972) *Nonstoichiometry, diffusion and Electrical Conductivity in Binary Metal Oxides*, Wiley, New York.
26. Dubok V. A. and Lashneva V. C. (1977) Electrical conductivity and nonstoichiometry of CaO-doped erbia. *Izv. AN SSSR, Ser. Neorg. Mater.*, **13** (9), 1636–9.
27. Zyrin A. V., Dubok V. A. and Tresvjatskij S. G. (1967) Electrical properties of REE oxides and some of their compounds, in. *Chemistry of High-Temperature Materials*. Nauka, Leningrad, pp. 59–62.
28. Kutsenok I. B., Kaul A. R. and Tretjakov J. D. (1974) Electrochemical production of solid electrolyte Ag-β-Al$_2$O$_3$. *Zhurn. Fiz. Khimii*, **48** (8), 2128–32.
29. Valeev K. S. and Mashkevich M. D. (1957) ZnO–TiO$_2$-based nonlinear semiconductors. *Zhurn. Teor. Fiz.*, **27** (8).
30. Valeev K. S. and Kvaskov V. B. 1983 *Nonlinear Metal-Oxide Semiconductors*, Energoizdat, Moscow.
31. Tresvjatskij S. G. (1960) High-temperature electric heating resistance. USSR Inventor's Certificate No. 132, 347. Publ. in *Bulletin of Inventions*, No. 19.
32. Rutman D. S., Toropov J. S., Pliner S. J. *et al.* (1985) *Zirconia-based High-Refractory Materials*, Metallurgija, Moscow.
33. Coble R. (1962) Transparent alumina and method of preparation. US Patent 3,026, 210. March.
34. Shevchenko A. V., Dubok V. A., Lopato L. M. *et al.* (1977) USSR Inventor's Certificate No. 563, **405**. Publ. 30 June.
35. (1985) Japan Pat. No. 58–237882. Publ. 13 July.
36. Ketron, L. Ceramic Sensors. *Am. Ceram. Soc. Bull.*, 1989, **68**, (4), pp. 860–8.
37. Kvantov M. A., Kostikov J. P., and Lejkina B. B. (1987) Nature of semiconductor properties of ceramic barium titanate. *Izv. AN SSSR, Ser. Neorg. Mater.*, **23** (10), 1722–5.
38. Takami A. (1988) Automotive oxygen sensors based on thick titanium dioxide films. *Ceramics (Japan)*, **23** (11), 1053–60.

FURTHER READING

McLachlan D. S. (1988) Measurement and analysis of a model dual-conductivity medium using a generalized effective-medium theory. *J. Phys. C: Solid State Physics*, **C21** 1521–32.

Dubok V. A., Lashneva V. V. and Uljanchich N. V. (1990) Electrical conductivity of gadolinium scandate-based solid solutions. *Izv. AN SSSR, Ser. Neorg. Mater.*, **26** (11), 2342–7.

Grebenkina V. G., Jusov J. P., and Sorokin V. N. (1976) *Bulk Resistors*, Naukova Dumka, Kiev.

Vinitskij I. M., Rud' B. M. and Telnikov E. J. (1987) Properties of BaB_6–LaB_6-based thick resistive films. *Elektronnaja Promyshlennost*, **6**, 5–10.

REFERENCES TO SECTION 1.2.2: FERROELECTRICS, PIEZOELECTRICS AND HIGH-TEMPERATURE SUPERCONDUCTORS

1. Burfoot J. and Taylor G. (1979) *Polar Dielectrics and Their Applications*, Macmillan Press, London.
2. Lines M. E. and Glass A. M. (1977) *Principles and Application of Ferroelectrics and Related Materials*, Clarendon Press, Oxford.
3. Rez I. S. and Poplavko J. M. (1989) *Dielectrics: Basic Properties and Applications in Electronics*, Radio i Svjaz, Moscow.
4. Vorotilov K. A., Petrovskij V. I. and Fedotov J. A. (1989) Ferroelectric films in microelectronics. Electronnaja tekhnika, Ser. 3. *Microelektronika*, **1** (130).
5. Likharev K. K., Semenov V. K., and Zorin A. B. (1988) *New Possibilities for Superconducting Electronics*, Itogi Nauki i Tekhniki, Ser. Sverkhprovodimost, Moscow.
6. Kurtz S. K., Cross L. E., Setter N. *et al.* (1988) Ferroelectricity in $ReBa_2Cu_3O_{7-\delta}$ superconductors. *Mat. Lett.*, **6** (10), 317–20.
7. Mattheiss L. F. (1985) Electronic structure of $BaPb_{1-x}Bi_xO_3$. *Jap. J. Appl. Phys.*, **24**, (Suppl. 24–2), 6–9.
8. Roleder K. (1989) Polar regions in the paraelectric phase in a $PbZr_{0.992}Ti_{0.008}O_3$ single crystal. *Phase Transitions*, **15**, 77–82.
9. Vugmeister B. E., and Glinchuk M. D. (1989) Dipole mechanism of attraction between carriers in high polarized crystals. *Phys. Lett.*, **139**, 471–4.
10. Fesenko E. G., Danziger A. J. and Rasumovskaja O. N. (1983) *New Piezoelectric Materials*, Rostov University Publications, Rostov.
11. Levanjuk A. P., Osipov V. V., Sigov A. S. and Sobjanin A. A. (1979) Changes in structure of defects and resulting anomalies in properties of substances near phase transition points. *Zhurn. Eksp. Teor. Fiz.*, **76**, 345–68.
    Levanjuk A. P. and Sigov A. S. (1985) Structural phase transitions in crystals with defects. *Izv. AN SSSR, Ser. Fiz.*, **49**, 219–26.
12. Glinchuk M. D. (1989) Impact of impurities of physical properties of ferroelectrics. *Visnik AN URSR*, **12**, 27–33.
13. Muller K. A. (1979) Intrinsic and extrinsic central-peak properties near structural phase transitions, in *Lecture Notes in Physics*, Springer, Berlin, pp. 210–50.
14. Altshuler S. A. and Kozyrev B. M. (1972) *Electron Paramagnetic Resonance*, Nauka, Moscow.
15. Glinchuk M. D., Grachev V. G., Dejgen M. F. and Rojtsin A. B. (1981) *Electrical Effects in Radiospectroscopy*, Nauka, Moscow.
16. Vugmeister B. E. and Glinchuk M. D. (1985) Cooperative phenomena in crystals with off-centre ions: dipole glass and ferroelectricity. *Usp. Fiz. Nauk*, **146**, 459–91.
17. Vugmeister B. E. and Glinchuk M. D. (1990) Dipole glass and ferroelectricity in random-site electric dipole systems. *Rev. Mod. Phys.*, **62** (4), 993–1026.
18. Vaks V. G. (1973) *Introduction to Microscopic Theory of Ferroelectrics*, Nauka, Moscow.

References

19. Van der Klink J. J., Rytz D., Borsa F. and Hochli U. T. (1983) Collective effects in a random site electric dipole system: $KTaO_3$: Li. *Phys. Rev.*, **B27** (1), 89–101.
20. Vugmeister B. E. and Glinchuk M. D. (1980) Features of cooperative behaviour of paraelectric defects in strongly polarizable crystals. *Zhurn. Eksp. Teor. Fiz.*, **79**, 947–52.
21. Vugmeister B. E., Glinchuk M. D. and Pechoniy A. P. (1985) EPR studies of dipole impurities in incipient ferroelectric $KTaO_3$: Li. *Jap. J. Appl. Phys.*, **24** (24-2), 670–2.
22. Vugmeister B. E., Glinchuk M. D., Pechonyj A. P. and Krulikovskij B. K. (1982) Interaction of dipole defects in ferroelectrics. *Zhurn. Eksp. Teor. Fiz.*, **82** (4), 1347–53.
23. Vugmeister B. E., Glinchuk M. D., Karmazin A. A. and Kondakova I. V. (1981) Electrodipole broadening of EPR lines in ferroelectrics. *Fiz. Tverd. Tela*, **23** (5), 1380–6.
24. Pechenyj A. P., Antimirova T. V., Glinchuk M. D. and Vugmeister B. E. (1988) Reorientation dynamics of Li^+ and Nb^{5+} impurity ions in $KTaO_3$: Li, Nb crystals. *Fiz. Tverd. Tela*, **30** (11), 3286–93.
25. Antimirova T. V., Glinchuk M. D. and Pechenyj A. P. (1990) Low-frequency dynamics of Nb^{5+} ions in KTN crystals. *Fiz. Tverd. Tela*, **32** (1), 208–11.
26. Glinchuk M. D. and Smolyaninov I. M. (1988) Features of interaction of elastic dipoles and induced transitions in virtual ferroelectrics. *Fiz. Tverd. Tela*, **30** (4), 1197–9.
27. Glinchuk M. D. and Smolyaninov I. M. (1990) Structural phase transitions induced by elastic fields of impurities. *Phase Transformations*, **29** (10), 95–103.
28. Jaffe B., Cook W. and Jaffe H. (1971) *Piezoelectric Ceramics*, Academic Press, London, New York.
29. Okazaki K. (1969) *Ceramic Engineering for Dielectrics*, Tokyo.
30. Glinchuk M. D., Bykov I. P. and Skorokhod V. V. The study of manganese admixture influence on the properties of PLZT ceramics. *Ferroelectrics*, in press.
31. Glinchuk M. D., Bykov I. P., Kurljand V. M. et al. (1990) Valency states and distribution of manganese ions in PZT ceramics simultaneously doped with Mn and Nb. *Phys. Stat. Sol. (a)*, **122**, 341–6.
32. Wersing W. (1978) Anomalous dielectric losses in manganese doped leadtitanate ceramics. *Ferroelectrics*, **22**.
33. Smolenskij G. A. et al. (1985) *Ferroelectrics and Antiferroelectrics*, Nauka, Leningrad.
34. Jushin N. K. and Dorogovtsev S. N. (1990) Acoustic studies of disordered ferroelectrics. *Izv. AN SSSR, Ser. Fiz.*, **51** (4), 629–36.
35. Tutsumi M., Hayashi J. and Hayashi T. (1985) *Jap. J. Appl. Phys.*, **24** 733.
36. Laguta V. V., Glinchuk M. D., Bykov I. P. et al. (1990) Study of microstructural features of lead magnoniobate. *Fiz. Tverd. Tela*, **32** (10), 3132–4.
37. Bishop A. R., Marlin R. L., Muller K. A. and Tesanovic Z. (1989) Superconductivity in oxides: toward a rectified picture. *J. Phys. B. Condensed Matter*, **76**, 17–24.
38. Bednorz J. G. and Muller K. A. (1988) Perovskite-type oxides – the new approach to high-T_c superconductivity. *Angewandte Chemie*, **5**, 757–70.
39. *Radiospectroscopy of Crystals with Phase Transitions*. (1990) Collection, IPM (Institute for Materials Science Problems), Kiev.

40. Pechoniy A. P., Glinchuk M. D., Mikheev V. A. and Babich I. G. (1990) ESR and magnetic susceptibility of localized Cu magnetic moments in YBCO ceramics. *Phase Transitions*, **29**, 105–14.
41. Glinchuk M. D., Tsendrovskij V. A. and Jakimenko J. I. (1987) Ferroelectric ceramic materials in technology of development of filtration elements of microwave electronic devices, in *Ferro- and Piezoelectrics in Acceleration of Scientific and Technical Progress*. MDNTP, Moscow. pp. 101–2.
42. Jakimenko J. I., Borisov V. N. and Klochko S. F. (1986) Features of application of piezoceramic materials for piezotransformer elements of bandpass filters, in *Abstracts of Reports to All-USSR Workshop on Ceramic Capacitor, Ferro- and Piezoelectric Materials*, Riga, p. 178.
43. Jakimenko J. I., Molchanov V. I., and Selivanov S. A. (1984) Application of piezoceramic converters of small displacements for functional electronic devices in *Abstracts of Reports to the Second All-USSR Conf. on Production and Application of Ferro- and Piezoelectric Materials*. MDNTP, Moscow, p. 20.
44. Andrushkevich S. V., Kovaleva L. S. and Tsendrovskij V. A. (1983) *Elektronika SVCh, Ser. I*, 6, 52–5.
45. Lavrinenko V. V., Kartashev I. A. and Vishnevskij V. S. (1980) *Piezoelectric Motors*, Energija, Moscow.
46. Bansjavichjus R. J., and Ragulskis K. M. (1981) *Vibromotors*, Mokslas, Vilnius.
47. Dzhagupov R. G. and Erofeev A. A. (1986) *Piezoceramic Elements in Instrument Engineering and Automation*, Mashinostroenie, Leningrad.
48. Jakimenko J. I. (1988) Piezoelectric ceramics and their application in electronics. *Ceramics for Electronics*, Proc. Int. Conf. Pardubice, Czechoslovakia, vol. 1, pp. 46–57.
49. Lavrinenko V. V. (1975) *Piezoelectric Transformers*, Energija, Moscow.
50. Kartashev I. A. and Marchenko P. B. (1978) *Piezoelectric Current Transformers*, Tekhnika, Kiev.
51. Erofeev A. A., Danov G. A. and Frolov V. N. (1988) *Piezoelectric Transformers and Their Application in Radioelectronics*, Radio i Svjaz, Moscow.
52. Smolenskij G. A., Isupov V. A., Jushin N. K. *et al.* (1985) Control drives from electrostriction ceramics. *Pis'ma Zhurn. Teor. Fiz.*, **11** (18), 1094–8.
53. Avdeev L. Z. *et al.* (1987) in: *Problems of high-temperature Superconductivity*. Publ. Board of Ural Branch of Ac. Sci. USSR, Sverdlovsk, Part 2, p. 242.

2
Glass ceramic-based composites

A.E. Rutkovskij, P.D. Sarkisov, A.A. Ivashin and V.V. Budov

High thermal shock resistance, fracture toughness, high-temperature strength, optimal tribological characteristics, chemical and erosion stability at a low density – such is a list, far from complete, of properties that should be offered by modern materials to provide serviceability of assemblies and mechanisms under extreme conditions of high temperatures, thermomechanical loads, erosion wear, corrosive environments, etc. An ever-wider application for such operating conditions is found by ceramic materials [1].

Glass ceramic materials, used in engineering, differ advantageously from oxide ceramic materials by the manufacturing technology, which allows high-density materials with properties similar to those of ceramic materials to be produced by powder metallurgy methods at relatively low temperatures (800–1000 °C) [2].

Glass ceramics, however, as ordinary ceramic materials, while having high hardness, chemical stability, wear resistance, etc., feature an inadequate impact strength, which restricts their wide use as structural materials.

Increasing the fracture toughness, strength over a broad temperature range, thermal shock resistance and erosion resistance and imparting special functional properties can be attained by reinforcing the glass ceramic matrix with various fibres (metallic and nonmetallic), whiskers, and discrete particles, i.e., by developing composites [3].

The principles and purposes of reinforcement of ceramic and glass ceramic materials are the same: increasing the strength properties (including those at high temperatures) along with upgrading the fracture toughness and thermal shock resistance. The reinforcement of glass ceramic matrices, however, is preferable from the following considerations.

1. Glass ceramic matrices offer broad and controllable ranges of the linear thermal expansion coefficient (LTEC) and elastic modulus, which makes it possible to increase either the level of strength of a reinforced composite, or the work of fracture, or, to a certain extent, both.
2. Temperatures of formation (hot pressing, sintering) of glass ceramic-based composites are within 800–1400 °C, i.e., much lower than those for most ceramic matrices.

Powder metallurgy methods allow products and semi-products from composites to be formed on the basis of both crystallized and amorphous powders with a subsequent ceramization of the latter, which also reduces both the pressing temperature and the degradation (recrystallization, brittle fracture, cracking) of reinforcing fibres [4, 5]. All this results in a homogeneous structure of a composite, a maximum utilization of the effect of reinforcements, and hence in higher strength and fracture toughness.

Research and development efforts along these lines in a number of industrial countries (USA, Japan, Germany, etc.) are very intensive [6]. Similar efforts are also under way in the former Soviet Union. This section endeavours to present all the available results of studies conducted in the former Soviet Union. The work on development of reinforced glass ceramics and glasses is primarily concentrated at the D.I. Mendeleev Moscow Chemical Technology Institute, 'Glass' Research and Production Association, Institute for Materials Science Problems, Academy of Sciences of the Ukraine (Kiev), VIAM (Moscow) and in some other organizations.

Despite a wide variety of the glass-ceramic composites being developed, they have in common that their development relies on powder metallurgy methods (pressing and sintering, hot pressing, etc.).

Since functional purposes of composite systems of one and the same group (of the same composition and starting constituents) may vary, it seems expedient to class them conventionally on the basis of the reinforcing element type:

- reinforced with metallic continuous fibres and nets;
- reinforced with inorganic fibres and whiskers;
- reinforced with discrete particles.

2.1 GLASS CERAMIC-BASED MATERIALS REINFORCED WITH METALLIC FIBRES AND NETS

The reinforcement of ceramics with metallic fibres is rather widely used for upgrading fracture toughness, strength and thermal shock resistance [7]. It is generally accepted that the concept 'reinforcement' is to be interpreted as the toughening of ceramics in the broadest sense of the word rather that solely as increasing the load-carrying capability of a material. It should be pointed out that only high-temperature, primarily refractory, metals can be used to reinforce most ceramic materials because of the high (over 1600 °C) temperatures of their sintering. However, a high-temperature heating of tungsten and molybdenum fibres (most widely used and commercially produced refractory metal fibres) involves their recrystallization and catastrophic embrittlement, which greatly impairs the effect of reinforcement and reduces the fracture toughness increase below the expected one [8]. The commercial production of fibres of tantalum, niobium, and tungsten – rhenium alloys is very limited and their cost is high [9].

In view of the above, development of glass ceramic-based composites reinforced with readily available plastic metals (steel, nichrome) is not only appropriate, but also very attractive. The use of metallic fibres having a high plasticity for reinforcement of brittle glass ceramic matrices provides localization of damages, forming at loading of the material, because of a high deformability of reinforcing elements.

A composite produced by reinforcement of glass with 50 µm diameter stainless-steel fibres (50% vol.) exhibited a bending strength of 254 MPa [10]. For a glass ceramic reinforced with nickel fibres (up to 40% vol.) the bending strength reached 182 MPa at a work of fracture of 15.7 kJ/m^2 [11].

The development of similar composites at the Institute for Materials Science Problems (Kiev) was based on the use of quartz glass and glass ceramics as matrix materials [5, 22]. Glass ceramics of the $BaO-B_2O_3-Al_2O_3-SiO_2$ system were selected, which have a high content of the glass phase and a relatively low softening temperature (800–900 °C); this made possible the use of relatively low-temperature materials – stainless steel and nichrome – as reinforcing elements.

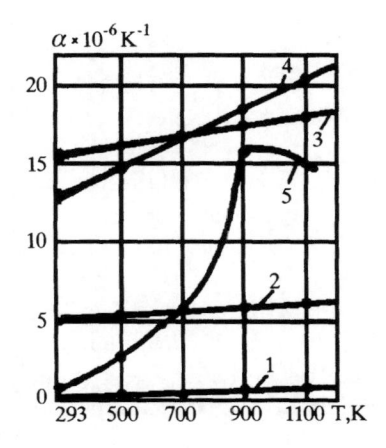

Figure 2.1 Temperature dependence of LTEC of starting materials: (1) quartz glass; (2) molybdenum; (3) stainless steel; (4) nichrome; and (5) glass ceramic STB-1.

The reinforcing fibres employed in structural composites should meet a set of operating and technological requirements, which include high strength and rigidity, stability of properties over a definite temperature range, chemical stability, etc. Reinforcing elements in composites are used in the forms of continuous wires, bundles, nets and fabrics. The possibility of developing a high-output process of manufacture of fibre-based products is determined by the adaptability of fibres to manufacture. An essential requirement is also the compatibility of fibres with the matrix material, i.e., the possibility of attaining a strong fibre–matrix bond under conditions ensuring the retention of initial mechanical properties of components.

Basic properties of fibres and matrix are presented in Figure 2.1 Table 2.1.

Table 2.1 Strength properties of starting components

Material	Density ($\times 10^3$ kg/m^3)	Elastic modulus (GPa)	Ultimate strength (MPa)
Glass ceramic STB-1	2.51	67	105
Quartz glass	2.2	74.5	110
Molybdenum	10.2	323	5740
Stainless steel	7.9	185	4200
Nichrome	8.2	205	1000

A unidirectional reinforcement results in the predomination of the strength properties of the composite in the direction of reinforcement. In the lateral direction, its strength properties are often, especially at a high volume fraction of the reinforcing phase, impaired rather than improved.

To reduce the anisotropy of properties, the manufacture of a reinforced material sometimes involves formation of unidirectionally reinforced monolayers, which are then composed in the required sequence to form complex plywood-like structures [12]. The anisotropy factor can in this case vary from one to values characteristic of unidirectionally reinforced monolayers.

Materials with a bidirectional orientation of reinforcing fibres can be produced with the use of woven nets and knitted nets (Figure 2.2) [13]. Woven nets, however, suffer from a number of serious drawbacks, the main one of which is that they can be made only from wires having a sufficient plasticity (about 20%), which are as a rule low-strength ones. The existence of contact between longitudinal and lateral wires in woven nets results in their pinching in the course of manufacture of composites, even at low pressures, One drawback, inherent also in plywood-like layerwise reinforced structures, is that reinforcement with woven nets does not improve the shear characteristics of the matrix material between layers. And, finally, the woven net reinforcement does not allow control of the material properties.

Glass ceramic-based materials

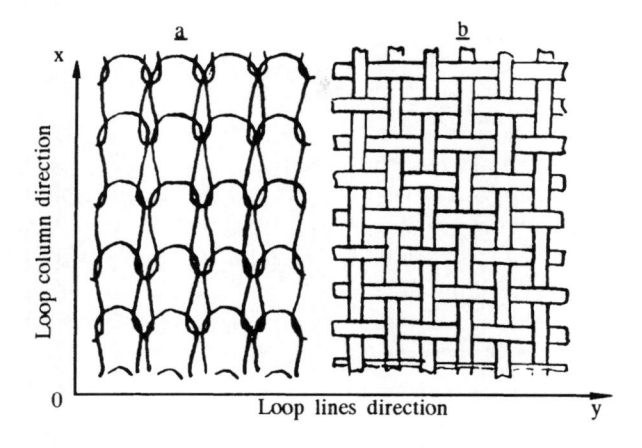

Figure 2.2 (a) Knitted and (b) woven metallic-fibre nets.

Knitted three-dimensional loop-type metallic nets, whose manufacturing technology has been developed and practically implemented at the Institute for Materials Science Problems (Kiev) [8], are free from the above disadvantages. They can be made from high-strength work-hardened metal wires having an elongation as little as 2–3% and of 0.02–0.2 mm in diameter. The mesh size and voluminosity of a net are controlled from 0.6–0.8 to 2.0–2.5 mm. A resilient three-dimensional looped structure of such nets provides lower contact stresses and a smaller number of contacts, especially with the use of powder matrices, with the result that fracture of a net occurs not in contact zones, but somewhere over the length of a wire.

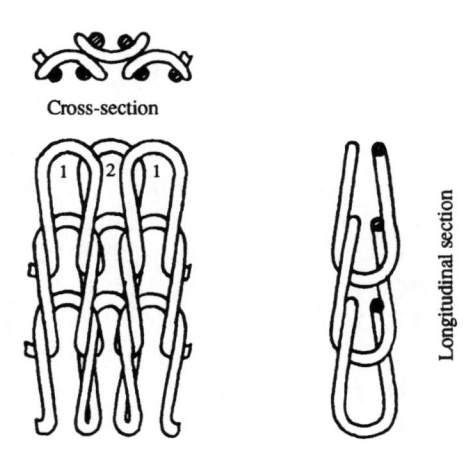

Figure 2.3 Lasting weave.

Knitted nets represent a system of interconnected loops, made by a type of knitting weave, such as 'plain', 'lasting', 'fang', etc. (Figure 2.3).

2.1.1 Moulding of materials

Hot pressing is most often used to manufacture metallic-and particularly ceramic-matrix composites [14, 15]. Its use for producing high-density glass ceramic-based composites is also attractive. Powder-based methods for manufacture of glass ceramics, which have gained a wide use, such as a semi-dry pressing, slip casting, hot die casting, and some other ones, are not very suitable for manufacture of dense, high-strength reinforced glass ceramics. The governing factor is that a pressureless sintering is incapable of producing in an acceptable time a nonporous composite, especially one reinforced with continuous fibres [16]. To increase the degree of compaction of the material in a 'cold' pressing, the pressing pressure has to be raised, but its rise above a definite value results in lamination of compacts because of an elastic after-effect of fibres.

Depending on the pressure application conditions, the pressing can be effected by a one-sided, a two-sided or an isostatic compression. A material can also be compacted by a dynamic pressing (closed-die or smith forging) or by explosion [17].

Depending on the pressure, temperature, and properties of the material being pressed, its compaction can occur through a quasi-viscous flow,

$$\frac{d\varepsilon}{d\tau} = \sigma^n,$$

where ε is the strain or shrinkage, τ the time, and σ the viscous flow stress (for exponent $n = 1$), or by a plastic deformation,

$$\frac{d\varepsilon}{d\tau} \simeq \sigma - \sigma_s,$$

where σ_s is the yield stress [18]. The plastic deformation is possible when metallic reinforcing fibres are used.

An important advantage of hot pressing over free sintering is the possibility of producing composites with the use of low thermomechanical stability fibres. When glasses or glass ceramics are used as matrices, composites can be produced at relatively low temperatures, which ensure retention of initial properties of reinforcing fibres.

The manufacture of composites by hot pressing includes the following production stages:

1. preparation and mixing of the matrix powder and reinforcing fibres;
2. preparation of the press mould and loading of the mix into it;
3. hot pressing;
4. machining of the pressed product.

Graphite is commonly used as the press mould material, as it exhibits good electrical characteristics and a higher mechanical strength in heating.

Disadvantages of the hot pressing method include a cyclic character of the process and the resulting relatively low output, a limited ability to produce complex-shape and large items.

2.1.1.1 Preparation of precursors for hot pressing

The starting powder for the matrix has to meet a number of requirements, including:

1. fine dispersity, needed for a better compaction of the material, for shortening the time and lowering the temperature of pressing;
2. homogeneity of the chemical and the size composition;
3. 'activity', which characterizes the powder's ability to fast compaction.

The manufacture of precursors of continuous and discrete fibre-based composites is substantially different. When short fibres or whiskers are used, the blend is prepared by a joint mixing of the matrix powder and reinforcing fibres; a wet or a dry mixing is used for this purpose. In the former case, suspensions based on organic or inorganic dispersion media containing the matrix material powder and fibres are prepared, while in the latter case the same components are mixed in a suitable mixer. The main criterion of quality of preparation of precursors of a composite is the uniformity of distribution of fibres throughout the matrix, which to a great extent determines the properties of the end-material.

The manufacture of a quality precursor of a composite with the use of continuous fibres depends greatly on the method of application of the matrix materials to separate fibres or to a bundle. The slip method is used most often, which consists in passing continuous fibres in the form of a bundle or band through a suspension containing the matrix material powder.

2.1.1.2 Hot pressing conditions

Selection of hot pressing conditions depends on physicochemical properties of starting components, product size and shape, and scheme of the stressed state in the pressing [15].

The process of hot pressing of a composite can be divided into several characteristic steps (Figure 2.4). At step I the blend is heated at a rate depending on the product size for a given time. As the temperature optimal for pressing the blend of the given composition has been reached, the material is heated through for a further time, needed for the temperature to equalize throughout the material (step II). Step III is characterized by a constant material loading rate. At this period there

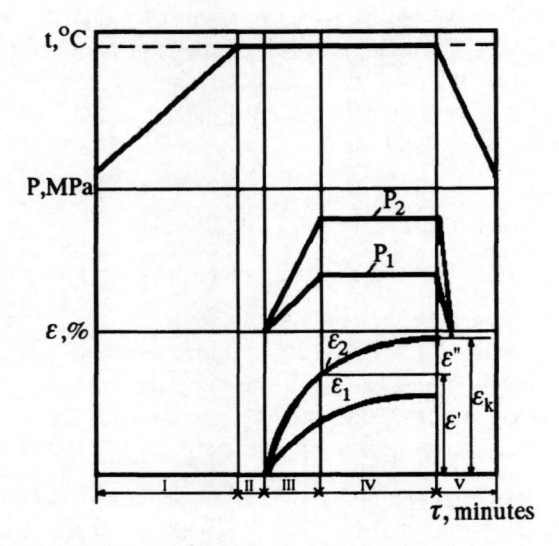

Figure 2.4 Stages and conditions of hot pressing of glass-ceramic composites.

occurs its considerable strain σ'', amounting to 60–70% of the total strain σ_h. As the loading rate is increased, the material strain rate rises. Increasing the pressing pressure to P_2 increases the material strain σ_2, especially at step III. The holding at the optimum pressing temperature (step IV) under a static pressure is needed to complete the matrix compaction process. The deformation of the material at this period continues (τ''), although at a slower rate. Step V involves removal of the external load and cooling at the required rate. Such conditions ensure a stable enough quality of products.

The composite based on a $BaO-B_2O_3-Al_2O_3-SiO_2$ glass ceramic matrix reinforced with metallic fibres (in the form of netted elements) was produced by a technology involving a layer-by-layer laying of netted semiproducts and glass ceramic powder into a graphite press mould [14]. Hot pressing was carried out in a unit with induction heating. Optimal pressing parameters were selected from results of studying the effects of temperature, pressure and pressing time on the density of pressed composites. The kinetics of compaction of reinforced glass ceramics was examined at the pressing in a temperature range of 750–1150 °C at isothermal holdings of up to 300 min under a pressure of 7.2–13.5 MPa [5]. The analysis of obtained data indicated the highest-rate increase in the relative density of materials under study to occur in a temperature range of 750–850 °C. A nonreinforced glass ceramic matrix reaches the maximum density (98% of the theoretical one) already at 850 °C. The presence of reinforcing fibres impedes the composite compaction to an extent depending on the nature and concentration of fibres. A

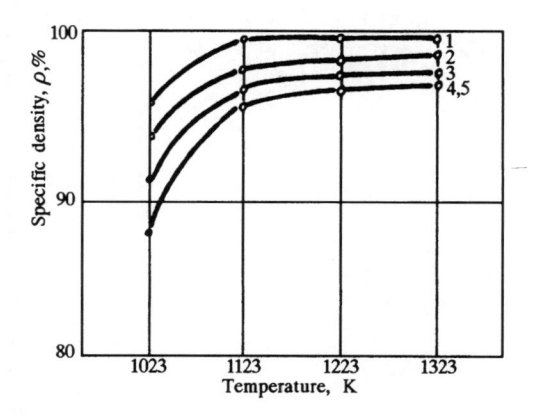

Figure 2.5 Effect of hot pressing temperature on relative density of glass-ceramic composites: (1) glass-ceramic matrix; (2) glass ceramic +2; (3) +5; (4) +10; and (5) +15% vol. metallic reinforcement.

further temperature increase does not change the relative density of the composite, but brings about intensification of the fibres – matrix interaction and fibre recrystallization processes. The dependence of density on the hot pressing temperature at a pressure of 10.8 MPa and a holding time of 60 min for composites with various fibre contents is shown in Figure 2.5.

Increasing the hot pressing pressure substantially increases the density of metallic fibre-reinforced glass ceramics, but the strength of graphite moulds restricts the possible pressures to 13.5 MPa. Figure 2.6 shows compaction curves for composites under study, having fibre concentrations from 2 to 15% vol., at various hot pressing pressures, a temperature of 950 °C,

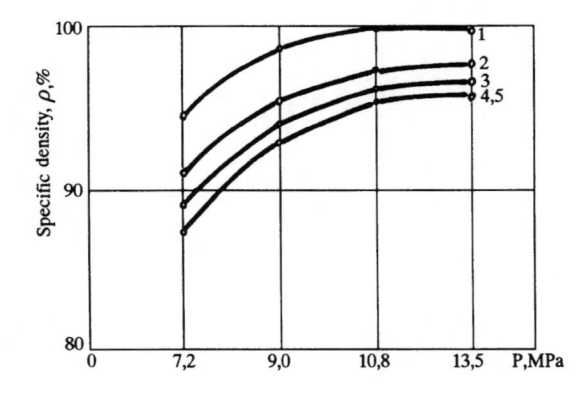

Figure 2.6 Effect of hot pressing pressure on relative density of glass-ceramic composites: (1) glass-ceramic matrix; (2) glass ceramic +2; (3) +5; (4) +10; and (5) +15% vol. metallic reinforcement.

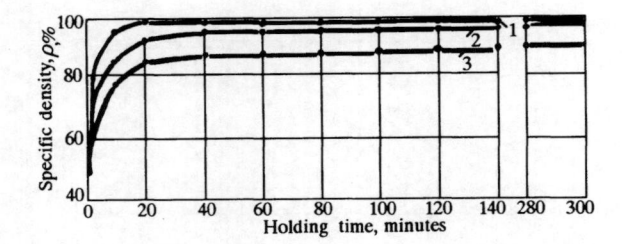

Figure 2.7 Effect of isothermal holding time on relative density of glass-ceramic composites: (1) glass-ceramic matrix; (2) glass ceramic +5; and (3) +10% vol. metallic reinforcement.

and a holding time of 60 min. A pressure increase from 7.2 to 10.8 MPa increases the composite density from 95.2 to 99.5%; a further pressure rise insignificantly affects the material density.

The effect of the holding time on the relative density of the composite is characterized by a steep increase of the density in the initial period (1–20 min), followed by its stabilization. The variation of the relative density with the holding time for composites containing 5–10% vol. reinforcing fibres is shown in Figure 2.7.

2.1.2 Thermomechanical properties of metallic net-reinforced glass ceramics

The main objective of reinforcement of brittle glass ceramic matrix with plastic metallic fibres at their volume content of up to 20% is to improve the thermal stability and impact strength of the material.

The dependence of impact strength of a composite on the volume content of metallic netted elements is shown in Figure 2.8. The initial

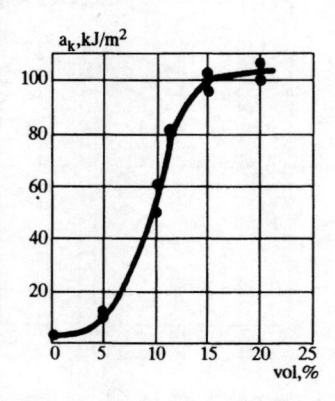

Figure 2.8 Effect of volume content of reinforcing metallic netted elements on impact strength of glass-ceramic composites.

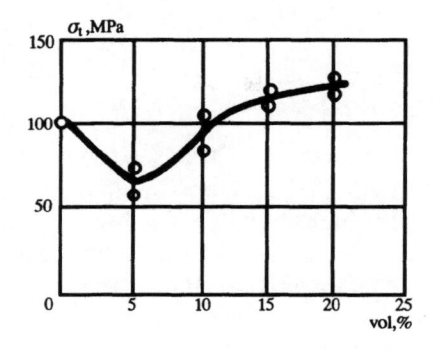

Figure 2.9 Effect of volume content of reinforcing metallic netted elements on bending strength of glass-ceramic composites,

impact strength of $BaO-B_2O_3-Al_2O_3-SiO_2$ glass ceramics is of 1.7–2.5 kJ/m^2. Addition of 14–16% reinforcing fibres raises it to 100 kJ/m^2, i.e., by a factor of 40–50. The thermal shock resistance of a composite containing 12–18% reinforcing nets amounts to 25–30 thermal cycles (800 °C – water) as against 1–2 thermal cycles for the initial glass ceramic [19].

The variation of the bending strength of the composite with the volume content of reinforcing elements (Figure 2.9) is such that at a reinforcement content of 5% the strength decreases somewhat and then rises (to 130 MPa) when up to 15% fibres are added [20]. A further increase in the content of reinforcing elements impedes production of a nonporous composite.

2.1.3 Investigation of interaction at matrix–fibre interface

The process of interaction between the matrix and fibres, proceeding both in the manufacture of a composite and after a long thermal treatment at temperatures equal to the hot pressing temperature and exceeding it was investigated by the local X-ray spectral analysis method.

The investigation demonstrated that in a hot-pressed steel fibre-reinforced glass ceramic composite, there occurs no interaction between its components [21]. A thermal treatment of the composite at 900–1000 °C for 100 h also brought about no changes in the interface. After a short-time (2–5 h) firing of the material at 1100 °C, the fibre outlines in Fe and Ni emissions were sharp, whereas the fibre–matrix interface in the Cr emission was smeared, which evidenced diffusion of chromium into the glass ceramic. According to some data, the forming interface can serve as a barrier layer, preventing a further degradation of fibres [22].

2.2 CERAMIC FIBRE-REINFORCED GLASS CERAMIC COMPOSITES

2.2.1 Reinforced glass ceramics

One of the most advantageous methods for increasing the strength, high-temperature strength and fracture toughness of glass ceramics is their reinforcement with high-modulus and high-strength ceramic fibres [5]. A composite produced from a lithium-aluminosilicate system glass ceramic ('Pyroceram 9608') as the matrix and continuous SiC fibres ('Nicalon') with a uniaxial reinforcement exhibited a bending strength of up to 400 MPa and a fracture toughness of up to 17 MPa m$^{1/2}$, i.e., severalfold those for a hot-pressed silicon nitride [5]. Essential for high-temperature applications of such materials is the fact that at high temperatures, their fracture toughness not only is not declining, but increases up to temperatures of 900–1000 °C and over.

The use as reinforcing elements of SiC whiskers, exhibiting the highest elastic modulus and strength, yields glass ceramic-based composites having high fracture toughness and strength. Thus, reinforcement of a cordierite-composition glass ceramic matrix with SiC whiskers (30% vol.) increased the bending strength of the matrix to 300 MPa and its fracture toughness to 4.37 MPa m$^{1/2}$. When a diopside matrix was used, the composite exhibited a bending strength of 550 MPa and a fracture toughness of 4.53 MPa m$^{1/2}$; for an anorthite matrix the characteristics were of 420 MPa and 4.15 MPa m$^{1/2}$ respectively [23].

Research in the field of composites in recent years has been aimed predominantly at studying the process of formation of precursors at an oriented reinforcement with continuous fibres, development of more complex two- or three-dimensional reinforcing structures, and also at the addition of whiskers and short fibres to the matrix [5]. Investigations into properties of the phase boundary, conditions of its formation and effect on the character of fracture of the composite are not less important for development of the composites under consideration. Some properties of starting constituents used to produce composites are presented in Tables 2.2 and 2.3.

Since the reinforcing fibre in a glass ceramic-matrix composite is the main load-carrying component, the strength characteristics of reinforcing elements are most fully realized at their uniform and controlled distribution throughout the matrix. The provision of such a regular structure is a complex technological task. One promising way for its fulfilment is the use of fibres pre-coated with the matrix material. For this purpose a continuous-action pilot plant for an electrophoretic deposition of powdery matrix materials, dispersed in an organic liquid, on continuous fibres, capable of varying the thickness of the coating on the fibres, has

Table 2.2 Mechanical and physical properties of reinforcing fibres

Fibre type	Melting point (K)	Density ($\times 10^3$ kg/m^3)	Bending strength (MPa)	Specific strength (MPa/kg)	Elastic modulus (GPa)	Specific rigidity (kN/m^2 kg)	Fibre diameter (µm)	LTEC ($\times 10^{-6}$ K^{-1}) at temperatures (K) 203	293
Monofilaments									
W/B	2573	2.60	2500–3500	961–1646	385–400	148–153	98–100	2.4	–
B/SiC	2573	2.63	3500	1330	385–400	146–152	104–145	–	5.22
W/SiC	2963	4.09	2100	513	490	120	100	3.3	3.24
Filamentary fibres									
Al$_2$O$_3$	–	–	1400–1800	–	–	–	12–15	–	–
SiC	–	–	1700	–	140	–	20	3.1	–
Whiskers									
SiC	–	3.20	10 000–15 000	–	500	–	0.1–0.25	3.0	3.2

Table 2.3 Physicomechanical properties of initial glass-ceramic matrices

System	Main phase	Grade	Density ($\times 10^3$ kg/m^3)	LTEC ($\times 10^{-6}$ K^{-1})	K	E (GPa)
MgO–Al$_2$O$_3$–SiO$_2$	Cordierite	STM-1	2.8	293–1.1 773–4.0 1273–5.7	1493	13.2
B$_2$O$_3$–BaO–Al$_2$O$_3$–SiO$_2$	Mullite	STB-1	2.5	293–2.7 323–6.0 1173–10.7	1593	6.7

been developed at the Institute for Materials Science Problems (Kiev). Its operating principle consists in that charged particles of one and the same sign, acted upon by a constant electric field, move to an oppositely charged electrode to form a deposit, whose structure and thickness depend on two parameters: the strength of the field and the duration of its action. The electrophoretic method provides a relatively simple control of the amount of reinforcing elements in the composite and their distribution.

Based on glass ceramic matrices and reinforcing fibres, whose properties are presented in Tables 2.2 and 2.3, two composite systems have been developed: $BaO-B_2O_3-Al_2O_3-SiO_2$ glass ceramic and continuous borsic fibres (on tungsten core); and cordierite-composition glass ceramic and continuous SiC fibres (on tungsten core) [2, 24]. The composites, containing 13–30% vol. reinforcing phase, were produced by hot pressing. Increasing the volume content of reinforcing elements raises the pressing temperature by 50–100 °C. The bending strength of the produced materials amounts to 170 and 350 MPa for composites with 10 and 20% vol. reinforcing fibres (B/SiC) respectively.

It should be pointed out that a well-founded selection of the chemical and phase compositions of a glass ceramic matrix can only be based on studies of the processes of its compaction and crystallization in the hot pressing. The expected toughening effect can in some cases, such as at the addition of continuous SiC fibres to a cordierite-composition glass ceramic matrix, be absent [25] due to a shift of the beginning of the process of separation of the principal crystalline phase, cordierite, at a simultaneous action of temperature and pressure to temperatures 100–150 °C lower than those in the thermal treatment of the same glass powder at a free sintering [26]. A rapid viscosity increase at crystallization of the matrix impedes impregnation of reinforcing fibres with the melt and densification of the composite.

2.2.2 Mechanical and corrosion properties of reinforced glass ceramics after one-sided radiant heating

Tests of materials at a simultaneous action on them of concentrated energy fluxes and corrosive media are of both a scientific and a practical interest [27]. The procedure and set-up for such studies have been developed at the Institute for Materials Science Problems (Kiev). The set-up is based on a radio-astronomic antenna, whose surface carries triangular mirrors. It can produce a radiant heat flux density of 900 kW/m^2 in the focal spot. The set-up has an automatic sun-aiming and tracking system. Samples under examination were quenched after every heating cycle by a natural cooling in the air or by spraying sea-water on the irradiated surface of the sample. A composite consisting of a

$BaO-B_2O_3-Al_2O_3-SiO_2$ glass ceramic reinforced with discrete SiC fibres and netted elements from stainless steel fibres (15% vol.) retained its strength after 450 'heating–cooling in the air' cycles [28]. When heated samples were quenched with sea-water, the strength declined by 4–5% after 300 cycles. Some increase in the strength of the composite after 70–100 cycles was noted, which is probably accounted for by 'healing' of surface defects of samples.

2.2.3 Chemical stability of glass ceramic-based composites at elevated temperatures

Glass ceramics, as ordinary ceramics, belong to materials having a high chemical stability. At a combined action of chemical reagents, elevated temperatures and frequent thermal cycles, however, the chemical stability of glass ceramic materials is inadequate. Studies of reinforced barium–boron–aluminosilicate glass ceramic composites were conducted with the aim to develop materials for operation in reaction chambers and pipelines at elevated temperatures and pressures, in chlorine-containing media [29]. Netted elements made from nichrome fibres, discrete SiC fibres (with a tungsten core), and discrete basalt fibres were used as reinforcing elements. Chemical stability of the composites was studied on a closed-cycle set-up at temperatures of up to 600 °C and a constant pressure of up to 100 atm; the test time was up to 100 h. The chemical stability of the SiC and nichrome fibre-based composite turned out to be 3–6 times that of chrome–nickel alloys used in the industry. The 'glass ceramic–nichrome' (8–12% vol.) system turned out to be the best.

2.2.4 Structure of composites

Structural studies of the prepared composites involved determining the uniformity of distribution of reinforcing fibres for various methods of their incorporation into the matrix and investigating the structure of the fibre–matrix interface. Structures of materials with different contents of reinforcing metallic fibres are shown in Figure 2.10. The composite was produced by hot pressing with a layer-by-layer laying of netted elements. It was ascertained that the distribution of reinforcing fibres is uniform enough and the degree of uniformity increases with their volume content [30]. The method of drawing continuous SiC fibres through a suspension containing the matrix material powder, followed by hot pressing of the precursor, also yielded a composite with a highly uniform distribution of fibres throughout the matrix (Figure 2.11). Optical and electronic microscope studies of the fibre–matrix interfaces in samples of the composites after the hot pressing revealed no appreciable indications of interaction between them (Figure 2.12).

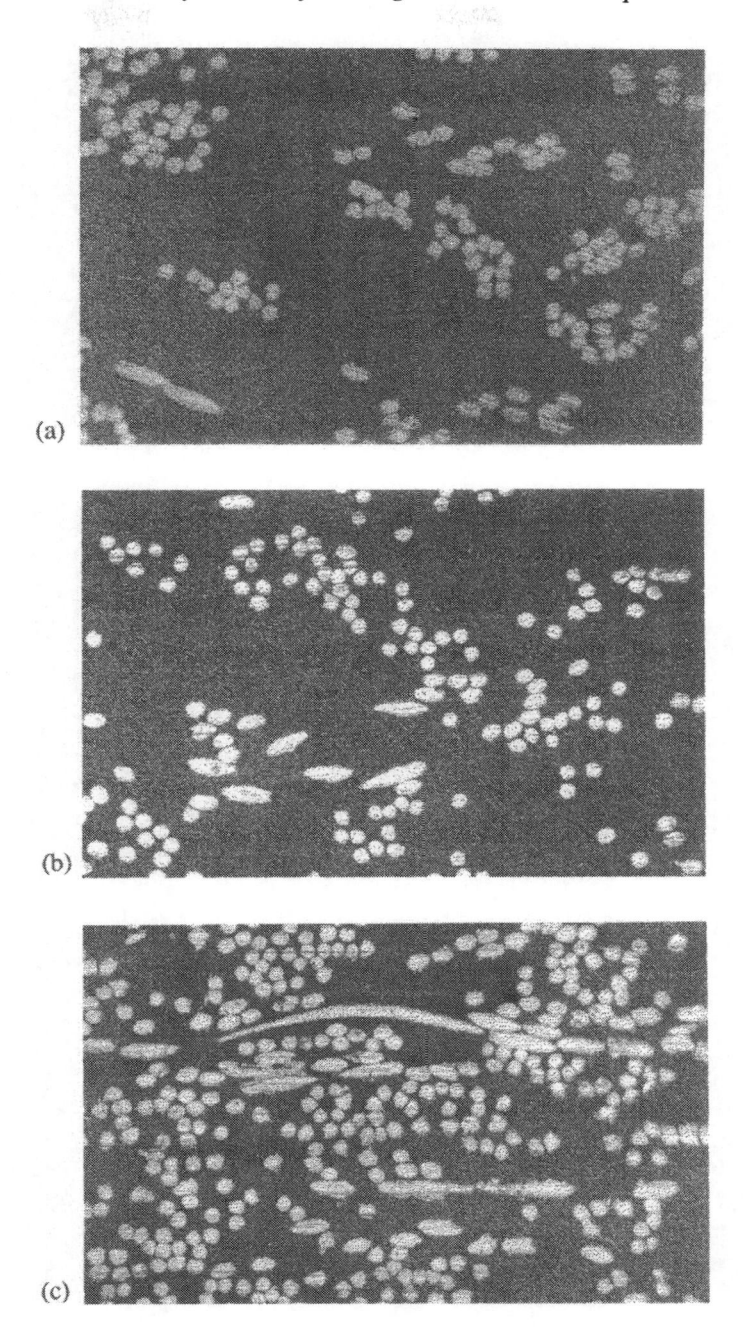

Figure 2.10 Structure of boron–barium glass ceramic reinforced with metallic-fibre nets; (a) 4–5% vol.; (b) 8–9% vol.; (c) 13–15% vol.

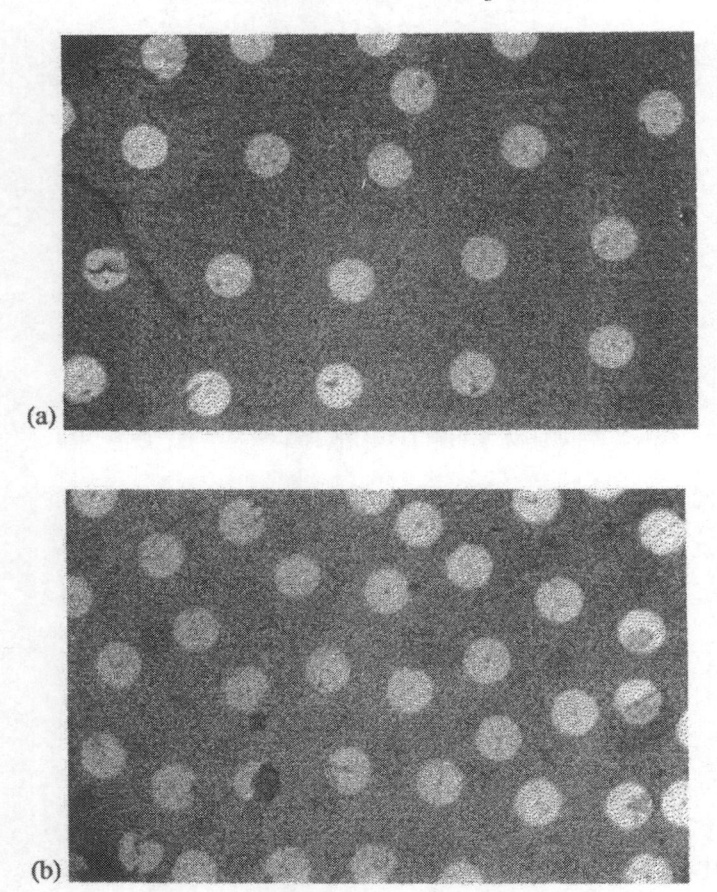

Figure 2.11 Microstructure of glass ceramic reinforced with continuous fibres; (a) 10% vol.; (b) 20% vol.

2.3 PARTICLE-FILLED GLASS CERAMIC-BASED MATERIALS

2.3.1 Filled glass ceramics

To meet various, often contradictory, requirements related to operating conditions of friction units, it is necessary to develop composite antifriction materials. Here the powder metallurgy methods often turn out to be the most effective, both in the technological and in the economic respect.

A limited field of application of liquid and grease lubricants spurred the search for antifriction materials capable of operation at high temperatures without a lubricant in the working zone. An extensive commercial application has been found by metal-based heterogeneous

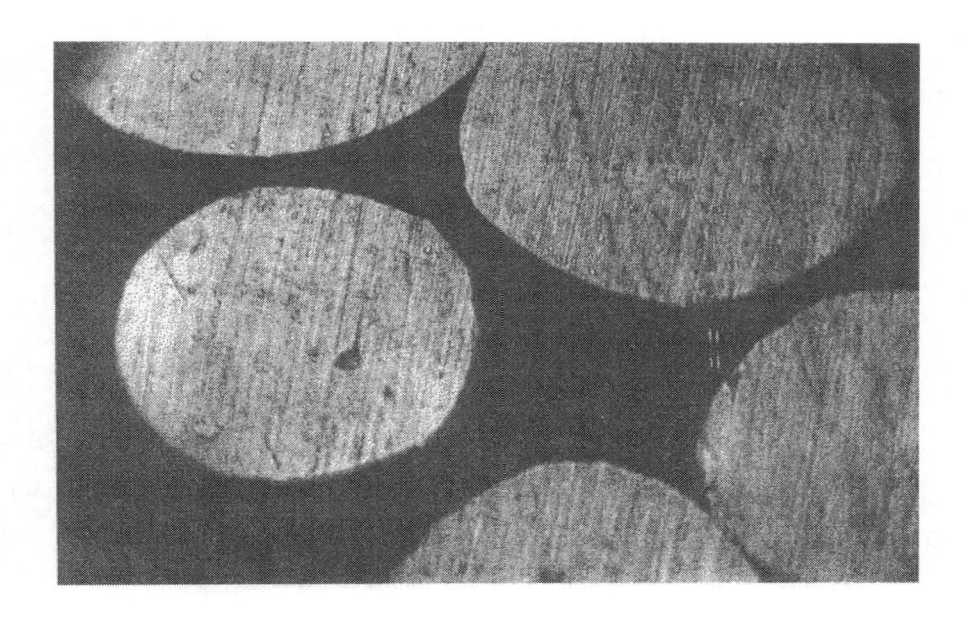

Figure 2.12 Fibre–matrix contact areas (boron–barium glass ceramic and stainless steel-fibre net).

antifriction materials [31]. At the same time the replacement of metals in bearings with ceramics increases their service life, obviates the need for lubrication, reduces the heat loss, allows the service temperature and rotation speed to be increased, and cuts down the fuel consumption, i.e., yields a great economic gain. An extensive application of ceramics in bearings, however, is retarded by their brittleness, inadequate thermal shock resistance and thermal conductivity.

Good candidates for this purpose are glass ceramic-based composites. The research conducted at the Institute for Materials Science Problems (Kiev) demonstrated the addition of plastic metallic powders to glass ceramic matrices to change qualitatively the behaviour of the composites under friction loading conditions as compared with that of ordinary ceramics. The incorporation of metallic fillers increases the thermal conductivity of composites, i.e., facilitates the heat removal from the contact zone within the bulk of material and also initiates formation of shielding films on friction surfaces, which prevents development of the seizure process and enhances the serviceability of the friction pair.

A developed triboengineering-purpose glass ceramic-based composite contains a metallic filler in the form of powder, discrete fibres, and also their binary mixture [5].

Reinforcement of a $BaO-B_2O_3-Al_2O_3-SiO_2$ glass ceramic with copper fibres (30–50% mass) increased the compressive strength from 800–900 to 3500–4000 MPa; the thermal conductivity in the direction perpendicular to the pressing one increased 1.3-fold. The friction coefficient was determined at tests in the air without lubrication at a constant pressure of 1 MPa and a sliding velocity varying from 6 to 12 m/s by a bearing-shaft scheme.

Comparison of friction characteristics of materials with various powder fillers demonstrated that the best properties at a sliding velocity of 6–8m/s in the air were exhibited by a sample containing 30% mass tin, but increasing the sliding velocity to 10–12 m/s brought about deterioration of its friction properties. The loss of serviceability by the composite resulted from melting of the filler (T_m = 232 °C) and its intense oxidation, since the temperature in the friction zone under such conditions amounts to 250–290 °C.

The most successful operation under severe friction conditions was displayed by samples of a composite containing 30% mass copper (friction coefficient f = 0.20–0.34). At friction in the air, it exhibited a stable performance at loads up to 3 MPa (f = 0.27–0.34). When this limit was exceeded, the wear resistance dropped steeply. Reinforcement of the glass ceramic with copper fibres in the same amount reduced the friction coefficient of the composite to 0.20–0.25.

Samples of the composite with a nickel filler had unsatisfactory properties, which apparently resulted from a much lower thermal conductivity of nickel ($\lambda = 1.428 \times 10^{-4}$ kcal/(cm s deg)) than that of copper ($\lambda = 0.984$ kcal/(cm s deg)) as well as from formation of a number of 'nickel–iron' solid solutions, causing seizure of such a material with a steel counterbody.

2.3.2 Composites with nonmetallic particles

The need for materials capable of operation under higher thermomechanical loads, including high temperatures, spurred the research on 'glass ceramic–crystalline filler' powder composite systems. Various refractory compounds and oxides – mullite, spinel, magnesium silicates, zircon, alumina, zirconia, etc. – can be used as the filler. When incorporated into the composition of a crystallizing glass, the filler does not remain inert with respect to the matrix, but interacts with it to form new crystalline phases, whose separation in the crystallization of the glass is impeded. The structure of such materials, forming in the sintering with the interaction, has been called 'reaction-formed structure' (RFS). Such an approach to developing new glass-ceramic materials has turned out to be fairly fruitful and is being successfully advanced at the D.I. Mendeleev Moscow Chemical Technology Institute [32].

Based on compositions consisting of a $MgO–Al_2O_3–SiO_2$ crystallizing glass and alumina of various modifications, a mullite-containing material has been obtained, exhibiting high fracture toughness ($K_{Ic} = 3.6$ MPa m$^{1/2}$) and bending strength over a broad temperature range (110 MPa at 1000 °C; 52–73 MPa at 1400 °C [33]. The thermal stability of the material amounts to 855 °C and the loss of strength at a thermal shock (1300–1320 °C) is not over 10–20%. The corundum content in the optimum composition with glass was of 50% mass, which corresponded to a full fixation of silica into mullite ($3Al_2O_3–2SiO_2$).

Studying the sintering of such composites demonstrated that it proceeded somewhat differently from that of the initial crystallizing glass (Figure 2.13) because of a different character of phase transformations. At a temperature over 1400 °C, mullite and spinel appear at a concurrent decrease in the corundum content. Such a change in the phase composition evidences an intense interaction between corundum and glass components. The maximum amount of mullite is formed at 1500 °C. Such an interaction can proceed in two ways: through dissolution of corundum in the melt, followed by crystallization of mullite, or at a direct contact of corundum grains with the liquid phase. A decline in the mullite yield and increase in the amount of the glassy phase with growing degree of interaction of corundum with the melt at its great excess with respect to glass (90%) indicates a significant impact of the former of the two processes. It is of interest that as the temperature exceeds 1550 °C, the mullite content declines once again at a concurrent increase in the

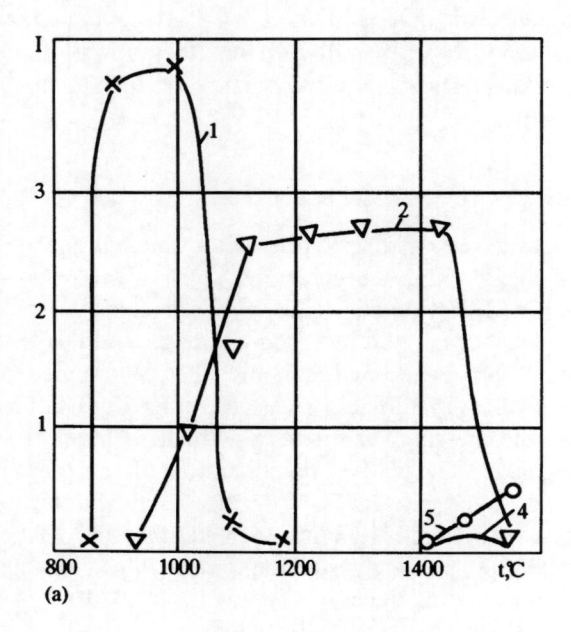

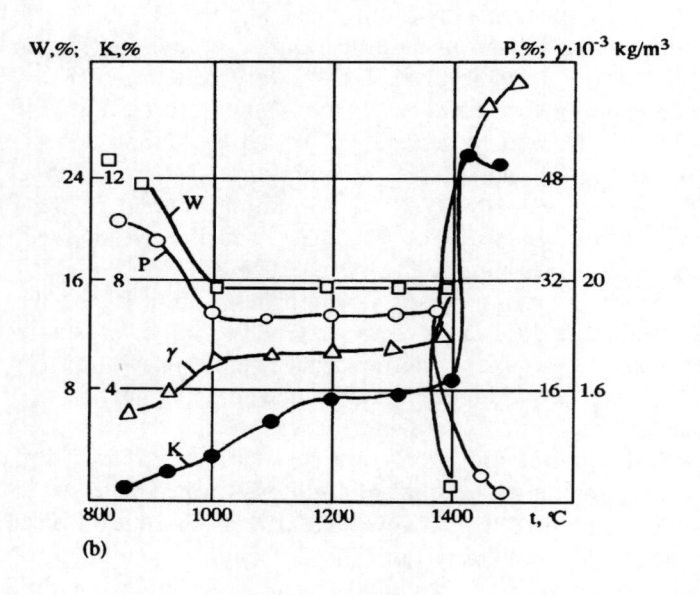

Figure 2.13 Phase transformations (a, c) and variation of properties of material (b, d) in the course of sintering of cordierite glass (a, b) and of its 1:1 mixture with corundum (c, d): (1) solid solutions with structure of β-quartz; (2) α-cordierite; (3) corundum; (4) spinel; (5) mullite; W is water uptake; P is open porosity; K is shrinkage factor; γ is apparent density.

(c)

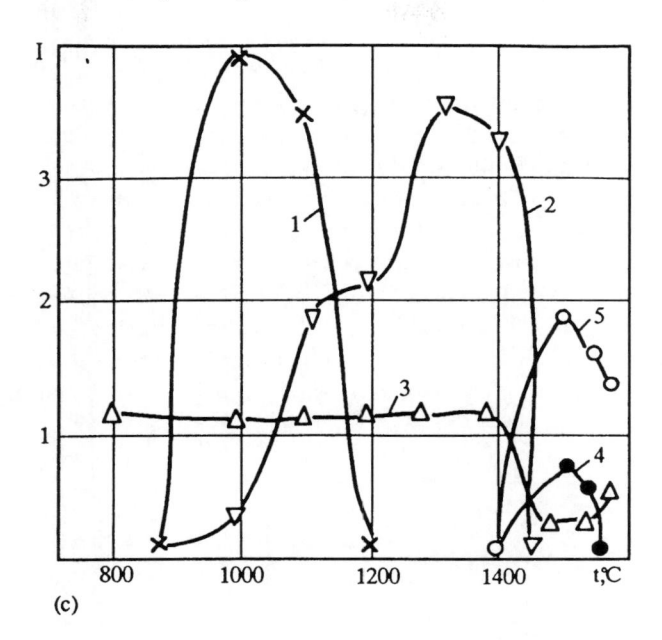

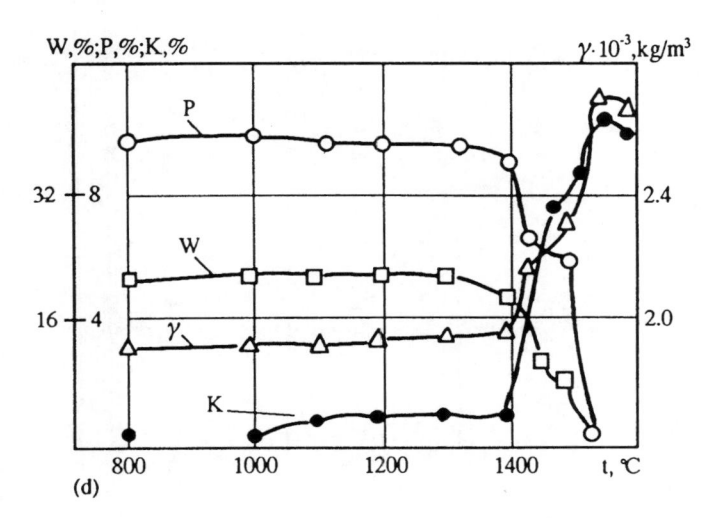

(d)

amount of corundum (Figure 2.13c), which evidences the incompleteness of the processes in the range of 1400–1550 °C and a nonequilibrium state of the system as a whole. As indicated by the X-ray phase analysis (Figure 2.14), variation of the amount of the added alumina affects the completeness of the mullite formation reaction and the mullite yield (with respect to the theoretically possible one), which amounts to 73–78%. The corundum–glass interaction reaction is also indicated by the fact that the degree of the interaction, characterized by the real-to-theoretical corundum consumption ratio, is maximal at a great excess of one of the components (Figure 2.14): glass (at 20% corundum) or corundum (90%).

A similar approach was employed to produce a thermal shock-resistant composite based on β-spodumenic solid solutions with the LTEC of

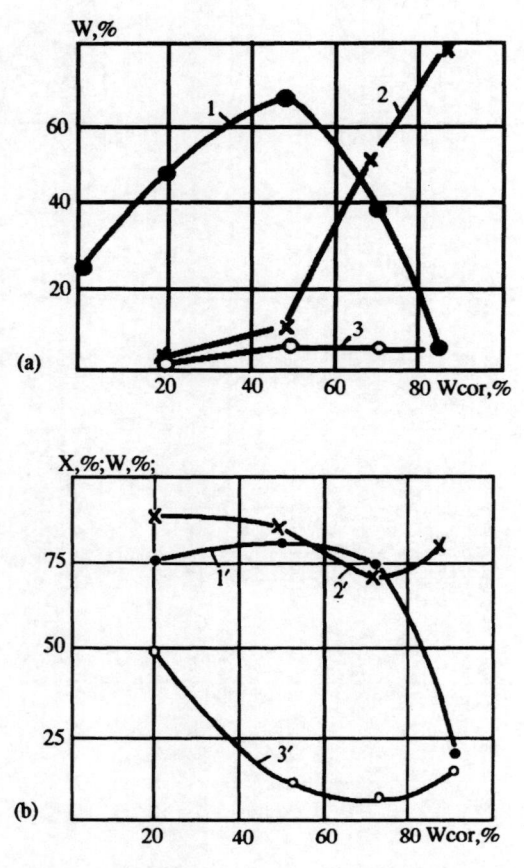

Figure 2.14 Dependence of phase composition of material (a) and of mullite formation process parameters (b) on corundum content in initial blend (sintering at 1500 °C for 1 h): (1) mullite; (2) corundum; (3) spinel; (1′) mullite yield (W); (2′) degree of corundum interaction with melt (X); (3′) glass phase content (C).

$5\text{--}15 \times 10^{-7}\,°C^{-1}$ by sintering of blends of a $Li_2O\text{--}Al_2O_3\text{--}SiO_2$ crystallizing glass with corundum [32]. In contrast to the above-described case, the alumina interaction with glass components occurs long before the melt appearance, at temperatures as low as 800–900 °C, where formation of solid solutions in the material along with decline in the lithium disilicate and quartz contents is observed. On the other hand, crystalline phases that appear in sintering of such blends are identical to those separating in sintered materials based on glass whose chemical composition corresponds to that of the composite under consideration. This confirms the governing role of the chemical composition of a composite, of its position in the phase diagram in formation of the structure of materials regardless of the method of introduction of components, either through glass or as a crystalline additive in the composite.

The use of the RFS method turned out to be effective also to produce glass ceramic-based high-temperature composite coatings. An example of such a composite system is $Bi_2O_3\text{--}P_2O_5\text{--}B\text{--}ZrO_2$, whose structure results from complex chemical and phase transformations in the thermal treatment and provides high performance characteristics of coatings: strength and heat-resistance.

The above-presented results are only individual examples, illustrating the great capabilities of glass-ceramic materials with a reaction-formed structure, which can be successfully employed also in developing highly refractory and highly strong matrices for reinforcing them with various fibres.

A significant progress in the development of ceramic oxide and nonoxide fibres produced by conversion of organometallic polymers has occurred in recent years, and therefore results of the research into 'glass ceramic–crystalline filler' composites may prove very useful for insight into complex physicochemical processes proceeding in 'glass ceramic–ceramic fibre' composites.

2.3.3 Impact of microstructure of glass-ceramic matrices on properties of composites

The strength of composites is greatly dependent on the level of microstresses arising at the phase interface as a result of different properties of the matrix and reinforcing material, first of all, the thermal expansion. When a glass-ceramic material is used as the matrix, the analysis of the stressed state and of its impact on physicomechanical properties of the composite is greatly complicated as the matrix itself is a heterogeneous, essentially composite material. The arising microstresses can, on the one hand, result in compression of the matrix, thereby enhancing its resistance to external loads, while, on the other hand, stresses at the interface can give rise to microcracks, which in most cases impair the material strength.

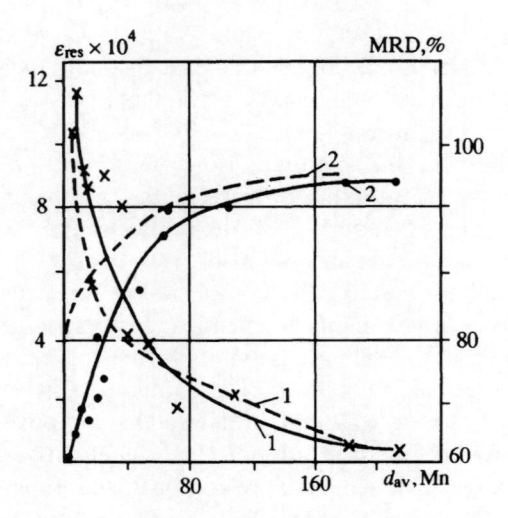

Figure 2.15 Dependence of residual strain (ε_{res}) (1) and of microstress relaxation degree (MRD) (2) on size of quartz crystals at their contents of 20% mass (solid lines) and 46% mass (broken lines) in glass matrix.

The doubt as to conditions where one or other impact of microstresses on the strength occurs, which so far exists, stems from absence of a unified theory of fracture of composite materials and a lack of studies dealing with a direct measurement of microstresses in heterogeneous materials.

In view of this, research on a radiographic determination of the degree of relaxation of microstresses and its relation with the strength of glass-ceramic materials [34, 35] is of great interest. The impact of three basic factors – crystal size, content of the crystalline phase, and difference of LTEC of constituent phases – on the degree of relaxation of the forming microstresses (MRD) was studied on model composite systems prepared by sintering of crystalline quartz, rutyl and zircon powders with various glass phases as well as on rutyl-containing cordierite glass ceramics. It was shown that as the crystal size rises from 3 to 300 μm, residual deformations decline, while the MRD grows (Figure 2.15). As the amount of the crystalline phases in composite samples containing quartz and zircon increases, the MRD decreases in spite of a growth of stresses in the glass phase (Figure 2.16). This fact shows retardation of propagating cracks at the phase interface.

Of great important for the development of glass ceramic-matrix composites are results of studies of the impact of different linear thermal expansion coefficients (LTEC) of composite constituents on the MRD. It was found that the MRD increase with increasing LTEC difference reduces the strength of a composite (more steeply if the LTEC of crystals is less than that of glass) because of formation of radial cracks, which are

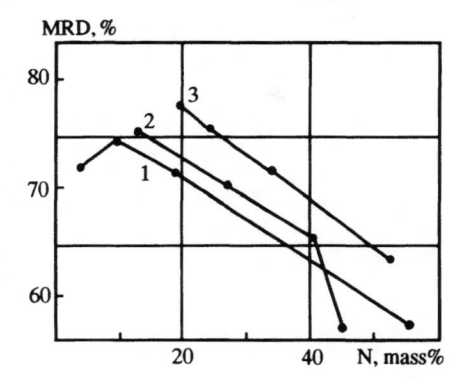

Figure 2.16 Variation of microstress relaxation degree (MRD) with crystalline phase content: (1) quartz; (2) zircon (glass phase with LTEC = $90 \times 10^{-7}\,°C^{-1}$); (3) zircon (glass phase with LTEC = $112 \times 10^{-7}\,°C^{-1}$).

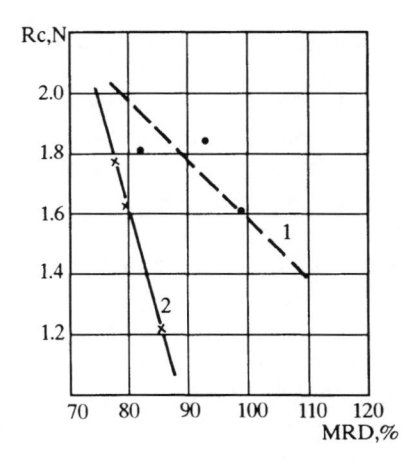

Figure 2.17 Variation of critical load, proportional to microstrength, with microstress relaxation degree (MRD) for 'glass-rutyl' composites (1) at $\alpha_{cr} > \alpha_{st}$ and (2) $\alpha_{cr} < \alpha_{st}$.

dangerous for the material (Figure 2.17). Studying the bending strength dependence on the MRD as well as on the residual deformation demonstrated the strength characteristics of investigated model composites to be governed not by theoretical or residual deformations, but by the degree of microcrack of the material, expressed in terms of the MRD.

It must be emphasized that the above-presented results have been obtained on model glass-ceramic materials containing up to 50% mass crystalline filler and may be employed in developing composites based on reinforced glasses and glass ceramics with a small content of the crystalline phase.

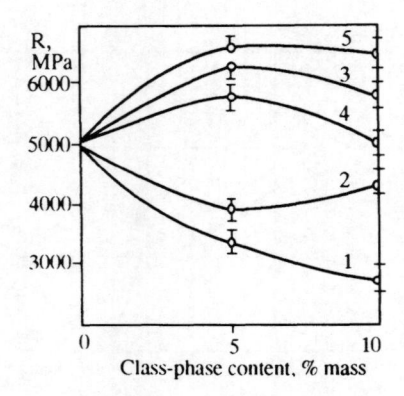

Figure 2.18 Variation of microbrittleness with calculated content of sodium-aluminosilicate glass phase with various LTEC ($\alpha \times 10^{-7}\,°C^{-1}$): (1) 96; (2) 70; (3) 42; (4) 27; (5) 5.8.

Analysis of the stressed state in glass ceramics of Li_2O–Al_2O_3–SiO_2 and MgO–Al_2O_3–SiO_2 systems, most attractive for development of composites, which contain up to 90% crystalline phase, presents certain difficulties stemming from their structural features. Due to this, findings of the study [36] dealing with determination of microdeformations in crystals which form the basis of glass ceramics with a high (over 90%) content of crystalline phase, as well as with regularities of the impact of structural parameters on the strength of glass ceramics, are of considerable practical and theoretical interest. Subjects of the study were model glasses of a cordierite composition, modified by additions of Na_2O–Al_2O_3–SiO_2 glasses with various LTEC (from 5.8×10^{-7} to $96 \times 10^{-7}\,°C^{-1}$). It was ascertained that the presence in a heterogeneous material of a glass phase with a LTEC much greater (over $70 \times 10^{-6}\,°C^{-1}$) than that of cordierite ($20 \times 10^{-6}\,°C^{-1}$) impairs mechanical properties (microbrittleness and microhardness) of glass ceramics (Figure 2.18). The addition of a glass ceramic with a low LTEC (up to $40 \times 10^{-7}\,°C^{-1}$), close to that of cordierite, improves the properties. The experimentally determined values of inhomogeneous microdeformations in cordierite crystals at high predetermined oriented stresses are due to relaxation of microstresses through microcracking which occurs as certain critical values of oriented strains (0.1×10^{-2}) have been reached.

The development of composites where the mechanism of microcracking only in the region of a high crack stress will be incorporated may be fairly promising, as this makes it possible to attain a relatively small power decrease, brought about by microcracking, and thereby to upgrade the crack resistance while retaining the strength of a composite [37].

A strong impact of the thermomechanical matching of constituents of a composite on the character of its fracture and level of its physicomechanical properties has been demonstrated with real 'LAS glass ceramics–mullite fibres' and 'borosilicate glass–SiC fibres' composites [38, 39].

REFERENCES

1. Hilling W. (1987) Strength and toughness of ceramic matrix composites. *Annual Review of Materials Science*, **17**, 341–83.
2. Berezhnoj A. I. (1976) *Glass Ceramics and Photoglass Ceramics*, Mashinostroenie, Moscow.
3. Strnad Z. (1989) *Glass-ceramic Materials*. Strojizdat, Moscow.
4. Prewo K. M. (1987) Glass and ceramic matrix composites: present and future. *Materials Research Soc.*, **120**, 145–6.
5. Rutkovskij A. E., Ivashin A. A., Alekseenko I. P. *et al.* (1990) Technology of structural glass ceramics-based reinforced composites. *Abstract of Reports to Moscow Int. Conf. on Composites*, Moscow, Part 2, pp. 78–9.
6. Kohuo Tohree and Nishino Yoshio. (1981) Development of sintered bearings for high speed revolution applications. *Mod. Dev. Powder Met.*, **12** (7), 855–70.
7. Walton J. and Corbett W. (1965) Metal fibre reinforced ceramics, in *Fiber Composite Materials*, Metals Park, Ohio.
8. Frantsevich I. N., Karpinos D. M., Rutkovskij A. E. *et al.* (1986) Tungsten- and molybdenum-based fibrous materials with oriented structure. Proc. of First All-USSR Conf. on Precipitation-hardened and Fibrous Materials. Kiev, pp. 13–25.
9. Savitskij E. M. and Burkhanov G. S. *Physical Metallurgy of Refractory Metals and Alloys*, Nauka, Moscow.
10. Gross R. (1972) FRG Patent No. 2,103,798. Attrition-, Impact-, and Corrosion-Resistant Material from Glass with Steel Inserts. Publ. 3 Aug.
11. Donald I. W. and McMillan P. W. (1977) The influence of internal stresses on mechanical behavior of glass-ceramic composites. *J. Mater. Sci.*, **12**, 290–8.
12. Karpinos D. M., Tuchinskij L. I., and Vishnjakov L. R. (1977) *New Composite Materials*, Vyshcha Shkola, Kiev.
13. Dalidovich A. S. (1970) *Fundamentals of Knitting Theory*, Legkaja Industrija, Moscow.
14. Ivashin A. A., Rutkovskij A. E. and Borshchevskij D. F. (1987) Hot pressing as method for producing technical glass-based reinforced composites, in *Manufacturing Technology and Properties of Powder Composites*, Penza, pp. 3–7.
15. Samsonov G. V. and Kovalchenko M. S. (1962) *Hot Pressing*, Gos. Izd. Tekhn. Literatury, Kiev.
16. Karpinos D. M., Tuchinskij L. I. *et al.* (1976) Sintering of two-phase composites with nondeformable inclusions. *Poroshkovaja Metallurgija*, **7**, 29.
17. Dorofeev J. G. (1972) *Dynamic Hot Pressing in Cermet Technology*, Metallurgija, Moscow.
18. Jones W. D. (1960) *Fundamental Principles of Powder Metallurgy*, Edward Arnold, London.
19. Tykachinskaja P. D., Budov V. V. and Khodakovskaja R. J. (1990) Effect of reinforcing fiber content on process of fracture of composites, in *Abstracts of Reports to Moscow Int. Conf. on Composites*, Moscow, Part 2, p. 194.
20. Cuper G. A. (1974) Micromechanical aspects of fracture, in *Composite Materials*, vol. 5, Fracture and Fatigue, (ed. Brouton L. I.), Academic Press, New York and London.

21. Rutkovskij A. E., Aleksenko I. P., Ivashin A. A. et al. (1988) Structure, strength and thermophysical properties of glass ceramics reinforced with continuous inorganic fibers, in *Application of Composites in Mechanical Engineering*, Minsk, pp. 170–1.
22. Rutkovskij A. E., Ivashin A. A. and Dzeganovskij V. P. Study of interaction between metallic fibers and glass-ceramic matrix after high-temperature annealings. *Poroshkovaja Metallurgija*, (in press).
23. Masahiro Ashizuka, Yoshinori Aimoto and Masahiro Watanabe. (1989) Mechanical properties of SiC whisker reinforced glass ceramic composites. *J. Ceram. Soc. Jap., Int. Ed.*, **97**, 789–90.
24. Antsiferov V. N., Sokolkin Ju. V., Pashkinov A. A. et al. (1990) *Titanium-based Fibrous Materials*, Nauka, Moscow, pp. 11–14.
25. Tykachinskaja P. D., Budov V. V., and Khodakovskaja R. J. (1990) Glass sintering and crystallization processes in hot pressing, in *Modern Problems of Powder Metallurgy, Ceramics, and Composites*, Institute for Materials Science Problems, Ac. Sci. Ukr. SSR, Kiev, pp. 61–4.
26. Khodakovskaja R. J. (1989) Glass-ceramic materials with reaction-formed structure. *Steklo i Keramika*, **6**, 36–9.
27. Frolov G. A., Polezhaev J. V., Pasichnyj V. V. and Zakharov F. I. (1981) Investigation of parameters of failure of thermomagnetic materials under nonstationary heating conditions. *Inzhenerno-Fizicheskij Zhurnal*, **40**, 608–14.
28. Aleksenko I. P., Podsosonyj V. V., Pasichnyj V. V. et al. (1988) Impact of cyclic thermal loads on structure and properties of reinforced glass ceramics, in *Application of Composites in Mechanical Engineering*, Minsk.
29. Rutkovskij A. E., Chekhovskij A. A. and Aleksenko I. P. Investigation of chemical stability of reinforced glass ceramics. *Poroshkovaja Metallurgija*, in press.
30. Rutkovskij A. E., Ivashin A. A., Artjukhova S. G. and Aleksenko I. P. (1979) Study of impact of reinforced glass ceramic-based structure on its physico-mechanical properties, in *Real Structure of Inorganic High-Temperature Materials*, Pervourlask, p. 291.
31. Ganz S. N. and Parkhomenko V. D. (1965) *Antifriction Chemically Stable Materials in Mechanical Engineering*, Mashinostroenie, Moscow.
32. Khodakovskaja R. J., and Tamm D. (1989) Mullite glass-ceramic materials based on magnesia aluminosilicate glasses–corundum compounds. *Proc. of D. I. Mendeleev MKhTI*, **157**, 67–75.
33. Levi E. A., Khodakovskaja R. J., Pobedimskaja E. A, and Belov N. V. (1980) Radiographic measurement of microstresses in glass-ceramic materials. *Dokl. AN SSSR*, **255**(3), 572–7.
34. Levi E. A., Khodakovskaja R. J. and Rudenko L. V. (1989) Radiographic determination of microstress relaxation degree and its relation to strength of glass-ceramic materials. *Izv. AN SSSR, Neorg. Mater.*, **25**, 496–501.
35. Khodakovskaja R. J., Baschenko J. V. and Gertsman A. F. (1987) Effect of glassy phase composition on fine structure of crystals and mechanical properties of $MgO-Al_2O_3-SiO_2$ system glass ceramics. *Proc. of D. I. Mendeleev MKhTI*, **146**, 117–26.
36. Rice R. W. (1981) Mechanisms of toughening in ceramic matrix composites. *Ceram. Eng. Sci. Proc.*, **2**(7–8), 661–701.
37. Borom, R. W. and Johnson C. A. (1987) Thermomechanical mismatch in ceramic fibre reinforced glass-ceramic composites. *J. Amer. Ceram. Soc.*, **70**(1), 1–8.

38. Desmukh U. V., Kanei A., Freiman S. W. and Craumer D. C. (1988) Effect of thermal expansion mismatch on fiber pull-out. *Materials Research Soc.*, **120**, 253–8.

FURTHER READING

Karpinos D. M. (ed.) (1985) *Composites*, Naukova Dumka, Kiev, pp. 212–4.
USSR Inventor's Certificate No. 623, 843. Ceramic Material. Karpinos D. M., Rutkovskij A. E., Kondratjev Ju. V., Ivashin A. A. *et al.* Publ. in *Otkrytija. Izobretenija*, 1978, No. 34, p. 71.
Rutkovskij A. E., Aleksenko I. P., Chekhovskij A. A., Furman V. V. *Plant for Powder Application to Continuous Fibers*, in press.
Rutkovskij A. E., Chekhovskij A. A., Ivashin A. A. *et al.* Investigation of corrosion, mechanical, and thermophysical characteristics of reinforced glass ceramics at one-sided cyclic radiant heating. *Poroshkovaja Metallurgija*, in press.
Kulik O. P., Denisenko E. T. and Krot O. I. (1985) High-temperature structural ceramics: manufacture and properties. Preprint No. 2, Institute for Materials Science Problems, Ac. Sci. Ukr. SSR, Kiev.
Nazarenko N. D., Juga A. I. *et al.* (1973) Effect of metallic fillers on friction properties of glass ceramics. *Poroshkovaja Metallurgija*, **7**, 51–4.
Nazarenko N. D., Juga A. I., Kolesnichenko L. F. *et al.* (1980) Investigation of friction properties of glass ceramic-metallic filler composites. *Poroshkovaja Metallurgija*, **7**, 75–89.
Catalogue of Engineering Glass Ceramics, (1967) Moscow.

3
Carbon-based composites

V. I. Kostikov

Nomenclature

T	temperature
τ	time
E_a	activation energy
l	linear dimension
S	surface area
V	volume
m	mass
d	density
d_{ap}	apparent density
d_p	pycnometric density
P	porosity
β	permeability coefficient
Q	gas flow rate
η	viscosity
K_f	filtration coefficient
k	proportionality factor
H	enthalpy
C_p	specific heat
λ	thermal conductivity
α	linear thermal expansion coefficient
σ	ultimate strength
E	elastic modulus
ε	ultimate strain
$\Delta\varepsilon$	elastic expansion
μ_T	Poisson's ratio
τ_{12}	tangential stress
τ_σ	shear strength

σ_{11}, σ_{22}	normal stresses
K_6	stress concentration factor
λ_1	dimensionless parameter
Σ	ultimate fracture strain
K_I	stress intensity factor
C_0	crack nucleus size
S_d	root-mean-square deviation
γ	surface tension
θ	contact angle of wetting
$\Delta H/\Delta g$	reduced width of EPR signal line
ρ	electrical resistance
q	EPR signal factor
$\Delta\rho/\rho$	magnetoresistance
χ	diamagnetic susceptibility
R	radius
r	pore radius
d_{ef}	effective parameter of pores
b_t	tape width
h_t	tape thickness
t	plate thickness
γ_g	graphitization degree
N	number of turn starts
K_{ov}	coefficient of overlap
α_w	winding angle
$[\Sigma]$	limiting stretching of taut tape bundles
λ_t	mandrel taper angle
δ	shell thickness
T_t	tape tension
L	grain size
L_a	crystallite size
K	coke yield
K	sinterability criterion
K_{ef}	effective reaction rate constant
C_G	volume concentration of reacting gas
δ_g	boundary gas layer thickness
η_g	amount of gas having reacted in pores
$a_l, a, b, n, D, \alpha, C, m_1, n_1, n_\theta\ \varphi_B$	empirical constants

3.1 THEORETICAL FUNDAMENTALS OF DEVELOPMENT OF CARBON–CARBON COMPOSITES

In a broad sense, an absolute majority of carbon-based materials are composites. Classification attributes of the materials are the composition of starting components, their processing method, macrostructure characteristics, thermal treatment type, and reinforcement method. There exists, however, no stable enough classification of carbon-based materials.

In the anthor's opinion, carbon-based materials can be classified as follows: coke–pitch compositions (graphites); pyrolytic graphites; vitreous carbon; carbon-filled plastics; and carbon–carbon composites.

The group of coke–pitch compositions is at present the most numerous one. It includes materials based on calcined and uncalcined cokes with matrices based on pitches, resins, varnishes and alcohols; materials with additions of metals, metal oxides, carbides, borides, nitrides and other refractory compounds to the starting coke–pitch blend; and materials based on porous coke–pitch compositions impregnated with liquid metals, alloys and organic compounds [1].

The group of pyrolytic graphites consists of carbon materials produced by the vapour deposition of graphite on a surface or into a porous carbon structure at pyrolysis of carbon-containing gases. These materials include pyrocarbon, pyrographite as well as pyrocarbon and pyrographite with inclusions of metals and of other compounds.

Vitreous (or glassy) carbon is a unique material produced by a special processing of cellulose. The average diameter of vitreous carbon pores is as small as of about 2 nm, and therefore it is practically impermeable for all gases [2, 3].

Carbon-filled plastics constitute a broad class of structural, heat-insulating and antifriction materials. They are produced on the basis of any types of carbon fillers and organic binders [4].

Carbon–carbon composites (CCC) represent a new type of carbon materials, based on fibrous carbon fillers and carbon matrices. Matrices consisting of carbon and carbides or a hybrid reinforcement of carbon or carbon–carbide matrices with both carbon and carbide fibres are employed in some cases to upgrade the heat resistance of such materials [5].

The CCC are, at present, extensively employed in all fields of technology; they are indisputably the materials of the future. They successfully combine a higher strength at high and superhigh temperatures, a low density, high corrosion and oxidation resistances, absence of ageing in storage, and a good physicochemical compatibility with polymeric, ceramic and biological systems. These CCC properties allow their effective use in development of parts and units of aerospace facilities, in manufacture of high-temperature production equipment for the metallurgy and chemical industry, in surface transportation and power facilities, for friction assemblies, prostheses of human supporting-skeletal system elements, and quite a number of other purposes in various fields. A vigorous advance in the CCC has been encouraged by a forestalling development of new carbon filler types: fibres, bundles, tapes, rods and fabrics. Basic methods for manufacturing these materials as well as their structure and properties are discussed in detail in Volume I of the present series.

Since many types of carbon-based composites have been described in adequate detail in the Soviet and foreign literature or in other volumes of this series, we will confine ourselves here to discussing just the carbon–carbon composites.

3.1.1 Structure of materials

3.1.1.1 Crystalline structure of carbon

Carbon exists in nature in two crystalline forms (diamond and graphite) and a number of amorphous modifications. The existence of the third form of carbon, carbin, has been reported [1]. Carbon can convert from one modification into another. The triple point in the phase diagram, corresponding to the diamond–graphite–liquid carbon equilibrium, lies at a temperature of 3800 °C and a pressure of 12.5–13.0 GPa. The graphite–vapour equilibrium at normal atmospheric pressure (0.1 MPa) occurs at 3270 °C. As the pressure increases to 10 MPa, the equilibrium temperature rises to 3700 °C. The temperature of the triple point (graphite–liquid–vapour) is of 3750 ± 50 °C at a pressure of 12.5 ± 1.5 GPa. Only one form of carbon, graphite, will be further discussed in this book.

Graphite has a hexagonal structure. The hexagonal graphite cell belongs to the spatial group C6/mmc-D_{6h}^4 with four atoms for an elementary cell. The latter is a 0.671 nm high prism with a rhombic base; sides of the rhombus are of 0.246 nm and the angle between them is 60 °C. Carbon atoms form an array of regular hexagons in every plane. The atoms are disposed in vertices of regular close-packed hexagons similar to that of a benzene molecule. Such planes are called basal planes. Every atom in a basal plane is bound to three adjacent atoms, spaced from it at 0.1415 nm. The energy of the bond between the atoms is of 710 kJ/mole.

Basal planes in a graphite crystal are parallel to one another, but can alternate in different sequences, which results in two crystalline modifications of graphite, hexagonal and rhombohedral. The former is characterized by displacement of layers by 0.1418 nm relative to one another, every third layer having the same arrangement of atoms as the first one. Adjacent layers are displaced so that under and over the centre of every hexagon there is a carbon atom in an adjoining layer.

The rhombohedral modification, which corresponds to the spatial group R3m-D_{3d}^5, has the parameter a = 0.246 nm and the angle of 39.49°. In the rhombohedral structure the layers are also displaced with respect to one another by 0.1418 nm, every fourth layer replicating the first one in the disposition of atoms. The rhombohedral structure is usually encountered in natural graphite, where its content can reach up to 30%. In man-made graphites it is practically not observed.

The forces of interaction between basal planes, van der Waals, are weak. The energy of bonds between the planes amounts from 4.2 to 18.2 kJ/mole and their spacing is of 0.3354 nm. Due to the weakness of bond forces between basal planes, the latter can be the principal shear planes, where laminations and cleavages of crystallites can occur. This results in formation of twins of a contact type with an axis parallel to the hexagon c-axis [6]. In the opinion of the authors of [6], twinning of single crystals is in some cases wrongly identified with the structure of the third crystalline form of carbon, carbin.

Various stable structural defects are inherent in all types of man-made and natural graphite. These include layer stacking faults, twins, screw and edge dislocations, 'hole' defects (absence of a group of atoms or of a single atom), interstitial atoms of elements. The defects bring about variations over very wide ranges of mechanical, thermophysical, semiconductor and other practically important properties of carbon materials. Some properties are also affected by heteroatoms, present in carbon materials as components of functional groups disposed on prismatic faces of graphite crystals.

Properties of man-made carbon are greatly dependent on the spatial distribution of its constituent atoms in the crystal lattice, which, depending on the nature of the substance and its manufacturing methods, varies over a broad range, from fully random to highly ordered.

Nongraphitized carbon consists of flat polymerized layers, similar to graphite layers. The layers are arranged in small stacks, within which there is neither a relative nor an azimuthal ordering between planes. In a nongraphitized material the interlayer distance is 0.344 nm and the diameter of layers (diameter of coherent scattering regions, CSR) is as small as 2 nm. The average spacing between stacks is about 2.5 nm. Some amount of disordered or amorphous carbon is present along with the ordered one.

Carbon materials with a lack of hydrogen and an excess of oxygen have in their structure highly developed cross-bonds between carbon networks, which impedes the rearrangement of structural elements, needed to produce a graphite structure, at a high-temperature treatment. A structural porosity retarding the growth of crystallites forms in the material. In nongraphitized material, only a part of the substance acquires the graphite structure at the thermal treatment. All this, of course, is conventional enough: in experiments on a forced orientation of graphite-like layers, at the stage of carbonization (under a pressure of 20 MPa), phenol–formaldehyde resin, which is considered to be nongraphitizable even at 3000 °C, was graphitized like a petroleum coke. Easily graphitizable carbon materials are as a rule obtained from low-oxidized, hydrogen-rich raw materials (petroleum and pitch cokes).

In the initial state, graphitizable carbon materials are structures consisting of large enough hexagonal carbon layers with a great number of bends, vacancies and their groups, as well as a considerable number of heteroatoms. There are separate flat regions of layers with an arrangement close to parallel. Such groups of planes can be arranged in the form of either linear-extended formations (jet-like components) or spherical ones (spherulitic structure).

The structural anisotropy, i.e., the relative arrangement of carbon networks in the starting material, predetermines its capability of a three-dimensional ordering or graphitization at a high-temperature thermal treatment. The parallel arrangement of layers in stacks and their small disorientation relative to one another, promotes their straightening and aggregation into groups. In this case a carbon material is easily graphitized.

The arrangement of atoms in graphitized carbons corresponds to the hexagonal system. Nuclei of hexagonal-structure crystals appear in graphitization of easily graphitizable carbon substances beginning from 1500 °C. Naturally, the formation of crystallites involves departures from a regular spatial distribution. Disturbances of periodicity of the spatial structure arise not only in the process of its formation; they can be created artificially at quench hardening, mechanical actions, etc., but, whichever the nature of atomic defects of carbon materials, they all result in distortion of the spatial disposition of atoms, change in interatomic distances in various directions, and distortion of the crystal lattice geometry.

3.1.1.2 Texture

Carbon materials feature anisotropy of physical properties, which is due to the hexagonal layered structure of graphite.

Because of this, properties of a graphite crystal in directions of crystallographic c- and a-axes are sharply different. The quantitative value of anisotropy of carbon materials can be characterized by the texture, determined by X-ray techniques.

The texture of materials can vary over a broad range depending on the raw stock and graphite production technology. Among structural graphite materials, the highest texturability and hence also anisotropy of properties are exhibited by pyrolytic graphite, whose texture changes strongly at a thermomechanical treatment (simultaneous action of temperature and load). This effect was ascribed to 'straightening' of graphite-like layers, resulting in remanent elongation of thermally treated samples [7]. On the other hand, a low anisometry of particles of domestic petroleum coke KNPS also predetermined a low anisotropy of the thermal properties and electrical resistance of graphite materials produced from this coke. The action of pressure and temperature on such a graphite, however, changes the X-ray index of the texture and hence also the anisotropy of properties of the material.

The texture of carbon materials can vary over a broad range depending on the type of the raw stock, method of moulding preforms, thermal and thermomechanical treatment.

3.1.1.3 Porous structure

Production of man-made carbon materials – coke, active carbons, electrodes, structural materials, carbon fibres, and many others – involves chemical processes, as a result of which individual atoms and their groups get detached from organic molecules and removed from the forming solid substance, which therefore gets enriched in carbon. The changes in the composition and structure of starting organic compounds are accompanied by their compaction, resulting in shrinkages and development of contraction cracks. As a result of the processes, porosity is formed concurrently with the atomic molecular structure in the solid carbon residue of carbonization, the pore size varying over a broad range, from a molecular to a relatively large size. Thus, poposity exists in all carbon materials, which significantly affects their properties.

Pores emerging in formation of a solid body can be connected with one another and with the surface, including the inside one. These are so-called channel or transport pores; they are responsible for the transport of the mass of a substance (filtration and diffusion) through a porous body. Some pores communicating with the surface cannot communicate with one another; these are blind pores. Transport and blind pores constitute an open porosity. Pores not communicating with one another and with the surface or communicating through passages that are smaller than molecules of a gas or a liquid, with whose aid the open porosity is determined, constitute a closed, or inaccessible, porosity. The division of porosity into total, open and closed is based on different methods of their determination. Most widely used is the calculation of the volume of open pores, which gives a general notion of the porosity of a body, from the apparent (d_{ap}) and pycnometric ('true') density (d_p) with the expressions

$$P = \frac{d_p - d_{ap}}{d_p} \quad \text{or} \quad P = \frac{d_p - d_{ap}}{d_p \, d_{ap}}.$$

The dimension of the former is m^3/m^3 (i.e., it is a dimensionless quantity), and of the latter, m^3/kg.

The porosity determined in this manner depends on the pycnometric fluid used. The larger the molecules of the pycnometric liquid or gas, the greater the size of open pores being determined, especially for bodies with molecular-size pores. The pycnometric liquids used for carbon materials include water (characteristic of porosity from so-called water uptake), methyl and ethyl alcohols, benzene, heptane, etc. Ethyl alcohol is most widely used to determine the pycnometric density of cokes and graphitized materials.

The data obtained in the measurements of d_p of carbon materials with the use of such substances as benzene, heptane, methyl and ethyl alcohols are very close to one another, which indicates the absence of the effect of molecular sieves. Determining the d_p with the use of water yields lower values than with the use of the above-listed liquids because of hydrophobicity of graphite, i.e., its poor wetting by water.

By analogy with the open porosity, the total porosity P_{tot} can also be found from the expressions

$$P_{tot} = \frac{d_r - d_{ap}}{d_r} \quad \text{or} \quad P_{tot} = \frac{d_r - d_{ap}}{d_r d_{ap}},$$

where d_r is the porosity calculated from X-ray structure analysis data from geometric dimensions of an elementary cell of the carbon material (the number of atoms in the cell and their mass are known).

Finally, the closed porosity (P_c), inaccessible for pycnometric fluids, is determined as

$$P_c = \frac{d_r - d_p}{d_r} \quad \text{or} \quad P_c = \frac{d_r - d_p}{d_r d_{ap}}.$$

It should be mentioned at once that structural defects make their contribution to the d_r value [8].

Determining the porosity by various methods makes it possible to find those pores which communicate with outside surfaces and with one another and play a very important role in impregnation and filtration processes. The simplicity of the pycnometric method and calculation of d_p gained a wide application for it. Gases used are as pycnometric fluids for more subtle investigation methods rather than liquids, in particular helium, which, owing to small sizes of its molecules, easily permeates into voids whose sizes are close to molecular ones and is practically not adsorbed on carbon at room temperatures.

An important characteristic of a porous structure is the size distribution of pores in it. According to the classification proposed by M. M. Dubinin, pores are classed into three categories; micropores, with an effective radius less than 1.5–2 nm; transition pores, from 1.5–2 to 100–200 nm; and macropores, with an effective radius over 100–200 nm. The classification of pores by sizes, proposed for carbon materials and carbons, is based on the mechanism of movement of gases in pores, and therefore the pore sizes falling into one or other group depend on the experiment conditions and nature of the gas.

The widest use in studying the size distribution of pores over a broad range of their effective radii has been found by the mercury porosimetry, adsorption methods, small-angle X-ray scattering and electron microscopy. The latter two methods are applicable to observation and study of distribution of micro- and transition pores. The volume of micro- and transition pores, their size distribution and specific surface can be calculated from isotherms of adsorption–desorption of various substances. The size distribution of pores is calculated with the use of the theory of capillary condensation of adsorbate in thin pores of adsorbent. The specific surface is computed on the basis of concepts of monomolecular adsorption by the BET method. Studying the small-angle X-ray scattering from carbon materials makes it possible to determine the size distribution of pores in a range 1–100 nm and the specific surface. The distribution of porosity of mainly large macropores (over 1 µm) can be measured by the continuous weighing method.

An essential characteristic of a porous structure and an important property of carbon materials is their permeability by gases and liquids. The permeability of a porous body is characterized by permeability or filtration coefficients. The filtration coefficient depends not only on properties of a porous body, but also on the substance being filtered. Since viscosities of gases most often dealt with are close to one another, filtration coefficients, whose determination in a number of cases seems to be preferable, can be used for comparative data.

The permeability coefficient (B, m^2), determined from the volume of gas flowing through the sample, is calculated with the Lejbenzon formula:

$$B = Q\eta L/(\Delta P S),$$

where Q is the gas flow rate (m^3/s); η the viscosity, (Pa s); ΔP the pressure difference (Pa); L the sample length, (m); and S the filtering area (m^2).

The filtration coefficient (K_f, m^2/s) is determined in dynamic tests from the gas inflow through the sample into an evacuated calibrated volume and calculated from the formula

$$K_f = \frac{\Delta P V L}{s} \frac{\Delta p}{\Delta \tau},$$

where $\Delta p/\Delta \tau$ is the pressure rise rate in the calibrated volume, Pa/s; V the calibrated volume, m^3; and ΔP the pressure difference across the sample (Pa).

With respect to the type of porosity formation in carbon materials they can be divided into two classes: materials where the binder fills all the gaps between filler grains and materials where the binder only forms bridges between the grains. The former undergo expansion in their calcination, and the development of porosity in them is primarily determined by the loss of mass because of destruction of the binder. In the latter, shrinkage develops, more or less intensively depending on the volume of pores existing in a green sample. Depending on the intensity of shrinkage, porosity can rise in the process of carbonization, remain at the same level, or decline, since shrinkage counterbalances the contribution of the mass loss.

In materials where gaps between grains are filled with the binder, the volume of developing pores can vary considerably in the absolute value, depending on the content of binder and the yield of coke from it. On introduction of two parameters, P_{max} – the maximum porosity after the thermal treatment, and P_0 – the porosity existing in the green preform, the porosity development can be described by the expression

$$(P_{max} - P)/(P_{max} - P_0) = K \log m,$$

where m is the mass loss and P the porosity at a given mass loss.

In the $(P_{max} - P)/(P_{max} - P_0)$–log m coordinates, the experimental points for various materials, produced both by forcing-through and by pressing, fall on one straight line, which indicates a common porosity formation mechanism at calcination of materials where gaps between filler grains are filled with the binder.

The existence of porosity in carbon materials affects many of their properties. The strength, elastic modulus, electrical resistance, thermal conductivity and other properties depend to one or other extent on porosity and its characteristics.

A power-law dependence of the bending strength on the overall pore volume has been found: $\sigma_{bend} = KP^{-1.6}$. For ceramic materials, a formula relating the porosity to the elastic modulus has been proposed by McKenzie:

$$1 - E/E_0 = kP,$$

where E_0 is the modulus for a nonporous solid body; E the modulus for a porous body; and k the proportionality factor.

A linear dependence of the elastic modulus on porosity within a pore volume range of 0.17–0.27, described by the McKenzie formula with $k = 2.8$, has been found for graphite. Its extrapolation to a zero pore volume yields $E_0 = 8.5$ GPa. For another carbon material, the parameters are different, namely $k = 2.15$, $E_0 = 22$ GPa. Thus, the manufacturing technology for a carbon material substantially affects the parameters of the equation, although the dependence remains the same. A linear equation describes also the thermal conductivity variation with porosity within the same pore volume range; at $k = 2.3$, $\lambda = 218$ W/(m K).

The electrical resistance variation with density can be presented by a power-law function with an exponent of, e.g., 3. It was also proposed to employ the Maxwell theorem to describe the electrical resistance of a body consisting of components with different electrical conductivities. Assuming a carbon material as consisting of two components, carbon particles and pores, i.e., spaces filled with air having an infinitely high resistance, we write the Maxwell equations as follows:

$$\rho = \frac{\rho_0}{2}\left(\frac{3d_0}{d} - 1\right),$$

where ρ_0 is the electrical conductivity corresponding to the material density d_0.

This expression has been checked for carbon materials differing not only in porosity, but also in the manufacturing technology. Differences in the technology can affect the electrical resistance, and therefore experimental data was adjusted to one constant density, selected at 1.62 g/cm^3. The empirical formula of the electrical resistance dependence on density is

$$\rho/\rho_{1.62} = 6.2/d - 2.8;$$

it was derived from a plot in the $\rho/\rho_{1.62} - 1/d_0$ coordinates, but the coefficients turned out to differ from theoretically predicted ones.

Having summarized results of numerous experimental studies, Virgilev [9] proposed the following empirical expressions:

$$c = c_0 \exp(-\alpha_c P), \qquad c = c_0 \exp(\alpha_c P),$$

where c is a property of a carbon material, c_0 the property of the nonporous material, and α_c a constant.

The expressions adequately describe the dependences of the thermal conductivity, compressive strength, elastic modulus, strength loss at oxidation and electrical resistivity, in particular its variation at oxidation, on the pore volume.

J. Hutchenson and M. Price indicated that porosity and permeability are related by a power-law dependence with an exponent of 3.5. However, a thorough examination of the graph in their publication shows the exponent at high porosity values to be much greater than 3.5. Most researchers concluded that permeability is proportional to the 3rd–6th power of porosity. Thus, in the theory by J. Kozeni, where a porous body is modelled by a set of parallel capillaries of the same diameters and lengths, this dependence has the following form:

$$B = cP^3/s^2,$$

where B is the permeability (m^2); P the porosity, (m^3/m^3); s the specific surface, (m^2/m^3); and c is a dimensionles constant depending on the geometry of the capillary cross-section.

As indicated by many researchers, carbon materials contain not only macropores, but also micropores, whose sizes have been estimated by various authors to be molecular (diameter of about 0.4 nm). The microporosity is formed at carbonization and has a maximum at 900 °C (from the adsorption data), after which the volume of micropores decreases steeply. Micropores are believed to be disposed at boundaries between crystallites. The calculation of the distance between adjacent crystallites showed it to be 0.3–0.8 nm, which coincides in the order of magnitude with pore sizes determined in carbon materials.

The comparison of data obtained in adsorption experiments and small-angle X-ray scattering studies demonstrates that micropores inaccessible for adsorbate molecules are formed in coke. This also follows from differences in specific surface values determined with the use of different adsorbates. Above treatment temperatures of 900–1000 °C the pores become practically inaccessible for all adsorbates. The number of such pores grows for nongraphitizable cokes, decreasing slightly only after 1500 °C. A decline in the volume of inaccessible pores in graphitizable and partly graphitizable cokes begins at lower temperatures than in nongraphitizable ones. Dependences of the inaccessible porosity on the treatment temperature for graphitizable and nongraphitizable cokes are different. An unfavourable effect of the microporosity development in the course of carbonization on graphitizability was noted.

The inaccessible porosity is determined from the theoretical (X-ray) and pycnometric densities; the calculation method does not allow it to be distinguished from structural defects. Thus, this porosity is a part of microporosity, which is not filled with the pycnometric fluid. It can be treated not only as isolated voids, i.e., zero-density spaces, but also as regions of a defective structure, filled with disordered carbon. The density of the regions is much lower than the carbon density, but more than zero. The calculation of the density of a carbon substance being in a disordered state yielded a value of $0.85–1.0 \text{ g/cm}^3$. In terms of the inaccessible porosity model, proposed in [8], the decrease of its volume with increasing treatment temperature can be considered as transition of disordered carbon, which has the form of deformed aggregates, to an ordered state.

3.1.2 Mechanics of carbon–carbon composites

The mechanics of composites of various natures and with various reinforcement types is given much attention elsewhere in this book, and therefore we will here confine ourselves to a very brief discussion of the mechanics of ceramic (carbon, carbide, oxide and combined) matrix, carbon fibre-reinforced composites.

Analysis of the structure of high-modulus filler-based carbon–carbon composites demonstrates their characteristic feature to consist in the existence in the matrix of pores and cracks, the average spacing between which is less than the effective carbon fibre length calculated for given filler and matrix type [10]. Based on the analysis, it is proposed to apply the Rosen model, with allowance for the matrix discreteness, to a description of the mechanism of fracture of a carbon–carbon material. The discrete-matrix model can account for the relatively low realization of the tensile strength of carbon–carbon materials, observed in practice, due to a multiple increase of the effective fibre length at fractionalization of the matrix as compared with a continuous-matrix model.

For carbon-ceramic materials with a carbide matrix or coating the presence of cracks in the matrix is noted as well, but the average distance between them is much greater than the effective length of the carbon filler, since other characteristics of the matrix in this case exceed similar characteristics of the filler (see Table 3.1).

Table 3.1 Physicomechanical characteristics of fillers and matrices of carbon fibre-based composites

Characteristic	Carbon-filled plastic		Carbon–carbon		Carbon–carbide	
	filler	matrix	filler	matrix	filler	matrix
Elastic modulus (MPa)	300 000	2400	300 000	20 000	300 000	400 000
Shear modulus (MPa)	40 000	800	40 000	8 000	40 000	80 000
Ultimate tensile strain (%)	0.6	5	0.6	0.15	0.6	0.05
Linear thermal expansion coefficient, ($10^{-6}\,°C^{-1}$)	0.5	50	0.5–1.5	3–5	0.5–1.5	4–8
Poisson's ratio	0.2	0.5	0.2	0.23	0.2	0.1

The data presented in the table indicate the possible difference in mechanical properties of carbon-ceramic materials differing by the E_m/E_f ratio, where E_m and E_f are elastic moduli of the matrix and fibre respectively.

Carbon–carbon materials feature a low shear resistance [11]. Stress–strain curves for such materials exhibit the pseudoplasticity, which is most marked at disorientation of the filler [12, 13] (Figure 3.1).

A considerable relation between the bending strength of carbon–carbon materials and the scale factor, due to the impact of a low shear rigidity on the effective strength, general rigidity, and character of the relative displacement of layers of reinforcing elements, was noted [14].

An elastic deformation of samples with various filler disorientation angles is observed for carbon-ceramic materials with a relatively low content of a carbide matrix. A complete substitution of a carbon matrix with a carbide one increases the effective elastic modulus, but in this case

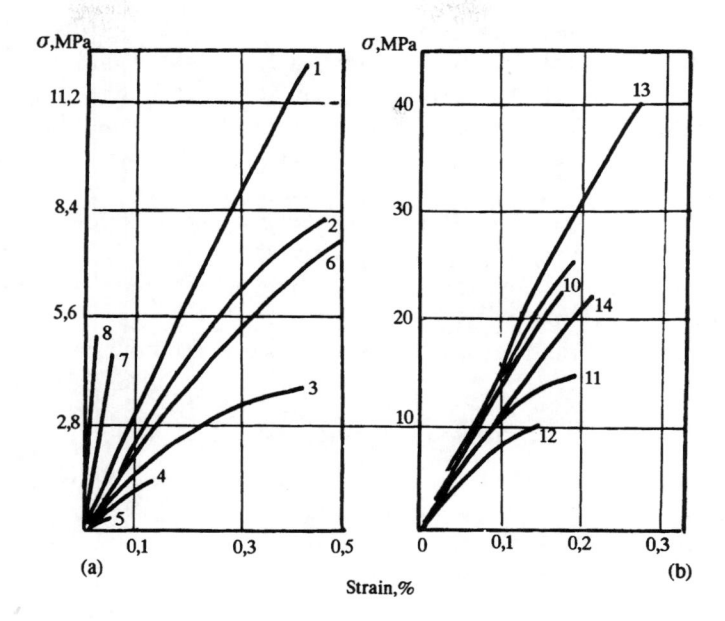

Figure 3.1 Stress–strain curves for composites based on (a) low-modulus and (b) high-modulus carbon fibres, produced from cellulose hydrate fibres: carbon matrices with filler disorientation angles of 0, ± 20, ± 45, ± 70 and 90° respectively (1–5); (6) fabric structure; (7, 8) carbide matrices; (7) composite after borosilicization; (8) silicon carbide; (9–12) carbon matrices with filler disorientation angles of 0, ± 7.5, ± 15, and 22° respectively; (1, 3) carbide matrices containing C, B_4C and SiC with disorientation angles of 0 and 15° respectively.

the strength and the limiting deformation of the composite approach their levels for an unfilled carbide. This also involves a change in the sample fracture character. For carbon–carbon composites with a filler disorientation angle not over 13–15° as well as for carbon-ceramic composites with any disorientation of the filler a cutting of samples by one or, less often, two cracks is observed predominantly.

For carbon–carbon composites with disorientation angles of up to 70–80°, there occurs a pulling-apart of fibre bundles, strands and braids. The sample stress–strain curve at a transversal loading is approximated by a straight line.

An important prerequisite for an effective utilization of the strength of carbon fibre in a ceramic-matrix carbon composite is the optimum relation between elasticities of the filler and matrix.

At an average filling of a composite of 0.5–0.6 only at $E_m/E_f = 1$ there occurs a 50% utilization of fibres, since in this case, at external loads of 100–150 MPa the load concentrated in fibres will amount to 200–300 MPa, which is comparable with the outer limit of fracture of fibres. Since E_m for man-made carbon materials is in practice within 10 000–30 000 MPa, it follows that fibres with E_f from 150 000 to 300 000 MPa and higher are needed to produce effective carbon–carbon composites. The upper limit of the preferable elastic modulus for a carbon fibre can be found, e.g., from the condition for decrease of the intensity of fracture of a carbon matrix under the action of thermal stresses.

At $E_f \approx 400\,000$–$500\,000$ MPa the brittle coating of a high-modulus substrate, carbon fibre, by a carbon matrix will crack with formation of continuous sections whose length is less than the effective length of the filler for the matrix used (l_{ef}).

The ultimate strength of a composite with a continuous matrix, such as of a fibre bundle, can be estimated from the expression [15]:

$$\sigma_c = V_f(al_{ef}m_1 e)^{-1/m_1}, \qquad (3.1)$$

where σ_c is the composite strength; V_f the volume filling of the composite with fibers; m_1 and a are parameters of the Weibull distribution of the strength of filler fibres.

The effective length of the filler for a porous composite, calculated from the above formula, can turn out to be greatly underestimated if fracture and pores in the matrix are encountered at every section of the calculated l_{ef} value.

Evaluation calculations indicate that the l_{ef} change due to the fractionality of the matrix can be as great as 10–50 times and amount to 10–20 mm [10]. This appears to be one of the causes of low effectiveness of utilization of the strength of starting fibres in composites with less than 10 mm long fillers [14].

In carbon-ceramic materials the filling of cracks and pores by a higher-modulus matrix increases the effective elastic modulus:

$$E_m = E_{carbon}V_{carbon} + E_{carbide}V_{carbide}, \qquad (3.2)$$

as well as reduces the relative fraction of porosity and the fractionality of the matrix:

$$K = V_p l_{ef}/d_{ef},$$

where K is the number of cracks in the matrix over a length of l_{ef}; V_p the relative fraction of porosity; and d_{ef} the effective diameter of pores disposed within the matrix between two nearest filler filaments.

Reducing the K and l_{ef} values will result in an accelerated increase in the strength at a decline in the porosity [13]:

$$\sigma_c = \sigma l_{ef} \exp\left[-\frac{\eta^l}{m_1} + \frac{\eta^l(1 - V_p)}{m_1 K}\right],$$

where σl_{ef} is the strength of the composite for an effective length of the filler in it, equal to l_{ef} from (3.1).

Along with a relative increase of the strength, the filling of pores and voids in a carbon material by a rigid ceramic matrix can reduce the standard deviation of the composite strength distribution in accordance with the expression

$$\bar{S}/\bar{S}^* = (l_{\text{ef}}/l^*)^{0.5},$$

where $\bar{S}$, $\bar{S}^*$ are standard deviations of the strengths of a fibre bundle of the starting carbon–carbon material and the composite with an elastic modulus calculated from formula (3.2).

Substitution of a carbon matrix with a more rigid ceramic one can give rise to the mechanism of fracture according to models of either 'single' or 'multiple' cracking at the propagation of the initial matrix crack [16]. Conditions for the manifestation of such a mechanism are relations

$$E_m > E_f; \quad \langle \Sigma_m \rangle \ll \langle \Sigma_f \rangle,$$

where $\langle \Sigma_m \rangle$ and $\langle \Sigma_f \rangle$ are ultimate fracture strains for the matrix and the filler respectively.

In the case of material reinforcement with continuous unidirectional fibres at a single cracking the composite strength will be

$$\sigma_c = \sigma_m \left(1 + \frac{V_f}{E_f/E_m}\right),$$

where σ_c, σ_m are the strengths of the composite and the matrix respectively.

If V_f is great enough for fibres to withstand the additional load after the matrix fracture, then the composite fractures after utilization of the strength of fibres, a multiple cracking of the matrix taking place. In this case the σ_c value is determined for reinforcement with continuous fibres [17]:

$$\sigma_c = \sigma_f V_f.$$

The matrix strength σ_m in this expression exceeds the strength of a nonreinforced brittle matrix, which can be estimated according to the Griffith criterion if the stress intensity factor ahead of the crack point (K_I) is known.

The increase in the strength of a ceramic-matrix composite is determined by a substantial increase in the work of fracture of the matrix as a result of a partial pull-out of fibres from it.

The strength of a composite in the case of a single cracking for reinforcement with discrete fibres is [16]:

$$\sigma_c = \frac{1}{K_I} \sqrt{\frac{2E_m}{C_0}\left(1 + \frac{\gamma_f}{\gamma_0}\right)} \left[1 + V_f\left(\frac{E_f}{E_n} n_\theta n_1^{-1}\right)\right],$$

where K_I is the factor of intensity of stresses ahead of the crack point; C_0 the crack nucleus size; γ_f and γ_0 are works of formation of a unit surface area for the reinforced and the nonreinforced matrix respectively; n_θ is a coefficient taking into account the orientation of fibres, equal to 1, 1/2 and 1/3 for one-dimensional, two-dimensional, and three-dimensional reinforcement respectively; and n_1 is a coefficient taking into account the utilization of fibres at a limiting loading of the matrix [18].

The transition from a single to a multiple cracking occurs at a critical volume fraction of fibres, determined from the condition

$$\sigma_f V_f = \sigma_m \left[1 + V_f \left(\frac{E_f}{E_m} - 1 \right) \right].$$

For carbon-ceramic materials with a rigid matrix featuring a smaller fracture strain as compared with the filler, the porosity of the matrix will exert the governing effect on the level of decrease of the composite strength being realized.

Experimental dependences of the strength and porosity of ceramics are most often described with the use of the Knudsen formula [19]:

$$\sigma_c = k L^{-a} \exp(-bP),$$

where σ_c is the strength of the composite; k the proportionality factor; L the ceramic grain size; P the porosity; a and b are empirical constants.

Experimental dependences of the strength and porosity for fibre-reinforced ceramics can be described by the equation presented in [20]:

$$\sigma_c = \sigma_{m_0} \sqrt{1 + \frac{\tau_1 V_f l l_{op}}{3 d \gamma_0}} \left[1 + V_f \left(\frac{1}{3} \frac{E_f}{E_m} l^{-bP} - 1 \right) \right],$$

where σ_{m_0} is the matrix strength at absence of porosity; τ_1 the shear strength of the fibre–matrix bond ($\tau \approx 1.5$ MPa); l the length of the reinforcing filler element; l_{op} the crack opening length; d the filler diameter; and V_f the fibre volume fraction.

An essential prerequisite for an effective performance of a composite is also the optimum filling.

The filling efficiency is determined not only by utilization of the strength and other properties of fibres to impart required qualities to a composite, but also by the impact of the filler on the matrix fracture mechanism, strengthening the matrix resistance to the crack growth, i.e., to its dispersion with formation of an ineffective discrete-matrix structure.

A high-rigidity stringer reduces the stress intensity factor (K_I) ahead of the point of a crack that cuts the matrix and extends to the stringer [21]. A decline in the coefficient K_I in this case is associated with an increase in the elastic modulus of the stringer and growth of its relative fraction in the composite proportionally to the decrease of the parameter λ_1:

$$\lambda_1 = \frac{4}{(3-\mu)(1+\mu)} \frac{E_m}{E_f} \frac{tl_{cr}}{A}, \qquad (3.3.)$$

where λ_1 is a dimensionless parameter; μ the Poisson ratio; t the plate thickness; l_{cr} the distance from the most remote crack point to a reinforcing element (fibre); and A the cross-section of the reinforcing filler element.

From formula (3.3), it follows that the stress intensity factor K_I can be reduced for a composite also reinforced with low-modulus fibres, but at a high filling degree

$$tl_{cr}/A \ll 1.$$

According to Korten [17], the zone of effect of a fibre on the surrounding matrix does not extends farther than $l_{cr} \approx 0.5 d_f$, which corresponds to a composite filling with fibres of about 25–30% vol. At great fillings the crack growth in the matrix under the action of an external load will be retarded by the decrease of the stress intensity factor at the crack point under the effect of the filler, which is equivalent to increase in the matrix fracture toughness.

The upper limit of the optimum filling with fibres is for carbon–carbon composites less than for carbon-filled plastics [22] because of a relatively high rigidity of a carbon matrix ($\sigma_m \approx 6000$–8000 MPa $\gg \sigma$ of epoxy resin) as well as of a complete absence of relaxation at temperatures below 1500–1800 °C.

3.1.3 Physicochemical features of carbon–carbon composites

The CCC manufacturing technology includes the following basic stages: formation of the filler reinforcement by winding, laying, assembling or other textile processing methods and formation of a carbon matrix by carbonization of binders or deposition of pyrocarbon in pores, followed by a high-temperature thermal treatment (Figure 3.2).

Methods for making spatial skeletons from carbon bundles, tapes, fabrics, and rods will be discussed below.

Here we only will note that after the formation of the skeleton with the specified reinforcement the porous precursor is impregnated with liquid organic binders – thermoplastic or thermosetting polymers as well as coal and petroleum processing products. The main requirement placed on raw materials for the matrix in ECC, in contrast to those for carbon-filled plastics, where various polymers and resins can be used to provide diversity of properties, is a high (at least 40%) coke yield at the thermal treatment.

The yield of carbon residue for a number of raw materials used to produce carbon matrices is given in Table 3.2. The greater the carbon yield, the smaller the disturbances occurring at the subsequent carbonization and

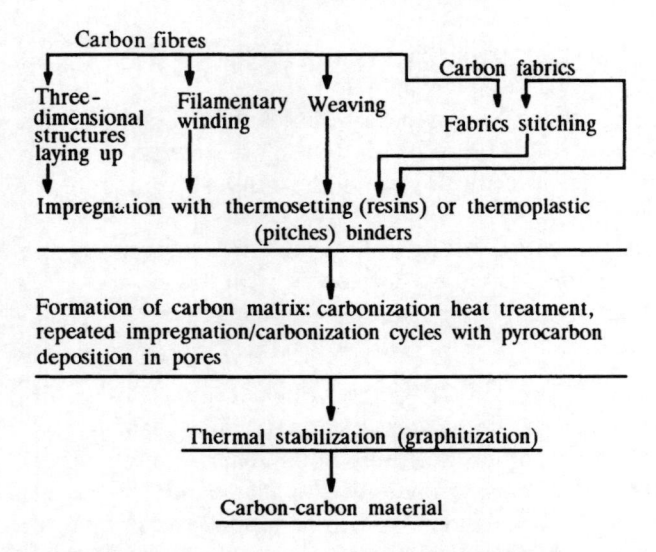

```
                    Carbon fibres
                                           Carbon fabrics
   Three-        Filamentary  Weaving
   dimensional   winding                   Fabrics stitching
   structures
   laying up

   Impregnation with thermosetting (resins) or thermoplastic
                         (pitches) binders

   Formation of carbon matrix: carbonization heat treatment,
   repeated impregnation/carbonization cycles with pyrocarbon
   deposition in pores

            Thermal stabilization (graphitization)

                  Carbon-carbon material
```

Figure 3.2 Flow-chart of manufacture of carbon–carbon composites.

its accompanying shrinkage; moreover, less compaction cycles are needed to fill pores in the matrix. The viscosity and wetting ability at the processing temperature are also important from the production standpoint.

Table 3.2 Carbon residue yield for various binders

Starting material	Carbon residue (%)	Carbon residue type
Polyesters	2–18	Amorphous
Epoxy resins	7–25	Amorphous
Phenol–formaldehyde resins	60–65	Glass-carbon

The CCC are most often produced with the use of pitches and phenol–formaldehyde resins.

3.1.3.1 Pitch-based matrices

Pitches, cheap and readily available raw materials featuring a high yield of the coke residue, hold a special place among substances that can be used as raw materials for CCC matrices [23]. The technology of pitches makes it possible to obtain a broad range of products differing in composition and set of physicochemical properties and thereby to prepare the raw stock for matrices with predetermined properties.

Both coal- and petroleum-origin pitches can be used as the raw stock for matrices [23]. Coal pitches are produced by distillation of coal tar and thermal oxidation. Petroleum pitches are produced from various residues: of cracking, pyrolysis tars, etc. Coal and petroleum pitches differ by the component composition, determined by a selective dissolution or chromatographically by means of a gel permeation chromatography and exography, and also by the molecular-mass distribution. Coal pitches contain a considerable amount of polycyclic aromatic compounds, which results in their poor solubility at a relatively small molecular mass (about 500 amu). Petroleum pitches have a much greater molecular mass (2000–3000 amu), but the presence of aliphatic hydrocarbons in them renders them more soluble. Thus, the content of substances insoluble in toluene for coal pitches amounts to 20% and more, whereas for petroleum pitches, it seldom exceeds 20%. Petroleum pitches practically do not contain the $_1$-fraction (the fraction insoluble in quinoline), whereas coal pitches contain from 4 to 40% of this fraction, depending on the softening temperature. This makes the petroleum pitches preferable for use as impregnates (Table 3.3). From the standpoint of the CCC technology, the pitches, as matrix materials, are evaluated from the softening temperature, wetting and sintering abilities, viscosity, surface tension and coke residue yield. The sintering ability and coke residue yield determine strength properties of CCC, while the wetting ability, viscosity and surface tension affect the blending and composite moulding stages.

Table 3.3 Comparative characteristic of coal-tar and petroleum pitches

Characteristic	Pitch type			
	coal-tar		petroleum	
Softening temperature (°C)	65–70	135–150	70[a]	94[b]
Solubility (%):				
in hectane	43–55	15–25	56.0	56.4
in toluene	68–75	25–55	86.5	80.4
in quinoline	93–97	60–80	99.5	100.0
Viscosity (conventional) (Pa s)	80–140 (at 130 °C)	75–125 (at 240 °C)	–	229 (at 150 °C)
Range of maximum destruction rates (°C)	380–460	390–480	375–425	430–490
Sinterability criterion (%)	15–20	5–8	10	6
Coke yield (%)	37–40	52–62	35	47

[a] From cracking residue
[b] From pyrolysis resins

As T_s increases, the coke residue (CR) of a pitch grows while the viscosity and wetting ability of the pitch decrease (Figure 3.3) [24], and therefore to process a filler-pitch blend the mixing is carried out at temperatures of 1.5–$2T_s$. A good wettability provides a uniform distribution of the binder, which forms the carbon matrix in the course of

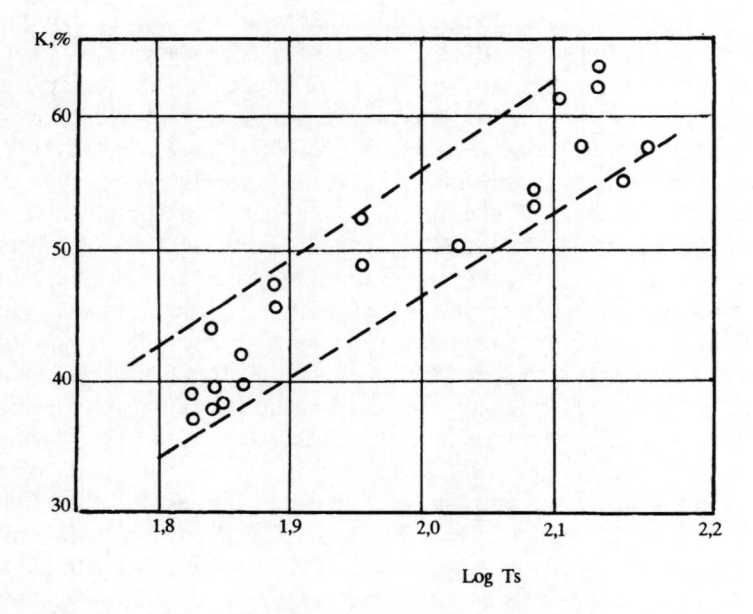

Figure 3.3 Dependence of coke yield (K) from pitch on softening temperature (T_s).

subsequent thermal treatment. The uniformity of mixing of a blend can be estimated from its plasticity. The plasticity of a blend is substantially affected by the content of fine fractions of the filler, especially of those less than 20 μm, as well as by the temperature and time of mixing.

As the content of the fraction less than 20 μm in the formulation increases, not only the plasticity of the mass decreases, but also the inverse elastic expansion grows, which exerts an adverse effect at formation of precursors. Increasing the content of the binder, while improving plastic and elastic properties of the composite, reduces the density and the properties of the end-material associated with it. The increase in the inverse elastic expansion with increasing content of fine fractions is caused by a growth of direct contacts between filler grains as the thickness of binder interlayers decreases [25]. This is confirmed by an increase of the elastic expansion of a blend at a temperature above the pitch softening one over that of the same blend at room temperature (Figure 3.4).

The carbon matrix is formed from an organic binder at the thermal treatment of a blend. The thermal analysis of pitches demonstrated that the run of the carbonization process is the same for all coal-origin pitches. Three stages can be distinguished in this process (Figure 3.5) [26]. Up to 360 °C, there occurs a mass loss, primarily due to a physical evaporation of substances with a low molecular mass, to which there

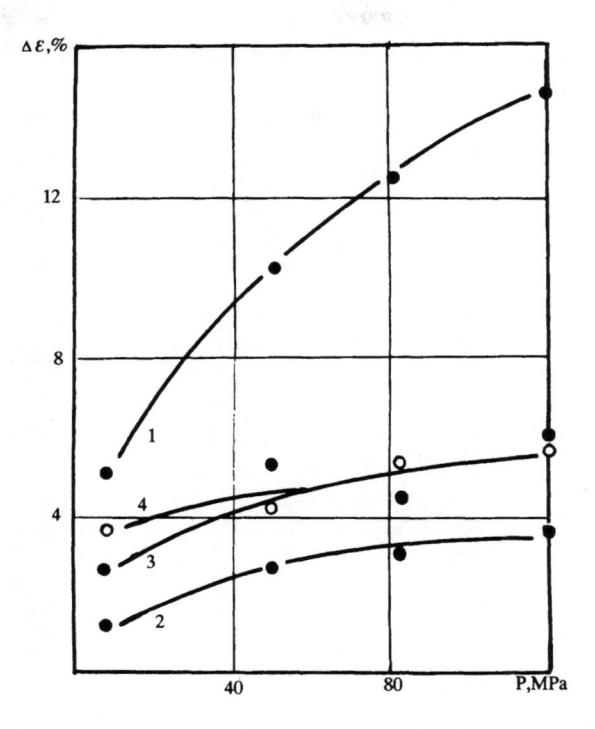

Figure 3.4 Effect of pressing pressure (P) on elastic expansion (ΔE): (1) compositions for binder; (2) compositions with 20% binder at (2) 20 °C; (3) 80 °C; (4) 150 °C.

correspond a low value of the effective energy of activation of this process stage and absence of significant thermal effects at an endothermal run of the thermographic curve. Within 360–500 °C destruction processes are occurring, which is confirmed by deep endothermal effects and a high effective energy of activation of this process stage (300–400 kJ/mol) as well as by changes in the component composition of pitch towards an increase in components with a high molecular mass through both the removal of destruction products and the beginning of the condensation process. At the third stage (above 500 °C) the coke formation starts through development of the process of the condensation of aromatic molecules, which is indicated by a steep increase in the hydrogen evolution, exothermal run of the process, increase in the microhardness, and decline in the effective activation energy down to a value characteristic of condensation processes.

Petroleum binders feature a much narrower range of the second stage, whose temperature maximum is shifted 20–30 °C towards lower temperatures. Such a behaviour of petroleum pitches is to be taken into account in the formation of the matrix: the heating rate over the

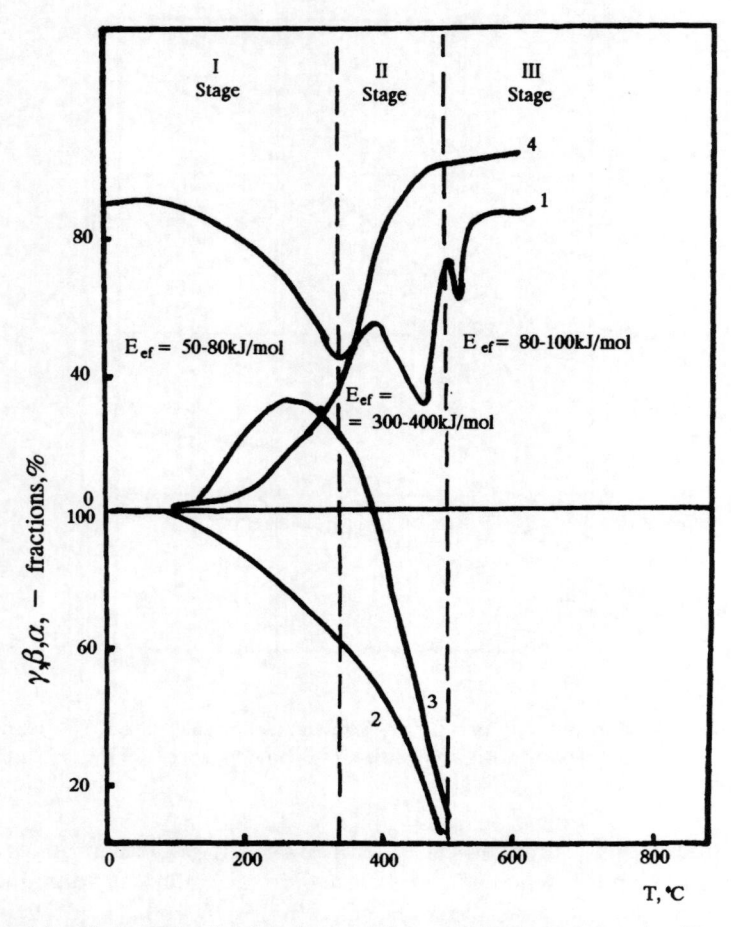

Figure 3.5 Pitch carbonization process diagram: (1) mass loss rate; (2) contents of γ-fraction; (3) β-fraction; (4) α-fraction.

temperature range where destruction processes are proceeding should be reduced.

To increase the density and the properties associated with it (strength, electrical and thermal conductivities, permeability), composite materials are impregnated (sometimes repeatedly) with pitch and then subjected to thermal treatment. The density increase is associated with the characteristic of porosity of the object of densification (volume and size distribution of pores), impregnation process parameters (vacuumization depth, pressure, holding time), and properties of the impregnate (viscosity, surface tension). The densification results in decrease of the volume of open pores, increase of the density, and decline in the permeability. In spite of a general reduction of the volume of pores, the impregnation with pitch, a thermoplastic substance cannot change considerably their size distribution because carbonization of the impregnate, due to evolution of volatiles, gives rise to pores comparable in the size with the spacing of the filler.

The use of fibrous materials as the filler brings about difficulties in the mixing. To improve the mixing and the binder distribution uniformity, use is made of the hydraulic mixing method, consisting in mixing the filler and the binder together in water [27]. In this process, carbon fibres are dispersed into separate 3–5 mm long filaments and mixed uniformly with the binder. After removal of the excess moisture and drying, the resulting powder is moulded into preforms or products. The moulding is carried out by hot pressing, since in pressing of a cold blend, there occurs a considerable elastic expansion after the load removal. The inverse elastic expansion can be reduced by firing a preform or products in a special accessory that fixes the dimensions [28]. The firing of fixed preforms improves the structure of CCC, reducing the porosity 2–3 times as compared with that at carbonization without the fixing.

The formation of the matrix is affected not only by properties of the binding or impregnating substance itself, but also by the matrix interaction with the filler. Studying the interrelation of individual characteristics of binders with properties of the end-material based on them failed to determine general regularities; as a rule, they extended only to one group of binders. The basic criterion for evaluating the quality of a binder is formation of a strong matrix that ensures the strength of a material as a whole. Suggested as such a criterion [29] was the index K (sinterability criterion), which represents properties of both a binder and a filler. The sinterability criterion, defined as the difference between coke residues from pitch in the presence of a filler (K_f) and without it (K_p), referred to the latter,

$$\Delta K = \frac{K_f - K_p}{K_p} \times 100,$$

correlates well with the compressive strength (Figure 3.6).

The obtained dependence makes it possible to predict the sintering ability of a binder with respect to a given filler and the strength of the end-material. Since ΔK depends on the binder content in a blend and has the maximum value at its optimal content, determining ΔK for blends with various contents of a binder makes it possible to estimate the optimal content of the latter.

Carbonization of pitch results in formation of polycrystalline carbon with a zero graphitization degree, $d_{002} > 3.44$ Å and crystallite size $L_a > 100$ Å, which is capable of graphitization (i.e., formation of a hexagonal layered structure) at a subsequent high-temperature treatment. To obtain dense precursors with a low residual porosity the

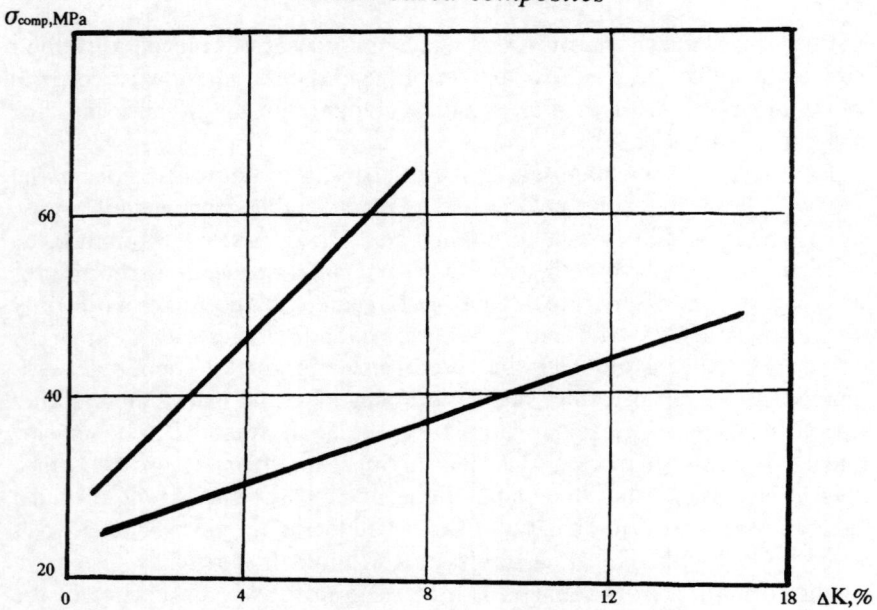

Figure 3.6 Dependence of material strength (σ) on binder sinterability criterion (ΔK) (1) after firing and (2) after graphitization.

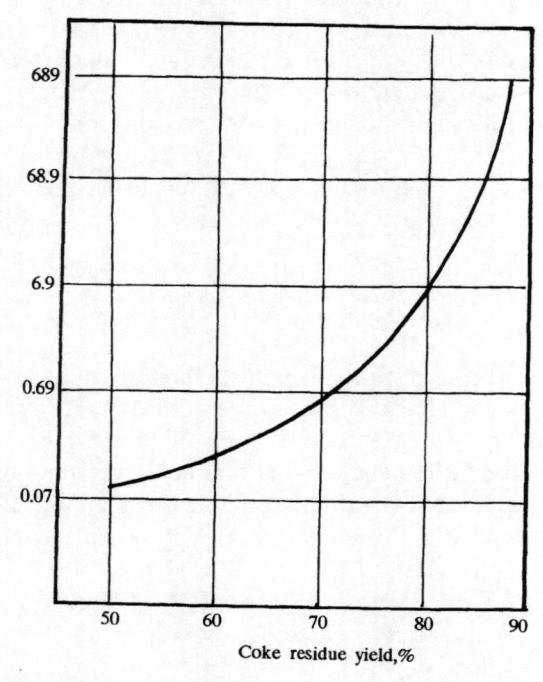

Figure 3.7 Effect of pressure on coke yield in process of carbonization of petroleum pitch.

impregnation–carbonization cycles are repeated several times. As follows from Figure 3.7, increasing the carbonization pressure increases the carbon residue yield and, as seen from Figure 3.8, reduces the number of impregnating cycles.

In some cases, porous skeletons are placed in a hot isostatic pressing vessel (hyperclave) together with pitch. Modern hyperclaves are capable of working temperatures up to 1000 °C and pressures up to 1000 MPa, which makes it possible to effect impregnation and carbonization in a single process. After such a treatment, preforms are subjected to a primary machining and then are heated (graphitized) at 2600 °C in an argon or nitrogen atmosphere. To attain a density of 1.9–2.0 g/cm^3, a complete impregnation–carbonization cycle (under a normal or a high pressure) or a graphitization cycle is repeated 4–6 times.

3.1.3.2 Resin-based matrices

The use of thermosetting synthetic resins for the formation of a carbon matrix makes it possible to attain a high coke residue and a more

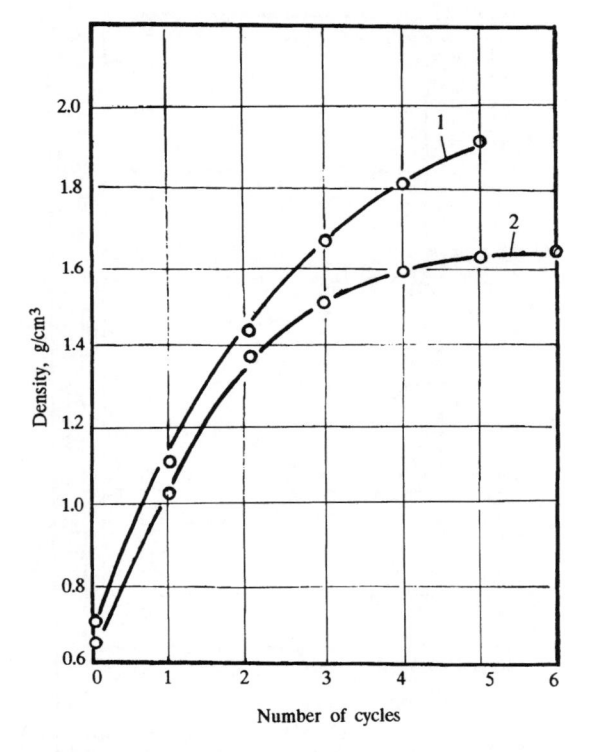

Figure 3.8 CCC densification kinetics: (1) carbonization under high pressure; (2) carbonization under atmospheric pressure.

uniform and finely dispersed distribution of pores forming at the thermal treatment. Of the greatest interest among synthetic resins are phenol–formaldehyde and furane resins, as the most available in terms of their commercial-scale production, wide possibilities for varying their properties, and high coke residues. Directionally changing the structure of an organic molecule through addition of functional groups capable of secondary reactions in the course of thermal treatment makes possible the transition of the carbonization process, where destruction processes predominate and the coke residue yield is low, to an exothermal process, where condensation reactions are prevailing and the coke residue is higher; in this case, however, the capacity of coke residues (matrix) to graphitization will decline. Study [30], using a model organic substance, polyvinyl alcohol (PVA), where hydroxyl groups were substituted with fural groups, capable at heating of forming strong carbon–carbon bonds through opening double bonds of the furan ring, attained a smooth transition from a well graphitizing carbon for PVA to practically non-graphitizing carbon at a content of fural groups of 41% and more (Figure 3.9). For polyvinylacetofural with a high content of fural groups, the

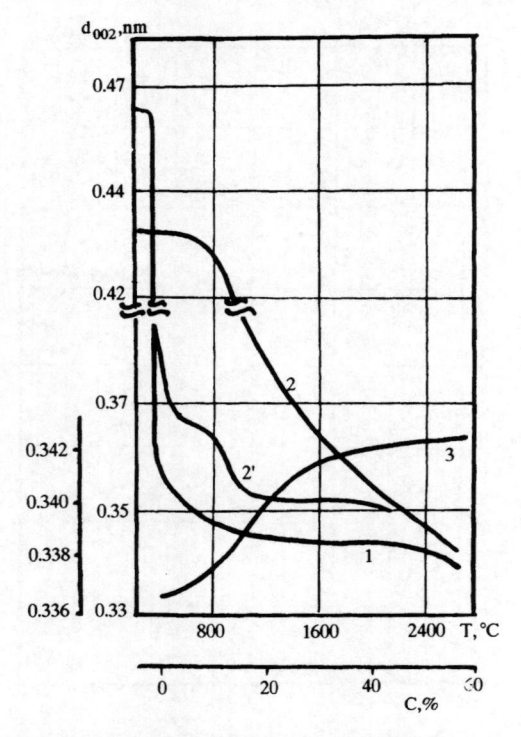

Figure 3.9 Dependence of interplanar distance d_{002} for (1) polyvinyl alcohol and (2) polyvinylacetofural with 41% substituted hydroxyl groups on treatment temperature and (3) content of polyvinylacetofural groups C.

crystalline structure of the matrix are formed heterogeneously, retaining elements of the initial structure up to relatively high treatment temperatures. Using synthetic resins for formation of the matrix, however, makes it possible to obtain more dense, and hence stronger, CCC than with the use of thermosetting binders. Formation of the matrix on the basis of synthetic resins provides for a homogeneous finely dispersed porosity. Table 3.4 presents the distribution of pores by dimensions of effective radii for cokes from pitch and synthetic resin, where it is seen that the coke from resin contains practically no pores larger than 1.0 μm. Accordingly, the permeability of a CCC with a synthetic resin-based matrix will be 1–2 orders of magnitude lower.

Table 3.4 Characteristic of porosity of cokes from pitch and synthetic resin (T_{treat} = 900 °C)

Coke	Open porosity (cm^3/g)	Volume of pores (cm^3/g) with effective radius (μm)			
		< 0.01	0.01–0.1	0.1–1.0	> 1.0
From pitch	0.135	–	0.005	0.003	0.127
From resin	0.119	0.026	0.068	0.025	–

Formation of a synthetic resin-based matrix is in the course of the thermal treatment affected by the degree of curing and the final temperature of curing of the resin, but this effect occurs up to 400 °C [31]. A significant impact at the same stage is exerted by the direction and magnitude of the applied pressure. These factors determine the shrinkage and LTEC of the matrix.

The set of physicochemical processes proceeding at the stage of carbonization of a carbon-fibre skeleton impregnated with phenol–formaldehyde resin is shown in Figure 3.10. At first stages of the process, the prevalence of the relative content of water in the composition of gaseous products indicates the completion of polycondensation processes. In the zone of 400 °C, there occur the maximum gas evolution rate (curve 3), increase in the density of phenol–formaldehyde resin (curve 7), steep decline in the specific hydrogen content in the binder, estimated from the H/C value (curve 8), beginning of a steep increase in the specific surface (curve 4), and beginning of detection of localized paramagnetic centres (curve 5).

The second stage of the process, beginning from 650–700 °C, involves intense sintering processes, accompanied by decrease in the specific surface, decline in the concentration of radicals, substantial increase in the binder density, and predomination of hydrogen in gaseous products.

The low-temperature stage of carbonization of phenol–formaldehyde resin terminates in formation of polycrystalline carbon with a zero graphitization degree d_{002} = 3.44 Å and crystallite size L_a = 20–30 Å.

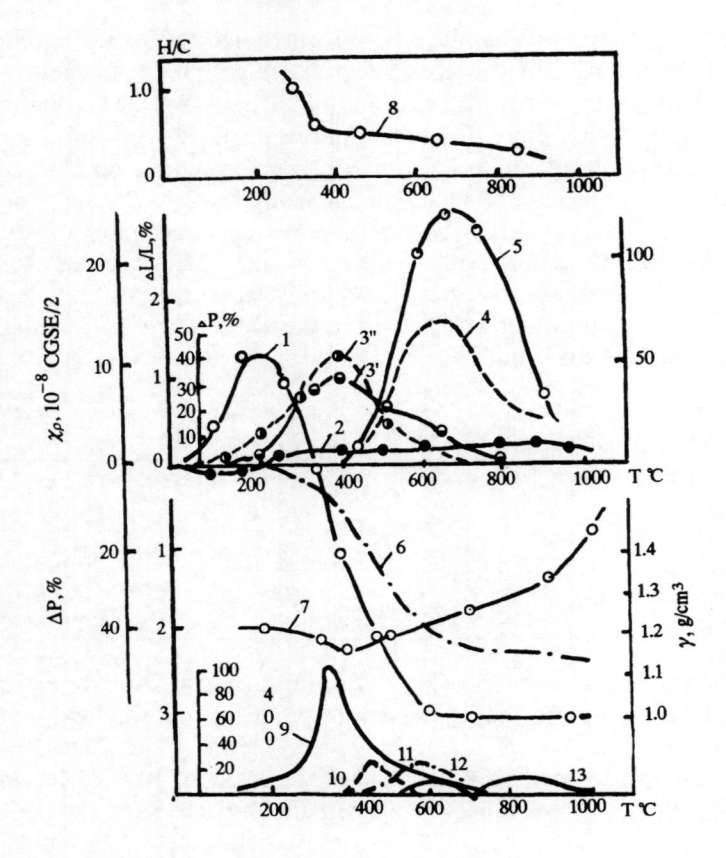

Figure 3.10 Variation of physicochemical and electronic properties of phenol–formaldehyde resin-based carbon-filled plastic: (1) dilatometric curve in transverse direction ($\Delta L/L$); (2) the same in axial direction of fibres; (3′) differential dependence of loss of mass (ΔP) of carbon-filled plastic (one-step phenolic resin); (3″) differential dependence of loss of mass (ΔP) of carbon-filled plastic (novolac phenol–formaldehyde resin); (4) variation of paramagnetism (χ_p) of binder; (5) variation of specific surface (S) of binder; (6) integral dependence of loss of mass (ΔP) of binder; (7) variation of binder density (γ); (8) variation of specific hydrogen content related to carbon content in binder (H/C); and variation of gas evolution rates (volume fractions) for (9) H_2O, (10) CO, (11) CO_2, (12) CH_4, and (13) H_2 in process of carbonization.

Instead of impregnation with a resin or pitch, followed by carbonization, the initial rigidification of a fibrous precursor can be provided by a chemical vapour deposition of carbon. In some cases a material can be produced solely by the vapour deposition of carbon in pores of a fibrous skeleton.

Formation of pyrocarbon (PC) on a surface occurs at participation of active centres, whose role, depending on the process conditions, can be played by carbon atoms, radicals, agglomerates, and other active particles [33]. The PC formation reaction rate at the initial period depends on properties of the substrate, after whose coating with a PC layer the reaction rate for given specific conditions becomes constant.

The interaction of the hydrocarbon being used with the substrate is accompanied by dehydration, cyclization and other reactions both in the case of a heterogeneous and in the case of a homogeneous–heterogeneous reaction. The rate of a homogeneous–heterogeneous reaction exceeds that of a heterogeneous one because radicals emerging in the homogeneous phase perish on the substrate surface, giving rise to new active centres of PC growth [23]. Hydrogen resulting from dehydration reactions inhibits the PC growth, bringing about perishing of radicals in the bulk or reacting with active centres of the substrate [34].

One of the most well-known models of PC formation from methane is represented in Figure 3.11 [35]:

According to this scheme, in a certain temperature range the reactions of decomposition of initial molecules and of synthesis of their fragments result in formation of various types of agglomerates in the form of more or less viscous drops [36]. The agglomerates, depositing on the surface, have a pseudo-liquid viscous structure and get deformed at their contact with the substrate, which results in formation of a PC layer with a high degree of crystalline orientation. As a result of dehydrogenation, the agglomerates lose a considerable amount of hydrogen, and the layer

```
   PC formation on                              PC formation in
     surface                  CH₄                gaseous phase
                               ↓
                              C₂H₄
                               ↓
                              C₂H₄
                               ↓
   Absorption on  ←─────────  C₆H₆  ─────────→  Condensation
     surface              and polyaromatic
                            hydrocarbons
         │                                           │
         ↓                                           ↓
   Dehydrogenation ←                             Dehydrogenation
                     Collision with
                        surface
         │                                           │
         ↓                                           ↓
    Anisotropic                                Isotropic low-density
    dense carbon                                   sooty carbon
```

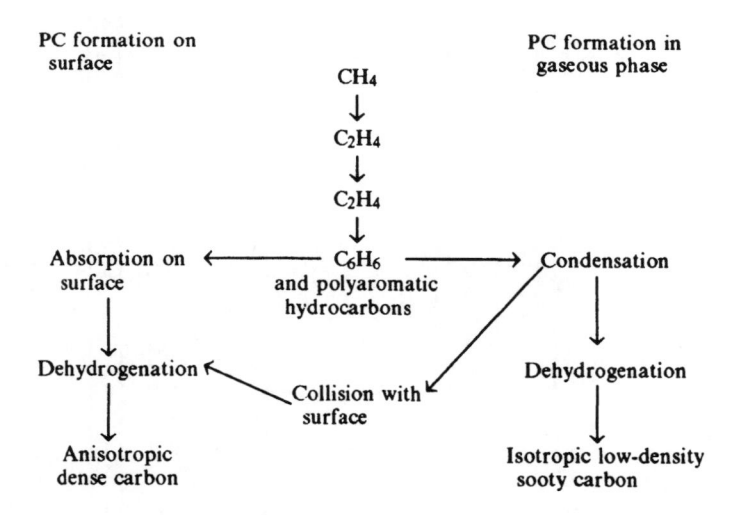

Figure 3.11 Model of PC formation from methane.

hardens to form a solid anisotropic dense carbon deposit. Various agglomeration and hardening schemes are possible depending on the temperature, gas concentration and surface state, which result in PC with various supermolecular structures and properties.

The process of PC deposition in pores is much more complex and can be qualitatively described as a function of the process temperature and hydrocarbon gas concentration in the reaction zone (Figure 3.12). The process is usually conducted within a temperature range of 950–1150 °C at pressures of $0.13–20 \times 10^3$ Pa with the use of methane or natural gas.

In the low-temperature zone, the gas conversion rate is determined solely by the reaction on the surface and in pores of a solid. The whole surface of pores of a solid participates in the reaction. The reaction rate is low; there is practically no gradient of the gas concentration in the volume and pores. As the temperature and the rate of chemical reaction increase, diffusion through pores starts affecting the conversion rate. The concentration of the reacting gas along pores decreases and the effective rate constant increases with temperature slower than in the low-temperature region.

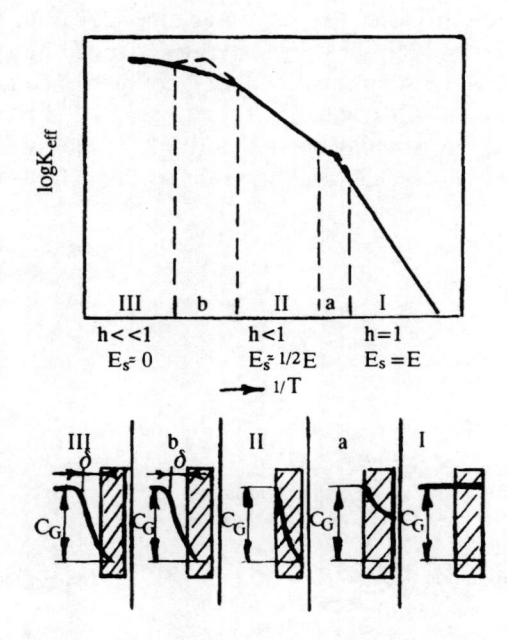

Figure 3.12 Graphic interpretation of Arrhenius' equation for reactions of gases with porous solids (as applied to densification by pyrolytic carbon): K_{eff} – effective reaction rate constant; T – temperature; C_G – concentration of reacting gas in volume; δ – thickness of boundary gaseous layer; η – amount of gas having reacted in pores; E_S – experimentally determined energy; E_A – true activation energy.

As the temperature increases further, the mass transport across the boundary gas layer at the substrate surface becomes the limiting factor for the reaction rate.

A theoretically ideal condition for PC formation in pores is the minimum deposition rate (low temperature and low pressure of gas), but in this case the real densification rate is low.

Three basic methods – isothermal method, temperature gradient method and pressure gradient method – are practically employed for densification of PC in fibrous carbon precursors; a combination of them is used sometimes. In all events, the principal process is the diffusion of an active hydrocarbon gas into a fibrous precursor.

The three principal PC deposition methods are schematically shown in Figure 3.13.

The isothermal process is conducted at a constant temperature (usually of 950–1000 °C), low pressures (1–15 mmHg) or pressures close to the atmospheric one, with considerable additions of an inert gas [37]. The starting hydrocarbon gas is either passed over the surface of parts or filtered through it. Conditions are selected so that the vapour deposition of PC should primarily occur in zone I (Figure 3.12). The isothermal process yields a fairly uniform PC distribution over an up to 12 mm thick wall; a PC concentration gradient across the thickness arises on thicker walls. Also, gradients along the parts can arise at a wrong organization of the overdensification process.

In the thermal gradient method, a fibrous preform with an inner hole is put onto a graphite or molybdenum rod within an induction coil or is heated by passing the current directly through the rod. As a result, the inner part of the fibrous preform, which adjoins the rod, has the highest temperature, whereas the outer part contacts a cooler environment; this gives rise to a temperature gradient across the preform

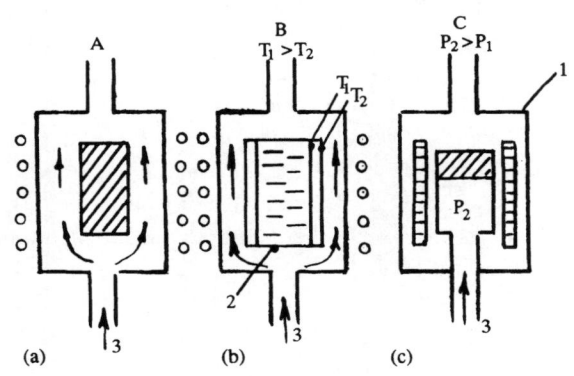

Figure 3.13 Pyrolytic carbon deposition methods: (a) isothermal method; (b) thermal gradient method; (c) pressure gradient method; (1) induction coil or resistance heating element; (2) receiver; (3) gas inlet.

thickness. PC first deposits on inner regions of the preform, and then, as the densified part of the preform starts heating up to the reaction temperature, its deposition advances radially. This method is ordinarily limited to manufacture of a single part and involves additional difficulties in attaining the homogeneity of PC being deposited. The thermogradient method, however, makes it possible to densify large crosssections and is in the PC formation rate greatly advantageous over the isothermal one.

The third method is a pressure differential one. It involves a a forced feed of the working gas into the inner region of the preform, just where the PC deposition occurs (Figure 3.13c). This process is superior to the isothermal one in the PC deposition rate.

The problem of reducing the temperature of processes of the vapour deposition of carbon has been actively studied in recent time. For example, the use of dichloroethylene as the working gas allows the PC deposition temperature to be lowered to 700 °C [38].

3.1.3.3 Carbonization of pitches at high static pressures

The process of carbonization under pressure is widely used in manufacture of carbon–carbon composites. This is associated with two factors: first, a high degree of impregnation of a carbon filler with pitch is provided under pressure and, secondly, the coke yield rises greatly, reaching 80–90% [39].

Cokes from pitches carbonized under pressure exhibit some specific features as compared with cokes produced at pressures close to the atmospheric one. The features as well as the mechanism of the process of carbonization under pressure can be revealed by studying the variation of physicomechanical and thermophysical properties, porous structure, structure of the produced coke at all levels: macrostructure, supercrystallite structure, and microstructure.

A direct study of the processes occurring at carbonization of pitch under pressure is difficult enough because of complexities involved in introduction of instrument sensors into the reaction chamber. In view of this, intermediate products of carbonization are studied, obtained as follows: when a predetermined temperature has been reached, a sample is cooled rapidly so as to 'freeze' the chemical composition and structure at a given carbonization stage.

Variation of EPR parameters with the temperature of carbonization at an atmospheric pressure and at a pressure of 60 MPa is shown in Figure 3.14, where it is seen that at carbonization under pressure the paramagnetic susceptibility curve shifts towards higher temperatures. Such changes result from the slowing of aromatic fragment formation processes under pressure. Examination of variations of the EPR signal linewidth,

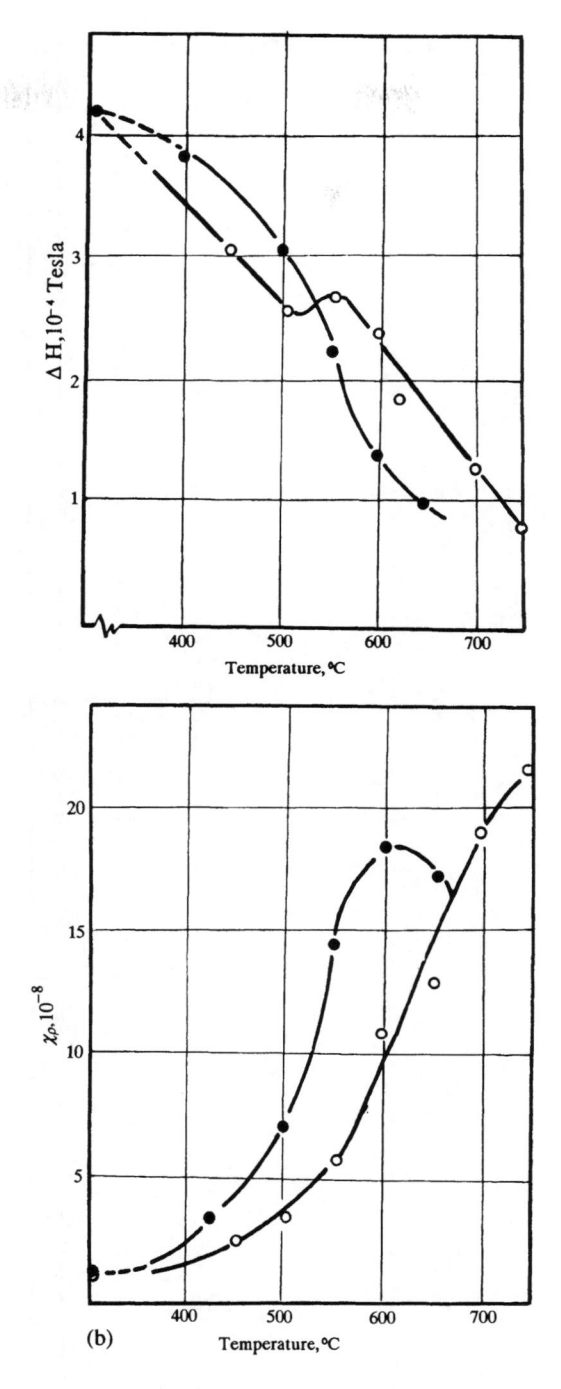

Figure 3.14 Variation of (a) width of line and (b) of paramagnetic susceptibility of EPR signal with carbonization temperature.

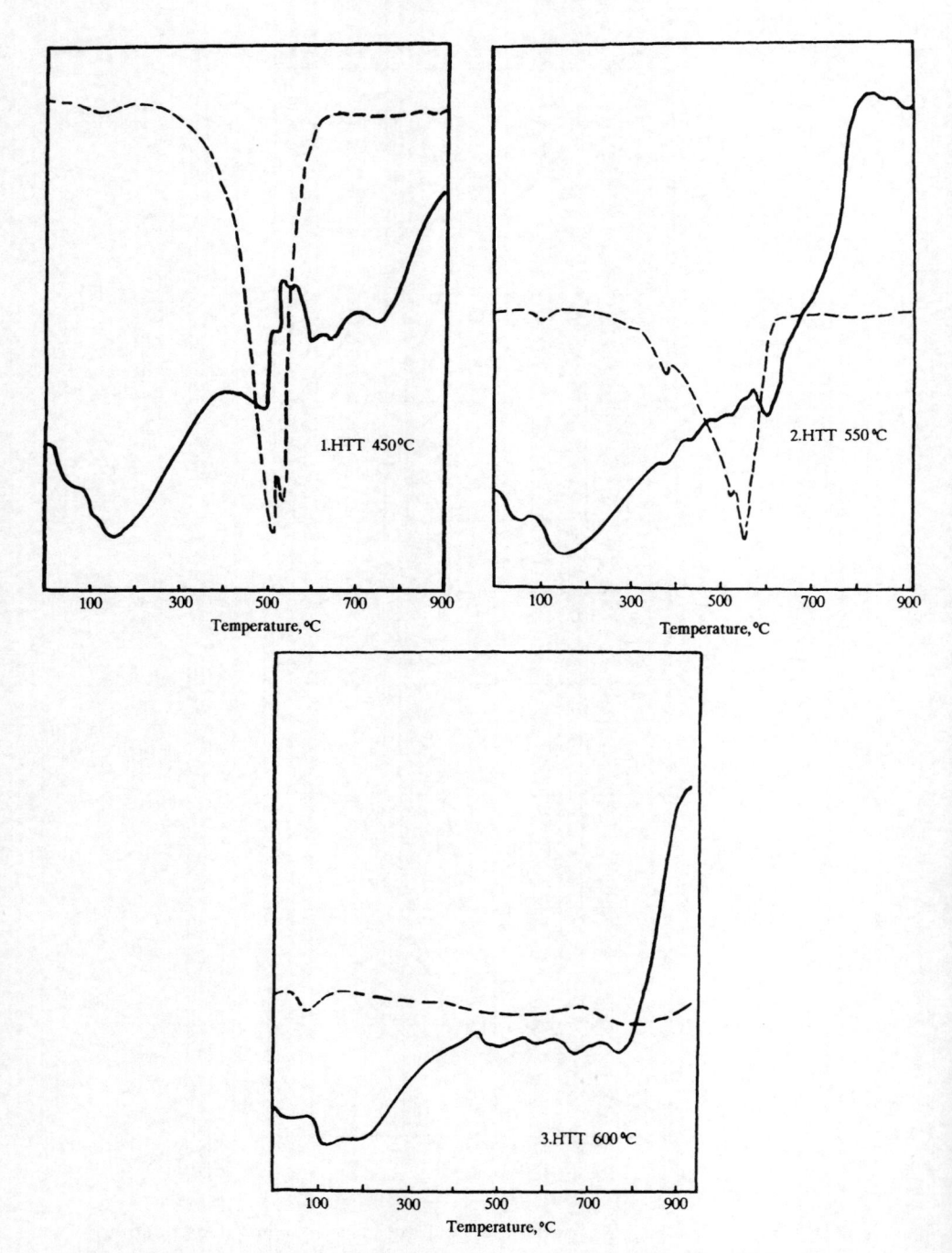

Figure 3.15 Derivatograms of products of pitch carbonization at atmospheric pressure (– – –) and pressure at 60 MPa (———).

Theoretical fundamentals 321

4.HTT 650 °C

5.HTT 750 °C

Temperature, °C

Temperature, °C

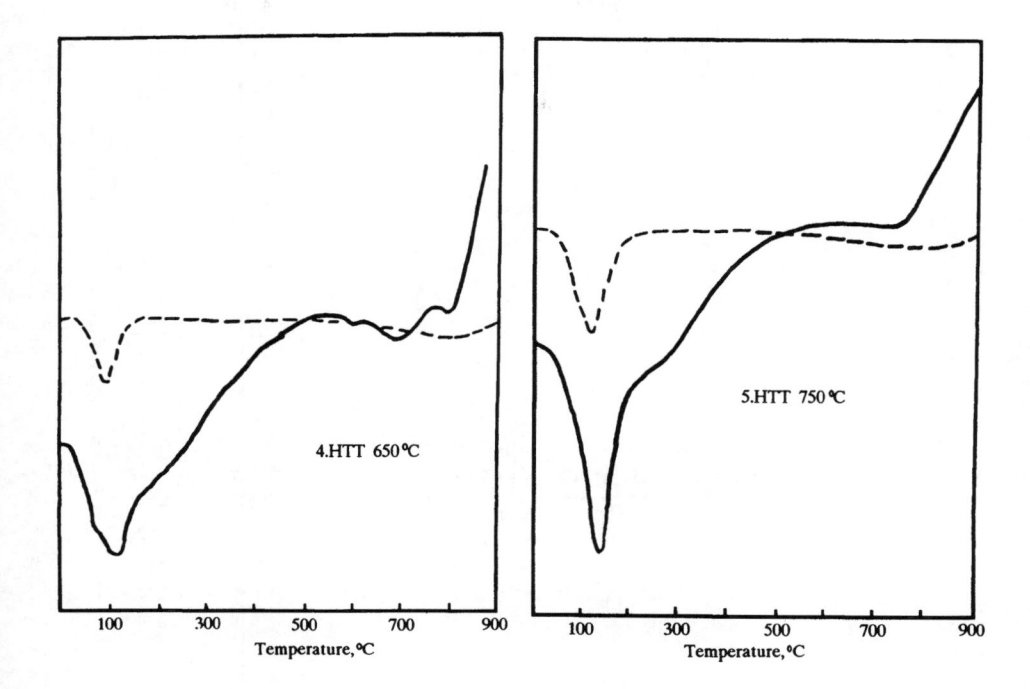

however, shows this to occur not in the whole range of temperatures under study. Thus, the change in the EPR signal linewidth up to 500 °C is greater for a sample carbonized under an atmospheric pressure, and over 550 °C, for pitches carbonized under pressure. This means that up to 500 °C, the growth of sizes of aromatic fragments is accelerated on account of the aromatic constituents existing in the pitch. The slowing-down of growth of aromatic fragments after 550 °C may result from a hydrating action of volatile carbonization products.

The thermal analysis, along with the EPR method, is informative in studying the carbonization process mechanism. Thermograms of pitches carbonized at the atmospheric pressure and at a pressure of 60 MPa are shown in Figure 3.15. At temperatures of 450–550 °C, thermograms of samples carbonized under pressure show a shift of thermal decomposition of pitch towards higher temperatures with respect to that of pitches carbonized under ordinarly conditions. Characters of thermograms for samples at treatment temperatures 450–550 °C are similar, but at higher temperatures the thermograms differ considerably. Pitch decomposition effects in DTA curves disappear, which is evidenced by flattening of the curves. DTG curves (Figure 3.16) also have no extremal regions, characteristic of samples with a treatment temperature of 450–550 °C. In the low-temperature region of DTG for samples with a treatment temperature of 600–750 °C, there appears an extremum of moisture release

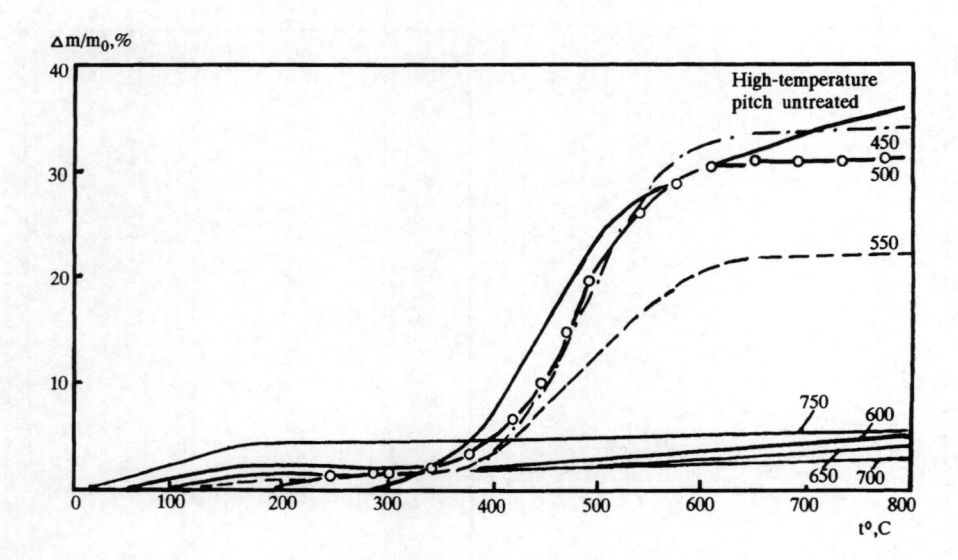

Figure 3.16 Thermogravimetry of products of pitch carbonization under pressure of 60 MPa.

(100–130 °C), increasing with the sample treatment temperature; an endoeffect in DTA curves corresponds to it.

Recalculating the loss of mass of samples in terms of a dry weight indicates decline in the mass loss at 800 °C with increasing treatment temperature, the most drastic decline occurring within a range of temperatures of treatment under pressure of 550–600 °C; a further temperature increase only insignificantly reduces the mass loss.

Cokes produced under pressure differ substantially from cokes carbonized under ordinary conditions in the porous structure, both by a lower porosity and by the form of the pore size distribution curve. The apparent and pycnometric densities and characteristics of the porous structure of cokes produced by carbonization of a high-temperature coal pitch under pressures of up to 100 MPa are presented in Table 3.5.

The pycnometric density (by ethyl alcohol) remains practically unchanged over the whole pressure range from atmospheric pressure ($d_p = 1.50 \text{ g/cm}^3$) and up to 100 MPa ($d_p = 1.49 \text{ g/cm}^3$), whereas the apparent density, directly associated with the porosity of a material, rises from 0.6 g/cm^3 at atmospheric pressure to 1.24 g/cm^3 at 100 MPa. The accessible (open) porosity as well as the average pore size decrease with increasing carbonization pressure. Also, as the pressure increases, redistribution occurs of the main forms of pores inherent in cokes. Thus, the amount of large spherical pores, traps, predominating in

Table 3.5 Density and characteristics of porous structure of cokes obtained under pressure

Specific pressure (MPa)	d_a (g/cm^3)	d_p (g/cm^3)	Porosity (%)	Specific volume of pores (cm^3/g)			r_{max} (μm)
				total	traps	capillaries	
60	1.13	1.53	18.8	0.164	0.142	0.022	1.26
80	1.20	1.55	16.5	0.140	0.102	0.038	1.00
100	1.24	1.49	15.9	0.126	0.085	0.041	0.89

cokes produced at atmospheric pressure, decreases progressively with increasing pressure, which indicates their receding from one another and decrease in their sizes, whereas the volume of capillary pores is growing.

It is to be noted, however, that a material with a high enough porosity (15.9%) is formed even at a pressure of 100 MPa.

The character of distribution of the porous structure practically does not change with pressure (Figure 3.17), only a shift of the main maximum to the left and, accordingly, decrease of the specific volume of pores with r_{max} with increasing carbonization pressure being observed in differential curves of the size distribution of specific volumes of pores. Cokes from pitches carbonized under pressure exhibit a number of structural features, distinguishing them from cokes produced at pressures close to the atmospheric one. The features can be revealed by examining the structure of a thermally treated material by methods relying on the analysis of electronic properties. Three structure levels – macrostructure, supercrystallite structure, and microstructure – are analysed.

The macrostructure is conveniently characterized by the coefficient of connectivity [40], defined as the ratio of the electrical conductivity

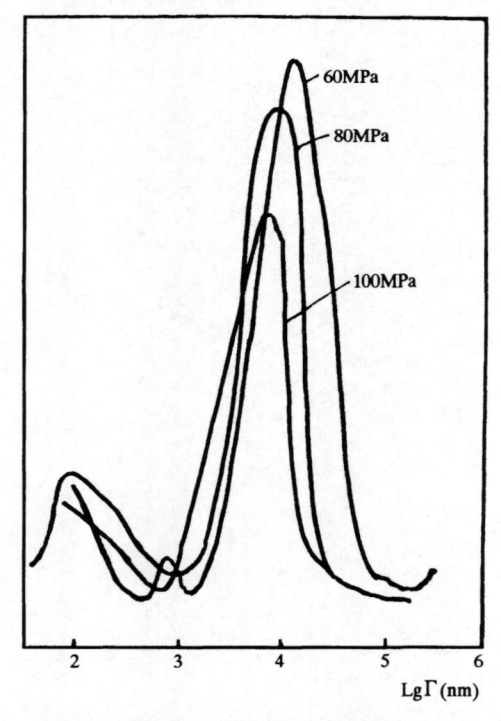

Figure 3.17 Differential curves of size distribution of specific volumes of pores in cokes of pitch carbonized under various pressures.

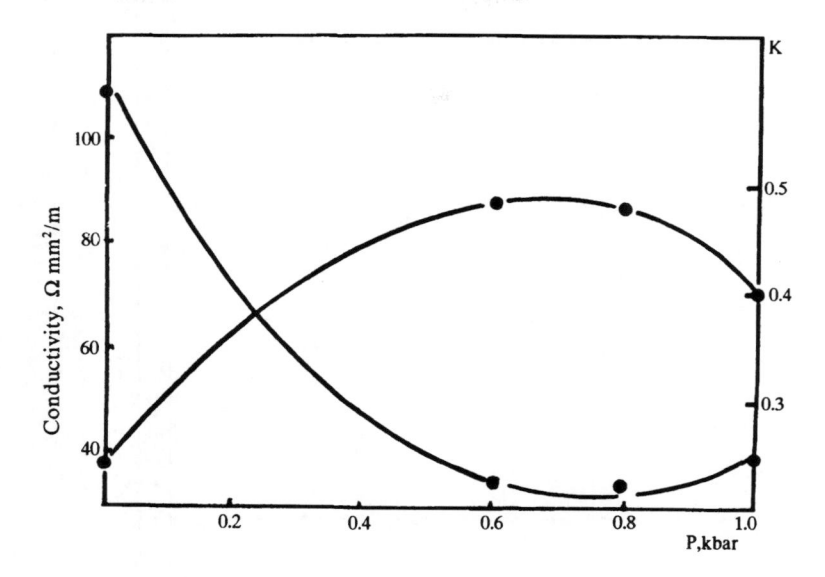

Figure 3.18 Effect of carbonization pressure on electrical conductivity and coefficient of connectivity of cokes produced under pressure.

of a sample to the electrical conductivity of its nonporous microvolumes, determined from the concentration and mobility of charge carriers.

The coefficient of connectivity (Figure 3.18) increases twofold, from 0.24 for a pitch carbonized at an atmospheric pressure to 0.49 for pitches carbonized under pressures of 60–80 MPa. This increase evidences ordering of the macrostructure of carbon produced under pressure.

Variations of supercrystallite structure parameters with pressure are shown in Figure 3.19. $\Delta H/\Delta g$, a reduced width of the EPR signal line, normalized to a change of the g-factor, characterizes the short-range order in the arrangement of crystallites in volumes at the level of fractions of a micrometer. This parameter grows with increasing preferable orientation of crystallites in said volumes. The second parameter, the angle of disorientation of crystallites in particles sizing up to 40 μm, decreases with growing degree of orientation. The analysis of Figure 3.19 shows that the variation of the two parameters unambiguously evidences a decline in the degree of preferable orientation of crystallites in microvolumes of the material.

The least changes under the effect of pressure were noted in the microstructure of investigated cokes; this is evidenced by the data of Table 3.6 [41], which characterize the microstructure by four parameters independent of one another.

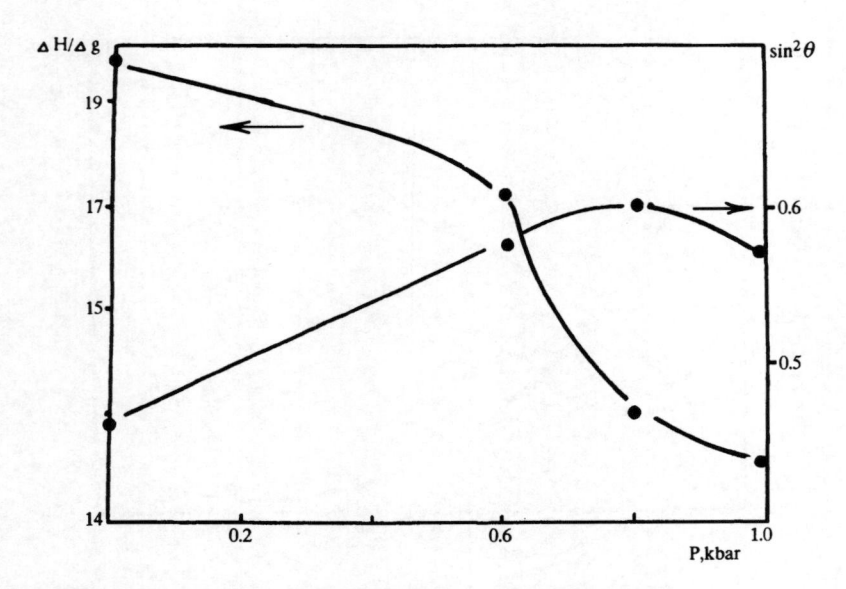

Figure 3.19 Variation of parameters of supercrystallite structure of cokes produced under pressure.

Table 3.6 Effect of carbonization pressure on microstructure of cokes (treatment temperature, 2700 °C)

Characteristic of microstructure	Pressure (MPa)			
	0.1	60	80	100
g-factor of EPR signal (arb. units)	2.0113	2.0109	2.0111	2.0110
Magnetoresistance $\Delta\rho/\rho$ (%)	3.0	2.9	2.5	2.5
Diamagnetic susceptibility $\chi(\times 10^{-6})$	7.1	7.5	7.2	7.0
d_{002}(Å)	3.398	3.402	3.396	3.395

Examining the changes in properties of cokes produced under pressure indicated that their character is close to that of changes in parameters of the supercrystallite structure and macrostructure. Thus, the electrical resistance curve, shown in Figure 3.18, has an extremum with minimum values in the pressure range of 60–80 MPa (similar to K). The compressive strength and Young's modulus curves, shown in Figure 3.20, have maxima in a pressure region of 80 MPa. The extremal character of the curves of variation of the properties of cokes produced by carbonization under pressure stems from build-up of defects like fine cracks as the pressure increases from 80 to 100 MPa, which is substantiated by the appearance of an additional maximum of the differential curve of the size distribution of specific volumes of pores (Figure 3.17) for coke produced under a pressure of 100 MPa.

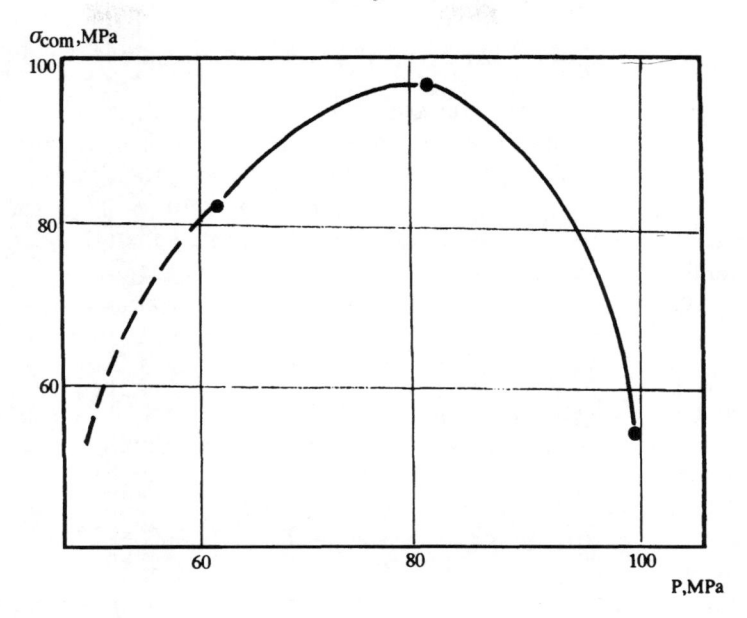

Figure 3.20 Effect of carbonization pressure on strength of cokes.

3.1.3.4 Graphitization

Graphitization is the process of a thermal transformation (ordering) of nongraphitic carbon materials into graphite. The rate and completeness of transformation of a starting carbon material into graphite depends on many factors: process temperature and time, nature of the raw materials used, composition and content of impurities, gas atmosphere, and applied pressure, as well as on features of the production equipment and process.

All carbon materials can be divided into graphitizable and nongraphitizable. There are many hypotheses about the structure of graphitizable carbon materials and mechanism of the graphitization process [1]. Most close to the reality are, in our opinion, Fischbach's concepts [42]. According to them, graphitization is a thermally activated process, consisting first of all in removal of a relatively small amount of defects that create and stabilize a disordered structure of graphitizable carbon materials. Graphitization of graphitizable carbon materials proceeds in two stages. The first of these terminates in a few minutes, and therefore an experimental determination of any physical properties, characterizing the degree of graphitization, at temperatures of 2300–3000 °C in such a short time is very difficult. It is most expedient to examine the regularities of variation of the interplanar distance of the lattice of carbon materials in graphitization, and therefore it is reasonable to examine the variation of the graphitization degree γ_g, determined from the interplanar distance:

$$\gamma_g = (0.3425 - \bar{d}_{002})/(0.3425 - 0.3356).$$

This characteristic varies from zero for a carbon material (semiproduct) to unity for a fully graphitized material. It was in [43] compared with the volume of material that had undergone the transformation in graphitization; the volume also varies from zero to 100%. Then for the two-stage process, $\gamma_g = \gamma_0 + \gamma_1$, where γ_0 and γ_0 are graphitization degrees corresponding to the above-mentioned graphitization stages.

In accordance with [1], for the first stage of graphitization,

$$\gamma_0 = B \exp(E_a/RT), \qquad (3.4)$$

where B is a constant allowing for features of the initial structure of carbon materials; E_a the activation energy; T the absolute temperature; and R the gas constant.

Most experimentally determined activation energy values are around 2.45 eV, which corresponds to the energy of activation of movement of displaced atoms along the crystallographic c-axis (2.80 ± 0.2 eV).

The second graphitization stage (γ_{g^1}) can be treated as transformation in the solid phase, similar to the recrystallization process. Such transformation are described by the Avrami equation:

$$\gamma_{g^1} = \gamma - \gamma_0 = 1 - \exp(-D\tau^n), \qquad (3.5)$$

where D is a coefficient depending on the temperature and material properties; τ the time; n a constant; and γ_0 from equation (3.4).

Equation (3.5) describes processes resulting in a complete transition of a substance from one state to another, so that γ_{g^1} varies from zero to unity.

The experimentally determined activation energy is close to the energy of the process of carbon self-diffusion in the graphite lattice (7.1–8.3 eV) and somewhat exceeds that presented in literature for the movement of vacancies along the c-axis (5.45 ± 0.05 eV).

The above-described allows the first stage of graphitization to be interpreted as displacement of some intermediate carbon atoms to the nearest equilibrium position. When displaced atoms move to the equilibrium position, there occurs some straightening-out of packets of crystallographic planes, which terminates at a given temperature at the moments when edges of the packets come into contact with one another. The decrease in the curvature (i.e., growth of texturization) of the planes allows a thickening of the packets in a direction perpendicular to the basal plane. It is effected through diffusion mainly of vacancies (which takes time). As a result, the defectiveness of the layer declines and dimensions of the b.c.c. lattice increase. The rearrangement terminates when the possibility of crystallite growth has been exhausted because of contact between individual packets.

A further graphitization requires a higher temperature. Then some of the remaining planes straighten out, thereby making possible a further thickening of packets. At temperatures above 2500–2700 °C, however, the ordering process can slow down, the slowing-down degree varying with the specific conditions.

The existence of the preferable orientation of carbon layers at regions a few micrometres in size determines graphitizability of the material at heating up to 2300 °C. At the absence of such regions, i.e., when there is a 'rigid' structure with strong side bonds, the material is not graphitized even at a thermal treatment of up to 3000 °C.

As noted above, graphitizable materials are produced from low-oxidized hydrogen-rich carbon substances which get softened at the initial stage of carbonization (petroleum and pitch cokes, coking coals, etc.). Nongraphitizable materials usually result from oxygen-rich substances; they do not soften at the initial carbonization stage. A high oxygen content (or lack of hydrogen) gives rise to cross-links between carbon networks, which create a 'rigid' structure. The creation of a 'rigid' structure, such as through a pre-oxidation of pressed samples in a temperature range of 200–300 °C, impairs the graphitizability of a material [44]. On the other hand, addition to the blend of a number of elements and chemical compounds, acting as catalysts, facilitates graphitization of a material [45]. Thus, the addition of silicon to materials with various graphitizabilities, produced from phenol–formaldehyde resin, improved graphite-like layers that form a crystal; this in turn shifted the process of graphitization in graphitizable materials towards lower temperatures [46].

A pretreatment through application of pressure at carbonization stage to a carbonaceous substance, nongraphitizable under ordinary conditions, made it possible to obtain a graphitizable coke. The applied pressure produced regions with a preferable orientation of aromatic macromolecules in the substance and thereby created the prerequisites for their ordering at a high-temperature treatment [7].

Carrying out graphitization under pressure should shift the process towards lower temperatures. Thus, the application of a pressure of 1 GPa lowered the temperature of graphitization of cokes from polyvinylchloride and samples of petroleum coke with a pitch binder by about 1000 °C. The authors ascribe this effect to the creep of the material, caused by the applied pressure. Conversely, reducing the pressure shifts the graphitization process towards higher temperatures.

The thermal treatment of carbon materials is always carried out in a gas atmosphere, whose composition affects not only the graphitization kinetics, but also the structure type. The presence of oxygen and (to a lesser extent) of carbon dioxide in the gaseous phase impairs graphitization [47]. Thermal treatment in a chlorine atmosphere speeds up graphitization, and further heating increases the disorder in the structure of the fired material. No impact of nitrogen or argon on graphitization was detected.

3.2 TECHNOLOGY OF PRODUCT MANUFACTURE FROM CARBON-BASED MATERIALS

As follows from the preceding discussion, methods of processing the starting materials (fillers and binders) in manufacture of carbon materials are fairly diverse. These include press moulding, followed by a thermomechanical treatment; lay-up and winding with impregnation, polymerization, and thermal treatment; high-pressure and thermal treatment; gas-phase and liquid-phase processes. These methods of manufacturing carbon composites call, as a rule, for unique equipment and special thermal treatment techniques; they are mostly multistage ones, involve emissions of toxic substances, are fire- and explosion-hazardous. The complexity of processes and high temperatures impede development and introduction of automated production-process control facilities. In spite of a vast experience in development of such materials, developing new carbon materials remains in a certain sense a semiempirical process.

A characteristic feature of modern production of carbon composites is the search for ways to prepare such a starting semiproduct that is most close in the shape and dimensions to the finished part. The manufacture of such materials for parts of various purposes includes essentially four stages.

1. Production of preforms of parts. This stage includes selection of the reinforcing component and binder, reinforcement type, moulding method, and hardening conditions.
2. Carbonization. Here the pyrolysis and then carbonization of hardened preforms in an inert atmosphere at a stepwise heating and holding are carried out and, when required, the formed porous structure is saturated with pyrocarbon.
3. Graphitization. Graphitization of carbonized preforms is carried out. When required, they are saturated with pyrocarbon or pyrographite.
4. Densification. Repeated or cyclic operations of impregnation of preforms with the binder, carbonization and graphitization are carried out with possible, when required, saturation with pyrocarbon or pyrographite.

Distinguishing features of the methods for manufacture of carbon-based materials are a prolonged production cycle and also sometimes continuity of processes, precluding a current control of the quality of materials.

Carbon composites lend themselves to practically all types of machining. Only limitations of the size imposed by working spaces of the thermal equipment, autoclaves, capabilities of the pressing equipment, volume of the reaction space at liquid- and gas-phase production methods are critical.

3.2.1 Winding of preforms

The formation of reinforcing structures of carbon composites by winding is widely used in the manufacturing technology for parts made from CCC as one of the most productive, mechanized and automated production techniques.

Winding is used to produce exhaust cones, e.g., for rocket engines, while the technology for winding is being intensively advanced. Winding methods and devices make it possible to manufacture hollow products of any complex configuration by winding of continuous fibres. Winding of parts such as a turboprop engine silencer, carbon fibre-reinforced continuously wound tubes, T- and I-beams from a prepreg where the fibre/resin ratio is maintained to within 1% and filler structures are laid accurately to within 0.002 inch have been effected. The automation of winding and laying of tapes makes it possible to vary the width of tapes, to unreel and compact the prepreg.

A wide development of the winding practice is promoted by an adequate assortment of carbon fibres, bundles, and tapes, manufactured in the former Soviet Union, USA, Europe and Japan.

New packing designs, allowing the transportation of easily damageable fibres from the manufacturer to the consumer, have been developed for processing of carbon fibres and tapes. When selecting fibres for winding, preference is given to bundles with a higher breaking strain. The winding makes possible processes of a continuous manufacture of varying-profile tubes, which practically cannot be effected by other production techniques.

The output of the winding equipment is raised when prepreg tapes are made beforehand. Sheet prepreg production methods yield thermally pretreated semiproducts, suitable for a subsequent hardening of carbon-filled plastics without a substantial release of volatiles or considerable shrinkages at the plastic formation stage.

Selecting the textile form – tape of radially uncoupled filaments or fabric – in winding is very important for carbon–carbon composites. The winding of a fabric reduces the shrinkage in the plane of the layer of a two-dimensionally reinforced material in carbonization and subsequent high-temperature procedures.

The use of prepreg makes possible the automation of tape-laying and even a changeover to application of industrial robots. Robotization of plants makes it possible to wind fibres on mandrels of the most complex configuration with six rotation axes for the feed of fibres. The seventh

332   *Carbon-based composites*

axis is intended to manipulate the mandrels. Figure 3.21 shows a working moment of winding a product on a multiaxis machine with formation of the tape being wound directly on a movable platform which carries the thread guide and other actuating mechanisms.

The methods for winding of products for a subsequent production by a high-temperature treatment of carbon–carbon materials in a predetermined shape take into account specific conditions of the further manufacturing procedures, apart from the traditional allowance for the composite-mandrel interaction, flowability of resin, subsidence in winding, and residual stresses [48, 49, 50].

These specific conditions are listed below.

1. Emission of a considerable volume of volatiles in carbonization of hardened polymeric binders [51].
2. Considerable radial shrinkages of transversally isotropic structures along with a positive linear elongation of a composite in the reinforcement directions.
3. Relative displacement of the reinforcement and hardening matrix in the microstructure of a composite, which exceeds ultimate deformations of fracture of brittle fibre-reinforced carbon matrices.
4. Development of residual deformations in surface layers as a result of anisotropy of a structure with reverse turns of tapes or bundles at a filamentary winding.

Figure 3.21 A winding bay.

Technology of product manufacture

The above features necessitate a departure from schemes of reinforcement without overlaps, traditional for carbon-filled plastics [52], which in plastics with a yielding binder provide the greatest utilization of the strength (elastic modulus) of the reinforcement in a composite [53].

Increasing the stability of a reinforced structure under such conditions calls for changeover to reinforcement schemes with a mutual crossover of tapes, formation of a relatively thermomechanically stable reinforcement pattern, geometrically closed in a plane, which is produced by a multistart winding. This reinforcement pattern, known earlier, but as a rule not used for plastic matrices, is formed when, after laying the preceding forward and then reverse turn, the laying of the next turn on the mandrel starts after the latter has been turned through some constant angle. At the greatest diameter of a part, the tapes or bundles are laid with the minimum overlap; the latter, of course, increases as the diameter decreases. The thickness of the layer being wound increases in proportion to the ratio of local diameters of the mandrel. This results in the winding pattern shown in Figure 3.22. It can also be seen in Figure 3.21. It is used by the Schunk company to manufacture tubular products [54, 55]. In Figure 3.22 the principal axes of the part are τ (circular direction) and θ (axial direction), and of the composite, '11' (reinforcement direction)

Figure 3.22 Winding pattern. τ, circular direction; θ, axial direction; l, free length of tape in reinforcement pattern, between interlaces of tapes; b, width; h_0, thickness of tapes; w, deflection of lifted-off tape (defect).

and '22' (transversal direction); a tape-reinforcement cell 'D' can be seen, where l is the free length of a tape between tape overlaps in the reinforcement pattern, b is the tape width, and h_0 the tape thickness. The region A, where the top tape is not strengthened by the interweave, contains a defect: a lifted off tape with a deflection w. In region C the tape at the edge of the part is the most free to move, while in region B the underlying tape is clamped by the interweave, and its movement is the most restricted.

The mutual restriction of interweaving tapes enhances, on the whole, the stability of the structure at high-temperature movements, shrinkages, and opposition to residual internal deformations.

Experimentally-determined absolute levels of residual strains in surface layers of carbon–carbon preforms are ten times lower than the ultimate fracture strain for shells made of these composites. At an inadequate adhesion bond between layers, which in carbon materials is strengthened with the use of viscose precursor-based carbon fibres, a carbon matrix containing pyrocarbon deposits, and at a lower composite treatment temperature [56], surface layers can separate off through the stability loss mechanism. The incidence of this phenomenon rises with increasing free length of fibres in the reinforcement pattern of Figure 3.22. Limits of the region of optimal values of the free length of the fibre and width of the tape are obtained from a modified model [50] of a longitudinal deformation on the surface of a tape of veneer, little resistant to shear, fixed in supports, and supported on one side by an elastic base of a wall (Figure 3.23).

The selection of l determines, in turn, the number of starts in the reinforcement pattern

$$N \geq \frac{\pi D}{l \cos \alpha_w},$$

where N is the number of starts of a forward turn and α_w the winding angle at the maximum diameter (D).

The tape width has a restriction to ensure its close fit to a curvilinear surface of the mandrel. Here the stretching of the tape can partly compensate for the difference in the lengths of paths of outer filaments in a b wide tape

$$b \leq \frac{[\Sigma] R}{\cos \alpha_w \times \tan \lambda_t},$$

where $[\Sigma]$ is the limiting stretching of bundles of a taut tape; λ_t the mandrel taper angle; and R the mandrel radius.

Principal stresses in shells of revolution fall at directions of the greatest local curvature of the surface. Carbon–carbon composites qualitatively differ from plastic-binder composites by an order of magnitude lower

Technology of product manufacture

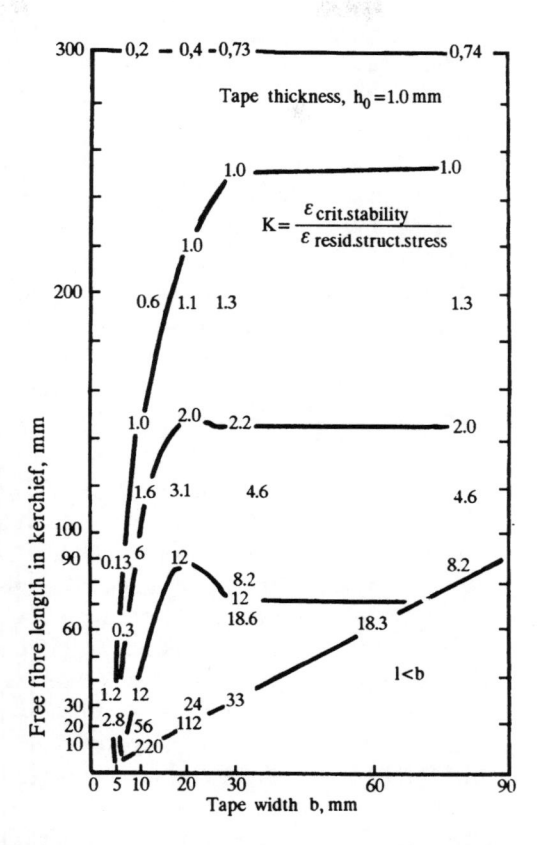

Figure 3.23 Dependence of free length of fibre in kerchief on tape width.

ultimate tensile strain. This decrease for CCC occurs relatively fast even at small deflections of the reinforcement direction from the breaking stress action vector. Optimal in this case are reinforcement patterns with geodesic paths of filaments, bundles, or tapes [50] according to the law

$$R \sin \alpha_w = R_2 \sin \alpha_{2w}, \qquad (3.6)$$

where R, R_2 are shell radii (current and maximum) and α_w and α_{2w} are the respective winding angles.

The constraints imposed on the winding pattern by expression (3.6) specify the optimal overall geometric dimensions of parts as well as permissibility of local deviations of the curvature. As the preform diameter decreases, a natural process of shell thickness growth is proceeding according to the equation

$$\delta_1 = \frac{R_2}{R_1} \times \frac{\cos \alpha_{2w}}{\cos \alpha_{1w}} \times \delta_2, \qquad (3.7)$$

where: R_2, R_1 are the maximum and minimum radii of the mandrel; α_{2w}, α_{1w} are the winding angles in zones of the radii; δ_2 is the tape thickness at the maximum diameter.

Nonuniform conditions for volume displacements at carbonization, high-temperature treatment, and then cooling down to room temperature are created on the part surface in all cases. Physical prerequisites for repeated spatial volume displacements are strongly pronounced anistropies of the chemical shrinkages and physical LTEC, which stem first of all from the known nearly tenfold anisotropy of thermal properties of the reinforcement. This results in a restriction to exclude emergence of tensile stresses in a shell of revolution in the form:

$$\frac{\delta}{R} \leq \frac{\varepsilon_{\tau,\theta}}{\varepsilon_r}, \qquad (3.8)$$

where: δ is the shell thickness; $1/R$ is the shell curvature, found from two local principal radii of curvature; $\varepsilon_{\tau,\theta}$ is the strain in a circular or an axial direction; ε_r is the strain in a radial direction.

Expressions (3.6)–(3.8) together give specific restrictions on the geometry of parts only for carbon–carbon materials with account of anisotropy of shape changes of a microcomposite, prepreg reinforcement pattern, binder composition, coke yield in the binder, and degree of texturization of the carbon filler and then of the composite as a whole.

In contrast to composites with plastic matrices, only for carbon–carbon materials the final formation of the matrix after arranging the reinforcement by winding on a mandrel in the geometry of the future part is also characteristic [57]. The tension in winding determines in a large measure the character of the pore structure of the carbon–carbon skeleton, since it can initiate squeezing of the binder out of the prepreg, especially at a 'wet' winding of an impregnated tape. The contact pressure in winding varies over the surface of the preform being wound with the angle of curvature of the part, winding angle, and tape tension in accordance with the expression

$$P = \frac{2K_{0v}T_t}{Rb} \sin^2\alpha \cos\alpha, \qquad (3.9)$$

where T_t is the tape tension; α the winding angle, λ the mandrel taper angle; b the tape width; K_{0v} the coefficient of overlap; and R the current radius.

The squeezing-out of the binder over the height of a preform, resulting from a greater local curvature, can increase drastically right up to the disappearance of the adhesive interlayer.

The thickness of resin interlayers remaining in the bulk of the material decreases proportionally to the contact pressure and stabilizes depending on the level of the surface tension:

$$r = \frac{2\gamma}{P}, \qquad (3.10)$$

where γ is the surface tension.

The maximum size of pores in places of resin interlayers is limited within the dimension 'r'. The binder squeezing-out develops with time in proportion to the resin viscosity. In the case of a 'wet' winding the binder squeezing-out can result in a 'sagging' of the filler in a radial direction until the tension relaxes to a level where the capillary pressure turns out to be sufficient to prevent further squeezing-out.

As a result, in different regions of a part, because of different conditions of the local contact pressure (3.10), the structure of binder interlayers becomes greatly dissimilar and a sharply different structure of the pore volume is formed after carbonization, which will lead to a substantial difference in physicomechanical properties both of the carbon-filled plastic-precursor and of the end carbon–carbon material. Conditions for creation of an inhomogeneous structure of the binder and then of the pore volume are prevented by using resins that are solid under the winding conditions in preparing prepregs. At a 'wet' winding, the winding conditions in preparation of carbon-filled plastic preforms, precursors of carbon parts, as well as the winding parameters appearing in expression (3.9) have to be defined more exactly by experiment.

3.2.2 Lay-up

To make skeletons of carbon fibres or fabrics by the lay-up method, use is made of a negative mould (female die) or a positive mould (male die), manufactured from metals, plastics, reinforced plastics or their combinations. Bundles, tapes or fabrics are laid, by hand, on the surface of the mould, coated beforehand with an adhesive layer, the binder is applied by brushes, and then the product being moulded is rolled over by a roller to remove air and compact the material.

The product wall thickness is varied by the number of filler layers arranged in the mould. A layer of the binder without the reinforcement is applied to the product surface. Next, polymerization is carried out, and a ready carbon-filled plastic preform is extracted from the mould. When pitch is employed as the binder, then metallic moulds are used, in which the subsequent carbonization is carried out.

3.2.3 Weaving of three-dimensional structures

One method of producing skeletons for three-dimensional structures is weaving on special machines [58, 59]. The widest use has been found by processes of a four-stage or Decart weaving and a two-stage weaving.

Carbon-based composites

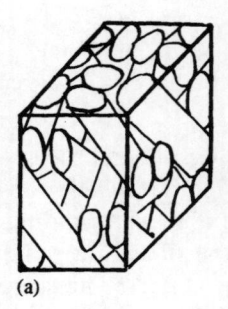

Figure 3.24 Structures of elementary cells of three-dimensional fabrics at (a) two-stage and (b) four-stage weaving methods.

In the four-stage process, the weaving is carried out with the use of at least four movements (strokes) per machine cycle, while in the two-stage process every machine cycle includes two different movements. Structures resulting from the processes are shown in Figure 3.24.

At the two-stage weaving the skeleton structure consists of two groups of fibres: axial and braiding. Axial fibres are laid in the fabric-forming direction and remain nearly straight without any interweavings. Braiding fibres pass between stationary axial ones to form a specific pattern that tightens up axial fibres and stabilizes the skeleton shape. When the arrangement of axial fibres has been determined, the number and arrangement of braiding fibres are laid down. The cross-section of the skeleton replicates the pattern of axial fibres on the machine frame. The ratio between braiding and axial fibres in fabrics being manufactured is always less than unity.

Figure 3.25 Schematic of two-stage weaving.

Figure 3.25 shows a 4×8 rectangular sample. The number of axial fibres in an $m \times n$ rectangle is $(2mn - m - n + 1)$, and of braiding fibres, $(m + n)$. The total number of fibres in the structure is $(2mn + 1)$.

The four-stage process makes it possible to produce several different woven configurations. Figure 3.26 schematically illustrates the four-stage process with the simplest and most popular weaving pattern 1×1. The cross-sectional shape of finished fabric replicates the shape of the main part of the general pattern of the arrangement of fibres in the weaving machine. The total number of fibres in a $m \times n$ rectangle is $(mn + m + n)$; it is always less than in the two-stage weaving of a sample. A four-stage sample contains no axial fibres.

Comparing the two methods reveals the following.

1. Both methods are capable of producing complex-section three-dimensional structures.
2. The minimum normalized length per machine cycle in a four-stage sample is 2.8, which corresponds to the maximum angle of orientation of internal fibres of 55°. Maximum values of these characteristics for two-stage samples are close to 2.0 and 90° respectively.
3. The maximum possible calculated volume content of fibres for a two-stage fabric is of 0.569–0.785, and for a four-stage one, of 0.685.
4. The normalized values of strength and elastic modulus at an axial tension are for two-stage three-dimensional fabrics, owing to the presence of axial fibres which are higher than for four-stage ones with the same fibre orientation angle.

The normalized weaving length is the ratio of the weave pitch length to the diameter of braiding fibres. The normalized strength and elastic

Figure 3.26 Schematic of four-stage weaving.

340   *Carbon-based composites*

modulus are the ratio of the strength to the number of fibres in the fabric, i.e., the strength or modulus of an individual fibre.

3.2.4 Braiding

3.2.4.1 Braiding of orthogonal three-dimensional structures

Some typical versions of orthogonal three-dimensional structures are shown in Figure 3.27. Numerals there indicate the numbers of strands of carbon fibres in every reinforcement direction, which correspond to relative fraction of fibres. These values can be compared, however, only taking into account that strands or bundles of fibres of different grades contain dissimilar numbers of filaments. Diameters and cross-sectional shapes of filaments are also dissimilar.

Only versions with an even numbers of strands along the z-axis can be made by braiding.

The braiding method includes the following main stages:

1. Assembling of the fixture for braiding;
2. Braiding of the skeleton along the x- and y-axes;
3. Threading of strands along the z-axis.

The stages are schematically shown in Figure 3.28. The initial stage, assembling the fixture, consists in installing metal rods that simulate skeleton components along the z-axis. The rods are inserted into sockets in the fixture, arranged so as to set the distribution of fibres in the x- and y-directions.

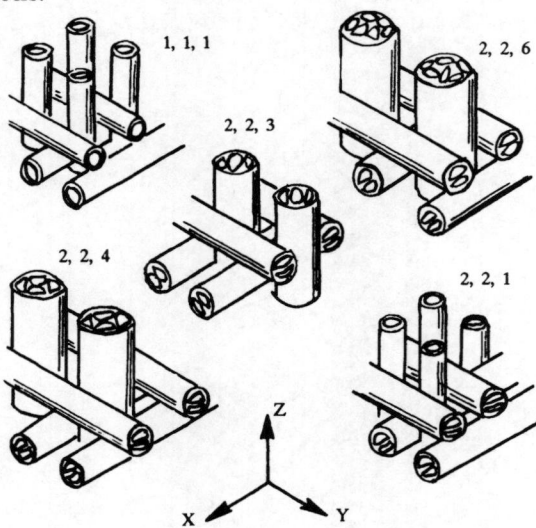

Figure 3.27 Typical versions of orthogonal three-dimensional structures.

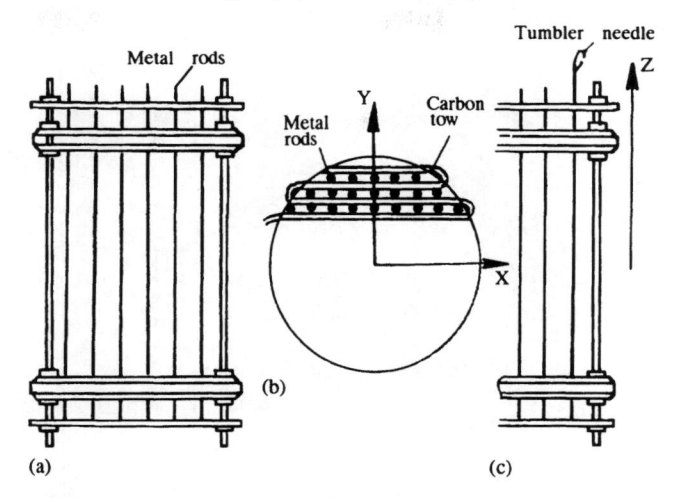

Figure 3.28 Schematic diagram of braiding of an orthogonal three-dimensional skeleton: (a) assembled fixture; (b) braiding along x- and y-axes; (c) replacement of metal rods with fibres along z-axis.

The braiding along the x- and y-axes is effected as follows. A strand of carbon fibres is drawn successively between rows of the metal rods so that a layer along the x-axis is formed. The next layer is laid after the fixture has been turned 90°, i.e., along the y-axis. In this manner the future skeleton of the required size is made layer by layer.

The threading along the z-axis consists in replacing the rods with carbon fibres. Carbon fibre strands are hooked on by hooks in which the rods terminate.

3.2.4.2 Braiding of cylindrical structures

The same braiding principle is applied to an automated manufacture of cylindrical preforms [60]. The manufacture of a cylindrical skeleton with reinforcing strands arranged in radial, circular and axial directions is schematically shown in Figure 3.29. Ends of thin metal rods rest on the foundation plate of a vibrator. Top parts of the rods extend beyond the top plate. The rods define between them radial and axial passages. Such a structure is placed in a braiding machine.

After the machine is started, the network base is continuously rotating, its axis of rotation corresponding to that of the machine. The braiding in the radial direction is effected as follows. A hook-like needle moves to and fro across the network base through radial passages between rods. It catches strands, fed by a special device, and draws them through the network structure so that in every passage is the same number of strands.

The braiding in the radial direction occurs concurrently with the radial braiding. Fibre strands are under an individual tension and are fed into

342   *Carbon-based composites*

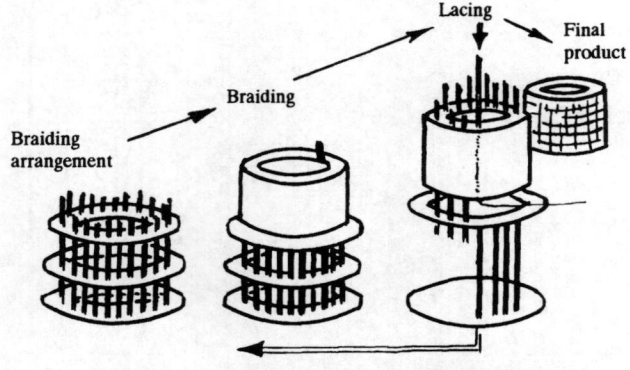

Figure 3.29 Main stages of automatic manufacture of three-dimensional skeletons.

circular passages with the aid of metallic tubes (one tube per passage). Fibres descend only under the tension resulting from rotation. Layers of circular fibres are wound continuously and slightly helically, so that they fill the framework from the bottom up to the top.

A horizontal metallic strap is provided for compaction in every radial passage. Near the place of introduction of circular fibres the straps are advanced from the network structure one after another as the structure rotates. They return back through radial passages and come over the next radial and circular layers of just-laid fibres. Next, the straps move back and assume a horizontal position to compact the layers.

For the final formation of the skeleton the metal rods are replaced with a taut fibre strand by means of a special needle. Every pass of the longitudinal fibre strand is followed by indexing of the preform for bringing the next metal rod to the needle.

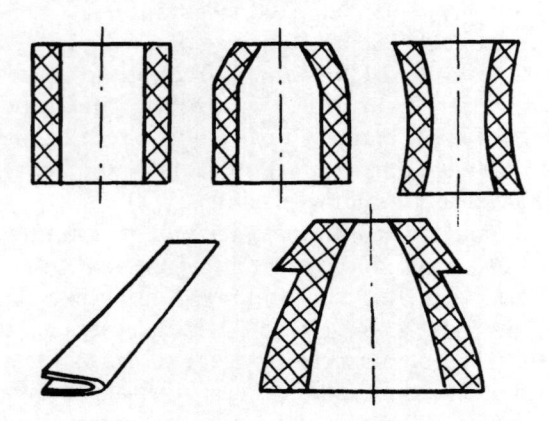

Figure 3.30 Types of skeletons produced by cylindrical braiding.

Such a skeleton braiding system allows the use of different fibres in each of the three main braiding directions as well as in different circular layers. The skeleton can be given the required outline by means of a special shaping device (Figure 3.30). One of advantages of such a braiding process is that radial fibres continuously form a loop around a metal rod at the inside and a stitch in the form of a chain at the outside. This makes it possible to cut the cylinder and to change the preform shape without destroying the structure.

Another method for braiding a skeleton of a three-dimensional structure has been developed by the AVCO company [61, 62]. The tooling for the skeleton braiding is a mandrel made of phenolic foam plastic, machined so that its inside diameter and configuration correspond to those of the finished part.

The pattern of arrangement of fibres in the filler structure is shown in Figure 3.31. Radial rods of a screw-like configuration are phenolic resin-based plastics, made as follows. Carbon fibre bundles, impregnated with resin, are twisted together to form a rod of the required diameter. The rigid material, polymerized in a furnace, is wound on a large-diameter spool. Next, when passing through a special head, the material is cut into rods of the required length, which are inserted into holes in the foam-plastic mandrel. Then the mandrel with the radial rods is transferred to an automatic system of axial braiding and circular winding.

When strands are passed in the axial and circular directions alternately, layers are formed which, growing, constitute a skeleton wall of a certain thickness. Dry fibres or prepregs are used for this purpose.

This braiding method yields a material with a volume fraction of fibres of 35–55%. The maximum dimensions of the produced skeletons are: length, 1.5 m; diameter, 2.1 m; and wall thickness, 210 mm.

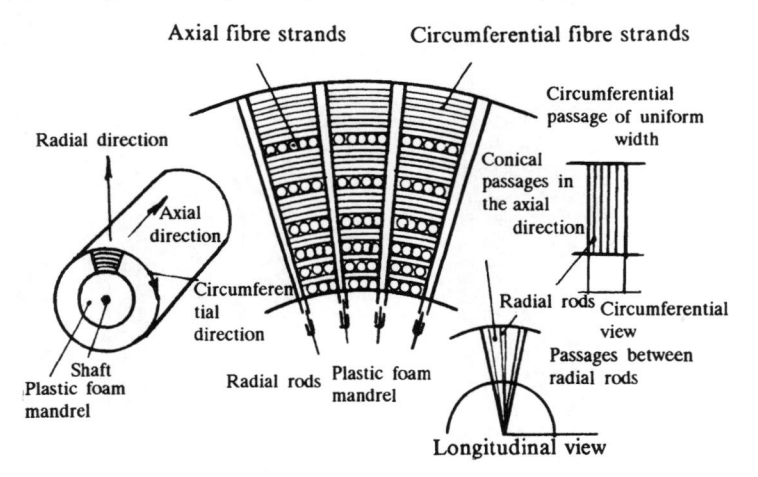

Figure 3.31 Manufacture of skeletons by method of AVCO.

3.2.5 Assembling of multidirectional skeletons

Skeletons of three, four, five dimensions, etc. spatial structures are assembled with the use of rods which are carbon-filled plastics based on thermoplastic or thermosetting resins [61, 62].

With regard to a further use of the rods for the assembling of skeletons and their saturation with a carbon matrix, two different approaches to selection of binders are employed. The first of these treats the binder as an auxiliary material serving solely to secure fibres together for the skeleton assembling time. One basic requirement for such a binder is the minimum coke residue after carbonization. The other approach to selecting the binder considers it as a part of the future matrix in a carbon–carbon material; then resins yielding a high coke residue are employed.

Resins used in the former case include polyvinyl alcohol, polystyrene, and some epoxy resins, and in the latter case, phenolic, modified phenol–formaldehyde, and epoxyphenolic resins.

Rods of a round cross-section are made as a rule. As known [64], the maximum volume filling with fibres in a composite is obtained with hexagonal-section rods, but spatial structures are today assembled just from cylindrical rods. The rod diameter depends on the diameter and number of carbon fibre bundles in it as well as on the diameter of the moulding drawing die, which in turn should be selected proceeding from the requirement for the fibre content in the rod: at least 60% vol. The rods used for the skeleton assembling should have the strength needed for this operation.

Figure 3.32 shows diagrams of tests of the rods for stability and end-face crushing. The requirements for the strength of rods are based on two possible loading versions: first, a rod is in its cell, a jamming has occurred at some depth; secondly, a rod, when being installed in its cell, has met a rod installed in another direction.

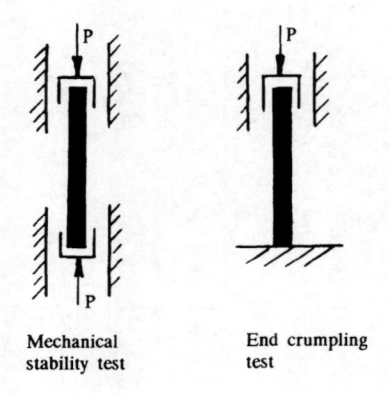

Figure 3.32 Schematics of mechanical tests of rods.

Technology of product manufacture 345

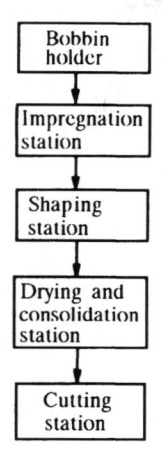

Figure 3.33 Block diagram of rod-making plant.

The block diagram of a plant for manufacture of rods is shown in Figure 3.33. It consists of several basic units: spool bank, impregnating unit, moulding unit, drying and hardening unit, and cutting unit.

The spool bank comprises spools with carbon fibres. It has a braking device to control the tension of a bundle unwinding from a spool.

Then the bundles pass to the impregnating unit: a bath with a solution or melt of the binder. In the impregnating bath, the binder combines with fibres. While being in the bath, fibres should get impregnated with the liquid; the time of residence in the bath needed for this depends on the wetting ability of the liquid and the fibre surface properties. In some cases the impregnating bath is heated to reduce the viscosity of the solution or melt and thereby to shorten the impregnation time. Types of impregnating baths are shown in Figure 3.34. Roller systems are used to reduce the

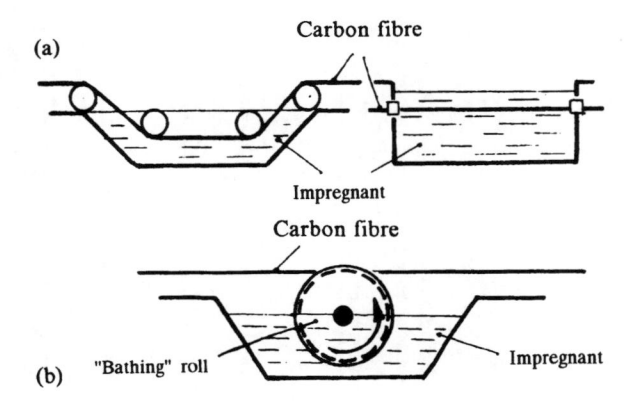

Figure 3.34 Impregnating bath types.

Preparative operations	Thermal stabilization of carbon fillers	Thermal activation of surface	Modification of surface properties by sizing	Preparation of prepregs including two-dimentional ones	Tape waving from carbon tows	Combining binders
Formation of preform	Winding of tapes and tows	Laying-up of fabric	Needle stitching reinforcement by pultruded carbon tows	Three-dimensional weaving of a part	Combination of discrete and continuous fabric structures	Aeration mixing to obtain non-woven fabrics
Formation of integrated porous structure and nominal volume of a part	Volume stabilization ↓ Carbonization of carbon plastic precursor including pitch-based ones	Confinement of dimensions in shape-forming tool ↓ Fixing of the preform via pyrolytic densification	Stitching of stacks ↓ High-temperature treatment of carbon–carbon preforms resulting in thermomechanical dispersion of the matrix	Fixing of the preform with a curable polymer matrix	Fixing of the preform in a thermoplastic (pitch) curable under cooling	
Perfection of porous structure to obtain CCCM of special purpose	Gas-impermeable chemically resistant materials ↓ Impregnation with thermosets and thermoplastics to minimize the porosity	Refractory high-strength materials ↓ Filling the pores with pyrolytic carbon via CVI process	Refractory power-intensive materials ↓ Impregnation with pitches (including isostatic pitch impregnation) and heat treatment	Erosion-resistant materials ↓ Deposition of pyrolytic liners (graphite, carbides, oxides)	Precursor of carbon-carbide and carbide-based CM ↓ Deposition of solid carbon into pores to reduce their sizes down to $R_{eq} = 50$–100 μm.	

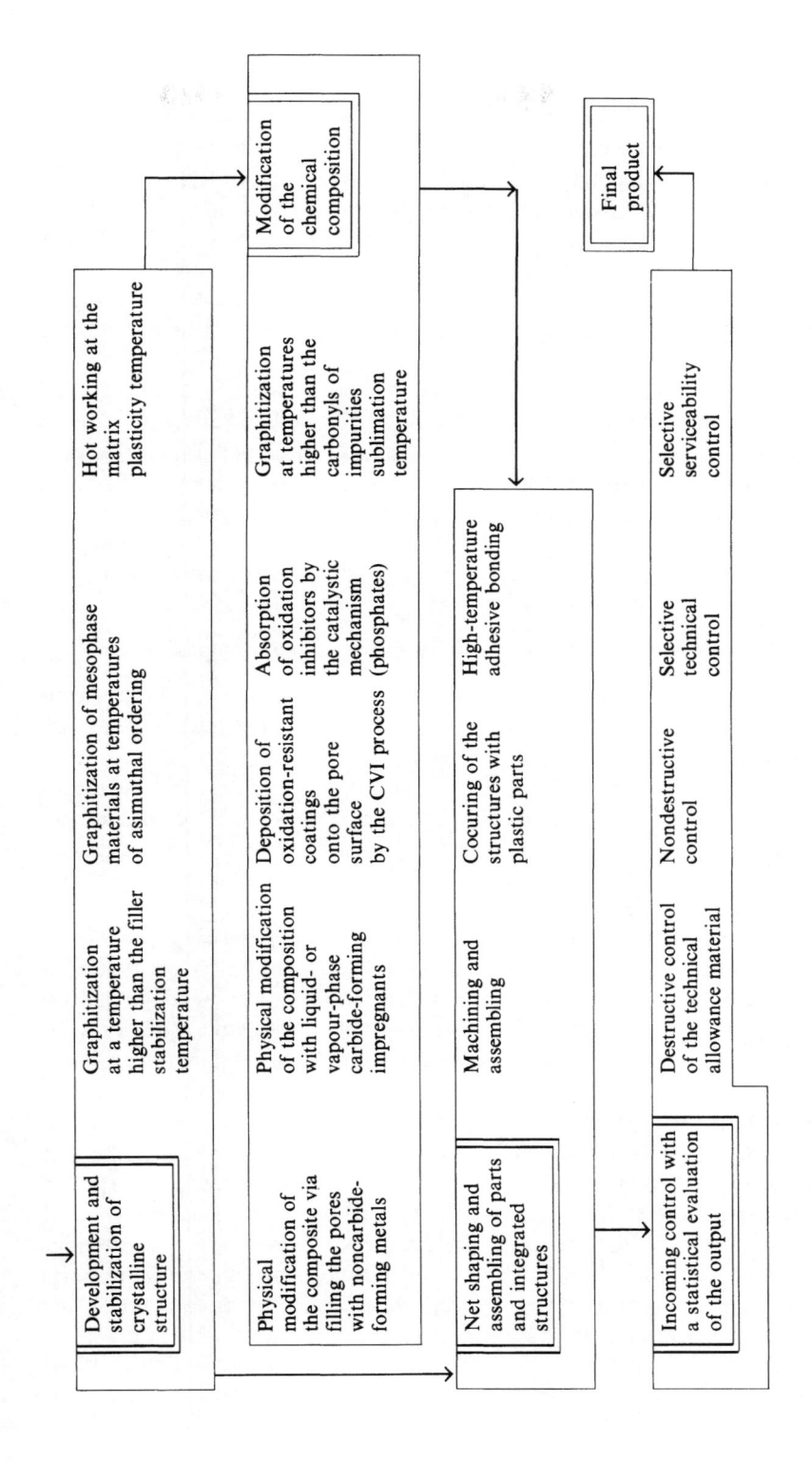

Figure 3.35 Basic flow-chart of manufacture of two-dimensionally reinforced CCC.

bath length; in this case the fibres bend when passing through the bath. Due to brittleness of carbon fibres, the bends result in fracture of individual filaments, which reduces the fibre strength, and therefore the radii of guide rollers should incommensurably exceed characteristic dimensions of fibres. The bath with a 'bathing' roller is free from such drawbacks (Figure 3.34b). A fibre bundle being impregnated in such a bath can be at one and the same level over the whole rod-making machine length, from its unwinding from a spool and up to the cutting unit.

The moulding unit, serving to shape the rod, is a drawing die or a set of drawing dies with an outlet orifice that ensures the predetermined rod diameter. Depending on the binder used, solvent type, drawing speed, etc., drawing dies can be disposed directly after the impregnating bath, within a furnace, or between the drying and the hardening furnace. Several drawing dies can be disposed at different points of the rod-making machine; in this case a gradual moulding of the rod is effected.

3.2.6 Manufacturing technology for two-dimensional composites

General processes of manufacturing two-dimensionally reinforced composites include basic traditional procedures of filler laying (winding, lay-up, formation of unwoven materials from staple) as well as traditional procedures of carbon matrix formation on the filler: carbonization and high-temperature treatment of hardened resins, pitches or deposition of carbon from hydrocarbon gases in pores. Practically, the processes consist of sets of the above procedures in various sequences depending on the tasks and technical capabilities.

The basic flow-diagram of manufacture of parts from two-dimensionally reinforced CCC, based on the structural analysis, is presented in Figure 3.35. Preparatory procedures serve to ensure the deformation stability of dimension of parts at subsequent high-temperature procedures. A dimensionally stable load-bearing skeleton of the filler, determining the rigidity of a composite and of a part as a whole, ensures intactness of the part, absence of initial distortions. The modification of surface properties serves to improve the wetting by the binder, for an effective interfilament impregnation of bundles and fibres, protection of the fibre surface from mechanical and thermo-oxidation damages. The formation of the load-bearing skeleton by woven tapes of prepregs containing weft filaments enhances the thermal stability at the thermal treatment stage (carbonization, graphitization). The provision of the load-bearing skeleton in the bulk of a part determines the main performance characteristics of products. The processes indicated in the flow-chart are at present mechanized, although they do not provide for an automated mass manufacture of products. The level of volumetric filling of the composite with fibres throughout the volume of the part is provided process-wise at this stage. The control parameters are the tension of tapes in winding, tension of fabric, warp and weft beams at a spatial weaving, pressure of pressing on filaments, fibres, and fabric at combined reinforcement patterns. In all these cases the resistance to the external pressure is provided by a squeezeable polymeric binder. The equilibrium of these factors produces the structure of the future material, including the maximum volume and maximum sizes of micropores.

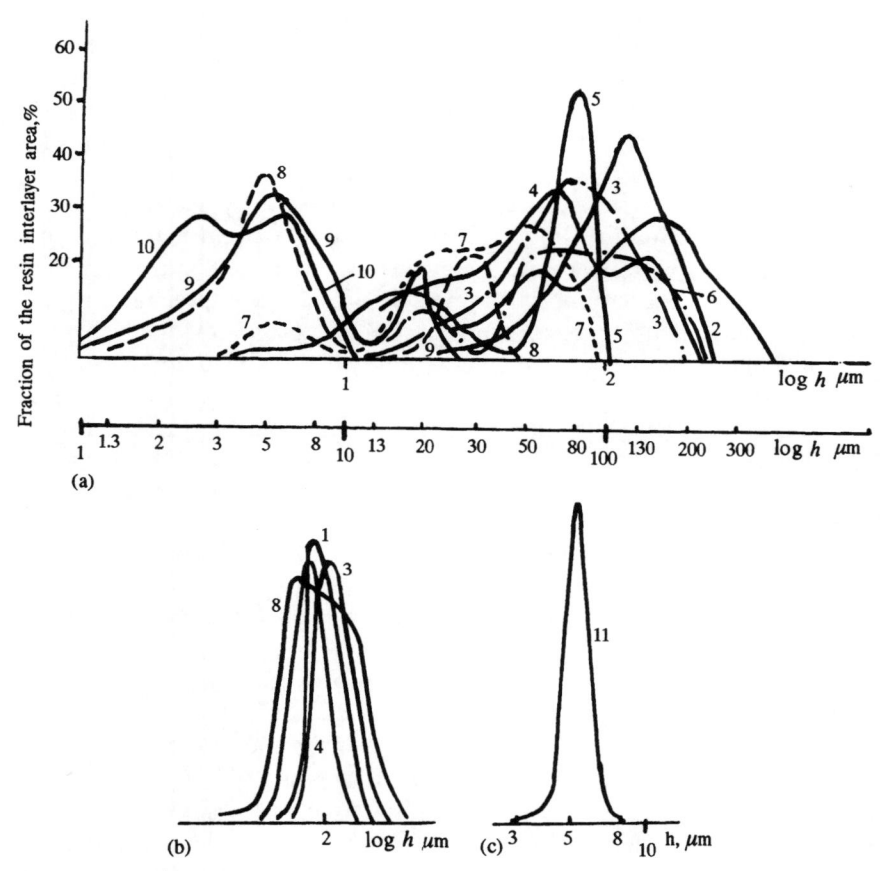

Figure 3.36 Results of fractographic analysis of sections of carbon-filled plastic, produced by winding and autoclave moulding: (1) tape with allowance for pores (binder content, 50%); (2) carbon-filled plastic with binder content of 47%; (3) the same with binder content of 45%; (4) tape without allowance for pores (binder content, 50%); (6) carbon-filled plastic with binder content of 33%; (7) 29%; (8) 27.6%; (9) 25.6%; (10) carbon-fibre bundle without binder (mercury program). (b) size distribution of fibres for specimens with nos given in (a). (c) size distribution of filaments.

Figure 3.36 shows the variation of the structure of polymeric interlayers in the bulk of the carbon-filled plastic-precursor with the binder content. The latter was reduced by squeezing in the winding, autoclave moulding, and control of the build-up in preparation of the prepreg. Carbon compositions subject to a finishing impregnation by solutions of polymeric binders, which wet the carbon surface, are more expedient to produce from carbon-filled plastics-precursors containing the minimum of polymeric interlayers. In this case the most highly dispersed porosity, most efficiently impregnated by such impregnates, is formed after high-temperature procedures. For compositions subject to a volumetric densification at a chemical carbon deposition in pyrolysis of hydrocarbons it is more expedient to select a carbon-filled plastic-precursor with a higher binder content. Then, as seen from Figures 3.36 and 3.37, resin interlayers of such thicknesses are retained which create prerequisites to emergence at their places of pores 4 to 50 μm in size, sufficient for a Poiseuille's diffusion of gaseous hydrocarbons with the highest absolute diffusion coefficient as compared with a Knudsen's or, the more so, Folmer's diffusion modes. At a subsequent physical modification of the composition by a liquid-phase impregnation with carbide-forming substances, from conditions of real rates of the impregnate–metal melt chemical interaction, rate of the liquid-phase impregnation, and change in the volumes of initial substances and interaction products, sizes of transport pores should be of the order of 50–1000 μm. In such a case, when materials are produced by winding, higher binder contents in the carbon-filled plastic-precursor are preferable (see Figure 3.36).

When the load-bearing skeleton of a part is formed by other methods indicated in the flow-chart (Figure 3.35), every technology contains its own effective techniques for controlling the structure and pore volume.

The total porosity volume is fixed by a subsequent stage of high-temperature procedures by a geometric restriction of thicknesses by means of piercing with a hardening thermoplastic or thermosetting binder. The next three process steps are aimed at improving the pore, crystalline structure and modifying the composition of the composite. The specific selection of manufacturing techniques is determined by the target-orientation of the technology as a whole. Individual processing types are presented in the flow-chart of Figure 3.35. These procedures are predetermined by physical properties: level of the initial temperature of three-dimensional crystallization; decrease of this boundary temperature at the thermomechanical action of internal stresses in the microstructure; level of the initial temperature of plastic deformations of the matrix; and level of the temperature of chemical transformation of gaseous or liquid impregnates.

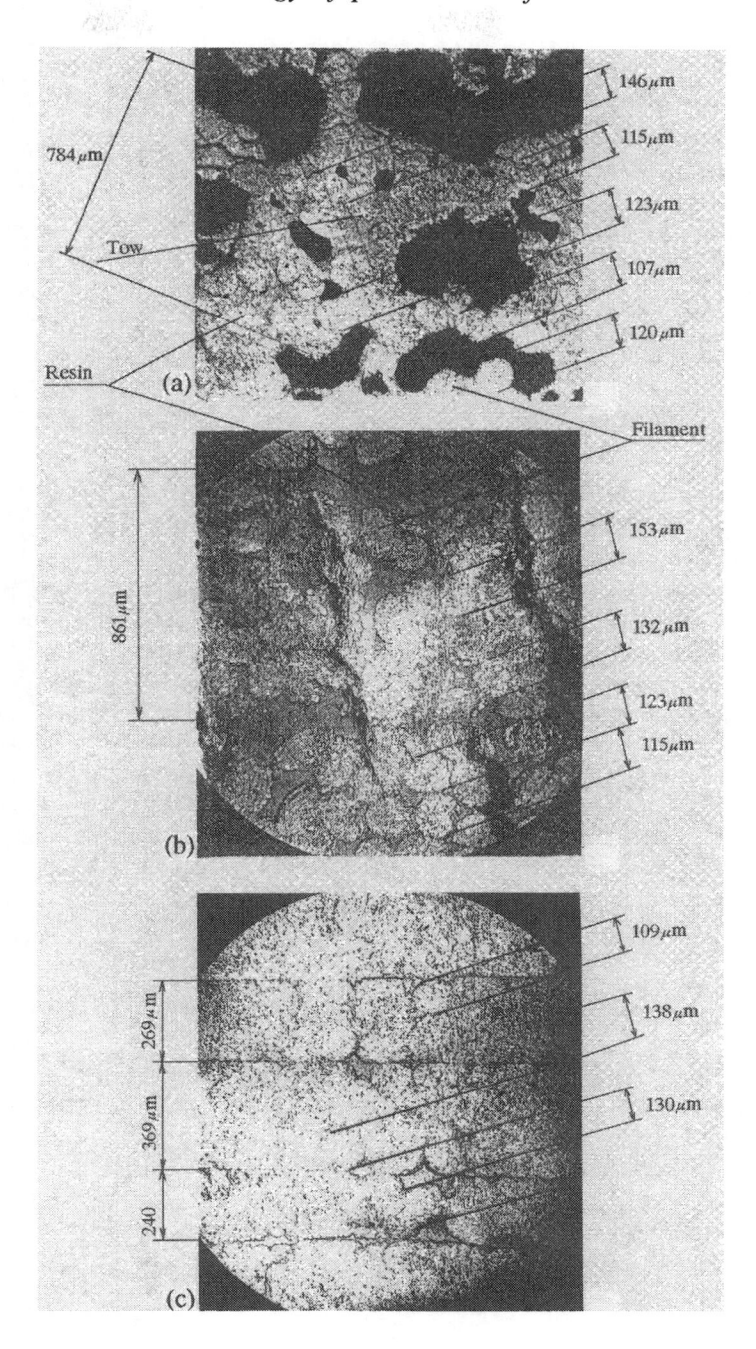

Figure 3.37 Sections of structure of starting tape and carbon-filled plastic with binder contents of 29 and 45%.

The machining of parts from CCC includes new production techniques and specialized high-quality tools. The assembling of units and integral structures from CCC is effected by hardening of a carbon part in the volume of plastics, which becomes possible owing to its full inertness and small changes of their dimensions over the temperature range of hardening of carbon-filled organoplastics and glass-reinforced plastics.

An indispensable component of the technology of parts from CCC, which as a rule perform most critical functions in high-temperature units and mechanisms, is the outgoing destructive and nondestructive testing and process inspection.

3.2.7 Manufacturing technology for three- and four-dimensional composites

The process of manufacture of multidirectional composites based three-, four-, etc., dimensional structure skeletons is a process of filling all the voids with a carbon matrix. The skeleton structure contains two levels of porosity: channel pores between filaments of a fibre and spaces between bundles. The end-face of a carbon fibre bundle, or strand, is shown in Figure 3.38. The size (diameter) of a filament is usually of 5–10 μm. Simple geometric considerations indicate that at the most close, hexagonal packing of filaments, the characteristic size of the channel between is at least 0.5 μm. The micrograph in Figure 3.39, however, shows the spacing between filaments in a real material to be much larger, up to

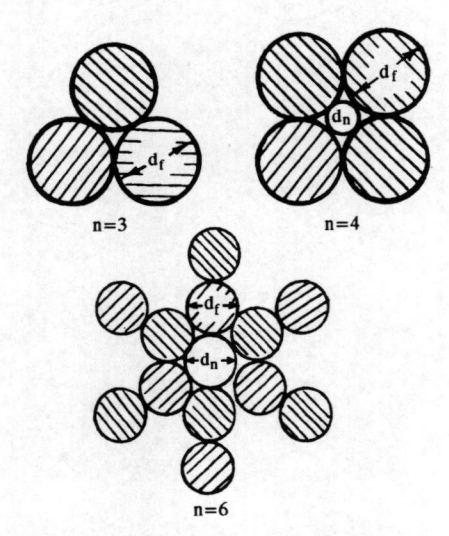

Figure 3.38 Possible arrangements of filaments in fibre bundle with various laying types.

Technology of product manufacture 353

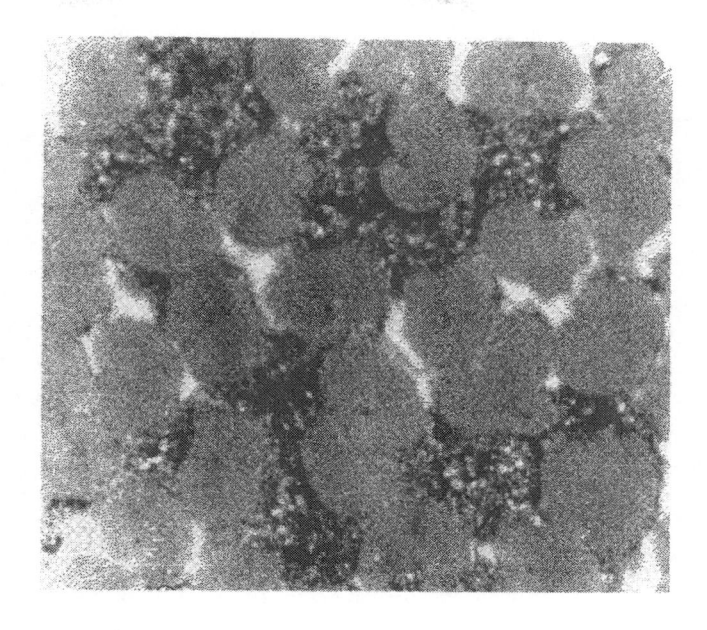

Figure 3.39 Microstructure of bundle end-face (× 2500).

10 μm. Thus, sizes of channel pores within a bundle of fibres range within 0.5–10 μm.

The second level of voids to be filled in the course of CCC manufacture is the interbundle porosity. An elementary unit of an orthogonal three-dimensional structure is schematically shown in Figure 3.40. A three-dimensional structure is characterized by the presence of a closed space whose characteristic size, several millimetres, is determined by the laying pitch and fibre bundle diameter. The skeleton of a four-dimensional structure features an open porosity and absence of local voids, which is the most favourable structure version for a subsequent impregnation. A level with millimetre-size pores, however, is inherent in this structure as well, since it is associated with characteristic sizes of the bundles that form the skeleton.

The main objective of using multidirectional fillers is to obtain CCC with a high density. Because of this, initial binders yielding a high coke residue are expedient to use for matrices in such composites. Coke residues of resins that may become attractive for the CCC manufacture are presented in Table 3.2. The greatest coke residue can be obtained by applying pressure at carbonization of pitches. Is it reasonable to employ them as a binder, foreseeing the complexity of carrying out the carbonization under pressure? To answer this question, one has to examine the quality of cokes obtained from pitches and from thermosetting resins.

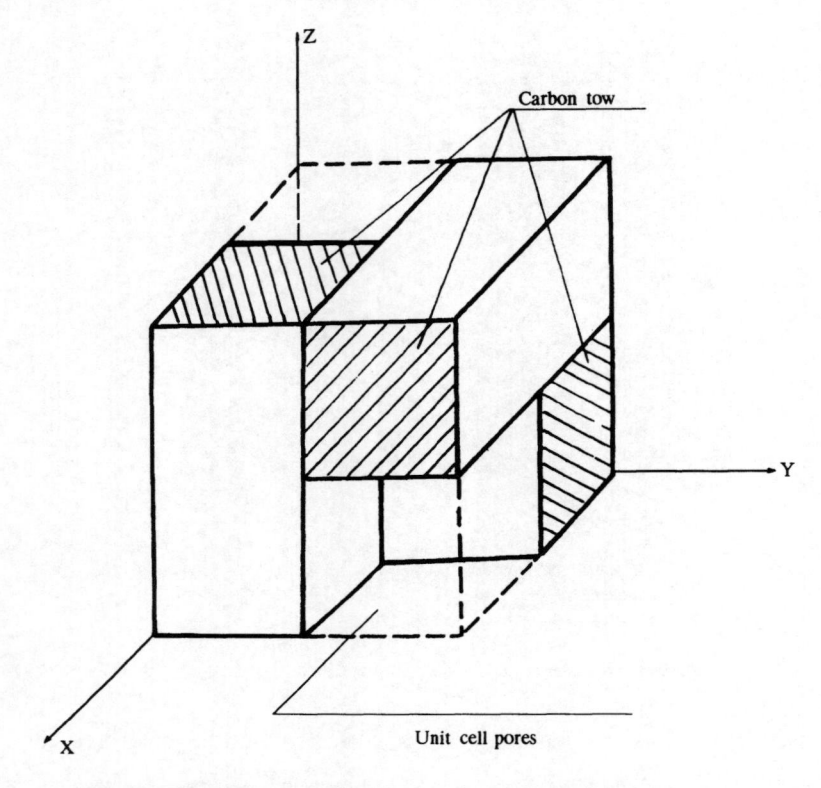

Figure 3.40 Elementary cell of a three-dimensional structure.

The variation in the true density and in the interlayer distance of products of carbonization of pitches and thermosetting resins in the course of heating is shown in Figure 3.41 [64]. Comparing the densities indicates that obtaining high-density CCC based on resin matrices is impossible since at their high-temperature treatment, the coke density is not over 1.5 g/cm^3 [64]. As regards the interlayer distance, its values show differences in structure of cokes: resin cokes have mainly a turbostratic structure, whereas pitch cokes treated at temperatures over 2500 °C feature a graphitic structure. Just the possibility of obtaining a high-density, graphitizable coke predetermined the use of pitches as a binder in manufacture of multidirectionally reinforced CCC.

Apart from a pitch coke, also a pyrocarbon matrix or combinations of coke and pyrocarbon matrices are employed in multidirectionally reinforced CCC.

The manufacture of a three- or four-dimensional structure CCC is illustrated in Figure 3.42. A skeleton made by braiding or assembling of rods is usually not rigid enough for carburization to be carried out under high pressures. In view of this, the densification process is started with a

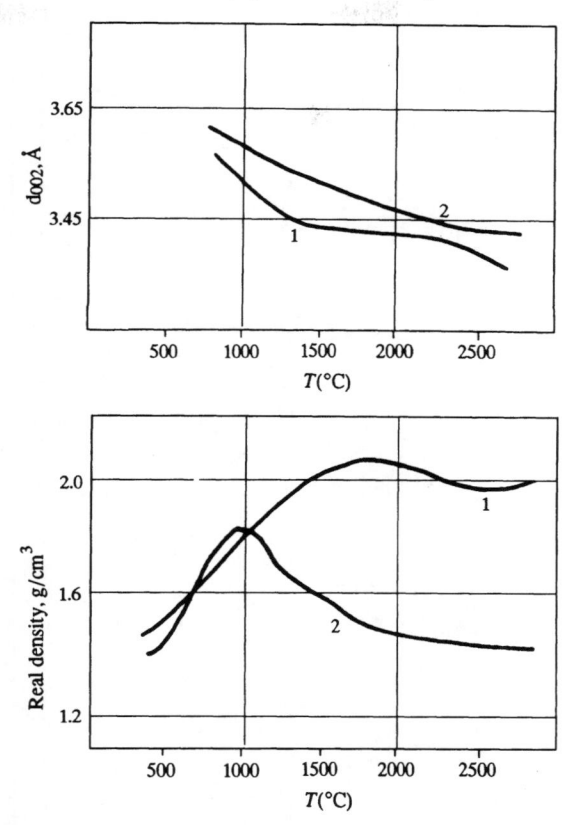

Figure 3.41 Effect of treatment temperature on true density of cokes and interlayed distance d_{002}: (1) pitch; (2) thermosetting resin.

preliminary stage whose aim is to increase the skeleton rigidity and to prepare for a further densification. Pitches with low softening temperatures (70 °C) are used at this stage as they wet carbon fibres much better than high-temperature ones do, which is evidenced by the data presented in Figure 3.43 [65].

A skeleton is impregnated with pitch heated to 200–250 °C, the minimum viscosity temperature [66], by the vacuum-compression method. The impregnation pressure is of 0.5–1.0 MPa. Impregnated preforms are carbonized at temperatures of 650–1000 °C [67]. The impregnation–carbonization cycle is repeated 2–3 times. The impregnation efficiency, however, declines with every next cycle.

Figure 3.44 shows results of studying the porous structure of preforms at the preliminary stage. Three maxima of differential curves of distribution correspond to three porosity levels in CCC. The right-hand maximum is the interbundle porosity; the middle maximum, a region of units of micrometres, represents pores between filaments; and the left-hand

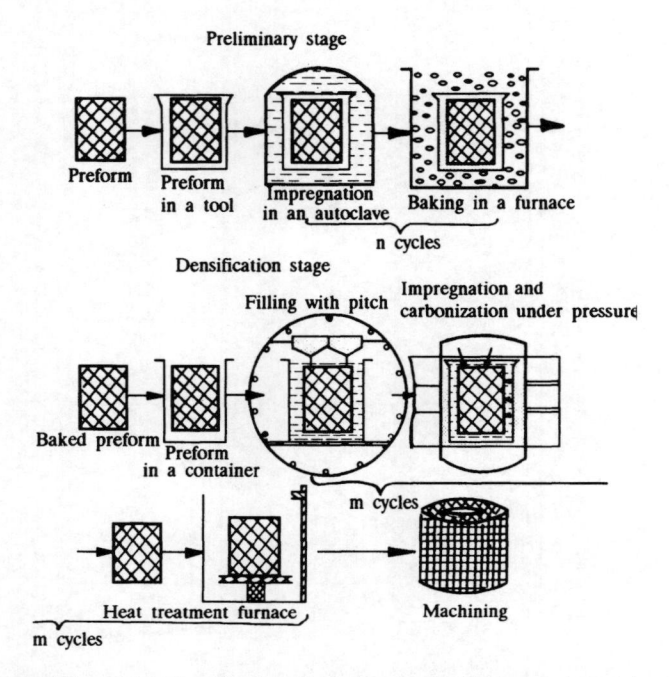

Figure 3.42 Schematic diagram of manufacture of high-density carbon–carbon material.

maximum corresponds to finer pores forming in the matrix in the process of carbonization. Analysis of variation of the porous structure in the course of the preliminary stage indicates that at impregnation under low pressures, pitch does not penetrate between filaments, while the decrease in the volume of medium pores after the second cycle evidences only that these pores change over to closed ones, which is confirmed by a steep increase of their volume after the high-temperature treatment (curve 3).

To increase the preform density in a single cycle, impregnation and carbonization are carried out in containers allowing carbonization to be carried out in a pitch layer (Figure 3.45). In this case the impregnate that has penetrated into the skeleton at impregnation cannot flow out as it is heated and melts at the initial stage of carbonization. Indeed, the use of containers shown in Figure 3.45 yielded in a single impregnation–carbonization cycle such preforms where the matrix mass was twice as great as that after two cycles without containers. Subsequent cycles of production of a high-density material are carried out with carbonization under pressure. The increase in density of a three-dimensional structure preform with number of cycles is shown in Figure 3.46. As seen, intermediate thermal treatments at 1000 °C retard the density increase. The apparent density attained after five impregnation–carbonization under

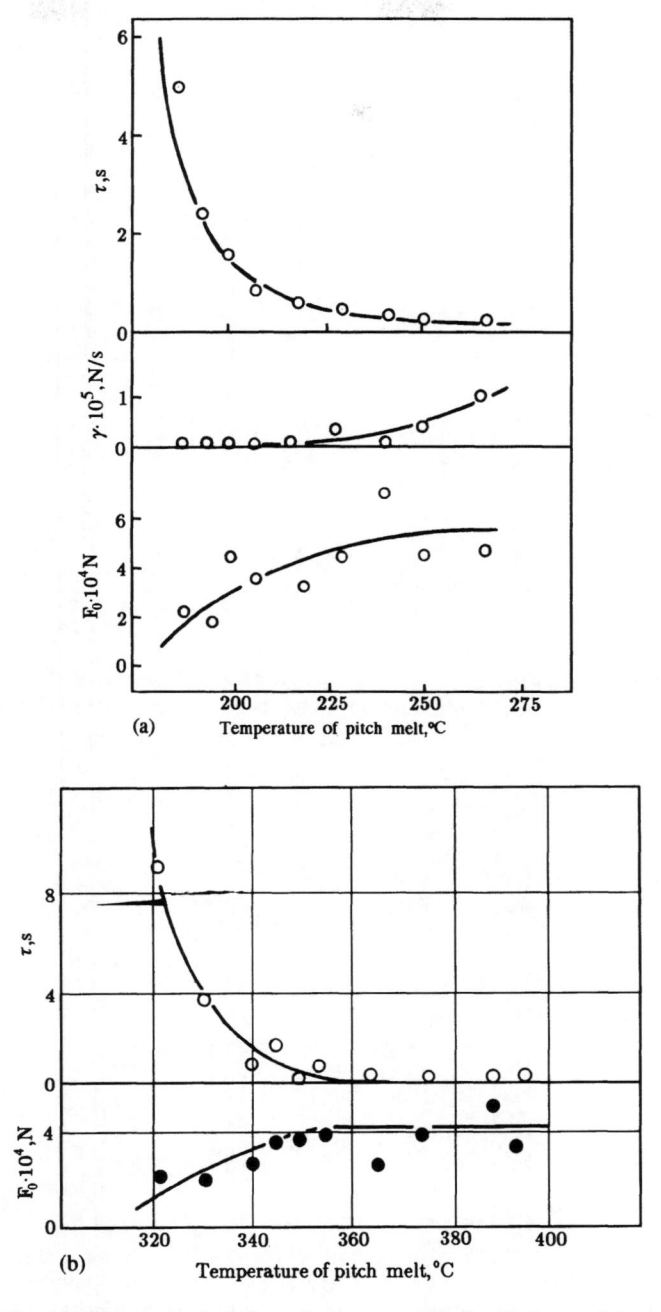

Figure 3.43 (a) Wetting relaxation time τ, wetting force F_0, and impregnation rate γ, measured by tendometering method in studies of wetting and impregnation of carbon fibres with pitches: medium-temperature coal-tar pitch; (b) high-temperature coal-tar pitch.

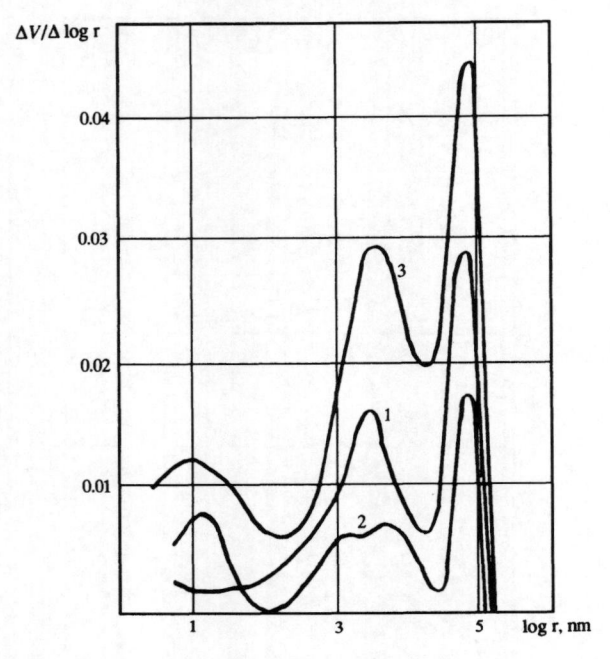

Figure 3.44 Differential curves of size distribution of specific volumes of pores at preliminary stage of CCC manufacture: (1) first impregnation–carbonization cycle; (2) second impregnation–carbonization cycle; (3) thermal treatments.

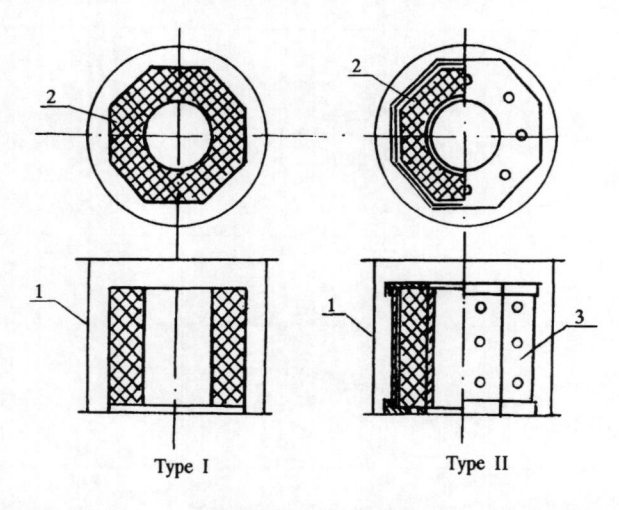

Figure 3.45 Tooling for preliminary stage of manufacture of high-density CCC: (1) steel container; (2) preform (skeleton); (3) perforated sectional container.

pressure-thermal treatment cycles is only of 1.8 g/cm^3, whereas the use of a high-temperature treatment raises this value to over 1.9 g/cm^3. The

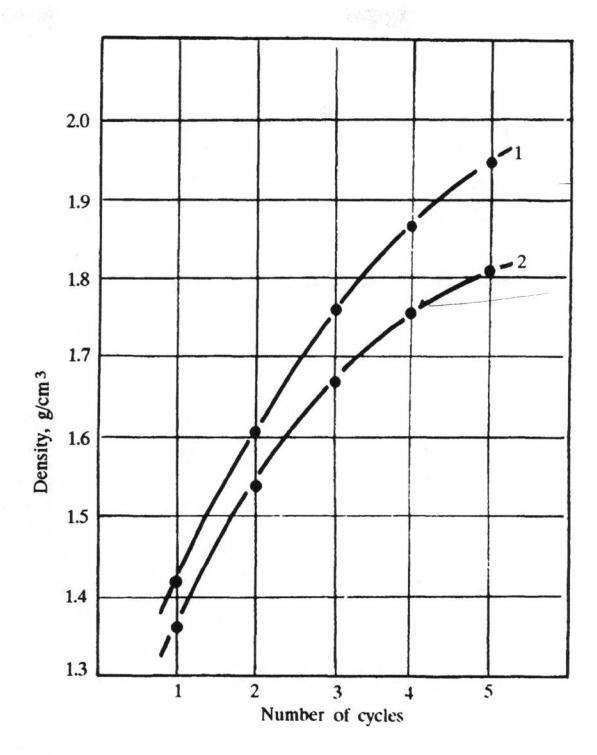

Figure 3.46 Increase in density of preforms from a three-dimensional structure material at intermediate thermal treatments at temperatures of (1) 2400 °C and (2) 1000 °C.

character of the curves indicates the tendency to saturation in the first case (curve 1) and to a further density increase at the sixth cycle in the second case. The analysis of differential curves of the size distribution of specific pore volumes, shown in Figures 3.47 and 3.48, demonstrated that a low-temperature thermal treatment involves no opening of porosity before the next cycle, and hence the latter cannot be effective. Moreover, the coke density also does not reach maximum values.

In various processes, the intermediate thermal treatment temperature is selected not only from the position of formation of a porous structure for a subsequent impregnation, but also considering the possibility of distortion of the preform with increasing temperature as well as the prospects for formation of properties of the produced material. The temperatures used are usually within 2200–2800 °C [61].

Density growth in the process of manufacture of materials with various treatment is shown in Figure 3.49.

The cycle diagram at the densification stage in the manufacture of CCC is shown in Figure 3.50. A preform after the preliminary stage,

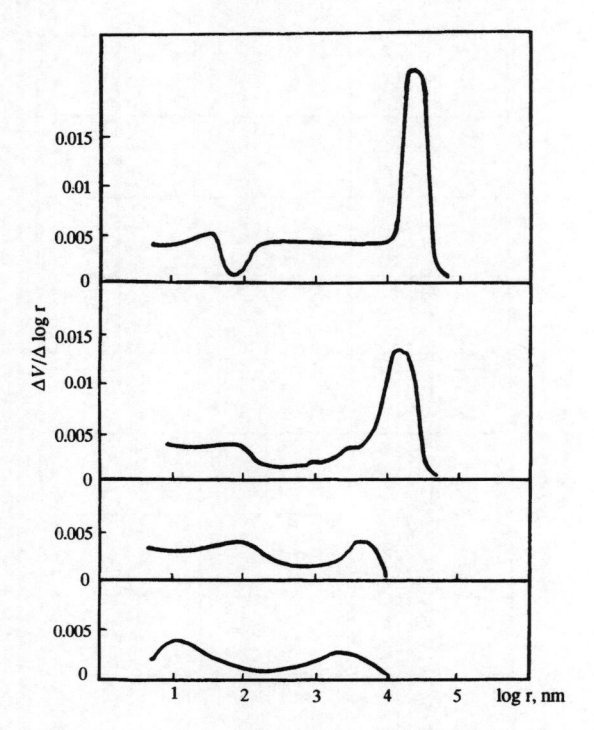

Figure 3.47 Differential curves of size distribution of specific volumes of pores in process of CCC manufacture with thermal treatment temperature of 1000 °C: 1–5 – Nos of impregnation – carbonization under pressure – thermal treatment cycles.

which is as a rule machined into a block close to the part to be produced, is placed into a steel container, which serves as the tooling for the impregnation–carbonization under pressure process. Next, pitch is poured into the container with the preform. High-temperature pitches are used as the impregnate at the densification stage. At the pouring, the preform is vacuumized and heated to 250 °C [68]. Molten pitch flows over into the container with the preform. The temperature at which the vacuum pouring is carried out is not optimal for the fluidity and wetting of the carbon preform by pitch, but cannot be raised because of the foaming featured by pitches at higher temperatures. Then the container with the preform, filled with pitch, is sealed.

As shown by studies of wetting and impregnation of carbon fibres by high-temperature pitch [65], no impregnation of the interfilamentary porosity occurs even at a temperature over 350 °C, and therefore pressure is to be applied to fill these and finer pores in the preceding coke layer. Thus, the application of pressure turns out to be necessary both for the impregnation process and for the carbonization process. Based

Technology of product manufacture

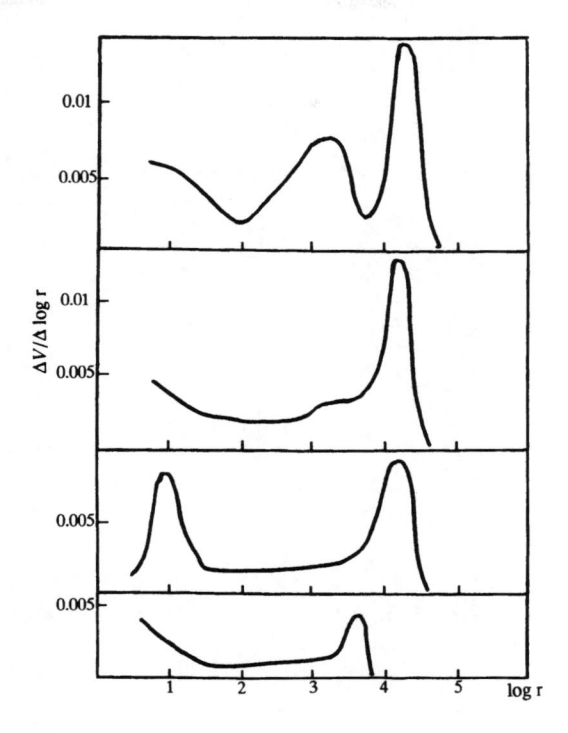

Figure 3.48 Differential curves of size distribution of specific volumes of pores in process of CCC manufacture with thermal treatment temperature of 2400 °C: 1–5 – Nos of impregnation – carbonization under pressure – thermal treatment cycles.

on this, a combined process of impregnation and carbonization under pressure has been developed. A typical graph of the impregnation–carbonization under pressure is shown in Figure 3.50. When the system has been heated to the temperature of the minimum pitch viscosity, the impregnation pressure is applied. Next, a holding at the impregnation temperature is carried out. A further temperature rise is slowed down since the carbonization process is proceeding. Thus, impregnation and carbonization are spaced apart in time. Impregnation and carbonization pressures can differ as they are determined by different factors. Thus, the impregnation pressure depends on the minimum size of pores in the preform, which should be entered by pitch.

Proceeding from the values given in Figures 3.47 and 3.48, this pressure at every cycle can be calculated from the Washburn's formula:

$$r_{eq} = \frac{2\gamma\cos\theta}{P},$$

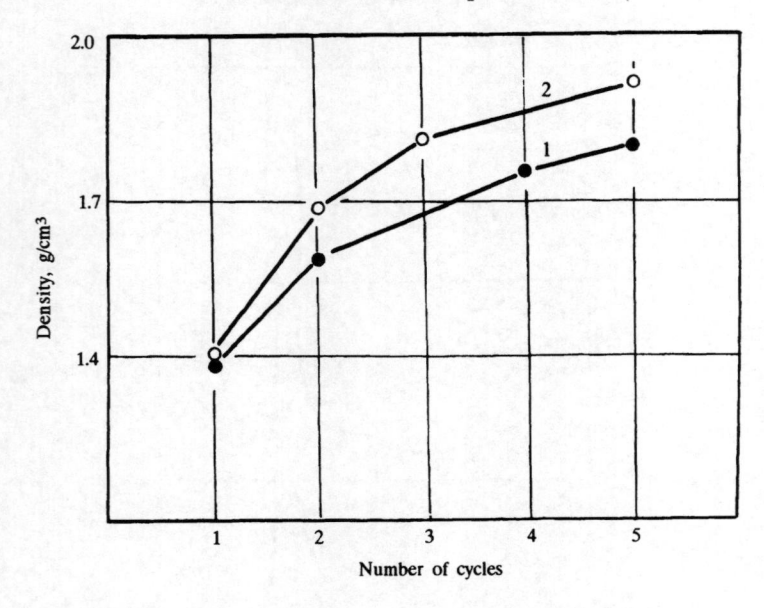

Figure 3.49 Density growth in process of manufacture of material with various treatment temperatures: (1) 1100 °C; (2) 2400 °C.

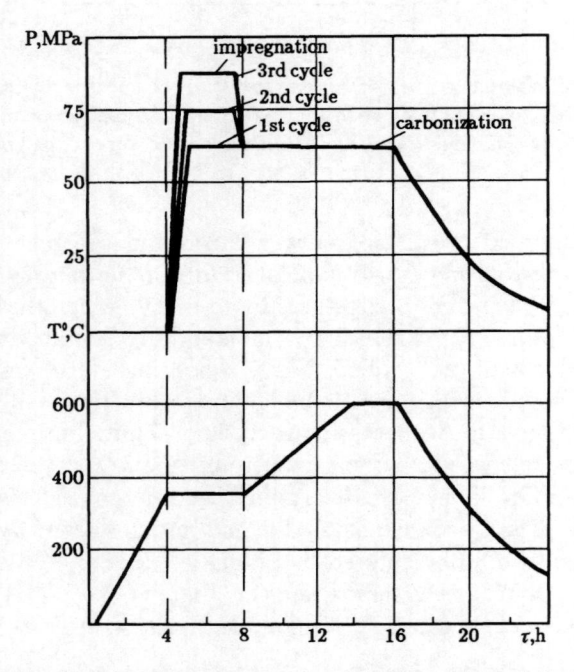

Figure 3.50 Typical diagrams of process of impregnation–carbonization under pressure.

where r_{eq} is the pore radius; γ the surface tension of the liquid (high-temperature pitch) at a given temperature; θ the contact angle of wetting of the solid with the liquid at the given temperature; and P the specific pressure of impregnation.

It follows that the impregnation pressure should change with increasing cycles for a more complete proceeding of this process, whereas the carbonization pressure is determined by the magnitude of the coke residue and the quality of the coke being obtained. The carbonization pressure in manufacture of multidirectional CCC is commonly on the order of 100 MPa [69].

The final carbonization temperature is about 600–700 °C; this is due to the fact that the pressure-transmitting medium in the course of carbonization is the carbonizing pitch itself, in which the preform is 'floating'. Such an organization of the process is therefore reasonable as long as the carbonizing body is liquid. Pitch is a thermoplastic system and passes in carbonization through the stage of mesophase transformations, being a liquid-crystalline system.

On completion of the mesophase transformations the carbonizing pitch goes over to a solid state. Just the temperature of completion of the mesophase transformations is the temperature of completion of the impregnation–carbonization under pressure.

The impregnation–carbonization under pressure is carried out on presses fitted with a special tooling, in gasostats or superclaves.

The initial rigidification of the skeleton is sometimes effected by a vapour deposition of pyrolytic carbon. In some instances, the matrix is fully formed by the vapour-phase pyrolysis.

Two basic methods are employed for densification of multidirectional structures with pyrocarbon: an isothermal method and a temperature gradient method, described above.

With the isothermal method the process is conducted at a constant temperature, a lowered pressure or a pressure close to atmospheric, with considerable additions of inert gases. The initial hydrocarbon is passed over the surface of a part or filtered through it. The isothermal process yields a fairly uniformly distributed pyrocarbon. This method is employed to increase the rigidity of the skeleton at the preliminary stage. The process is conducted at temperatures of 950–1150 °C and pressures of 1–150 mmHg with the use of methane or natural gas. The densification cycle time, depending on the furnace size, is 60–120 h.

The temperature gradient method is restricted to manufacture of a single part, but is greatly advantageous over the isothermal one in the densification rate.

The thermal conductivity and structure of the skeleton greatly affect the pyrocarbon matrix manufacture process. A carbon felt is a better heat-insulating material, and therefore higher temperature gradients can be obtained in it than in a high-thermal conductivity fibrous skeleton. The temperature differential is in this case controlled with the aid of a high-velocity gas flow. A higher temperature gradient at deposition of a pyrocarbon matrix is obtained by increasing the isotropy of thermal properties of the substrate. This condition is better met by a four-dimensional structure than by a three-dimensional-one. To obtain a pyrocarbon-matrix composite with the maximum density and high physicochemical properties the initial density of the skeleton or preform should not exceed 0.5–0.6 g/cm^3.

The reagent used for densification of fibrous substrates by pyrocarbon is commonly a methane–carbon tetrachloride mixture with a ratio CCl_4: CH_4 = 0.05–0.1. The pressure of the reagent mixture (CH_4 + CCl_4) varies from 10–20 to 800 mmHg, which determines the matrix formation time. The time of reagent residence in the reaction zone should not exceed 0.1 s. There is a trend to provide a high temperature gradient, on the order of 20 deg/mm, for the best run of the process.

In order to increase the temperature differential in the saturating skeleton and to reduce the nonuniformity of concentration of reagents across its thickness it is expedient that the reaction chamber be made exactly to the dimensions of the preform. In this case the temperature of the skeleton outside can be of about 100 °C due to a water cooling. A filtration principle of the feed of reagents along the substrate axis at a high flow-through velocity makes it possible to reduce the radial concentration gradient.

3.3 STRUCTURE AND PROPERTIES OF COMPOSITES

3.3.1 Properties and application of two-dimensional composites

Physical, mechanical and thermophysical properties of two-dimensionally reinforced CCC are determined by interaction of the anisotropic reinforcement with the isotropic carbon matrix. The main factors taken into account in analysing the properties of composites and the functional suitability of parts made from CCC are the reinforcement pattern, volumetric filling with fibres, and porosity of the matrix.

Specimens and parts undergo a brittle fracture at tension along the reinforcement and in the transverse direction. Deformation curves at bending, compression or at fracture of orthotropic specimens are nonlinear. The degree of nonlinearity increases with the relative thickness of the product, filler disorientation degree, and matrix porosity level.

Figure 3.51 shows stress–strain curves for grade KUP-VM-PU carbon–carbon composite [70, 71], obtained in tension of 200 × 10 × 3 mm plates and compression of 36 × 6 × 6 mm parallelepipeds. Similar dependences have also been obtained in tension and compression of 312 mm diameter shells.

Structure and properties of composites

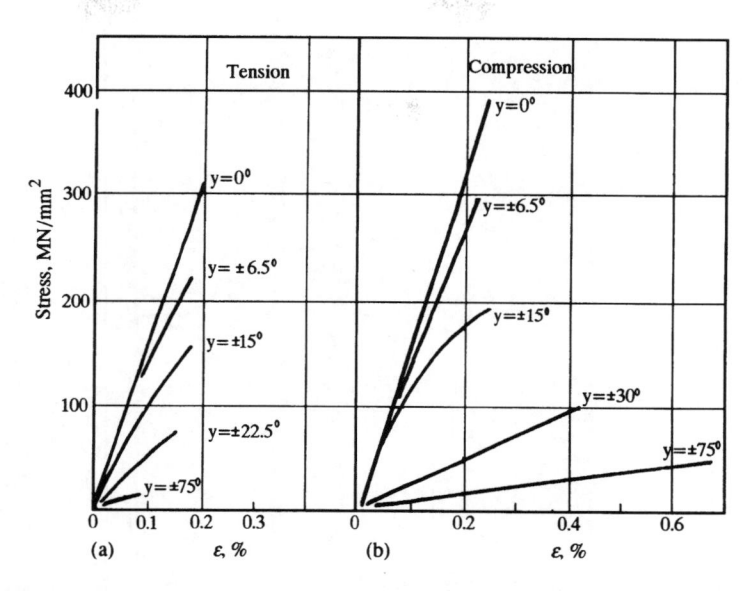

Figure 3.51 Stress–strain curves for KUP-VM grade CCC in (a) tension and (b) compression of plates and cylindrical shells.

A testing machine for cylindrical shells is shown in Figure 3.52. Tested and initial 312 mm diameter shells are shown in Figure 3.53. Testing such shells depending on their manufacturing method, matrix composition for various filler types with determination of the Young's and shear moduli, Poisson's ratios, ultimate fracture strains has become a mandatory step in solving the problem of the transfer of results of tests of specimens of various shapes to the stage of designing of products from carbon–carbon materials. The conducted tests confirmed that the CCC fracture is best described by the criterion of strength, based on maximum normal and tangential stresses:

$$|[\sigma_{11}]| \leq \sigma_{11,\,act.}; \quad |[\sigma_{22}]| \leq \sigma_{22,\,act.}; \quad |[\tau_{12}]| \leq \tau_{12,\,act.}.$$

A low ultimate strain of carbon matrix, brittle as compared with textured fibres, results in a drop of the ultimate strain of a composite specimen at disorientation of the filler. When compressed in a transverse direction, specimens get strongly deformed with pressing until fracture cracks become visible. In all cases the fracture of anisotropic specimens involves formation of fractures with an interlayer cleavage. When the angle of mismatch between the vector of external forces and the principal axis of reinforcement is small (Figure 4.51, curves 1 and 2), a brittle fracture with multiple cracks over a considerable length of specimen is predominating. Thus, an increase in the fracture toughness of two-dimensional CCC is demonstrated, which is a very important advantage of these materials in the series of known high-temperature structural materials.

Figure 3.52 Test of cylindrical shells from KUP-VM.

Bending deformation curves are shown in Figure 3.54. The relative thickness of the specimen is most important here. As it increases, the deformation curves indicate a decrease in the ultimate breaking load and an ever more nonuniform joining of layers in the resistance to fracture (curve 4). The picture being observed exhibits the impact of edge effects, which for CCC, as compared with carbon-filled organoplastics, can turn out to be more substantial because of a solely brittle fracture of the matrix. Due to this, the designing of parts as well as transfer of results of measurements of the ultimate resistance to fracture of specimens through known calculational relations to fracture of parts call, as a rule, for an extensive and thorough elaboration.

An increase in porosity of brittle carbon materials results in manifestation of a pseudoplastic character of fracture because of an effective reduction of the shear and Young's moduli of the matrix, which results in an incomplete utilization of reinforcement properties in an imperfect

Structure and properties of composites

Figure 3.53 312 mm diameter shells before and after test.

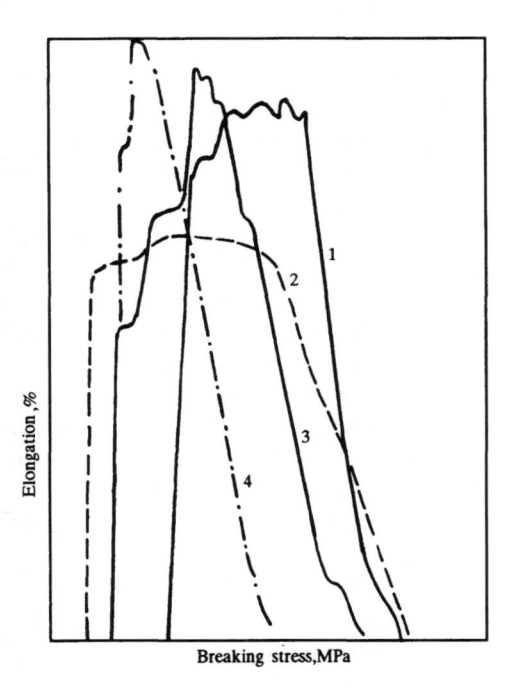

Figure 3.54 Deformation curves of material KUP-VM during three-point bending of short-beam and full-size specimens.

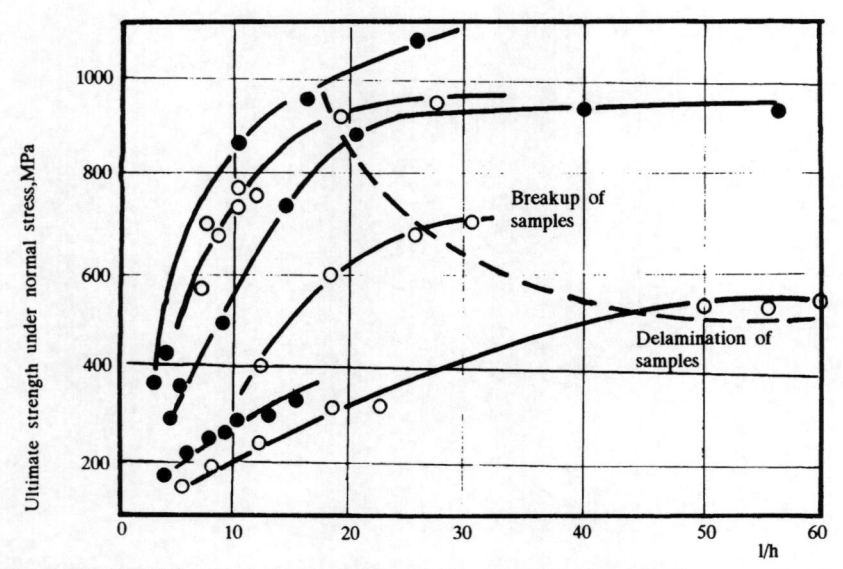

Figure 3.55 Ultimate resistance to normal stresses during bending as a function of geometric dimensions of specimen and porosity of material KUP-VM: (1, 2) 5%; (3) 8.7%; (4) 12%; (5) 15% and (6) 21%.

carbon matrix. The variation of the ultimate resistance of the material KUP-VM-PU to calculated normal stresses with the relative thickness for various porosities of the composite is shown in Figure 3.55.

Boundary curve 7 can serve for selecting the optimal type and size of specimen and, in designing of parts, for restricting the dimensions of fastening segments, where termination of the weakening impact of edge effects can be relied upon.

Figure 3.56 shows the variation of the ultimate resistance to tension and bending of two-dimensional CCC specimens with the volumetric filling with fibres. The region of optimal values is relatively small. The Young's modulus of the composite continues increasing according to the mixture rule at a higher volumetric filling. The levels of the ultimate resistance to fracture, calculated from the mixture rule, are not attained, which demonstrates an inadequate effectiveness of a uniform redistribution of load in a brittle porous carbon matrix throughout the specimen volume. The level of the elastic modulus, calculated from the mixture rule, agrees with experimental data. It can be assumed that transfer of the stress to the whole reinforcement volume has occurred in the carbon matrix, at least at the initial stage of loading, when, strictly speaking, the elastic modulus is determined from the deformation curve.

The above-listed factors mean that a nonuniform deformation of the specimen volume, concurrent development of normal and shear deforma-

Structure and properties of composites 369

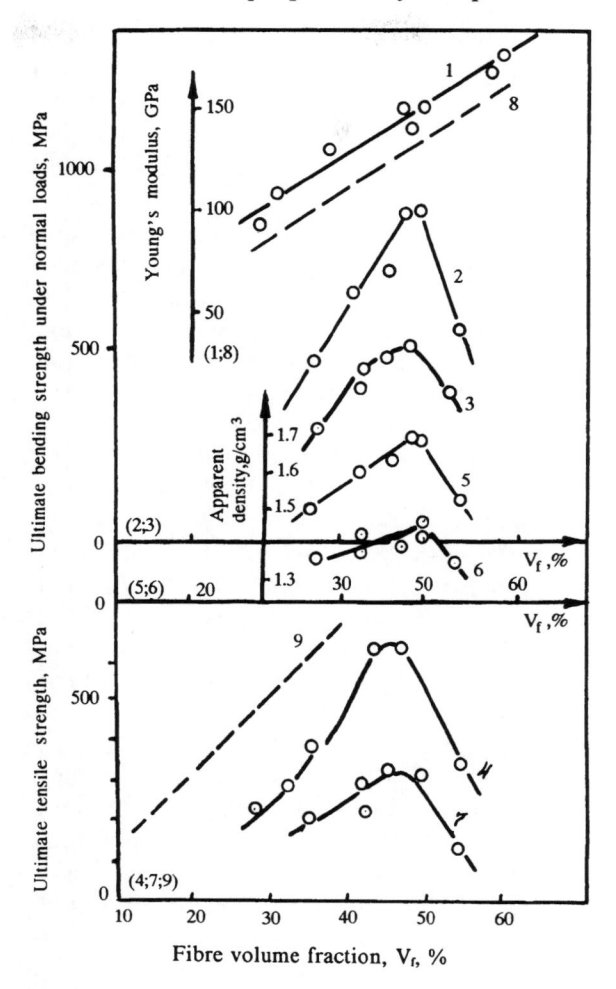

Figure 3.56 Ultimate resistance to normal loads in bending and tension as a function of volumetric filling with fibres for CCC specimens with density of 1.5–1.65 g/cm^3 ($P = 5\%$) (1, 2, 4, 5) and of 1.3–1.35 g/cm^3 ($P = 21\%$) (1, 3, 6, 7); (8, 9) corresponding calculations by rule of mixture.

tions and of other forms of a complex stressed state occur even at simple loading types. The governing importance for the level of strength attained under such conditions is displayed by the matrix porosity and the scale factor of the specimen or part being tested.

The commercial and scientific literature presents numerous results of investigations of the characteristics, although standards or procedures for tests of carbon–carbon materials have not been worked out and are not indicated. A designer or other user has to analyse and use such an information on physicomechanical properties, comparing them with the above-presented information. An analysis of the elasticity of CCC, their thermophysical properties, and retention of physicomechanical properties over the service temperature range before the beginning of sublimation is facilitated by the fact that all CCC are textural modifications of a commercial polycrystalline carbon, whose base is a common hexagonal crystalline lattice.

The main thermal and mechanical properties up to 3000 K were measured [72] on cylindrical or rectangular specimens cut out of large layered fillers. For compositions A and B, a plain-weave fabric was made from nongraphitizable viscose; for composition C the fabric of the same weave was obtained from graphitizable polyacrylonitrile. The compositions were densified to densities of 1.500–1.550 g/cm^3 (A), 1.625–1.875 g/cm^3 (B) and 1.825–1.875 g/cm^3 (C) with a weakly graphitizable coke, product of a high-temperature treatment of furfuryl alcohol, which was repeatedly impregnated into a successively densified carbon–carbon skeleton, two-dimensionally reinforced by the 'rosette' pattern with a displacement of every next layer by 22.5°. The volumetric filling with fibres was in the specimens as follows: A, 50%; B, 65%; and C, 65%.

An experimental determination of the enthalpy and specific heat demonstrated their complete predictability, and for composition C, a complete coincidence with graphite (see Figure 3.57). The linear thermal expansion coefficient and thermal conductivity also fully agree with the theory if the experiment is analysed jointly with results of thermophysical tests of fibres of various chemical natures [73, 74], taking into account the relation of these parameters to texturization of carbon-graphite materials [71] and bearing in mind the revealing of a physical relation between the crystalline structure of graphite and the thermal and mechanical properties [75]. The LTEC of a high-density composite in the transverse direction is very high over the whole temperature range and in a plane is negative at low temperatures, which is similar to the behaviour of anisotropic graphite. Medium- and low-density specimens have the same LTEC as those of isotropic carbons, i.e., not differing in various directions more than three times, whereas for graphitizable fibre-based specimens the LTEC in the reinforcement plane and across layers differ by an order of magnitude.

The graphite elasticity matrix is inversely related to anisotropy of the LTEC of a hexagonal carbon structure [75], and the tendency of relation of energy parameters of the crystal lattice of carbon to temperature affects results of high-temperature tests. Results of determination of strength, elastic and deformation characteristics of a two-dimensional CCC are shown in Figures 3.58 and 3.59. The ultimate resistance to compression and tension grows in all specimens and in all measurement directions, which is typical for carbon-graphite materials. The values of these characteristics for high-density specimens (C) are at least twice as

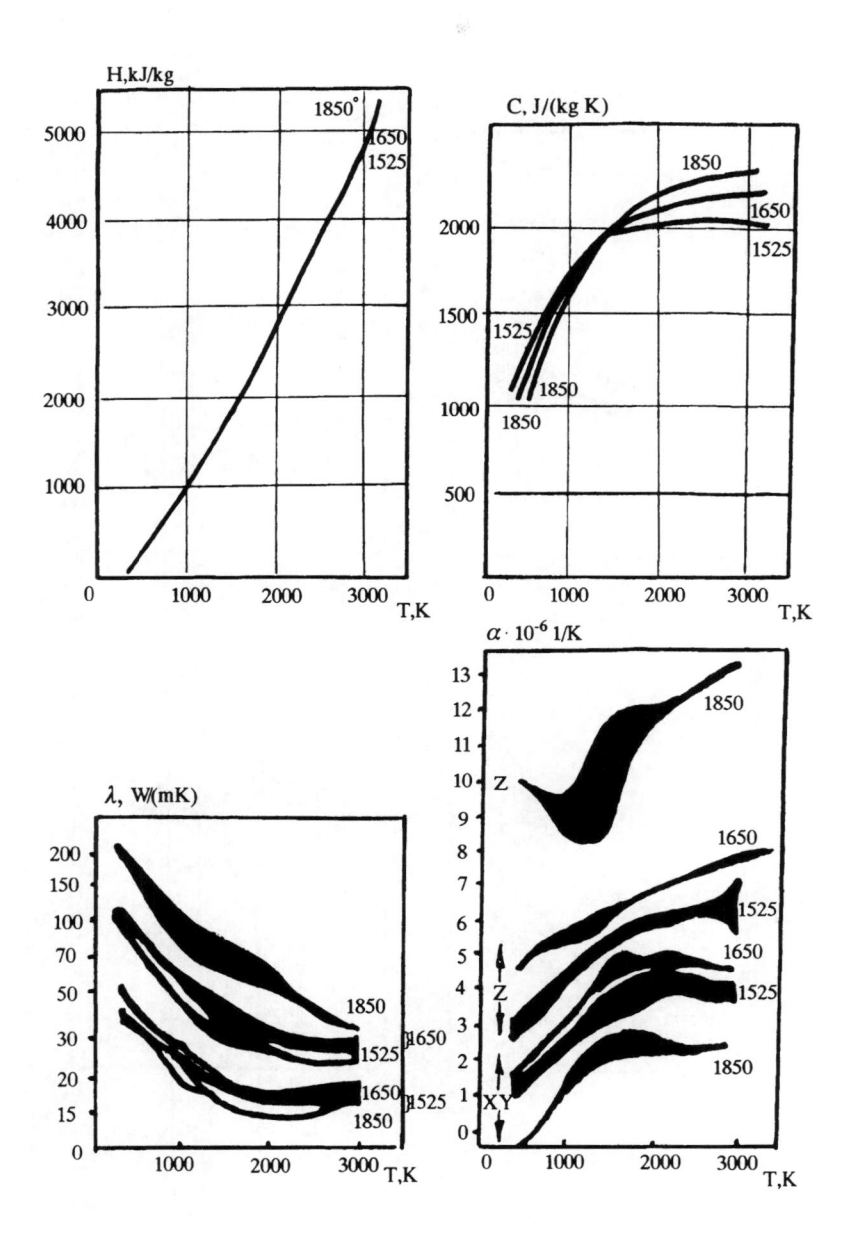

Figure 3.57 Thermophysical properties of materials: H, enthalpy; C_p, specific heat; λ, thermal conductivity; α, linear thermal expansion coefficient (LTEC).

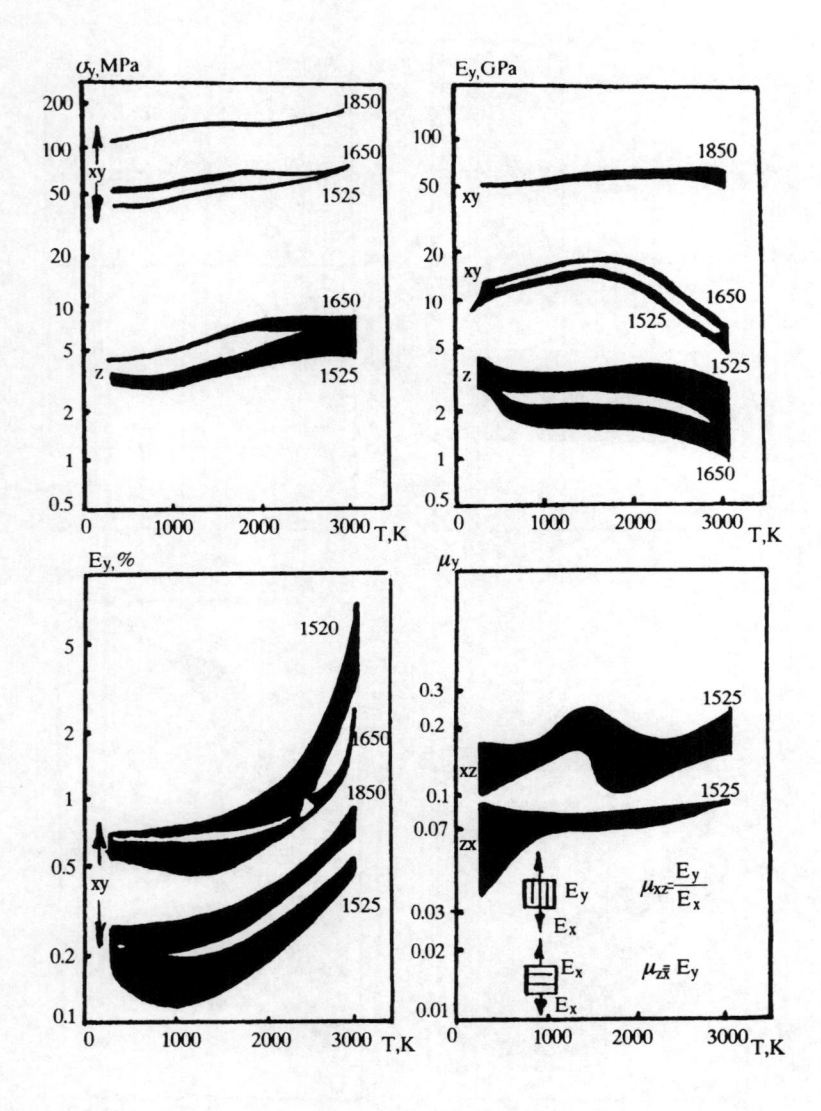

Figure 3.58 Investigation of CCC plates under tension.

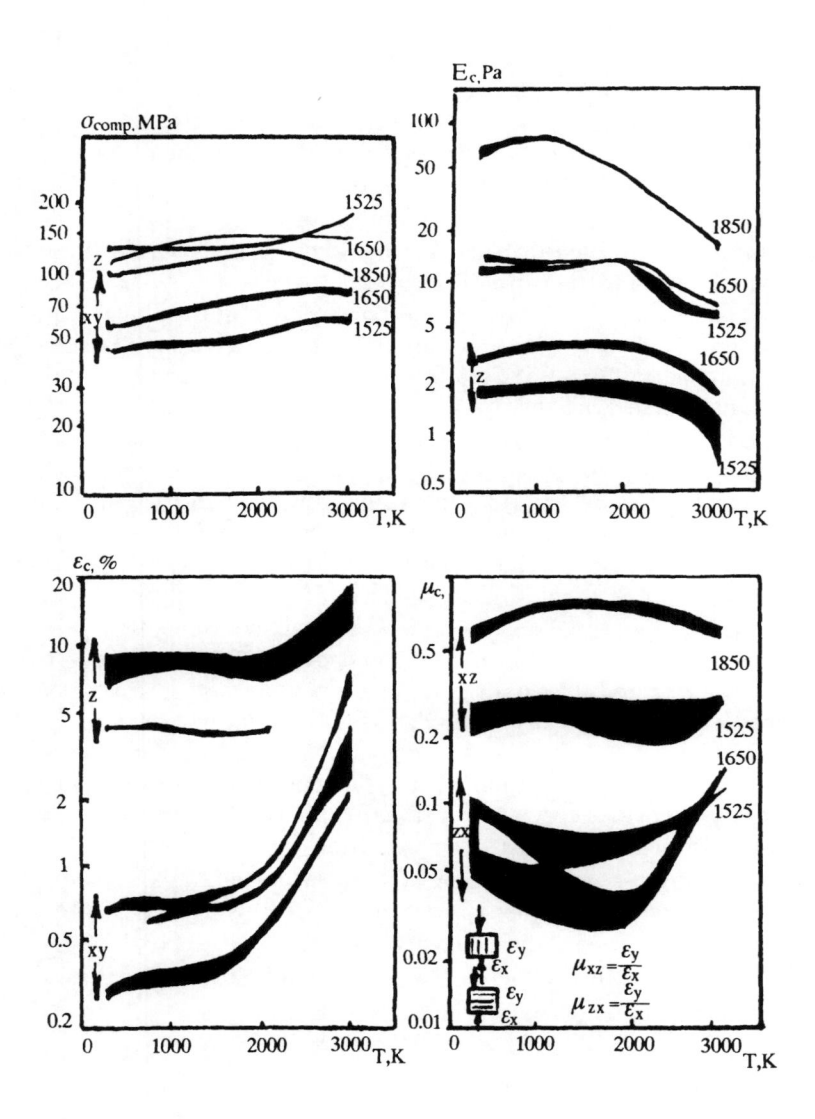

Figure 3.59 Investigation of CCC plates under compression.

high as those for medium-density specimens and much higher than for low-density ones (A). The same behaviour is exhibited by elastic moduli. In the plane of reinforcement of high-density specimens the modulus is five times as high as that across the layers. The moduli remain practically constant up to a temperature on the order of 2000 K, when plastic deformation of carbon fibres starts if they have been pretreated to about this temperature. The fracture strain measured in the reinforcement plane is not over 1% for low- and medium-modulus materials and is twice as low for a high-density one. In the transverse direction the ultimate strain under a tensile load is very low (Figures 3.51 and 3.58) and is more marked at compression. At low temperatures, it is within 5% for medium-density and within 10% for high-density materials with respect to its value in the plane of reinforcement.

Poisson's ratios in compression and tension take into account both the thermal and the deformational expansion. For a high-density material with a high modulus in the plane of reinforcement the Poisson's ratio in the same plane is more than twice that for a low-density material, where deformative characteristics of the reinforcement and matrix are closer to one another.

The thermal expansion component makes its great contribution to an anomalous increase of the Poisson's ratio above 2000 K. The decrease of this parameter at lower temperatures does not contradict the known data for composites with a similar reinforcement pattern.

The behaviour of the moduli at shear and torsion and of the strength at shear along and across a layer were studied simultaneously [72]. The character of main dependences up to and over 2000 K is similar to the variation of the moduli and ultimate resistance to fracture at tension and compression and is largely dependent on the adhesion interactions of carbon surfaces [76, 77].

The commercial information on physicomechanical and thermophysical properties of basic two-dimensional CCC types does not contradict the results presented in scientific publications.

Table 3.7 presents properties of carbon–carbon materials for exhaust parts of nozzles of solid-fuel rocket engines [78]. In the course of technological improvement, their density was increased from 1.36 to 1.8–1.9 g/cm^3 along with decline in the porosity. The following regularities were revealed.

1. Carbon reinforcement from better graphitizable and most texturized polyacrylonitrile fibres provides the highest elastic modulus.
2. Carbon mats containing randomly distributed fibres in the plane of reinforcement exhibit a higher ultimate fracture strain, but a much lower strength.
3. The highest compressive strength is exhibited by composites with a matrix formed by chemical vapour deposition of pyrocarbon in pores at pyrolysis of hydrocarbon.

Table 3.7 Physicomechanical properties of CCC for exhaust parts of nozzle of solid fuel rocket engines

CCC grade	Manufacturer	Material	Manufacturing method	Density (g/cm³)	Ultimate strength (MPa)				Elastic modulus in tension (MPa)	Ultimate strain (%)
					tensile	shear	bending	compressive		
Pyrocarb 903	Hitco	Polyacrylonitrile (PAN) fabric	Impregnation with resin	1.95	172.5	8.4	94.31	106.9	7790	0.207
Carbon 608	Kaiser	PAN fabric	Impregnation with resin	1.88	170.0	5.3	192.8	59.0	8620	0.220
Pyrocarb 901	Hitco	Viscose fabric	Impregnation with resin	1.78	100.0	4.1	112.4	95.8	2440	0.403
MK 4929	Kaiser	Material from pitch	Impregnation with resin	1.78	32.7	5.3	48.0	44.6	2130	0.150
Pyrocarb	Hitco	PAN fabric	Chemical vapour deposition	1.76	225.3	7.8	187.6	113.2	5740	0.401
MO HX 5120	Heyweg	PAN fabric	Chemical vapour deposition	1.67	40.4	4.8	60.4	41.1	1290	0.554
Standard CCC	Heyweg	PAN fabric	Chemical vapour deposition	1.45	94.4	–	107.9	78.1	1500	0.147
MO HX	Heyweg	PAN fabric	Impregnation with resin	1.36	27.0	2.1	37.1	30.7	870	0.447

Table 3.8 Physicomechanical characteristics of Sepcarb materials of type ST

Characteristics	Two-dimensionally reinforced T_1		Two-dimensionally reinforced T_2	
Filler orientation relative to load	Parallel	Perpendicular	Parallel	Perpendicular
Density (g/cm³)	1.55		1.7–1.8	
Tension:				
resistance (MPa)	45	–	40–90	–
modulus (MPa)	12 800	–	25 000–50 000	–
Compression:				
resistance (MPa)	70	105	55–110	50–120
modulus (MPa)	13 200	3 000	25 000–50 000	4000–7000
Linear thermal expansion coefficient (10^{-6} °C^{-1}):				
at 1000°C	2.0	5.5	1.0	11
at 1500°C	2.4	6.0	1.2	12
at 2200°C	2.8	6.8	1.4	13
Thermal conductivity W/m K:				
at 100°C	78	23	210	70
at 500°C	47	17	120	40
at 800°C	42	14	110	32

The CCC produced by SEP [79] and Aérospatiele [80], subsidiaries of the Le Carbone-Lorrain, which have found wide application in aerospace and medical engineering, confirm a high anisotropy of LTEC for composites based on more textured carbon fabrics from polyacrylonitrile (Tables 3.8 and 3.9). The same compositions have a several-fold higher elastic modulus in the plane of reinforcement. The commercial data on the LTEC, thermal conductivity, and elastic moduli ratio, presented in Table 3.8, are consistent with the above-discussed results. At the same time a higher ultimate resistance to compression in the transverse direction than in the reinforcement direction is reported. This results from a nonoptimal selection of the specimen for the test, since it means an incomplete utilization of properties of the high-strength and high-modulus reinforcement. Properties of unidirectionally reinforced and fabric-reinforced Aerolor materials are compared in Table 3.9. At the same density level, the attained elastic modulus and fracture resistance for the fabric-reinforced material are several-fold lower than for the unidirectional one. A lower elastic modulus in bending than in tension is reported for the material Aerolor-223, which may result from a nonoptimal selection of specimen dimensions and distortion of the physical result by an interlayer shear. Commercial publications of the NIIgrafit (Graphite Research Institute) report properties of two-dimensionally reinforced

Table 3.9 Physicomechanical characteristics of materials made by S.N.I. Aérospatiale

Characteristics	Aerolor II		Aerolor 223	
Filler orientation	Parallel	Perpendicular	Parallel	Perpendicular
Density (g/cm^3)	1.3–1.5		1.3–1.5	
Tension:				
resistance (MPa)	800–1100	–	40–70	10
modulus (MPa)	140 000–160 000	–	20 000–30 000	–
Bending strength:				
resistance (MPa)	110–1350	–	80–200	–
modulus (MPa)	150 000–170 000	–	10 000–20 000	–
Shear strength:				
resistance (MPa)	–	–	20–30	–
Compressive strength:				
resistance (MPa)	–	–	120–200	120–200
Linear thermal expansion coefficient ($10^{-6}\,°C^{-1}$):				
at 1000 °C	–	–	1.5	6
at 2500 °C	–	–	2.5–3.5	8
Thermal conductivity (W/(m K))	–	–	30–40	10

composites based on high-modulus polyacrylonitrile [81] and low-modulus viscose [82] fibres; they are presented in Tables 3.10–3.14. The materials are intended for use in electric furnaces as electric-heating [83] and structural elements. As seen from the tables, they offer high specific strength characteristics, superior to those of high-temperature metals, stable thermophysical and electrical properties. Parts made from the materials operate in vacuum and an inert or reducing atmosphere at temperatures up to 2000 °C. An electric heater which has a longer service life than heat-resistant metals, a higher structural strength than graphite heaters, and a low specific mass is shown in Figure 3.60. The structure is assembled on bolted graphite elements with application of heat-resistant adhesives [84] between to exclude electric-arc discharges in the fastening assembly. The rated power of such three-phase electric heaters is of hundreds of kilowatts. Electric heaters from these materials are manufactured in the form of tubes or thin-walled cylindrical shells and are superior to graphite in the structural strength. Sigri [85], Schunk [86] and Polycarbon [87] manufacture two-dimensionally reinforced carbon–carbon composites (Tables 3.15, 3.16 and Figures 3.61–3.63) for applications as electric furnace linings, heat-resistant structural elements, chemically resistant parts, high-temperature heat insulation and heaters. Performance characteristics of all the materials have close orders of magnitude. On the whole it can be stated that two-dimensionally reinforced materials exhibit a low density, typical for plastics, a fracture resistance at the level of that of aluminium materials, an elastic modulus at the level of that of titanium structural alloys and materials, a heat resistance of tungsten, a thermal expansion in the plane of the layer at the level of that of silicate glasses, and a thermal conductivity of steel. In the transverse direction their thermal conductivity equals that of oxide ceramics, and the thermal expansion coefficient, that of copper.

Table 3.10 A comparison of physicomechanical properties

Characteristics	KUP-VM-2	KM-5415
Density (g/cm^3)	1.35	1.25
Ultimate strength (MPa):		
in tension	370	70
in bending	550	150
in compression	25	100
Thermal conductivity (W/(m K))	2.9	6.5
Linear thermal expansion coefficient (10^{-6} °C^{-1}):		
at 20–2000 °C	3	
at 20–1500 °C		2.8
at 20–2000 °C		3.1
at 20–2500 °C		4.6
Specific heat (kJ/(kg K))	0.68	0.65

Table 3.11 Variation of ultimate strength (MPa) of composite with reinforcement angle

Reinforcement angle (deg)	Tension	Compression	Bending
0.0	370	200	550
22.5	253	137	376
45.0	105	57	156
67.5	75	40	110
90.0	1.5	11	5.3

Table 3.12 Variation of thermal conductivity of composite (W/(m k)) with reinforcement angle

Reinforcement angle (deg)	Along	Across	In thickness
15	23.0	4.6	2.9
30	19.1	8.5	2.9
45	13.8	13.8	2.9
60	8.5	19.1	2.9
75	4.6	23.0	2.9

Table 3.13 Variation of thermal conductivity with temperature

Temperature (°C)	1200	1500	1800	2100	2400
Thermal conductivity (W/(m K)):					
along warp	7.5	7.7	8.3	8.8	9.5
along weft	7.5	8.8	9.9	11.4	13.1
across thickness	4.4	4.6	4.9	5.3	5.6

Table 3.14 Physicomechanical characteristics of carbon materials* (strength characteristics are stable at temperatures up to 2000 °C)

Characteristics	KUP-VM-PU Winding of high-modulus PAN-based fibres	Tapir Lay-up and sewing-across of viscose fabric	KM-VM Lay-up of high-modulus PAN-based fabric
Volumetric density (g/cm^3)	1.4	1.3	1.5
Tensile strength, kPa:			
longitudinally	37 000	6900	16 500
transversely	150	3700	5000
Compressive strength (kPa):			
longitudinally	20 000	11 000	15 000
transversely	1100	5100	5000
Elastic modulus, (10^5 kPa):			
longitudinally	170	21	80
transversely	3	16	50
Shear strength (kPa)	1800	3000	2700
Shear modulus (10^5 kPa)	4.8	–	6
Poisson's ratio (%):			
longitudinally–transversely	0.45	0.16	0.12
transversely–longitudinally	0.05	0.16	0.12
Bending strength (kPa):			
longitudinally	55 000	13 000	29 500
transversely	5300	7200	10 000
Linear thermal expansion coefficient (10^{-6} °C^{-1}):			
at temperatures up to 200 °C:			
longitudinally	– 0.4	0.8	– 0.4
transversely	3.2	1.2	0.7
at temperatures of 20–1500 °C	–	–	–
at temperature of 20–2000 °C	–	–	1.3
Thermal conductivity (W/(m K)):			
longitudinally	15	4.4	45
transversely	3.2	4.7	25

Figure 3.60 Three-phase heater from material KM-5415.

Such a specific set of properties opens up unique potentialities for designing machine elements and high-temperature units.

The CC-class materials offered by the Sigri company are produced by manufacturing techniques that are most readily mechanized and are therefore most economical: winding and pressing of cloth laminates. Volumetric densification methods and maximum processing temperatures also involve no extreme conditions. As a result, cheap carbon–carbon materials with a medium level of strength and rigidity, but with high enough oxidation and erosion resistances and a high heat resistance are manufactured and offered to users. A temperature increase to 2000 °C reduces strength characteristics by not more than 10%, except for the shear resistance, which is reduced to a greater extent. However, the ash content of the material decreases more than by half and the electrical resistance decreases appreciably, which opens up broad possibilities for using thermally treated materials in electrothermal equipment designs.

Commercial data on variation of thermophysical properties of materials made by Sigri are presented in Figures 3.61–3.63. They feature considerable thermal expansion coefficient and thermal conductivity, which exhibit a general character of texturization of the composite as a result of its laminated structure. The thermal treatment increases the

Table 3.15 Physical properties of various two-dimensionally reinforced carbon–carbon composites

Characteristics	CC 1001D	CC 1001G	CC 1501D	CC 1501G	CC 1501D	CC 1501G	CC 2001D	CC 2001G
Treatment temperature (°C)	1000	2000	1000	2000	1000	2000	1000	2000
Volumetric density (g/cm^3)	1.35–1.40	1.35–1.40	1.40–1.45	1.40–1.45	1.30–1.40	1.30–1.40	1.50–1.60	1.50–1.60
Open porosity (%)	20–25	25–30	15–20	20–25	20–30	25–35	10–15	15–20
Bending strength (N/mm^2)	140–160	135–150	230–280	210–250	140–160	130–150	300–400	280–350
Dynamic module (kN/mm^2)	30–35	35–40	55–60	60–65	50–60	50–60	100–120	110–130
Tensile strength (N/mm^2)	90–110	85–105	280–350	260–330	170–200	160–200	400–500	380–470
Electrical resistance in direction of fibres (ohms)	35–40	25–30	30–35	25–30	30–35	25–30	35–40	30–35
Coefficient of permeability (cm^2/s)	4.4×10^{-2}	6.8×10^{-2}	5.9×10^{-3}	4×10^{-2}	2×10^{-3}	1×10^{-2}	10×10^{-1}	1.2×10^{-1}
Interlayer shear strength (N/mm^2)	13–18	10–15	10–15	9–12	10–13	9–10	10–15	10–12
Content of impurities (%)	0.14	0.07	0.20	0.08	0.25	0.10	0.10	0.06

Table 3.16 Physical properties of materials made by Polycarbon and Shunk

Material	Treatment temperature (°C)	Content of fibres (% vol.)	Density (g/cm^3)	Porosity (%)	Bending strength (N/mm^2)	Young's modulus (GN/m^2)	Ultimate strain (%)	Shear strength (N/mm^2)	Torsional strength (N/mm^2)	Torsion modulus (GN/m^2)
CF 113	1100	55–65	1.50–1.55	5	340–420	90–130	0.3–0.5	22–28	28–33	2.8–3.4
CF 122	1100	55–65	1.50–1.55	5	240–275	65–85	0.3–0.5	20–25	25–30	2.5–3.2
CF 222	1700	55–65	1.60–1.65	10	200–240	70–90	0.2–0.4	15–18	20–25	1.8–2.2
CF 322	2500	55–65	1.60–1.70	10	150–200	75–95	0.2–0.4	12–15	15–20	1.8–2.2

Structure and properties of composites

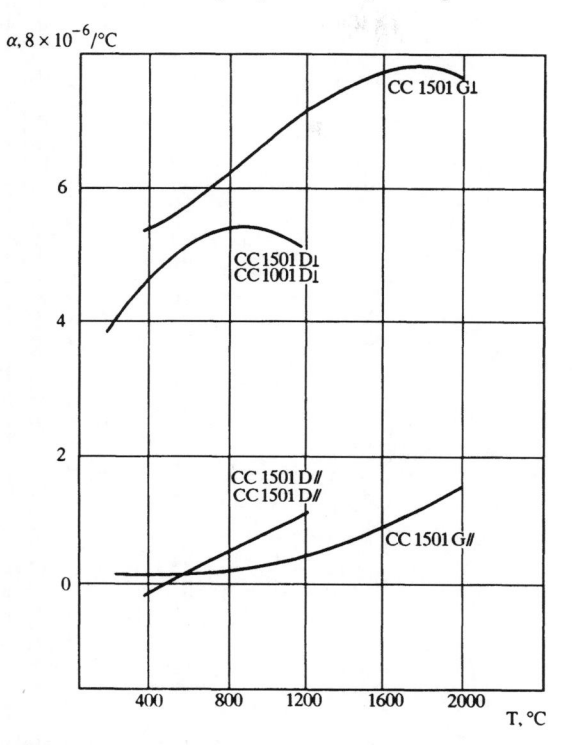

Figure 3.61 Linear thermal expansion coefficient of materials made by Sigri.

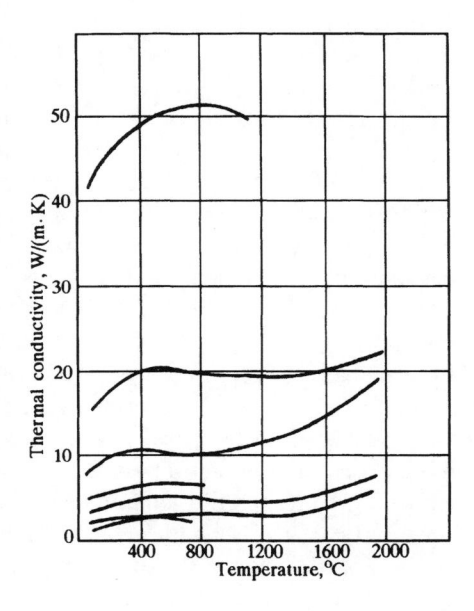

Figure 3.62 Thermal conductivity of materials made by Sigri.

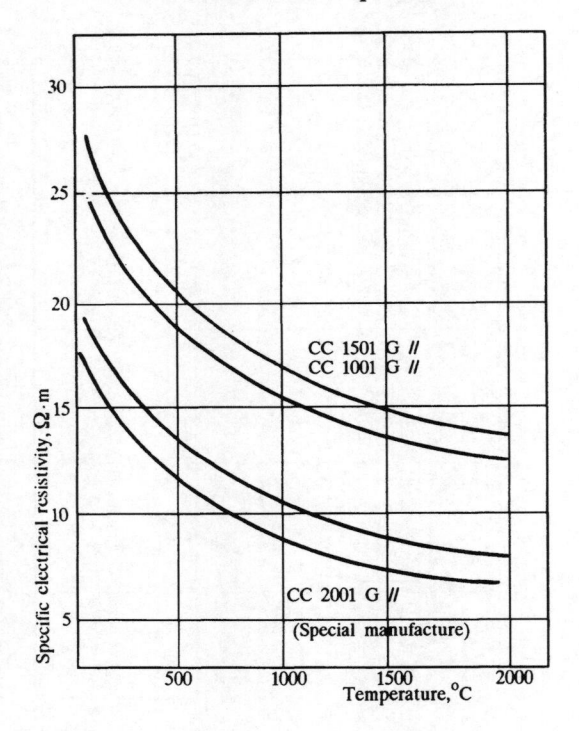

Figure 3.63 Electrical resistivity of materials made by Sigri.

LTEC anisotropy. The highest thermal conductivity is along the layer in a composite produced by winding of high-modulus, most textured fibres. This thermal conductivity level corresponds to that of the material KUP-VM-PU, also produced by winding of high-modulus high-temperature-treated fibres. The electrical resistance decrease with increasing temperature is typical of KUP-VM-2, KM-5415 materials and CC materials, which are similar in composition and structure. This is a feature of materials with a low graphitization degree, where a perfect enough structure in the coherent scattering region layer has been attained, but the azimuthal ordering of layers in the height is absent.

The carbon–carbon materials made by Polycarbon and Shunk (Figure 3.64 and Table 3.16), offered in the commercial information, also exhibit physicomechanical properties typical of this class of structural materials [72] (see Figures 3.14–3.15). Their properties at room temperature (Table 3.16) indicate a high degree of completion of the manufacturing densification cycle, since the open porosity is maximally lowered. Levels of physicomechanical properties vary with the initial filler types and, of course, decline after the high-temperature treatment. The latter reduces the strength of the carbon matrix, which affects the shear strength and the torsional strength and modulus. The composite's elastic modulus for

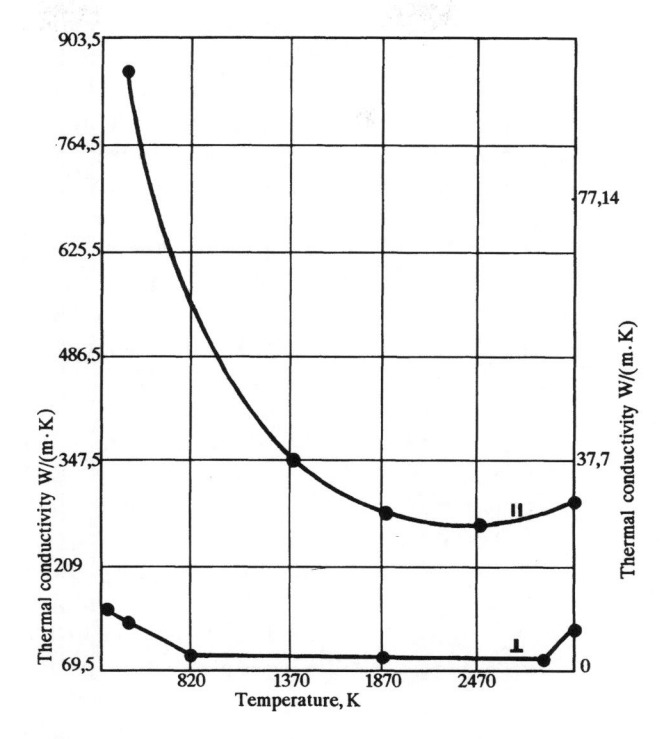

Figure 3.64 Thermal conductivity of materials made by Polycarbon.

normal loads (bending) grows somewhat. These commercial results are in a good accord with scientific results of studying the interlayer strength at cleavage [76] and the tear-off resistance of carbon surfaces with varying processing techniques [77].

3.3.2 Friction-purpose carbon–carbon composites

Intensive efforts on developing carbon–carbon friction materials were started in the late 1960s and early 1970s by the largest western companies, suppliers of aircraft engineering products, such as Dunlop (Great Britain), Bendix (USA), Goodyear Aerospace (USA), Goodreach (USA), A Bex (USA), SEP (France), etc.

The work along this line turned out to be so promising that as early as 1970 the aviation division of the Dunlop company informed government representatives in Great Britain, France and the USA that the company had ceased experimenting with beryllium and other materials for brakes and had fully focused the research on the CCC. The company abandoned beryllium brakes because carbon ones provided a greater mass reduction, a lower wear, and a higher reliability under extreme braking conditions.

As a result, most west European aircraft operated by Air France (France) and British Airways (Great Britain) have been fitted with CCC brakes.

In 1977, the US Air Force started testing CCC brakes on the F-15 aeroplane. In 1979, the CCC brakes were installed on the F-15, F-16, F-18 and B-1 aeroplanes. CCC brakes made by Goodyear were used on the *Challenger* space shuttle. In all, this company has already fitted more than 2500 aeroplanes with such brakes.

A high interest in friction-purpose CCC has recently been shown by the Toho Beston and Kobe Steel companies of Japan.

An extensive application of CCC-based braking devices in railway transport, tanks, and other mechanical engineering products is predicted.

Despite numerous advertisement messages, publications on friction-purpose CCC are practically absent in the scientific and technical literature.

In the USSR, intensive work on development of friction-purpose CCC has been conducted since 1973. As a result, novel methods for the manufacture of such materials of a high quality have been devised [88, 89]. They rely on special techniques for orientation of continuous and discrete carbon fibres having different elastic moduli and provide stable values of the friction coefficient and minimum wears of discs even at particularly severe brake system operating conditions. The main characteristics of braking wheels with carbon discs, manufactured in the USSR, are presented in Table 3.17.

Table 3.17 Basic technical characteristics of brake wheels with carbon discs

Tyre size	Wheel index	Initial pressure on wheel (kPa)	Static load on wheel (kN)	Landing speed (km/h)	Work of braking (10^6 kgf m)	Mass (kg)	
						wheel	brake
950 × 400	KT 206	120	130	420	1.60	39	67
1030 × 350	KT 213	260	166	360	1.90	66	66
1300 × 480	KT 204	115	186	330	2.97	79	91
1260 × 425	KT 19117	170	222	420	2.45	91.5	77

3.3.2.1 Factors governing friction-wear characteristics of carbon–carbon composites

Carbon discs are tested in the brake, whose components they are, on an inertial test-bench; friction-wear characteristics of the material are evaluated on a friction machine. Rings of 75 and 53 mm outside and inside diameters respectively and 14 mm thickness are tested under conditions simulating those of operation of aircraft brakes.

The influence of properties of reinforcing fibres and of the reinforcement type on the friction coefficient and wear resistance of carbon–carbon materials has been studied. Figure 3.65 shows the variation of the

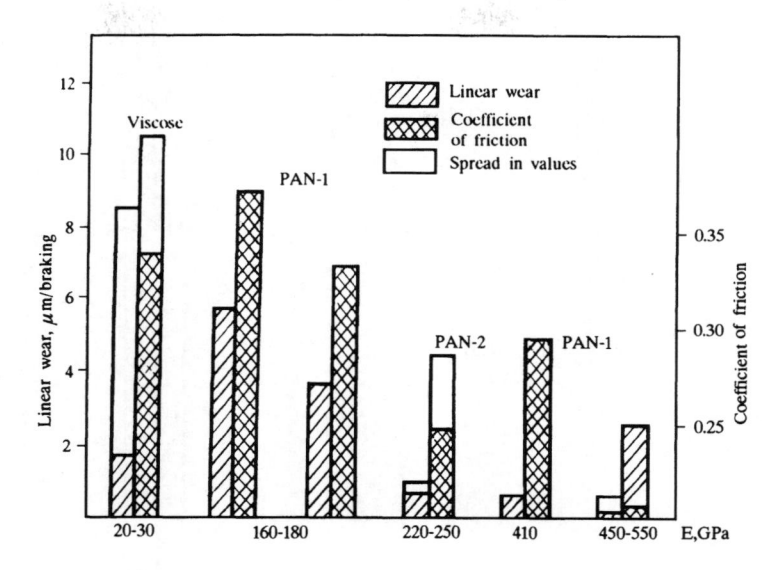

Figure 3.65 Variation of friction coefficient and linear wear per braking with elastic modulus of fibres for discrete fiber-reinforced materials.

friction coefficient and the linear wear per braking for discrete fibre-reinforced materials with the elastic modulus of fibres, the latter value being most sensitive to changes in the conditions of processing and structure of fibres. As can be seen, both friction characteristics decline with increasing elastic modulus of fibres; the friction coefficient can decrease 1.5–2 times, and the wear, by an order of magnitude.

The variation of friction–wear characteristics of the material with the reinforcement type (discrete fibres or fabric) is shown in Figure 3.66. They are in principle identical for materials based on fibres of various nature.

The wear of discrete fibre-reinforced materials is always within a range of values smaller than the wear of a material with a fabric filler. For materials with a combined filler, i.e., a combination of one or other fabric with discrete high-modulus fibres, the wear value range extends as compared with that for a fabric reinforcement, chiefly through reduction of its lower limit.

The dependence of the friction coefficient on the reinforcement type is much less pronounced than its relation to fibre characteristics. At a combined reinforcement, the lower limit of the range of values decreases, as does the wear.

Both the friction coefficient and the wear are practically independent of the material porosity over a relatively broad range of its values (up to 20–25%), while at a porosity over 27% the two characteristics, especially the wear, increase considerably (Figure 3.67).

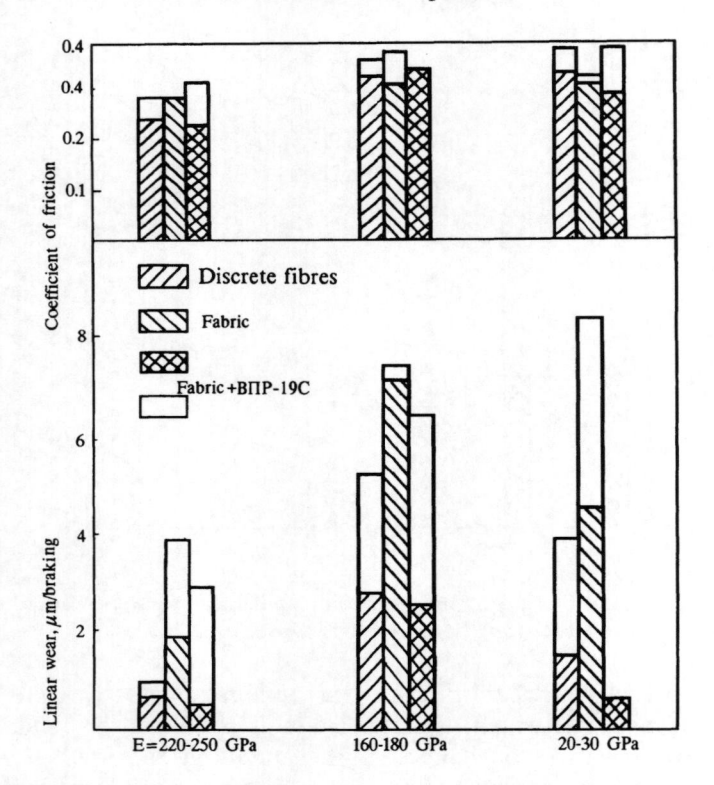

Figure 3.66 Variation of friction-and-wear characteristics of materials with reinforcement type.

Thus, of the above-discussed factors, the strongest impact on the friction–wear properties of carbon–carbon materials is exerted by the value of the elastic modulus of reinforcing fibres. Its change results in change in dissipative properties of the materials and hence also in temperature-and-force conditions at the contact spots. The interrelation of friction and wear characteristics with the elastic modulus of fibres, however, may also stem from the influence of other parameters of fibres (e.g., density or thermal conductivity), which change along with the elastic modulus.

The decline of wear resistance at replacement of discrete reinforcing fibres with fabrics appears to be caused by strengthening of the influence of technological-origin defects: pores, cracks, various structural inhomogeneities, resulting from difficulties in impregnating fabrics with the binder.

The release of the braking energy brings about an intense thermal and force action on the friction surface, which cannot but give rise to changes in the structure of surface layers. The structure of surface layers of

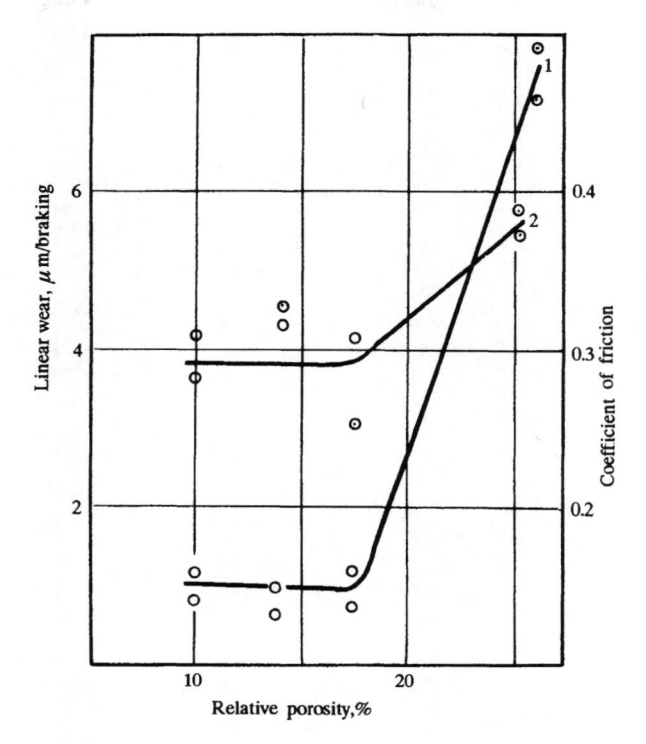

Figure 3.67 Effect of material porosity on (1) linear wear and (2) friction coefficient.

samples of various materials after friction tests were studied with the aid of a scanning electron microscope.

The micrograph of the friction surface of a sample of a material having a relatively small wear (2 µm/braking) is shown in Figure 3.68a. A dense homogeneous smooth film is formed in the course of friction on the surface of the material, which is heterogeneous in the initial state.

A similar film is formed on the friction surface of a material with a greater wear (8 µm/braking), but it is loose, consisting of particles weakly bonded to one another (Figure 3.68b). For a material that undergoes a catastrophic wear (linear wear of about 40 µm/braking) the above-described film does not form at all, and the 'third body' is a compacted conglomerate of particles not bonded to one another.

The obtained results made it possible to form a notion about the mechanism of wear of carbon–carbon composites (Figure 3.69).

Primary products of wear in the form of highly disperse particles are formed at initial stages of friction as a result of attrition or fatigue fracture of the surface. Under the action of high temperatures and pressures at contact spots (which can exceed 3000 °C and 10 MPa), the

(a)

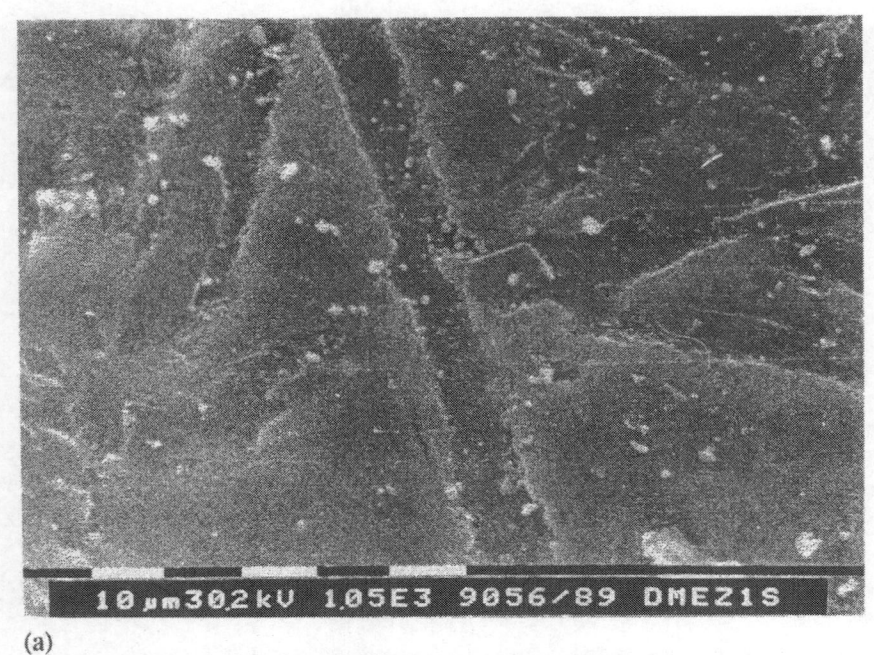

(b)

Figure 3.68 Micrographs of friction surfaces of samples of materials with (a) small and (b) increased wear.

Structure and properties of composites

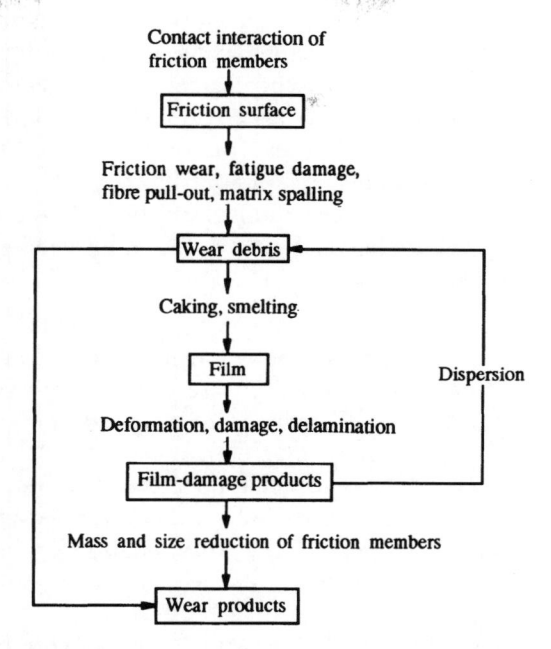

Figure 3.69 Scheme of processes proceeding under friction of carbon–carbon materials.

particles, having a high surface energy, are sintered to form a film observed in micrographs. Elastically and plastically deforming, the formed film takes up the friction load and protects the base material from wear. When the deformability of the film has been exhausted, it breaks down or separates off from the base. Fragments of the film get partly dispersed and participate once again in its formation and partly thrown out by centrifugal forces beyond the friction unit. The cycle repeats many times at various regions of the friction surface.

In the light of the suggested mechanism, it can be assumed that the wear resistance of a material is associated with the efficiency of the protective action of the surface film, which depends on its deformation-and-strength properties and adhesion to the base material. These characteristics appear to be governed by the size, shape and surface properties of wear particles, which depend on the composition and structure of the material, as well as on the structure and properties of its components, carbon fibres and coke matrix.

3.3.2.2 *Demands placed on materials and designs of braking devices*

At initial stages of the research, the Dunlop, Bendix and Goodreach companies tried to use for aircraft brake systems the CCC developed earlier for other purposes, where carbon fibres, various fabrics, felts, etc. were reinforcing components, but the results were, as a rule, negative. A new approach to selecting the reinforcement patterns, raw materials, equipment and processing conditions was needed. The lack of information on requirements to be met by friction-purpose CCC had to be compensated for by testing tens of versions [90, 91].

Based on the analysis of patent publications [92, 93] and of the efforts conducted in the USSR, the following basic requirements for friction-purpose CCC can be formulated.

1. The material density should be not less than 1.7 g/cm^3.
2. A high interlayer strength of the composite is needed, since discs are in a complex stressed state in the course of brake operation.
3. The friction coefficient of the friction pair should be not less than 0.2 and not more than 0.4, and stable enough.
4. The wear of friction elements should not exceed specified values under any brake operating conditions (usually 2–3 µm/braking).
5. The material should have a reliable antioxidation protection at temperatures over 1000 °C, exerting no adverse effect on friction-and-wear characteristics.
6. The material should have a high thermal conductivity, including that in the direction perpendicular to the friction surface.
7. The material should meet additional requirements in resistance to damages from vibration of the brake, thermal compatibility with the housing parts, etc.

3.3.2.3 Features of manufacturing technology for friction-purpose materials

The manufacturing technologies for friction-purpose CCC can be conventionally divided into two groups; the first including the methods used to manufacture nozzles, nose cones, thermal protection elements, and other aerospace parts, and the second including special methods and techniques for directional control of friction and wear characteristics (FWC) of composites.

Various fibrous materials (tapes, fabrics, short and continuous fibres, multidimensional three- and four-dimensional reinforcement blocks) from viscose, pitch and polyacrylonitrile raw materials are used as the reinforcing filler, but preference is given to the latter, which provides the highest wear resistance and stability of friction characteristics of CCC.

Cokes of phenolic and furane resins, of pitch as well as pyrocarbon serve as matrices for friction composites. Best FWC are attained with the use of pitch cokes and pyrocarbon, whose advantages for developing high-density CCC are clear from the values of matrix density after graphitization (Table 3.18).

Table 3.18 Matrix density of cokes after graphitization

	Density (g/cm^3)
Phenolic resin coke	1.65
Furane resin coke	1.80
Pitch coke	2.10
Pyrocarbon	1.6–2.10

Every variety of friction CCC of the first group has its technological features stemming from the binder and filler types and the reinforcement pattern, but, on the whole, the manufacture of such composites includes the following steps.

1. Carbonization of premoulded preforms at 900–1000 °C in nitrogen atmosphere.
2. Densification by a repeated impregnation with resin or pyrocarbon.
3. Densification with pitch and carbonization under pressure in high-pressure units (hyperclaves).
4. Graphitization in induction furnaces with purging by nitrogen at temperatures up to 2600 °C.

Special processes for manufacture of CCC for use as brake discs on aircraft, high-speed railway transport facilities and other high-energy structures, such as brakes and clutches of high-speed motor vehicles, have been developed most intensively in recent times.

Any unidirectional reinforcement versions turn out to be inadequate to ensure serviceability of CCC for the purpose in question. With two-dimensional-type reinforcement patterns the interlayer strength of composites is, in a number of cases, inadequate to prevent lamination of products under complex-stressed conditions of operation of heavily loaded brake systems. Transition to three- and four-dimensional-type materials is not very promising, as it involves complication of the technology and a higher cost of products. Moreover, the presence in this case of a large number of fibres oriented at right angles to the friction surface results in a greater wear of friction parts in service.

New methods for production of friction-purpose CCC, relying on the use of combined reinforcement patterns, employing both continuous and discrete fibres (CCD reinforcement patterns), have been developed in the former Soviet Union (at the Graphite Research Institute). The essence of such a reinforcement is that layers of continuous fibres (tapes, fabrics, special weaves) are combined with discrete fibres, partly oriented at right angles to surface layers formed by continuous fibres.

Such a reinforcement is much simpler than a three-dimensional one, but provides a much higher strength of a two-dimensional composite and improves its FWC.

3.3.2.4 Properties of friction-purpose carbon–carbon composites

Designs of brake discs are most diverse [94, 95]. Two designs are most popular: segmental, where carbon–carbon parts have the form of linings on strengthening metallic discs, and monodiscs, all-CCC ones.

Basic characteristics of friction-purpose CCC from leading manufactures and made in the former Soviet Union are presented in Tables 3.19–3.22.

Table 3.19 Physicomechanical characteristics of material Termar-ADF, manufactured by different technologies

Characteristic	Impregnation with pitch in autoclave and firing in gas furnaces		Impregnation and firing under pressure in high-pressure apparatus	
	average	variation (%)	average	variation (%)
Density (g/cm^3)	1.75	22	1.95	8
Bending strength (MPa)	95	20	130	10
Compressive strength (MPa)	105	28	150	12
Shear strength (MPa)	15.0	9	–	–
Tensile strength (MPa):				
parallel to fibres	4.9	38.0	–	–
perpendicular to fibres	44.5	13.0	–	–
Thermal conductivity (W/(m K)):				
parallel to fibres	20	–	65	
perpendicular to fibres	50	–	165	

Table 3.20 Basic properties of materials Sepcarb A2 and Aerolor

Characteristic	Load direction w.r.t fibres	Temperature (°C)	Sepcarb A2	Aerolor	
				41	42
Density, g/cm^3	–	–	1.7–1.85	1.6–1.8	1.9–2.0
Tensile strength (MPa)	Parallel	200	–	40	40
	Perpendicular	–	–	10	10
Elastic modulus in tension (GPa)	Parallel	–	25.0–45.9^a	25.0–35.0	20.0–35.0
Bending strength (MPa)	Parallel	–	41–194	60–150	90–150
Elastic modulus in bending (GPa)	Perpendicular	–	–	10.0–30.0	20.0–30.0
Compressive strength (MPa)	Perpendicular	–	56–153	80–220	80–100
	Parallel	–	56–255	80–220	60–100

Characteristic	Load direction w.r.t fibres	Temperature (°C)	Sepcarb A2	Aerolor 41	Aerolor 42
Thermal expansion coefficient ($10^{-6}\,°C^{-1}$)	Parallel	1000	1.0	0.7	0.5
	Parallel	2500	1.4[b]	3–3.5	2.5
	Perpendicular	1000	11	6–7	3–4
	Parallel	2500	13[b]	5–8	5–6
Thermal conductivity (W/(m °C))	Parallel	100	210	–	–
	Parallel	800	110	–	–
	Perpendicular	100	70	–	–
	Perpendicular	800	32	–	–

[a] Elastic modulus in compression
[b] At 2200 °C

Table 3.21 Basic properties of materials made by 'Dunlop'

Characteristic	Direction relative to layers	CB5	CB7
Density (g/cm^3)		1.70–1.75	1.70–1.75
Tensile strength (MN/m^2)	Parallel	80	70
Modulus in tension (GN/m^2)	Parallel	40	35
Compressive strength (MN/m^2)	Parallel	170	130
	Perpendicular	110	100
Modulus in compression (GN/m^2)	Parallel	20	15
	Perpendicular	2.0	1.9
Bending strength (three-point bend) (MN/m^2)	Parallel	135	120
Modulus in bending (three-point bend) (GN/m^2)	Parallel	30	29
Interlayer shear strength (MN/m^2)	Parallel	17	12
Thermal conductivity (W/(m K))	Parallel	25	150
	Perpendicular	8	40

3.3.2.5 Oxidation protection methods

The CCC used as friction and heat-absorbing elements are in the course of braking acted upon by an elevated temperature, which, depending on the energy load on the brake, properties of the material itself, and external heat-exchange conditions, can be as high as 2000 °C on the friction surface and in a thin surface layer and 1000 °C in the bulk of material, at a distance over 1 mm from the friction surface (Figure 3.70). Since CCC designed for a dry friction are as a rule operated in the air,

Table 3.22 Summary table of properties of Termar materials (former Soviet Union)

Material grade	Density (g/m³)	Bending strength (MPa)	Compressive strength (MPa)	Impact strength (kJ/m²)	Thermal conductivity (W/(m K))		Friction coefficient[a]	Linear wear (μm/braking)
					Perpendicular to pressure axis	Parallel to pressure axis		
Termar-DF	1.75–1.85	80–90	120–150	3.6–5.2	70–90	21–28	0.20–0.29	0.3–1.9
Termar-TD	1.65–1.71	90–100	130–160	4.6–9.8	60–70	15–20	0.25–0.30	2.5–3.0
Termar-STD	1.68–1.75	150–170	120–150	15.1–18.0	50–70	13–18	0.28–0.35	2.0–2.5
Termar-DNV	1.85–2.20	170–200	140–250	–	–	–	–	–
Termar-ADF	1.75–1.98	160–180	150–160	9.6–12.2	40–60	25–30	0.25–0.45	1.0–3.0

[a] Tests conducted on an IM-58 machine.

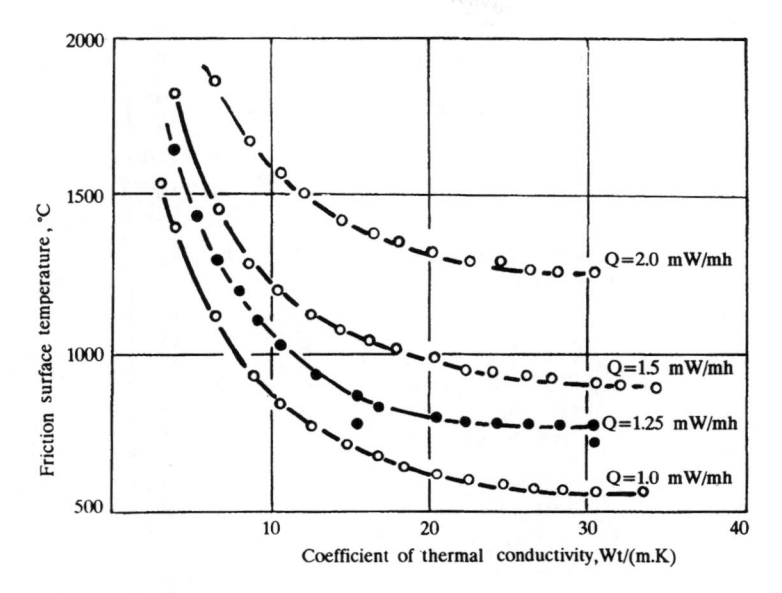

Figure 3.70 Dependence of maximum temperature of friction surface on thermal conductivity of carbon material for various friction powers (Q).

carbon gets oxidized at heating in the course of operation. Due to this, in tests of friction CCC on an inertial test-bench the material ablation determined from the mass loss exceeds, as a rule, the ablation calculated from changes in linear dimensions of a sample.

From the difference between these values the contribution of oxidation to the wear, called 'oxidation wear', is calculated. Depending on the test temperature, the oxidation component of wear can vary from 10–15% to 90–95%, reaching its maximum at 800–900 °C. A further temperature increase leads to a very rapid breakdown of the material structure, and the mechanical component of wear rises once again.

In the low-temperature region (temperature in bulk of material up to 600 °C) the carbon oxidation reaction is controlled by the stage of chemical interaction proper. The oxidation wear rate is, in this case, primarily determined by the degree of purity of the carbon material. By reducing the content of impurities of metals-oxidation catalysts from 10^{-3}% to 10^{-5}–10^{-6}% the oxidation rate can be lowered by 1–1.5 orders of magnitude and thereby the service life of a friction-purpose product can be extended accordingly. Treatment of products with phosphorus-containing reagents is also very effective. According to the model adopted at present, the mechanism of phosphorus action consists in blocking of active centres, in particular of impurity atoms. At temperatures up to 600 °C the oxidation component of wear is relatively small, less than half the total wear, reducing the oxidizability of a material little affects its wear resistance. A good correlation between the apparent activation energy of carbon oxidation in the transition zone, i.e., at temperatures of 600–1000 °C (30 kcal/mole), and the activation energy for the loss of mass of nonfriction surfaces of samples tested on the inertial test-bench (27 kcal/mole), was shown. Since at such temperatures the oxidation reaction proceeds in the intrakinetic region, then the stage controlling the mass loss is the diffusion of gases in pores of the material. The most effective means for reducing the oxidation component of wear is the use of a barrier layer that prevents diffusion of gases; since the oxidation wear amounts to up to 95% of the total one, the use of coatings upgrades substantially the serviceability of CCC.

In view of the specifics of operation of friction materials, protective coatings are applied to nonfriction surfaces; this is most justified for two-dimensional structure composites having channel pores which are mostly parallel to the friction surface, and to a lesser extent, for isotropic materials having a developed porous structure also in the direction perpendicular to the friction surface. The use of coating for the Termar laminated composite drastically changed the character of variation of the linear wear with the number of braking cycles (Figure 3.71). Quite a number of antioxidation coatings for friction CCC have been developed and are being used; they can be divided into the following classes.

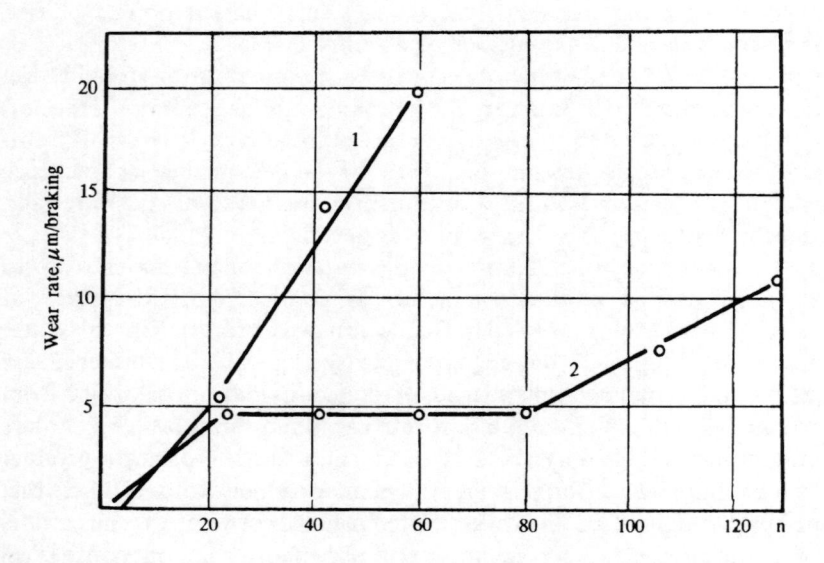

Figure 3.71 Typical dependences of wear rate on number of brakings (on inertial test-bench) for specimens of carbon friction material Termar-STD: (1) without coating; (2) with coating.

1. Coatings based on an organic binder and a dispersed filler-antioxidant (boron and phosphorus compounds). Their operating temperature is determined by the thermal stability of the binder and is generally not over 500 °C.
2. Metallic and cermet coatings based on silicon and/or silicon carbide, coated with layers of nickel, chromium, etc. Their operating temperature is limited by the initial temperature of an active oxidation of the metals (700–750 °C).
3. Coatings based on inorganic binders (solutions of polyphosphates, silicates, etc.), containing a filler-antioxidant. For heat-resistant binders (chromophosphates, alumophosphates, borophosphates) the operating temperature can be of 900–1000 °C.
4. Composite coatings based on inorganic glasses and crystalline heat-resistant fillers. The operating temperature is determined primarily by the selection of glass and can be as high as 1300–1400 °C, particularly with the use of intermediate ceramic layers applied to a carbon substrate.

3.3.3 Three- and four-dimensional composites

The set of properties of multidirectional CCC is determined by a large number of factors, usual for the formation of properties of any composite: the fibre type, the matrix type, and manufacturing technology features. An important role in CCC is played by the structure of the matrix part in regions adjoining carbon fibres. It was shown [96, 97] that CCC, even those produced with the use of only one binder type, contain two modifications of the matrix: that carbonized under the influence of the fibre surface and that carbonized in a free state. The structure of the latter matrix type corresponds to the coke structure, described in section 3.1.3.1. Such a matrix, produced under pressures on the order of 60–80 MPa, is characterized by isotropy of the disposition of crystallites. A matrix with crystallites oriented parallel to the reinforcing element of the composite is formed in carbonization on fibres. The preferable orientation of the matrix makes it possible to increase the elastic modulus, thermal conductivity, tensile and compressive strength. The impact of the matrix orientation on the bending strength is less clear, although it was noted in [100] that with a fully oriented matrix (unidirectional composite) the bending strength increases by 10%. The orientation of matrix crystallites along the fibre surface can be reduced by using fibres with a thin isotropic pyrocarbon coating [97], which prevents the fibre surface from being wetted with the pitch melt. Varying the conditions of the process of pyrolytic deposition of carbon on fibres makes it possible to attain formation of different structures of it and hence also of the matrix in the interfilamentary region of the composite: from anisotropic to isotropic,

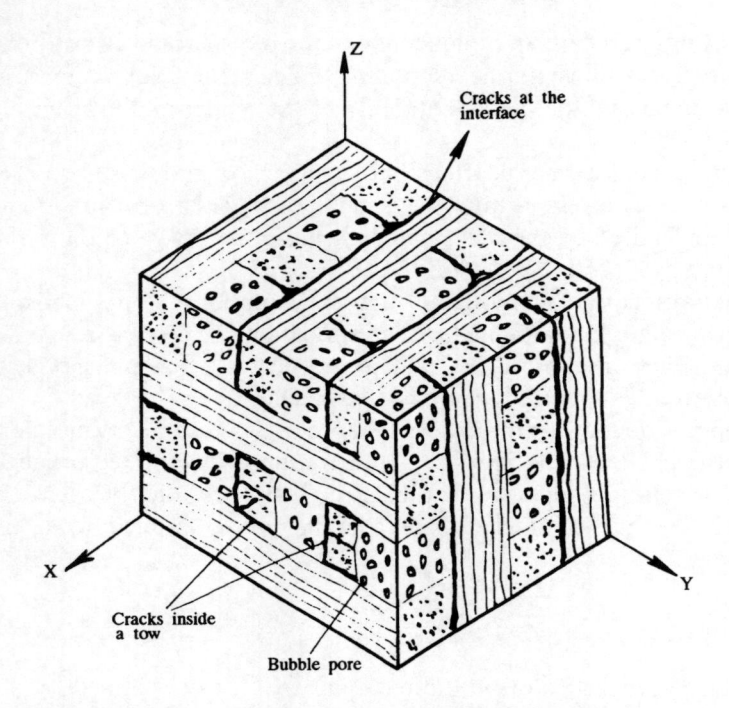

Figure 3.72 Pores and cracks in three-dimensional CCC.

similar to the matrix formed between reinforcing filler elements (a space in a three-dimensional structure, Figure 3.72). An increase in the isotropy of the whole matrix leads to a decrease of the elastic modulus by 16–41% [99] (depending on the initial binder type); the tensile strength changes much less, by 6–7%, but such properties as the resistance to thermal stresses and impact strength of the composite are improved.

3.3.3.1 Density and porosity

The development of multidirectional high-density CCC predetermined considering the density as one of the most important characteristics of such materials. According to the rule of additivity, the density of a composite consists of the density of the matrix and the density of fibres with account of their volume ratio. The porosity, inherent in CCC, however, necessitates considering two density types: true density and apparent, or volume, one. The densities and porosities of high-density three- and four-dimensional CCC are presented in Table 3.23. An open porosity predominates in materials with a four-dimensional skeleton, which is probably inherited from its initial structure. The porosity of a material, however, is also substantially affected by the temperature of treatment in its manufacture (Table 3.24).

Table 3.23 Density and porosity of three- and four-dimensional CCC

Characteristic	Material		
	Three-dimensional CCC based on PAN fibres	Three-dimensional CCC based on pitch fibres	Four-dimensional CCC based on PAN fibres
Apparent density (g/cm^3)	1.96	1.99	1.94
Density by mercury (g/cm^3)	2.016	2.042	2.01
Density by helium (g/cm^3)	2.125	2.171	–
Open porosity (cm^3/g)	0.025	0.029	0.032
Open porosity (%)	5.1	5.9	6.2
Closed porosity (cm^3/g)	0.029	0.019	0.080
Total porosity (cm^3/g)	0.054	0.048	0.040
Total porosity (%)	11.0	10.0	7.7

Table 3.24 Effect of treatment temperature on porosity of four-dimensional CCC

T_{treat} (°C)	d_a (g/cm^3)	P_{wat} (%)	P_{mer} (%)	V (cm^3/g)	P_{tr} (%)
2100	1.79	3.0	2.8	0.015	6.2
2400	1.93	3.6	6.2	0.032	55.5
2700	1.98	4.0	7.2	0.036	48.6

P_{wat} – porosity by water uptake
P_{mer} – porosity by mercury
P_{tr} – transport porosity

According to their origin, the basic types of porosity in multidirectional CCC can be divided into three groups: bubble pores, formed in the process of matrix carbonization; shrinkage cracks; and cracks resulting from thermal stresses (Figure 3.72). Shrinkage cracks and bubble pores weaken the matrix phase, while cracks resulting from thermal stresses play a more complex role in the thermomechanical behaviour of a composite in service, since they form at the fibre–matrix interface and can open and close at heating and cooling.

Interfaces between a bundle and the matrix and between two bundles are often not flat, but distinctly bent due to a wave-like expansion of bundles into the bulk of the matrix. The formation of waves is ascribed to the creep or plasticity, brought about by compressive thermal stresses at the thermal treatment [100]. Tensile stresses at cooling in combination with cracking within fibre bundles can also contribute to formation of wave-like bends. The bends produce a mechanical bond at the interfaces where cracks exist and prevent rupture of the composite in the event of a complete break of the bond at the interfaces.

3.3.3.2 Mechanical properties

The traditional examination of the mechanical behaviour of CCC on specimens cannot yield true values of characteristics, realized in a structure.

Figure 3.73 shows typical stress–strain curves for compression of a material with an orthogonal three-dimensional structure of the skeleton for various patterns of specimen cutting-out with respect to the x-axis. Specimens cut out parallel to the reinforcement axis (X) exhibit a brittle fracture. The ultimate strain in this direction coincides approximately with the ultimate strain of unidirectional composites. For specimens cut out at an angle to the fibre axis the character of curves changes. From an angle of 15° to an angle of 30° there occurs a change in the mechanical behaviour of the material. No brittle fracture occurs. After the proof stress has been reached, stresses in the material remain practically constant while the strain increases.

Data indicating that the most probable mechanism of fracture of a three-dimensional CCM in compression is by microbending were presented in [101]. At the initial period, when the compressive load is relatively low, the fibre bundles and matrix are deformed elastically, without fracture; linear segments of stress–strain curves correspond to this stage. Here the main elements experiencing the load are longitudinal

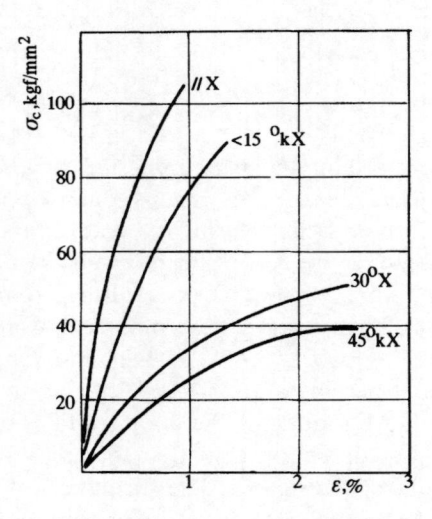

Figure 3.73 Variation of stress–strain curve in compression with specimen cut-out angle for three-dimensional structure material.

fibre bundles. As the load increases, the matrix, especially the part which contains a large number of pores and cracks, is deformed with damage. Fibre bundles near the deformed zone get bent. The material behaviour changes from a linear elastic to an elastoplastic one. As a result of bending of bundles, their ability to transmit the load declines, but they none the less continue to withstand compressive loads.

As the stresses increase further, the matrix gets crushed and disintegrates, longitudinal bundles undergo an ever greater bending, while transverse ones bend and slide, coming closer to one another. The deformation increases and a 'quasi-plastic' stage of the process sets in, at whose end the maximum load is reached and the material fractures.

Typical stress–strain diagrams for tension and compression of specimens of a material with a four-dimensional reinforcement structure are shown in Figure 3.74 for extreme cases of cutting-out of specimens: along the diagonal of a cube, which corresponds to the fibre-laying direction, and along an edge of the cube; in the latter case, fibres have been cut symmetrically in the four directions.

Analysis of the curves of Figure 3.74 shows that at tension the material behaviour under load is practically elastic and linear and the fracture is brittle. At compression the character of material deformation is nonlinear, 'quasi-plastic'. The material is anisotropic. Thus, depending on

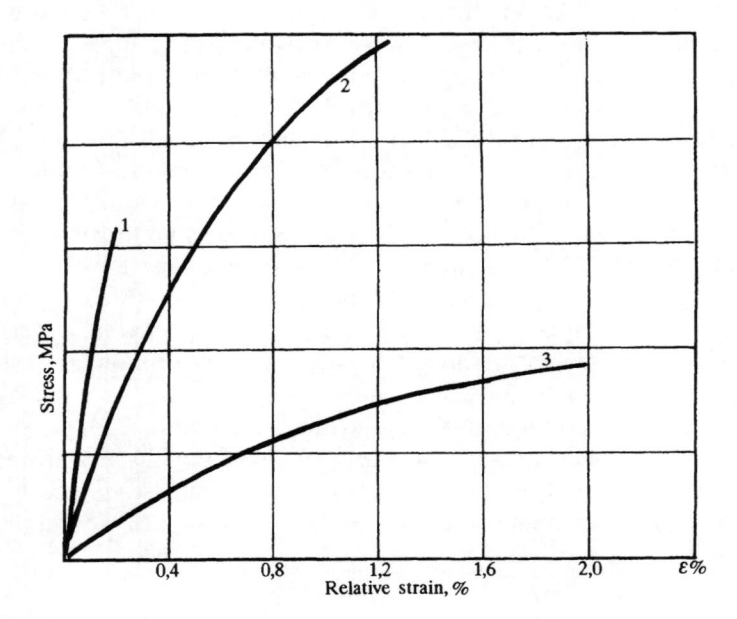

Figure 3.74 Typical stress–strain curves for three-dimensional CCC: (1) tension in x (y) direction; (2) compression in x (y) direction; (3) compression at angle of 45° to x (y) direction.

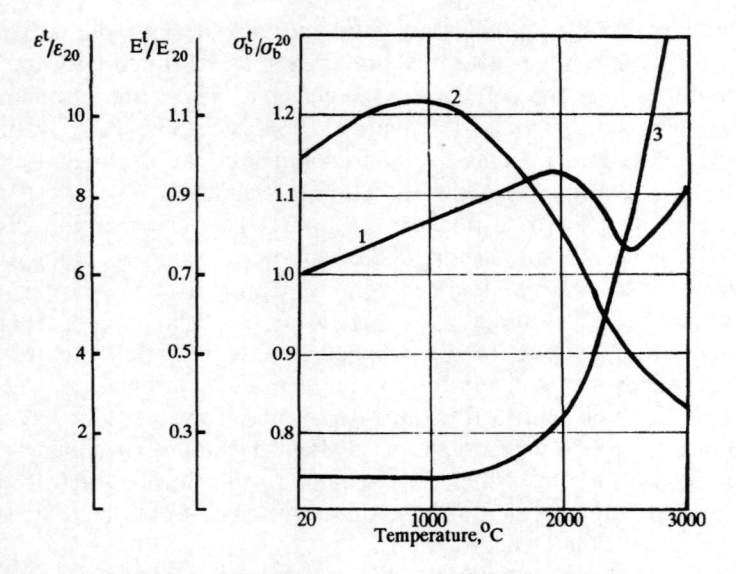

Figure 3.75 Relative variations of (1) ultimate compressive strength, (2) elastic modulus, (3) and ultimate strain of three-dimensional-structure materials with test temperature.

the cutting-out direction, the elastic modulus changes by an order of magnitude, the tensile strength decreases twofold at changeover from a direction along the fibre axis to a complete disorientation of fibres, and the compressive strength decreases nearly threefold.

CCC exhibit a feature typical for all carbon materials: increase in the strength with increasing test temperature. Typical curves of variation of the compressive strength, elastic modulus and strain with the test temperature are shown in Figure 3.75. The strength rises up to a temperature of 2000–2200 °C and then drops to values somewhat lower than the initial level (at room temperature), and rises again by 3000 °C. Such a behaviour of strength in carbon materials is usually ascribed to healing of minor cracks at the thermal expansion of fragments of the material under the action of temperature. The elastic modulus also increases to maximum values at 1200–1500 °C and then decreases steadily. At temperatures over 2000 °C; its values become less than the initial one (at room temperature). The ultimate fracture strain increases in accordance with the decrease of the module, exceeding 8–10 times the initial value.

3.3.3.3 Cylindrical three-dimensional composites

An extensive application of three-dimensional CCC [102] in cylindrical structures necessitated studying the behaviour of cylinders reinforced in three directions – radial, circular and axial – at heating. Such composites

Structure and properties of composites

Table 3.25 Tensile stresses in three-dimensional CCC at thermal treatment of up to 2000 °C

Volume fraction of fibre bundles	Axial stress (MPa)
0.05	1600
0.10	1000
0.15	700
0.20	550
0.40	270

offer advantages over two-dimensionally reinforced cylinders because radial fibre bundles improve the properties of a composite in this direction, reducing the risk of cracking of a structure in the course of thermal treatment and imparting resistance to lamination. The advantages are attained at the cost of a deterioration of properties in the circular and axial directions because of radial bundles. The volume fraction of radial bundles generally amounts to 10–15%. If it is too low, then a risk of their rupture in the course of thermal treatment exists, which is evidenced by high axial stresses arising in this case (Table 3.25). The presence of radial fibres reduces peak circular stresses in thick-

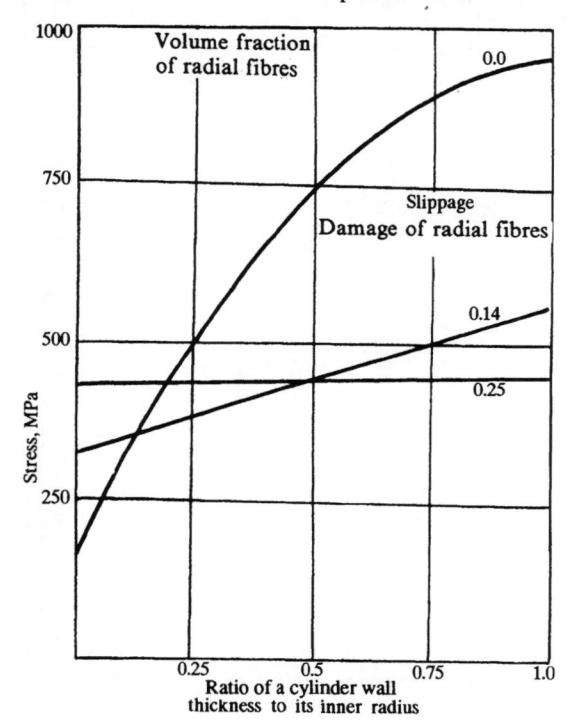

Figure 3.76 Maximum stress in circular bundle at 2200 °C.

406   *Carbon-based composites*

walled cylinders, owing to which radial and circular thermal expansions of the composite are nearly identical; in thin-walled cylinders the radial effects promote increase of stresses in circular fibre bundles through reduction of the volume fraction of the latter (Figure 3.76).

Between the inside and the outside surface of a cylinder, there exists a zone of an inertial shear along the diameter, where radial bundles are not absolutely effective. If the shearing force is high enough, then a radial bundle will lose the connectedness at a distance determined by the friction force and the strength at the interface (Figure 3.77).

In thin-walled cylinders the wall thickness can be the same as or less than the inertial shear distance. The sliding of radial bundles, caused by shearing forces, can result in their fully tearing off the remaining part of the composite. In a thick-walled cylinder, where connectedness of radial bundles is retained, a system of cracks around them is observed after cooling.

Thus, the presence of radial bundles in thick-walled cylinders of three-dimensional CCC reduces the risk of a catastrophic fracture of peripheral bundles, which can be caused by appearance of stresses in the course of thermal treatment. The sliding and damage of radial fibre bundles do not prevent a successful use of cylindrical three-dimensional CCC in high-temperature structures.

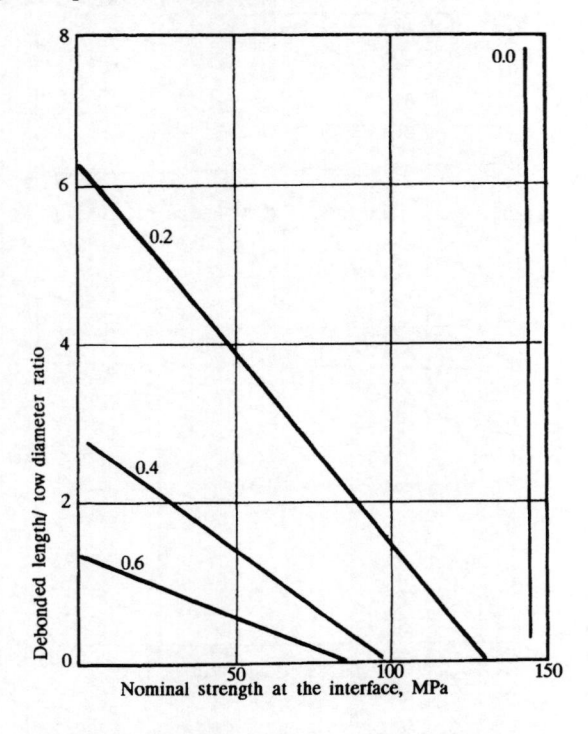

Figure 3.77 Length of broken bond at 2200 °C.

3.3.3.4 Thermophysical properties

The linear thermal expansion coefficient, as mechanical properties, depends on the pattern of cutting-out of the specimen under study. The LTEC increases with temperature both for three- and four-dimensional structure CCC (Figures 3.78 and 3.79). A negative LTEC is observed for a four-dimensional structure in reinforcement directions at low test temperatures. This is typical of carbon fibres, i.e., the behaviour at heating is in this case determined by the filler. In directions not coinciding with the fibre axis, the LTEC values are higher: the influence of the matrix shows up, since all fibres are disposed at an angle to the specimen axis. For three-dimensional specimens the difference is greater, since such a material is more anisotropic than a four-dimensional CCC [103].

The temperature dependence of the thermal conductivity, shown in Figure 3.80, indicates its decline at the initial segment of the curve, whereas at temperatures over 1500 °C, it remains practically unchanged up to 3000 °C.

The combination of thermophysical and mechanical properties of CCC is very important, as in many cases their application involves high heating rates and high temperatures. The following relation is used as a comparative characteristic of the thermal shock resistance:

$$K = \frac{\lambda \cdot \sigma}{\alpha \cdot E},$$

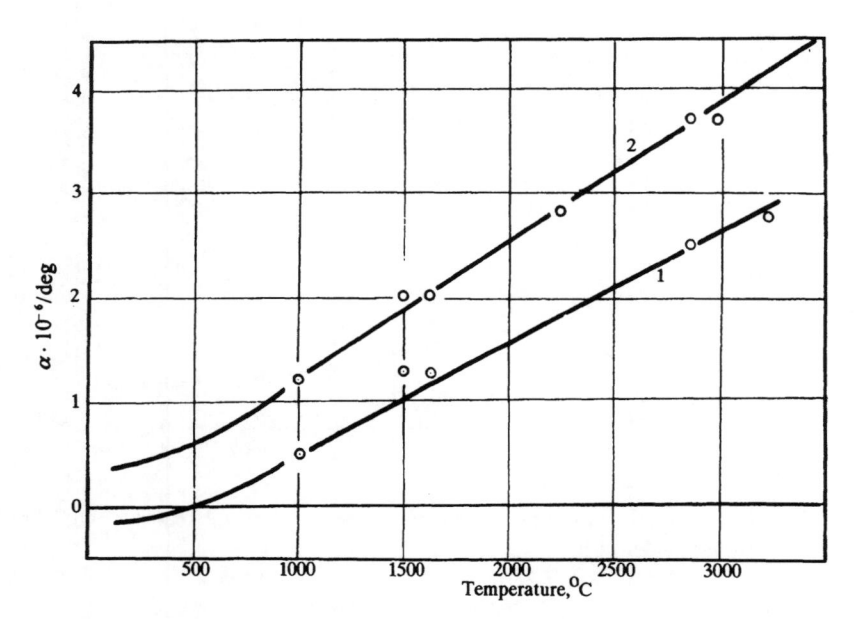

Figure 3.78 Linear thermal expansion coefficient in (1) direction of reinforcement and (2) at 45° to reinforcement axis.

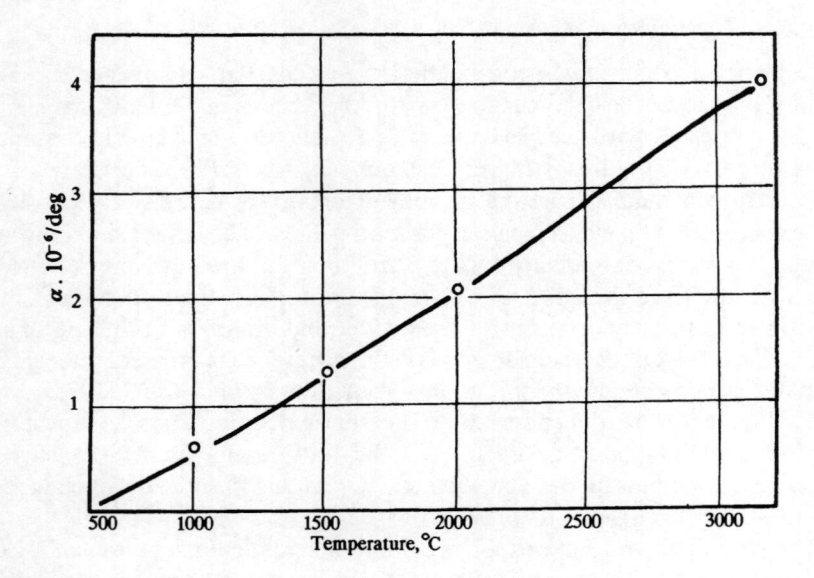

Figure 3.79 Linear thermal expansion coefficient of 4D-CCC parallel to reinforcement axis.

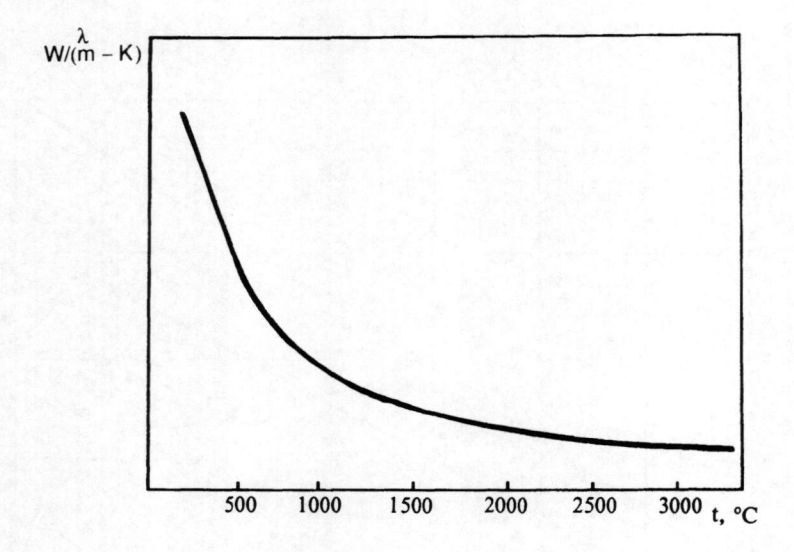

Figure 3.80 Typical temperature dependence of thermal conductivity for multidirectional-skeleton materials.

where λ is the thermal conductivity; σ the ultimate strength; α the LTEC; and E the elastic modulus.

3.3.3.5 Reactivity

A linear dependence of erosive ablation of carbon materials on the density is known [104], and therefore the aspiration to obtain such materials of the maximum density is justified. The behaviour of a composite at its use in parts of a jet engine, however, depends on the reactivity. The CCC operating in the critical and supercritical zones of rocket nozzles undergo the action of severe conditions arising in fuel combustion. The combustion creates an oxidizing atmosphere, and a nozzle of a composite is subjected to gasification: conversion of a solid or liquid fuel into combustible gases through an incomplete oxidation at a high temperature. One possible method for modelling the behaviour of a material in an engine is studying its reactivity at gasification by vapour, since vapour is the most important component of the gas flow, participating in gasification of a CCC nozzle in a rocket engine. Results of studying the reactivity of a three-dimensional CCC with polyacrylonitrile (PAN)- and pitch-based fibres [105] are shown in Figure 3.81. The difference in behaviours of the composites appears to be associated with the filler. PAN-based fibres have a bean-shaped cross-section of a filament, whose average size is of 5–6 μm, whereas pitch fibres are of a round cross-section, with a smooth outside surface, the filament diameter being of 8–10 μm. The difference in the structure of the matrix parts adjoining the fibres is in accord with the difference in the fibres. Thus, the matrix in the pitch fibre-based composite, as shown by studies [97],

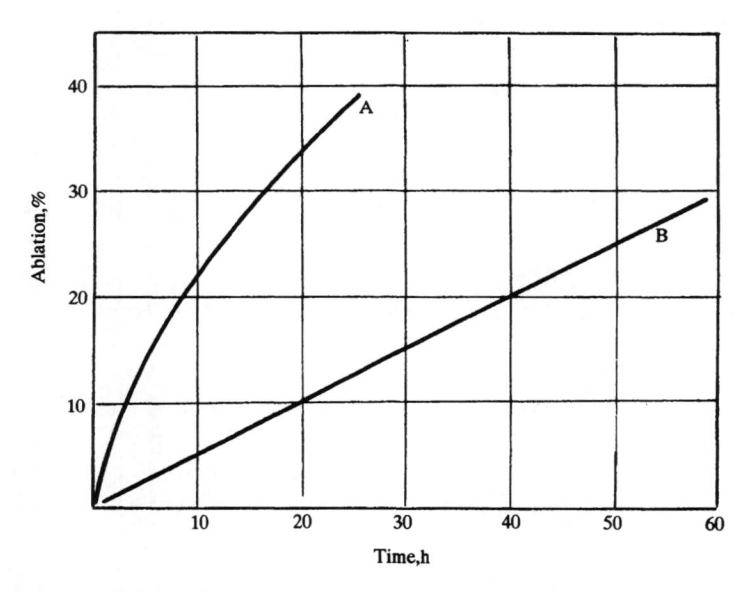

Figure 3.81 Characteristics of reactivity of three-dimensional CCC: (A) composite with PAN-based fibres; (B) composite with pitch-based fibres.

has a structure where crystallites are disposed parallel to the fibre axis, whereas no such a distinct orientation of crystallites was observed in the three-dimensional CCC with PAN fibres. Thus, the morphology of carbon fibres exerts a strong influence on the structure of the adjoining matrix, which inherits the fibre structure, and therefore the reactivities of composites based on different fibres differ three-fold. The specific surface of the PAN-based CCC, measured by the oxygen chemisorption method, turned out to be of $0.63 \text{ m}^2/\text{g}$, and of the pitch-based one, of $0.14 \text{ m}^2/\text{g}$. The active specific surface is primarily concentrated on the fibre end-faces disposed on the surface of the composite. Under conditions of engine operation the fibre ends undergo erosion much more rapidly than the remaining part of the material. Thus, edge parts of fibres as well as matrix regions accessible for gases are, first of all, responsible for the erosive ablation of the CCC. Most important in the matrix at ablation are the amount and sizes of pores.

3.3.3.6 Application

Composites of a three-dimensional structure were developed more than 15 years ago for manufacture of nose cones and thermal protection of rockets. Due to this, principal requirements in their development included such properties as resistance to thermal and mechanical shocks and to ablation, high density and minimal change in the product geometry in the course of operation.

In the past years, the three-dimensional structure CCC showed themselves to advantage as rocket and space engineering products, such as nose parts of rockets, rocket engine elements (nozzle unit with the inlet part and throat, exhaust flares and adapters from the housing to the nozzle), and fasteners for parts attached to large reservoirs and cowls of launch rocket vehicles. There occurs, however, nearly no extension of the scope of application, since the materials continue to be costly despite the development of automated processes of manufacture of reinforcing skeletons and improvement of the equipment for impregnation and high-temperature treatments. Typical products from three- and four-dimensional CCC are shown in Figure 3.82.

Nozzles, most important elements of rocket engines, have over the last 15 years undergone considerable changes in the dimensions, design, shape and materials. The nozzle is progressively shifted within the engine, characteristics of materials used are improved, and the shape changes depending on the engine control principle. There occurs a progressive transition from pivoted nozzles through fixed nozzles controlled by gas injection to flexible-stop nozzles. Jointly with the changes in the shape and design of engines, there occurs an unavoidable evolution of the materials used towards a higher erosion resistance, lower weight

Figure 3.82 Typical products from three- and four-dimensional CCC.

and cost. The development of materials for nozzles has resulted in structures from multidirectional CCC, integrating the inlet part, the throat, and the exhaust part of a nozzle in a single element (Figure 3.83).

Basic properties of a number of three- and four-dimensional CCC, used to manufacture nozzle parts, nose parts, and other parts of rocket bodies, are presented in Tables 3.26 and 3.27.

Le Carbone-Lorraine manufactures an Aerolor series of CCC. The Aerolor 30-group materials are based on skeletons of a three-dimensional structure with orthogonally disposed fibres. The matrices are produced by combination of methods of densification by pyrocarbon with impregnation by resin or pitch.

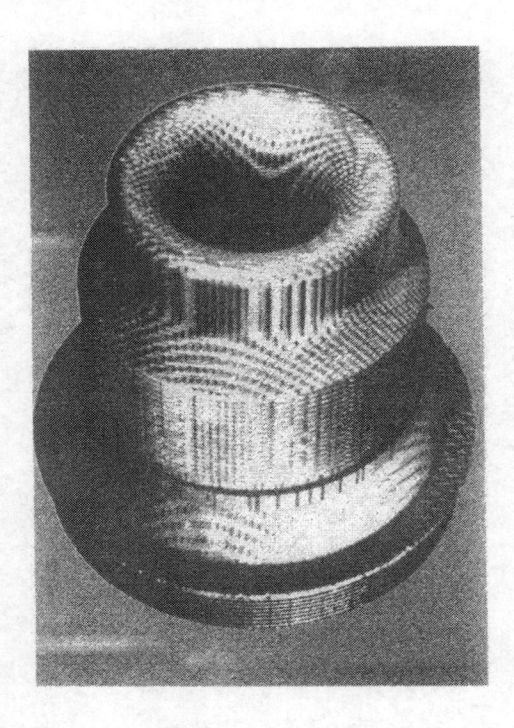

Figure 3.83 Nozzle from multidirectional CCC.

Table 3.26 Properties of three-dimensional CCC

Characteristic	Direction	Material		
		Aerolor 32	Aerolor 33	3KMP
Apparent density (g/cm^3)		1.90–1.95	1.80–1.90	1.90–1.92
Tensile strength (MPa)	$x(y)$	170	60–100	100–120
	z	170	60–100	–
Elastic modulus in	$x(y)$	100	40	40–55
tension (GPa)	z	110	–	–
Compressive strength	$x(y)$	140	100	130–150
(MPa)	z	120	100	–
Elastic modulus in compression (GPa)		80	40	30–35
Linear thermal expansion coefficient ($10^{-6}\,°C^{-1}$) at:				
1000 °C	$x(y)$	–	–	0.6
1500 °C	$x(y)$	–	–	1.3
2000 °C	$x(y)$	–	–	2.1
2500 °C	$x(y)$	1.2	2.3	3.0
1000 °C	z	–	–	0.6

Characteristic	Direction	Aerolor 32	Aerolor 33	3KMP
1500 °C	z	–	–	1.3
2000 °C	z	–	–	2.1
2500 °C	z	–	2.0	3.0
Thermal conductivity	x(y)	120	260	50–70
W/(m K)	z	180	140	50–70

Table 3.27 Properties of 4D-CCC

Characteristic	Sepcarb-500			4KMS			4KMS-L	
Density (g/cm³)	2.0			1.90–1.95			1.88–1.92	
Tensile strength (MPa)	F	100		F	120		Z	80
	Z	–			–			–
	X	30		X	45		X	50
Modulus in tension (GPa)	F	62		F	50		Z	45
	X	18						
Compressive strength (GPa)	F	80		F	140		Z	110
	Z	50		Z	80			
	X	50		X	40		X	80
Linear thermal expansion coefficient ($10^{-6}\,°C^{-1}$) at:	F	Z	X	F	Z	Z		X
1000 °C	0.6	0.5	0.5	0.6	0.5	0.6		0.6
1500 °C	1.0	1.8	1.4	0.7	1.3	0.9		1.4
2200 °C	1.4	4.0	2.0	2.6	1.7	1.1		1.9
2500 °C				3.6	2.5	1.4		2.4
2800 °C				3.8	2.7	1.8		2.95
Thermal conductivity, (W/(m·K)) at:	X			F	Z		Z	X
100 °C	180			55	20		53	46	20
500 °C	85			44	200		30	28	1000
				37	600		26	24	1500
				34	1000		24	22	2000
800 °C	70			31	2000		22	20	2500
				30	3000		22	20	2800

The material 3KMP is in the skeleton structure close to the Aerolor 30 CCC; its matrix is a coal-tar pitch coke. Skeletons of materials Sepcarb 500, manufactured by SEP, and 4KMS are produced in a similar manner: by laying rods in directions of the four spatial diagonals of a cube. Rods for the material 4KMS are based on a thermoplastic resin, polyvinyl alcohol; rods for Sepcarb 500, on thermosetting phenolic resins. Matrices of the two composites are made with the use of impregnation and carbonization under pressure. The material 4KMS-L differs in the structure of the skeleton: one of reinforcement directions coincides with the z-axis of a part or preform, whereas the three other ones, disposed perpendicular to the z-axis, are turned 120° with respect to one another.

3.4 CARBON- AND CARBIDE-MATRIX COMPOSITES

A low heat resistance of carbon–carbon materials imposes constraints on their use in a considerable number of constructions which operate for a long time in the air or other oxidizing atmospheres at temperatures up to 2500 °C.

The development and improvement of composites serviceable at superhigh temperatures in an oxidizing environment is one of the basic problems of modern materials science.

New high-temperature, heat-resistant materials will provide for development of integral constructions, offering a high degree of the weight perfection, for hypersonic aircraft, reusable spacecraft and aerospace planes with a level of performance characteristics unattainable for traditional materials.

Materials of this class will make it possible to solve the problem of development of more economical high-temperature aircraft and automobile gas-turbine engines and urgent materials-science problems having arisen in development of laser designs, especially mirrors.

Composites of such a type are undoubtedly needed for improvements in the chemical industry, metallurgy and electrical engineering. The use of new composites in these industries will upgrade the reliability and cost-effectiveness of production processes, their ecological cleanliness and the quality of products.

The information on materials of the carbon fibres-refractory compound type is rather limited. Classed with refractory compounds are carbides, borides, nitrides, oxides and silicides, whose melting point exceeds 2500 °C; about eighty such compounds are known.

The development of materials of this class should take into consideration the physicomechanical compatibility of their constituents; the adhesion interaction at the phase interface; and the chemical and diffusion interaction, which can be very active both in the process of manufacture of a composite and in the course of its high-temperature service.

The first two factors govern the level of physicomechanical properties of a composite, while the last one determine the maximum possible operating temperature of the material. A high-temperature interaction between carbon fibres and refractory compounds can give rise to eutectics, whose melting points decrease on the average from 3000 °C to 1000 °C in a carbide → boride → nitride → oxide series.

In view of the above, as well as of a higher refractoriness of carbides as compared with other compounds, their use for developing superhightemperature composites is more preferable.

When, however, the operating temperature is not over 2000 °C, borides and nitrides may be used, and at temperatures up to 1000 °C, oxides can be employed.

The latter compounds are the best candidates for obtaining especially heat-resistant materials, but research and development along this line is limited.

The greatest difficulty in development of a carbon fibre-based heat-resistant composite is the provision for its oxidation protection. The principal deteriorating factor, impairing the durability of carbon–carbon materials, is their oxidation, accompanied by development of a porous structure of the carbon matrix as well as by a burning-up of reinforcing carbon fibres, which eventually reduces appreciably the material strength. The major contributor to the mass rate of ablation of a carbon–carbon material oxidized by atmospheric oxygen at gaseous atmosphere pressures of 0.01 to 1.0 atm is the intraporous carbon–oxygen interaction.

Two versions of the oxidation protection of carbon materials are possible. The first one consists in deposition of a protective film of heat-resistant refractory carbides, e.g., of silicon carbide, on the surface of a carbon–carbon composite. Such films retard the oxygen diffusion to carbon and thereby extend the composite life.

At elevated temperatures, however, the front of a high-rate carbon oxidation shifts to the 'coating–substrate' interface, which results in peeling of the coating. The peeling of coatings occurs also due to considerable thermal stresses arising at the 'coating–substrate' interface because of the difference in elastic and thermal properties of the substrate and coating materials. Besides, protective coatings are in the course of service easily damaged by impacts of stones, sand, hail, etc.

The second version is to provide a volumetric oxidation protection of reinforcing carbon fibres by distributing them throughout a heat-resistant, e.g., carbide, matrix, or, in other words, utilizing the high heat-resistance of carbide matrices, to increase their strength, fracture toughness, and thermal shock resistance through reinforcement by carbon fibres. High-strength carbon fibres are used for this reinforcement produced from cellulose hydrate (viscose) and also high-strength high-modulus polyacrylonitrile (PAN) fibres.

In view of a low reliability of protective coatings, the latter version is more preferable.

3.4.1 Requirements for structural constituents of three-dimensionally reinforced high-temperature heat-resistant composites

The development of strong and light composites for operation at super-high temperatures in oxidizing atmospheres calls for the fulfilment of the following requirements.

1. Reinforcing fibrous fillers should have the maximally high strength and retain it up to a temperature of at least 2100 °C. The variation in the linear dimensions of fibres with increasing temperature should be minimal.
   Coefficients of variation of strength and elastic properties of reinforcing fibres should be minimal. Continuous reinforcing fillers in a composite should get deformed and fractured at a simultaneity factor close to unity.
2. The matrix should have a high heat resistance and reliably protect the reinforcing filler from oxidation, retain its strength characteristics, first of all the rigidity, to a temperature not lower than that to which this is done by reinforcing fibres.
3. The matrix properties and the character of interaction at the filler–matrix interface should provide the maximum degree of utilization of the reinforcing filler strength.
   Ultimately, the connection between reinforcing fillers should be arranged so that rupture of one of them is not transmitted to others, i.e., a fracture point that has arisen should be localized within the section of a single structural component (fibre).
4. The coefficient of the volumetric filling with the reinforcing component should be optimal.

Any departures from the above requirements result in reducing the coefficient of utilization of strength of the reinforcing component and the strength of composites.

Thus, the mechanical behaviour of a composite is determined by three basic parameters: strength of reinforcing fibres; rigidity and strength of the matrix; and character of bond at the fibre–matrix interface.

Along with this, properties of composites are substantially dependent on the arrangement of the reinforcing component in the volume of a composite, i.e., on the reinforcement pattern.

The selection of a high-temperature and heat-resistant matrix, capable of protecting carbon fibres from oxidation, is one of the basic problems in developing materials of this class.

Strength properties of carbides are much dependent on the existence of homogeneity regions, phase transformations, crystalline structure, grain size, material purity, and also the presence of pores, cracks and other defects.

A characteristic feature of carbide materials is brittleness. An appreciable macroplasticity (departure from linearity in the stress–strain diagram) appears at temperatures of about 1200–2200 °C.

Pores, cracks and a greater grain size increase the stress concentration, restrict the plasticity and increase the temperature of the brittle–ductile fracture transition.

The process of fracture of carbide materials includes the formation of fracture nuclei and extension of cracks, both having existed and formed at the loading. As the external load is increased to a certain limit, the process of fracture proceeds slowly. There occurs an equilibrium propagation of cracks, whose length at tensile and bending loads is, in carbide materials, much smaller than in metals.

At a brittle fracture, opening-mode cracks are always observed, while sliding- and tearing-mode cracks are not formed.

The features of crack propagation in brittle materials promote the fact that the changeover from a uniaxial tension to bending and compression is accompanied by increase in the strength in the ratio 1:(2–2.5):(10–15). The difference in the strength is retained up to brittle–ductile transition temperatures.

As the temperature is increased to 1500–1600 °C, the strength of carbides first falls and then, as for carbon materials, rises, reaching the maximum value within a temperature range of 1700–2500 °C. Such a strength behaviour is associated with redistribution of stresses and decline in the role of stress concentrators.

Most important in developing a reinforced composite is the knowledge of the level and character of variation of elastic properties of its constituents. The modulus of normal elasticity of carbide materials is within $(39–55) \times 10^4$ MPa.

At temperatures of 1800–1900 °C, the modulus decreases steeply due to relaxation processes, viscous flow at grain boundaries, and transition to a ductile fracture. The hardness of carbides varies in a similar manner; at temperatures over 1000 °C, it becomes commensurable with that of pure metals. Exceptions are SiC and B_4C, which at 1500 °C retain a strength over 1500 MPa.

Comparing the strength and elastic properties of fibrous carbon materials and carbides shows a substantial difference between their levels.

A volumetric combination of carbon fibres and carbides results in materials whose matrix, in contrast to classical composites, exhibits no high enough plastic properties.

Composites based on carbon fibres combined with silicon, titanium and zirconium carbides have been developed and studied in sufficient detail by the present time. Carbon fibre–silicon carbide materials are most widely known.

Interaction with oxygen at elevated temperatures results in formation of a protective oxide film on the silicon carbide surface. Strength characteristics of silicon carbide are retained up to a temperature of at least 1500 °C. The thermal expansion coefficient of silicon carbide is the closest to that of carbon fibre.

A covalent atomic bond provides the minimal diffusion interaction in the silicon carbide–carbon system, which ensures stability of the composition of a composite in the course of its long-time service at high temperatures.

3.4.2 Manufacturing technology for carbon fibre–refractory carbide composites

Several methods for manufacturing composites of this class are known, such as hot pressing of a carbon fibre–carbide blend or extrusion of such a blend, followed by sintering; plasma spraying of carbide coating on carbon fibres, followed by their hot pressing or sintering, etc.

Numerous papers [106–108], describing properties of carbon-ceramic materials, indicate that refractory compounds are introduced into the volume of a carbon–carbon material by vapour deposition. This method is attractive enough, but its features such as toxicity, necessity of a strict observance of component consumption rates; pressure and temperature of the process render it very complex as to the equipment needed, and it is labour-intensive.

In our opinion, the best method for manufacturing carbon fibre–refractory carbide composites is impregnation of the starting fibrous carbon base, having a certain porosity, with a carbide-forming metal melt.

The pore space of the initial preform is filled with the melt owing to a capillary impregnation process, whose rate depends on such factors as the wetting, spreading and viscous flow of the melt. At the same time, there occur chemical and diffusion interactions of the melt with the carbon base, which results in increase in the melt viscosity, slowing-down of the impregnation process, and ultimately to formation of a carbide phase.

The greatest impact on the impregnation process is exerted by temperature. At high temperatures, the melt spreads rapidly. The limiting stage of impregnation is the movement of liquid in pores of the body being impregnated under the action of the capillary pressure.

The kinetics of impregnation is determined primarily by the porous structure of the base (effective radius of pores, their shape and amount) and the melt viscosity.

An appreciable influence on the impregnation process is exerted by desorption of gaseous impurities from a carbon preform as well as by redistribution of metallic impurities from the bulk of the melt to the phase interface. Certain impurities markedly increase the rate of spreading and work of adhesion, improve the wetting and facilitate the impregnation process.

Accordingly, the process of manufacture of the materials consists of three basic stages: moulding of a carbon-filled plastic preform; its conversion into a carbon–carbon base suitable for impregnation; and its capillary impregnation with melt of a carbide-forming metal with an appropriate thermal holding for conversion of the metal into carbide.

Carbon- and carbide-matrix composites 419

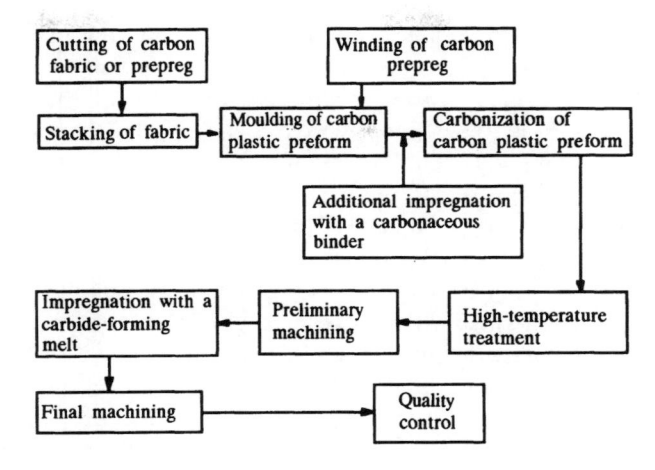

Figure 3.84 Flow-chart of manufacture of composite.

The flow-chart of manufacture of three-dimensionally reinforced carbon fibre–carbide matrix composites is shown in Figure 3.84.

Specific technological approaches are illustrated by manufacture of the nose fairing and parts of the front wall of the wing of the *Buran* space shuttle. Carbon fabrics made from viscose were used as the starting stock.

Carbon-filled plastic preforms of the nose fairing, up to 1.5 m in diameter, were moulded by the method of elastic punch made from a

Figure 3.85 Press mould for moulding of fairing.

twill-weave carbon fabric. The press mould for moulding of the fairing is shown in Figure 3.85.

Carbon-filled plastic preforms for the leading edge of the wing were moulded from a three-dimensional-weave carbon fabric by impregnation under pressure. The press mould for moulding the parts is shown in Figure 3.86. The carbon-filled plastic preforms had in all cases to meet definite requirements as to the content of the binder, degree of its hardening and porosity. The preforms underwent nondestructive testing to find out laminations, regions not filled by the binder, and other defects.

The carbonization stage involves formation of the porous structure of a carbon–carbon preform and its phase composition, i.e., the relation between carbon fibres and coke in the material structure.

The high-temperature treatment aims to stabilize the structure of carbon–carbon preform material components (fibres, coke). At the firing and high-temperature treatment stages, measures have to be taken to preserve the required aerofoil profile of preforms; this is attained by the use of special tooling and conditions of heating and cooling the preforms.

After the high-temperature thermal treatment, the preforms are machined and then checked for their mating with one another on the check-completing jig.

Next, the preforms are subjected to a volumetric liquid-phase impregnation with molten silicon. This stage involves formation of the final composition and structure of the material.

Figure 3.86 Press mould for moulding of preform of leading edge of wing.

Figure 3.87 Part of a leading edge of wing.

The above-described manufacturing process results in high-temperature heat-resistant composites where carbon fibres are the reinforcing component and silicon carbide is the matrix.

The uniformity of silicon carbide distribution throughout the volume of material of parts is checked by nondestructive testing. After the liquid-phase impregnation, the parts undergo the final fitting on a jig.

A high-temperature glass-silicide coating, protecting the parts from moisture and extending their service life, can be applied to their working surfaces when required.

One of the parts of the leading edge of the wing is shown in Figure 3.87.

3.4.3 Mechanical properties of carbon fibre–silicon carbide materials

The changes in the bending strength of a material-semiproduct after successive manufacturing steps and the interrelation between the strength of the end-material and the strength of the carbon-filled plastic are shown in Figure 3.88. The material subjected to carbonization exhibits a steep decline in the strength, which is brought about by destruction of the polymer, its transformation into coke, and formation of a considerable porosity. Next, as a result of an additional impregnation with the binder and carbonization, the strength rises somewhat, and then reaches

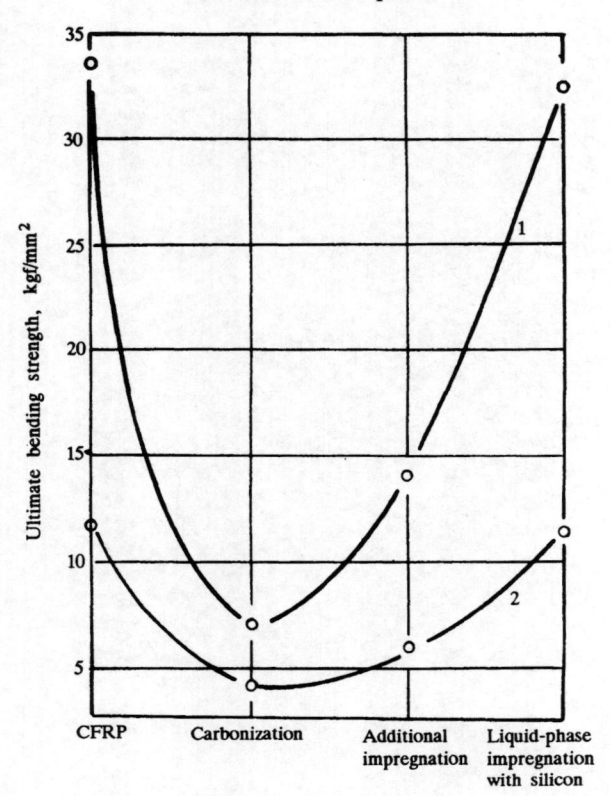

Figure 3.88 Changes in material strength at production process stages: (1) material based on carbon fibres from TNM and (2) on carbon fibres from cellulose hydrate (2).

the maximum value when pores have been filled with silicon carbide. The strength of a material reinforced with high-strength high-modulus fibres is much higher in all cases.

A further increase of the material strength occurs in silicization because of formation of a carbide matrix.

Physicomechanical properties of a composite depend not only on the ratio between carbon fibre and silicon carbide fractions, but also on the type of carbon fibres and on the structure into which they have been formed.

Table 3.28 presents basic characteristics of the following composites: Gravimol, based on carbon fabric TKK-2; Gravimol-V, based on fabric TNU-4; and Karbosil, whose reinforcing component is carbon bundle of grade VMN-4.

Karbosil was tested with specimens cut out by two different patterns: (1) 9 layers of a thickness of 2.7 mm and (2) 13 layers of a thickness of 5.4 mm.

Table 3.28 Physicomechanical characteristics of carbon fibre–silicon carbide composites

Material	Number of monolayers in specimen N	Specimen thickness (mm)	Material density (g/cm³)	SiC content, % mass	Strength characteristics (MPa)					Impact strength (kJ/m²)	Fracture toughness K_{Ic} (MPa m^{1/2})	Thermal expansion coefficient at 1000–2000 °C ($10^6\,°C^{-1}$)
					Tension	Compression	Bend	Shear	$E_t \times 10^{-3}$			
Gravimol (twill)	26	8.0	1.7–2.0	40–60	40	90	100	26	32	18	0.045	3.5
Gravimol (three-dimensional)	5	10.0	1.7–2.0	40–60	40	120	110	38	24	22	0.175	3.0
Karbosil (unindirectional)	9	2.7	1.74–1.76	12.0–14.0	220	170	300	12	70	35	0.560	1.5
Karbosil pattern (1)	9	2.7	1.84–1.98	33–34	150	200	350	13	80	30	0.450	1.6
Karbosil (unindirectional)	13	5.3	1.86	17	180	170	320	19	52	34	0.540	1.9
Karbosil pattern (2)	13	5.3	1.93	31	100	230	320	11.5	55	32	0.500	1.9
Karbosil (fabric TVMP-U, cloth)	9	3.8	1.76	21	100	160	150	9.5	70	36	–	1.6
Karbosil (fabric TVMS-U-2, sateen)	9	3.8	1.88	27	110	200	160	11	75	30	–	1.6

Strength properties of composites reinforced with bundle VMN-4 greatly exceed those of materials reinforced with carbon fabrics TKK-2 and TNU-4, which results from a higher strength of the carbon fibres making up bundle VMN-4. No appreciable influence of the reinforcement pattern, with the laying coefficient varying from 0.556 to 0.472, on strength characteristics of a material was found. At the same time, strength characteristics of materials reinforced with plain- or satin-weave fabrics from bundle VMN-4 are markedly inferior to those of a material reinforced by pattern (1).

There are two causes of the lower strength with reinforcement by fabrics from carbon bundles VMN-4.

1. Any woven structure restricts the laying of carbon fibres in a layer, which reduces the coefficient of filling with the reinforcing material, and also contains in the weft the fibres that are essentially worthless as they are laid in directions not corresponding to acting stresses.

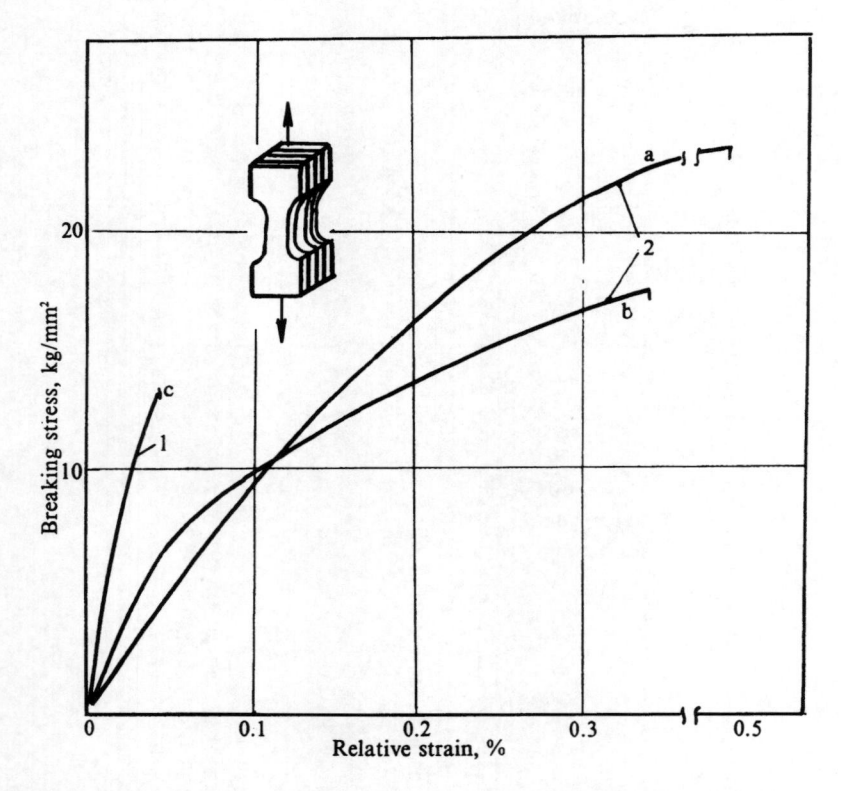

Figure 3.89 Stress–strain curves for tension of (1) Gravimol of density 1.70 g/cm^2 and (2) Karbosil of volumetric densities of (a) 1.85, (b) 1.75.

2. The weaving of high-strength high-modulus bundles results in fracture of a fraction of their constituent fibres (filaments).

Tensile stress–strain diagrams for Gravimol and Karbosil specimens are shown in Figure 3.89. Karbosil exhibits a low ultimate strain, but its strength is 4–5 times superior to that of Gravimol materials. The elastic modulus increases with the density of materials; for Karbosil it is twice as high as for Gravimol.

The coefficient of utilization of the strength of reinforcing carbon fibres for Karbosil is 20–33%. Such a low value stems not only from the traumability of the material, but also from imperfection in organization of interaction at the fibre–matrix interface as well as from meandering of fibres in their laying and existence of numerous defects in the material structure (cracks, laminations, cavities, large pores, etc.). The microstructures of Gravimol-V and Karbosil are shown in Figures 3.90 and 3.91 respectively. For Karbosil the maximum tensile strength of 250 MPa was reached.

Tests for the effect of a stress concentrator were conducted with specimens with a 6 mm hole, prepared in accordance with the above-presented lay-up patterns. The effective stress concentration factor was determined from the formula

$$K_\sigma = \frac{\sigma_b}{\sigma_{bf}},$$

where σ_b is the strength of a specimen without hole and σ_{bf} of a specimen with hole. For specimens with lay-up of the first type K_σ was 1.15, and of the second type, 1.11, i.e., the sensitivity of the material to a stress concentrator was low in both cases.

The fracture toughness was studied with specimens having a special slit (Figure 3.92), prepared by the same lay-up patterns. The slitted specimens were found to have a high enough residual tensile strength, amounting to 58–61% of the strength of an unslitted specimen.

The critical stress intensity factor, K_{Ic}, was determined from the formula

$$K_{Ic} = \sigma_b \sqrt{\pi a_l},$$

where a_l is the correction factor for the cracking zone,

$$a_l = \frac{l}{\left(\dfrac{\sigma_b}{\sigma_{res} \cdot \varphi_B}\right)^2 - 1},$$

where: l is slit length; σ_b strength of unslitted specimen; σ_{res} strength of slitted specimen; $\varphi_B = \sqrt{B/\cos \pi \cdot l}$ – correction for specimen width; and B specimen width. Results of the tests are presented in Table 3.29.

(a)

(b)

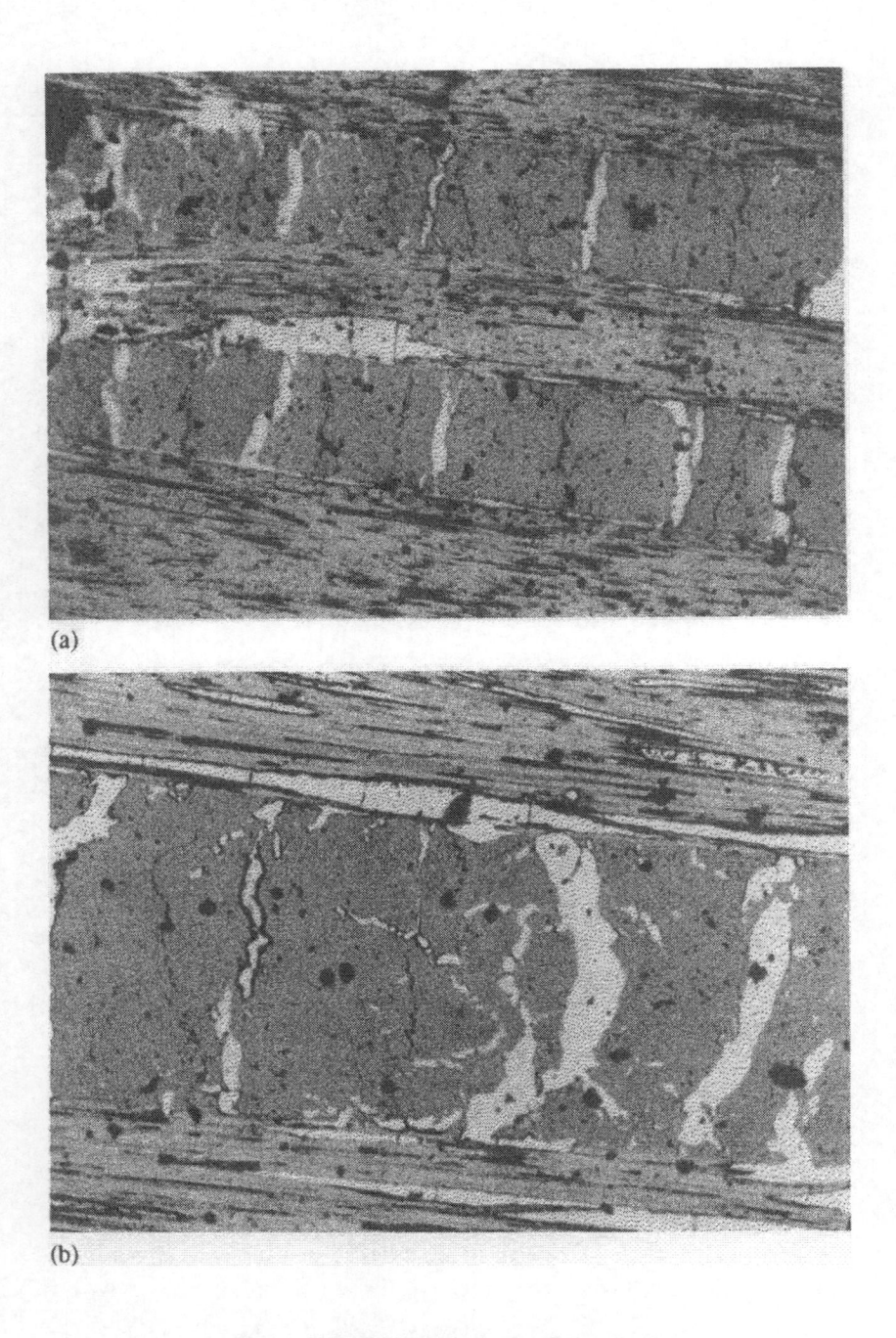

Figure 3.90 Microstructure of Gravimol-V.

(a)

(b)

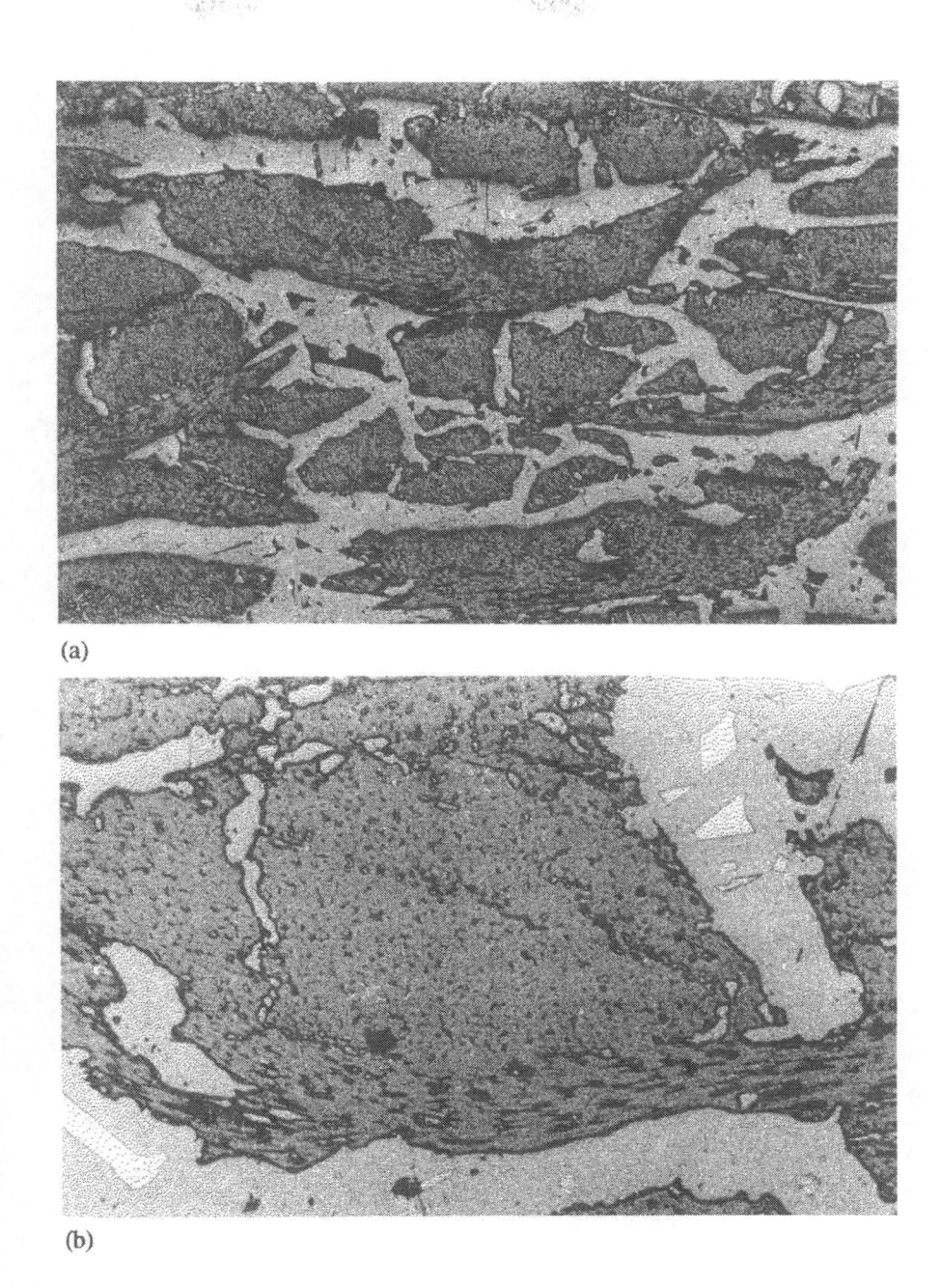

Figure 3.91 Microstructure of Karbosil: (a) × 63; (b) × 200.

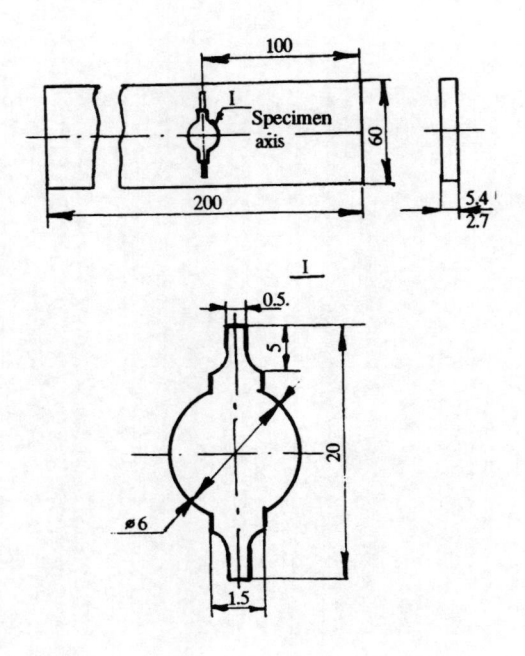

Figure 3.92 Specimen for fracture toughness studies.

Table 3.29 Strength of specimens and stress intensity factor

Lay-up pattern	Carbide phase content (%)	Density (g/cm^3)	σ_{res} (MPa)	a_1	K_{Ic} (MPa m$^{1/2}$)
I	27.5	1.85	101.0	7.61	0.81
	13.1	1.63	85.0	7.34	0.67
II	15.2	1.67	82.3	6.66	0.64
	15.2	1.68	84.6	6.90	0.65

Table 3.30 Loss of mass of specimens depending on initial density

Material density (g/cm^3)	Temperature on specimen surface (°C) centre	edge	Test time (min)	Mass loss (% mass)
1.72	1540	1630	20	2.30
2.07	1520	1570	20	0.25
2.09	1420	1500	20	0.25

The tests demonstrated that Karbosil has a high enough fracture toughness. The stress intensity factor was within $K_{Ic} = 0.64$–0.67 MPa m$^{1/2}$ and for a conventional monolayer it was of $K_{Ic} = 0.14$–0.20 MPa m$^{1/2}$.

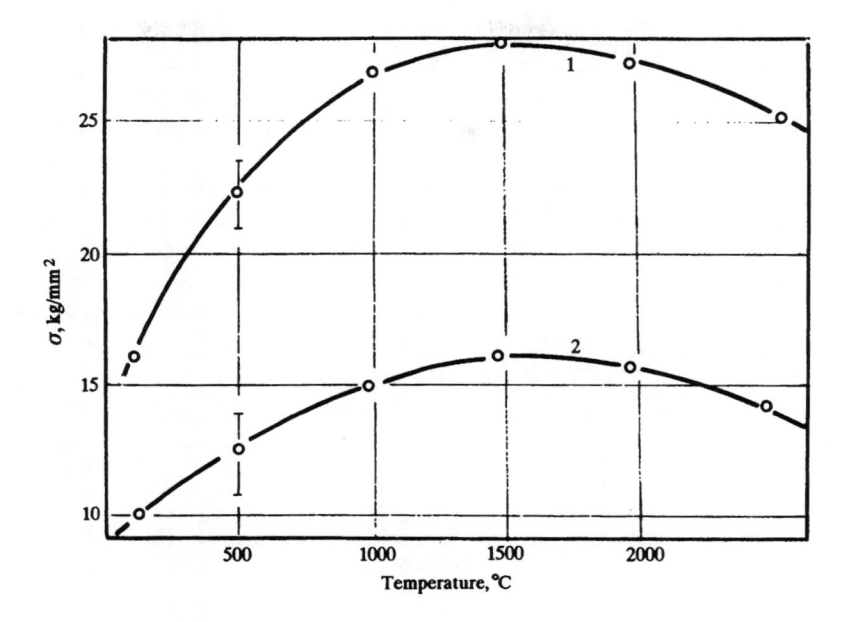

Figure 3.93 Temperature dependence of bending strength of (1) Karbosil and (2) Gravimol-V.

The bending strength was studied at temperatures up to 2500 °C. The samples were heated in an inert atmosphere. Results are presented in Figure 3.93. Over the whole studied temperature range, the level of strength of a composite based on high-strength high-modulus carbon fibres (bundle VMN-4) was 1.5–2 times as high as that of a composite based on high-strength fibres (fabric TNU-4). As the temperature rose, the strength of composites increased by 60–80% and reached the maximum value at 1500–2000 °C. At a further temperature increase to 2500 °C the strength decreased from the maximum level by 12–14%, i.e., the ultimate fracture stresses remained to be much higher than at room temperature. The strength increase with temperature is characteristic of carbon and carbide materials. Main factors resulting in such an increase are appearance of plastic properties in carbon fibres and silicon carbide, redistribution or relief of internal stresses in the bulk of materials.

Thus, it has been shown that carbon fibre–silicon carbide composites retain their strength characteristics and can be operated at temperatures up to 2500 °C.

3.4.4 *Heat resistance of carbon–carbide matrix materials*

Tests were conducted by two procedures. Specimens in the form of 30 mm diameter, 5–9 mm thick discs were tested on a plasma set under the following conditions: working medium, air with a nitrogen dissociation degree of 60%, and oxygen 100%; static pressure, 0.1–0.3 atm (abs.); gas flow velocity, 0.1 M; test-cycle time, up to 20 min. The specimen was installed perpendicular to the gas flow. Tests under the above conditions are most close to operating conditions of materials in parts of the nose fairing and leading edge of the wing of the *Buran* space shuttle. Results of tests of materials with various densities are presented in Table 3.30.

A material density increase from 1.72 to 2.07 g/cm^2 reduces its mass loss by a factor of 9. This is due to increase in the fraction of the carbide phase, whose oxidation results in formation of a heat-resistant self-healing oxide film that covers the surface of specimens and inhibits the access of oxygen to carbon components of the composite.

Results of studies of the temperature dependence of the specific rate of ablation of Gravimol-V and Karbosil are shown in Figure 3.94, where the rate of ablation of a carbon–carbon base with a density of 1.4 g/cm^3 is also shown for comparison.

As the temperature on the specimen surface rose from 1300 to 1550 °C, the rate of ablation for Gravimol-V increased from 10^{-4} to 10^{-3} kg/m^2 s, whereas for Karbosil it tended to remain at a level of 3×10^{-4} kg/m^2 s. The rate of ablation of the carbon–carbon material under the same conditions was 1–2 orders of magnitude greater.

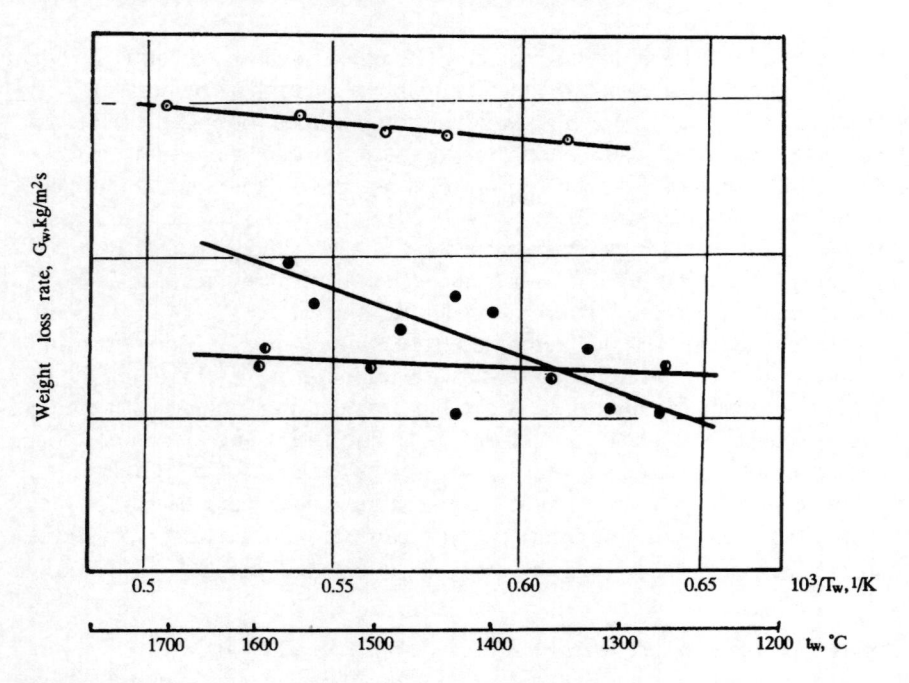

Figure 3.94 Temperature dependence of ablation rate for (1) CCM, (2) Gravimol-V, and (3) Karbosil.

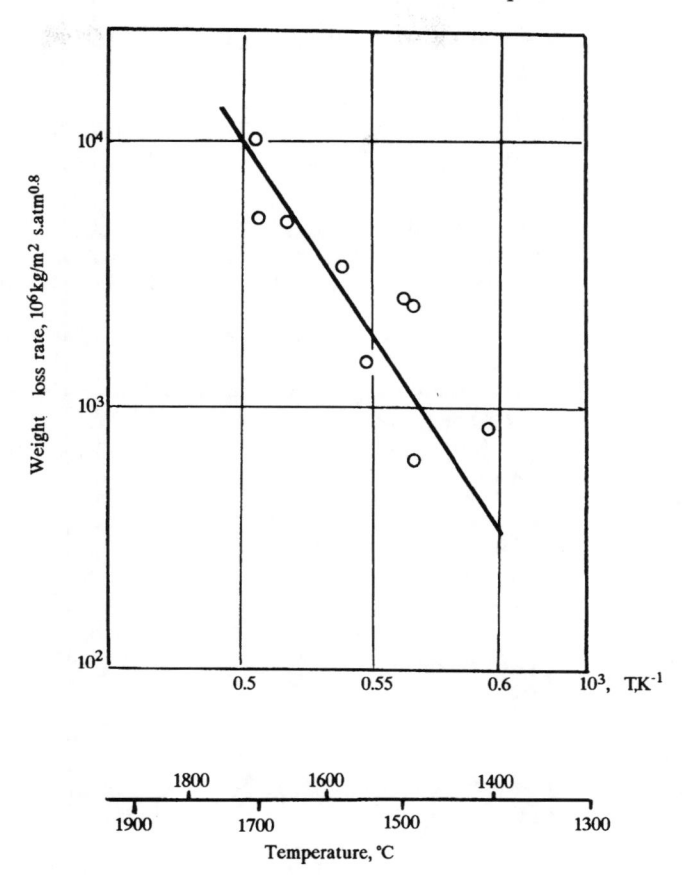

Figure 3.95 Temperature dependence of ablation rate for Gravimol-V.

Disc-shaped specimens of 50 mm in diameter and 5–9 mm in thickness were tested on a plasma set under the following conditions: working medium, air; static pressure, 0.12 atm (abs.); gas flow velocity, 3.5 M; test-cycle time, 10 min. Specimens were installed in silicized graphite holders parallel to the gas flow. Temperature dependences of the ablation rate are shown in Figure 3.95. The Gravimol-V ablation rate linearly increased from 0.5×10^{-3} to 1.0×10^{-2} kg/m^{2}s as the temperature rose from 1400 to 1700 °C.

The conditions of material testing on plasma sets in the thermochemical action and the static pressure at the specimen surface are more severe than conditions of the actual action of the oxidizing atmosphere at the landing of the *Buran* space shuttle.

Results of the tests demonstrated that the maximum operating temperature for a composite based on carbon fibres, carbon matrix and silicon carbide is of 1650 °C.

Similar results were obtained for specimens of composites based on carbon fibres, carbon matrix and titanium (or zirconium) carbide, which can be manufactured by the above-described technology.

REFERENCES

1. Ostrovskij V. S., Ostrovskij Ju. S., Kostikov V. I. and Shipkov N. N. (1986) *Synthetic Graphite*, Metallurgija, Moscow.
2. Vjatkin S. E., Deev A. N., Nagornyj V. G. *et al.* (1967) *Nuclear Graphite*, Atomizdat, Moscow.
3. Sosedov V. P. (ed.) (1975). *Properties of Carbon-based Structural Materials*, Metallurgija, Moscow.
4. Karpinos D. M., Tuchinskij L. I. and Vishnjakov L. R. (1977) *New Composite Materials*, Vyshcha Shkola, Kiev.
5. Konkin A. A. (1974) *Carbon-based and Other Heat-resistant Fibrous Materials*, Khimija, Moscow.
6. Nagornyj V. G., Nabatnikov A. P., Frolov V. I. *et al.* (1975) On existence of new crystalline form of carbon. *Zhurn. Fiz. Khimii*, **49** (4), 840–9.
7. Vinnikov V. A., Kotosonov A. S., Frolov V. I., Ostronov B. G. (1971) Variation of graphitizability of carbon material because of mutual orientation of hexagonal layers at carbonization stage, in *Carbon-based Structural Materials*, Proc. of NIIGrafit, No. 6, Metallurgija, Moscow, pp. 14–16.
8. Nagornyj V. G. and Ostrovskij V. S. (1970) On effect of structure on some physical properties of carbon materials. *Khimija Tverdogo Topliva*, **1**, 110–17.
9. Virgilev J. S. (1973) Porosity and some other physical properties of structural graphite. *Khimija Tverdogo Topliva*, **5**, 102–5.
10. Kolesnikov S. A. (1973) Discrete-matrix model for carbon–carbon compositions. *Mekhanika Polimerov*, **3**, 387–93.
11. Donald A. (1975) *Mater. Sci. and Eng.*, **17** (1), 139–52.
12. Rabotnov J. K., Kolesnikov S. A., Matytsin V. S. *et al.* (1976) Mechanical properties of composite with carbonized matrix. *Mekhanika Polimerov*, **2**, 235–40.
13. Brasell G. W., Horak I. A. and Butler B. L. (1975) Effects of porosity on strength of carbon–carbon composites. *J. Comp. Mater.*, 9 July, 288–96.
14. Granoff B., Pierson H. O. and Schuster D. M. (1973) Carbon-fert, carbon–matrix composites: dependence of thermal and mechanical properties on fiber volume percent. *J. Comp. Mater.*, **7**, 36–52.
15. Rosen B. W. (1965) in *Fiber Composite Materials*, ASM, Metals Park, Ohio, pp. 37–76.
16. Shchetanov B. V., Prilepskij V. N. and Chernjak A. M. (1977) Strength of high-strength fiber-reinforced ceramics. *Mekhanika Polimerov*, **6**, 994–7.
17. Korten T. (1967) *Fracture of Reinforced Plastics* Khimija, Moscow.
18. Jackson P. W. Some studies of the compatibility of graphite and other fibers with metal matrices, *Metals Eng*.
19. Coble R. L. and Parikh N. M. (1976) in *Fracture*, vol. 7. Mir, Moscow, 223–99.
20. Shchetanov B. V., Prilepskij V. N., Lapidovskaja L. A. *et al.* (1978) Toughening of phosphate foam ceramics by silicon carbide whiskers. *Mekhanika Polimerov*, **2**, 253–6.
21. Korten (1972) in *Fracture*, (ed. H.Liebowitz) vol. 7, Fracture of Nonmetals and Composites, Academic Press, New York and London, pp. 367–471.

22. Andrianov K. A., Kolesnikov S. A., Resanov V. I. et al. (1978) Mechanical properties and durability of thermally treated carboplasts. Plast. Massy, 5, 21–3.
23. Bejlina N. J., Kozhueva E. N., Golubkov O. E. et al. (1990) Study of coal-tar and petroleum pitch compositions by extrography method. Khimija Tverdogo Topliva, 5, 132–6.
24. Ostrovskij V. S. and Mechnikova O. E. (1964) On plasticity of coke-pitch masses, in Structural Carbon-graphite Materials, No. 1, Metallurgija, Moscow, 250–5.
25. Zhurid Ju. P., Popov V. L. and Ostrovskij V. S. (1966) On elastic expansion of coke-pitch masses, in Structural Graphite-based Materials, No. 2, Metallurgija, Moscow, pp. 12–16.
26. Ostrovskij V. S. (1987) Influence of filler and binder properties on quality of structural carbon material, in Structure and Properties of Carbon Materials, Metallurgija, Moscow, pp. 7–16.
27. Kulakov V. V., Pavlova A. I., Zlatkis A. M., et al. (1975) On use of coal-tar pitches as binders of carbon fiber-reinforced composites, in Carbon-based Structural Materials, No. 10, Metallurgija, Moscow, pp. 201–6.
28. Kulakov V. V., Pavlova A. I., Judin V. P. et al. (1976) Study of compositions based on discrete carbon fibers and coal-tar pitches, in Carbon-based Structural Materials, No. 11, Metallurgija, Moscow, pp. 148–53.
29. Lapina N. A., Starichenko N. S., Ostrovskij V. S. et al. (1975) Evaluation of sintering ability of pitches. Tsvetnye Metally, 12, 39–42.
30. Ostrovskij V. S., Lapina N. A., Khakimova D. K. and Ionova E. M. (1973) Study of chemical and structural transformations of solid residues of carbonization of polyvinylacetofurals. Vysokomolek. Soed., 15 (1), 217–23.
31. Lukina E. J. and Fokin S. I. (1990) Thermal expansion of binders for matrix of carbon composites. Tsvetnye Metally, 8 70–2.
32. US Patent No. 3914500. Tungsten wire reinforced silicon nitride articles and method for making the same.
33. Tesner P. A. and Shein O. G. (1987) Kinetics of pyrocarbon formation from acetylene, vinyl acetylene, and diacetylene, in Results of Science and Technology. Kinetics and Catalysis, VINITI, Moscow, 28 (1), pp. 747–9.
34. Makarov K. I. and Pechik V. K. (1975) Investigation into kinetics of thermal transformation of methane. I. Heterogeneous reaction. II. Homo-heterogeneous reaction, in Results of Science and Technology. Kinetics and Catalysis, VINITI, Moscow, 16 (6), pp. 1484–90, 1491–500.
35. Japan Patent Application No. 60–33263. 20 Feb. 1985. Structural Material.
36. Lahayc J., Badil P. and Dicret J. (1977) Mechanism of carbon formation during steam cracking of hydrocarbon. Carbon, 15 (2), 87–93.
37. Kotlensky W. N. (1973) Deposition of pyrolitic carbon in porous solids, in Chemistry and Physics of Carbon, 9, pp. 173–262.
38. Peacock S. J. and Reuben B. G. (1980) The pyrolysis of ethylbenzene: kinetics and carbon deposition. Third Intern, Conf. Carbon. Ext. Abst., pp. 659–62.
39. McAllister L. E. and Lachman W. L. (1983) Multidirectional carbon–carbon composites, in Handbook of Composites, 4, Elsevier. 1983.
40. Kotosonov A. S. (1982) Characteristic of macrostructure of synthetic polycrystalline graphites in electrical conductivity and magnetoresistance. Dokl. AN SSSR, 262 (1), 133–5.
41. Kostikov V. I., Kotosonov A. S., Ostronov B. G. and Kholodilova E. I. (1989) Structure of carbon material produced from coal-tar pitch, carbonized under pressure, in Carbon Materials. Metallurgija, Moscow, pp. 76–80.

42. Fischbach D. B. (1971) Kinetics and mechanism of graphitization, in *Chemistry and Physics of Carbon*, vol. 7, (ed. Walker P. L.), pp. 1–105.
43. Virgilev J. S. and Pekaln T. K. (1976) Carbon material graphitization process stages. *Izv. AN SSSR, Ser. Inorganic Mater.*, **12** (10), 1791–5.
44. Kondratjev I. A., Ostronov B. G., Rozenman I. M. and Kotosonov A. S. (1975) Production of single-component carbon materials with different graphitization degrees, in *Carbon-based Structural Materials*, Proc. of NIIGrafit, No. 10. Metallurgija, Moscow. pp. 66–70.
45. March H. and Warburton A. P. (1970) Catalysis of graphitization. *J. Appl. Chem.*, **20** (5), 133–7.
46. Ostronov B. G., Kotosonov A. S. and Polozhikhin A. I. (1974) Effect of silicon on variation of electronic properties and structure of graphitizable and nongraphitizable carbon materials at thermal treatment, in *Carbon-based Structural Materials*, Proc. of NIIGrafit, No. 9. Metallurgija, Moscow, pp. 134–9.
47. Chalykh E. F. (1972) *Technology and Equipment of Electrode and Electrocarbon Production Facilities*, Metallurgija, Moscow.
48. (1988) How to filament wind thick-wall composite cylinders. *Comp. Adhes. Newsl.*, **5** (2), 6.
49. French Patent Application No. 2,584,652, Int. Cl. B29C 67/14. Method for manufacture of tubular elements from composites.
50. Tarnopolskij J. M. (1969) *Features of Calculation of Parts from Reinforced Plastics*, Zinatne, Riga.
51. Kolesnikov S. A. (1980) Thermal stabilization and carbonization of plastics, in *Thermal Stability of Structural-Purpose Plastics*, (ed. Trostjanskaja E. B.), Khimija, Moscow, pp. 213–40.
52. Japan Patent Application No. 62–171871 (A), Int. Cl. B65H 65/04. Apparatus for rewinding of carbon-fiber thread.
53. French Patent Application No. 2,584,341, Int. Cl. B29C 67/14. Method for manufacture of shell wound from fiber-reinforced composite material.
54. Shunk Carbon Fiber Reinforced Carbon. Schunck Kohlenstofftechnik GmbH, Postfach 6420, D-6300, Giessen, I.
55. Fitzer E. The future of carbon–carbon composites. *Carbon*, 1987, **25** (2), pp. 163–190.
56. Davis H. O. (1976) Material and process effects on carbon–carbon composite shear strength. *J. Spacecraft and Rockets*, **13** (8), 456–60.
57. Butler B., Duliere S. and Tidmore J. (1971) The relation between thermal expansion and preferred orientation of carbon fibers. *Tenth Biennal Conf. on Carbons*, Bethlehem, Pennsylvania 27 June–2 July, p. 171.
58. Li W. and Shiekh A. B. (1988) The effect of processes and processing parameters on 3D braided preforms for composites. *33rd International SAMPE Symposium and Exhib.*, 7–10 March, pp. 104–15.
59. Ko F. K., Soebroto H. B., Lei Ch. 3D Net Shaped Composites by the 2-Step Braiding. Ibid., pp. 912–921.
60. Bruno P. S., Keith D. O. and Vicario A. A. (1986) Automatically woven three-dimensional composite structures. *31st International SAMPRE Symposium and Exhibition*, Las Vegas, Nevada, USA, 7–10 April, pp. 103–16.
61. Grenie J. (1984) Les composites multidirectionnels dans les vehicules balistiques et spatiaux. *Mater. et Techn.*, **72** (1–2), 33–41.
62. Rolincik P. G. (1987) Autoweare^{TM} – a unique automated 3D weaving technology. *SAMPE Journal*, **23** (5), 40–7.
63. Maistre W. A. (1976) Development of a 4D reinforced carbon–carbon composite. *AIAA Paper*, No. 605, p. 6.

64. McAllister L. E. and Lachman M. L. (1983) Multidirectional carbon–carbon composites, *Handbook of Composites*, vol. 4, Elsevier.
65. Grigorjev G. A., Borodin A. N., Sologubov A. I. *et al.* (1988) Wetting and impregnation of carbon fibers with finishes, in *Development and Study of Structural Carbon Materials*, Proc. of NIIGrafit, Metallurgija, Moscow, pp. 117–22.
66. Privalov V. E. and Stepanenko M. A. (1961) *Coal-tar Pitch*, Metallurgija, Moscow.
67. Thomas C. R. and Walker E. J. (1986) Carbon–carbon composites. *Mater. Aerosp. Proc.*, London, 2–4 April, **1**, pp. 138–65.
68. Stoller H. M. and Frue E. R. (1972) Processing of carbon–carbon composites – an overview. *SAMPE Quart.*, **3**, (3), 10–22.
69. Bauer D. W., Kotlensky W. V., Warren I. W. and Gray E. (1971) Fabrication and CVD carbon infiltration of carbon and graphite filament filament wound cylinders, in *Summary of Papers Tenth Biennial Conference on Carbon*, Bethlehem, Pennsylvania, 27 June–2 July, pp. 57–8.
70. Kostikov V. I., Kolesnikov S. A. and Shurshakov A. N. (1980) Carbon composites with ceramic matrices, in *Graphite-based Structural Materials*, Proc. of NIIGrafit, Metallurgija, Moscow, pp. 78–88.
71. Sosodov V. P. (ed.) (1975) *Properties of Carbon-Based Structural Materials*, Metallurgija, Moscow.
72. Goetzel C. G. High-temperature properties of some carbon–carbon composites. *High Temp.-High Pres.*, 1980, **12**, (1), pp. 11–22.
73. Fitzer E. (1987) The future of carbon–carbon composites. *Carbon*, **25** (2), 163–90.
74. Fitzer E. (1984) Carbon-based composites. *J. Chim. Phys. et Phys.-Chim. Biol.*, **81** (11–12), 717–33.
75. Shulepov S. V. (1972) *Physics of Carbon-graphite Materials*, Metallurgija, Moscow.
76. Devis H. O. (1976) Material and process effects on carbon–carbon composite shear strength. *J. Spacecraft and Rockets*, **13**, 456–60.
77. Kojiona A. and Otani S. (1982) Adhesive properties between graphite substrate and the low temperature pyrolytic carbon deposited on it. *Carbon*, **20** (2), 121–3.
78. Promising materials for solid-fuel rocket engine nozzles. (1978) Sci. Techn. Collection. *Inter-Industry Technology and Economics. Series T*. Inter-Industry Problems of Technology, **1**, 24–8.
79. Matériaux Composites Carbon/Carbon Sepcarb. Advertisement publication by SEP Company.
80. Carbon/Carbon Materials. Advertisement publication by S.N.I. Aérospatiale, Le Carbone-Lorraine.
81. 2D Reinforced Carbon Composite KUP-VM-2. Advertisement publication of NIIGrafit. Vneshtorgizdat, USSR, publ. No. 6677MV, 1989.
82. Carbon Composite KM5415. Advertisement publication of NIIGrafit. Vneshtorgizdat, USSR, publ. No. 6678MV, 1989.
83. Materials for electric heaters and their manufacturing technology. Licenses now. *New Engineering*. TsNIItsvetmetinformatsija, Moscow.
84. Heat-Resistant Adhesive for Joining Carbon-Graphite Materials. Advertisement publication of NIIGrafit. Vneshtorgizdat, Moscow, publ. No. 6675MV, 1989.
85. Carbon Fiber-Reinforced Carbon for High Temperature Technology SIGRI GmbH Productbereich, Vc-2.
86. Schunck Carbon Fiber Reinforced Carbon. Schunck Kohlenstofftechnik GmbH. Postfach 6420. D 6300 Giessen 1.

87. Fibergraph Carbon–Carbon Composites. Advertisement publication of Polycarbon Inc.
88. Kulakov V. V., Pavlova A. I., Zlatkis A. M. and Bagrov G. N. (1975) On use of coal-tar pitches as binders of carbon fiber-reinforced composites. *Carbon-based Structural Materials*, Proc. of NIIGrafit, No. 10. Metallurgija, Moscow, pp. 201–6.
89. Kulakov V. V., Samsonova L. S., Judin V. P. *et al.* (1975) Production of composite based on carbon fiber and thermosetting resin by hydromixing method, in *Carbon-based Structural Materials*, Proc. of NIIGrafit, No. 10. Metallurgija, Moscow, pp. 206–10.
90. Kulakov V. V., Pavlova A. I., Judin V. P. and Zhak I. V. (1976) Investigation of composites based on discrete carbon fibers and coal-tar pitches, in *Carbon-based Structural Materials*, Proc. of NIIGrafit, No. 11. Metallurgija, Moscow, pp. 148–52.
91. Carbon Friction Materials. Advertisement publication of NIIGrafit. Vneshtorgizdat, USSR, publ. No. 6679MV, 1989.
92. Awashi S. and Wood J. L. (1988) Carbon–carbon composite materials for aircraft brakes. *Adv. Ceramic Materials*, **3** (5), 449–51.
93. US Patent No. 4,339,021, Nat. Cl. 191/50. Friction Material Based on Carbon Matrix and Carbon Fiber and Its Manufacturing Technology. Kosuda Niima.
94. US Patent No. 4,119,179. Carbon Friction Disk.
95. French Patent Application No. 2,506,672, Int. Cl. B32B 31/20, F16D 69/00. Method for Manufacture of Friction Disks.
96. Fitzer E., Geigl K. H. and Huttner W. (1980) The Influence of carbon fiber surface treatment on the mechanical properties of carbon–carbon composite. *Carbon*, **18** (4), 265–70.
97. Zheng G. and Dong-Hua J. (1983) The microstructure of 3D carbon–carbon composite. *16th Bien. Conf. on Carbon*. Extended Abstracts and Program. Univ. of California, San Diego, 18–22 July, pp. 507–8.
98. Fitzer E., Huttner M. and Manocha L. M. (1979) Influence of process parameters on the properties of carbon–carbon composites with pitch as matrix precursor. *14th Bien. Conf. on Carbon*. Extended Abstracts and Program. Univ. Park, PA, USA, 25–29 June.
99. Rowe C. R. and Lander L. L. (1977) Mechanical properties of pitch fiber carbon–carbon composites. *13th Bien. Conf. on Carbon*. Extended Abstracts and Program. Iroine, Calif., USA, pp. 72–3.
100. Jortner J. (1986) Macroporosity and interface cracking in multi-directional carbon–carbons. *Carbon*, **24** (5), 603–13.
101. Jing Dong. *18th Bien. Conf. on Carbon*. Extended Abstracts and Program, 19–24 July, 1987, pp. 436–7.
102. Jortner J. (1987) The role of radial yarns in 3D carbon–carbon cylinders. *18th Bien. Conf. on Carbon*. Proc. and Program. 19–24 July, pp. 436–7.
103. Tarnopolskij Ju. M., Poljakov V. A. and Zhigun I. G. (1988) Composites reinforced on diagonals of cube. *Mekhanika Kompositnykh Materialov*, **6**, 1020–7.
104. Choury J. J. Carbon-Carbon Materials for Nozzles of Solid Propellant Rocket Motors. AIAA Paper, 1976, No. 609, p. 7.
105. Jones L. E., Thrower P. A. and Walker P. L. (1986) Reactivity and related microstructure of 3D carbon–carbon composite. *Carbon*, **24** (1), 51–9.
106. Emjashev A. V. (1989) Effect of silicon chloride additions on pyrographite deposition rate. *Khimija Tverdogo Topliva*, **6**, 127–9.

107. Emjashev A. V., Slavgorodskaja Z. V. and Stepanova A. N. (1980) Production and some properties of silicon pyronitride films. *Izvestija AN SSSR, Neorg. Mater.*, **16** (2), 293.
108. Emjashev A. V. (1987) *Gas Metallurgy of Refractory Compounds*, Metallurgija, Moscow.

INDEX

Page numbers appearing in **bold** refer to figures and page numbers appearing in *italic* refer to tables.

3KMP *412, 413*
3KMP *412, 413*
4KMS *413*

Activation energy, covalent crystals 9, *10*
Activation volume, diamond **15**
Adhesion
  ceramic-matrix composites, wetting with metallic melts 76–86
  mechanical strength of adhesive (brazing) contact 86–90
Aerolor *377, 394, 395, 412, 413*
Avrami equation 328

Boron carbide-based materials 4
Boron nitride-based materials
  applications 4, 143–9
  mechanical properties 143–9
Braiding
  cylindrical structures 341–3
  orthogonal three-dimensional structures 340–1
Braking devices, materials and design 391–2
Brazing
  ceramic-matrix composites 73–95
    joining methods and brazed products 92–5
    mechanical strength of adhesive contact 86–90
    stresses in brazed joints 90–2
    wetting with metallic melts 76–86
Brazing alloys 93–5
Brittleness 97

Carbon
  crystalline structure 289–91
  texture 291–2
Carbon fibre–silicon carbide materials, mechanical properties 421–9
Carbon residue yield for various binders *304*
Carbon- and carbide-matrix composites 414–32
  carbon–carbide materials heat resistance 429–32
  carbon fibre–silicon carbide materials 421–29
  manufacture of carbon fibre–refractory carbide composites 418–21
  requirements for high temperatures 415–8
Carbon-based composites 286–437
  applications, two-dimensional composites 364–85
  carbon- and carbide-matrix composites 414–32
    heat resistance 429–32
    requirements for high temperatures 415–8
    carbon fibre–silicon carbide materials 421–9
  classification 287–8
  manufacture 330–64
    assembling of multidirectional skeletons 344–8
    braiding 340–43
    lay-up 337

Index 439

refractory carbon fibre–carbide composites 418–21
three- and four-dimensional composites 352–64
two-dimensional composites 348–52
weaving of three-dimensional structures 337–40
winding of preforms 331–7
mechanics 297–303
physicochemical features 303–30, 374–85
  carbonization of pitches at high pressure 318–27
  enthalpy 371
  expansion coefficient 371
  fillers and matrices 298
  graphitization 327–30
  pitch-based matrices 304–11
  resin-based matrices 311–18
  specific heat 371
  thermal conductivity 371
properties 364–414
  carbon fibre–silicon carbide materials 421–9
  friction-purpose carbon–carbon composites 385–99
  three- and four-dimensional composites 399–414
  two-dimensional composites 364–85
structure 289–97
  crystalline structure of carbon 289–91
  porous structure 292–7
  texture 291–2
see also Composite materials
Carbon–carbon composites 288
cylindrical three-dimensional composites 404–6
manufacture 304
physicochemical features 303–30, 374–85
  carbonization of pitches at high pressure 318–27
  enthalpy 371
  expansion coefficient 371
  fillers and matrices 298
  graphitization 327–30
  pitch-based matrices 304–11
  resin-based matrices 311–8
  specific heat 371

thermal conductivity 371
three- and four-dimensional composites 399–414
  applications 410–14
  density 400–2
  mechanical properties 402–4
  porosity 400–2
  reactivity 409–10
  thermophysical properties 407–8
  two-dimensional composites 364–85
see also Carbon-based composites
Carbonization
  high pressures 318–26
  pitches 308–9
  resin-based matrices 312–13
  resin 313
Ceramic composites for electronics 158–216
  dielectric polarization and loss 190–6
  ECC-based devices 204–16
  electrical conductivity 179–90
    calculation 167–9
  nonstoichiometry and doping 196–204
  physical properties, theory 159–79
    classification 160–2
Ceramic fibre-reinforced composites 266–72
  chemical stability 270
  corrosion properties 269–70
  mechanical properties 269–70
  reinforced glass ceramics 266–9
  structure 270
Ceramic-matrix composites
  applications 131–56
    boron nitride-based materials 143–9
    quartz ceramics-based materials 153–6
    silicon carbide-based materials 141–3
    silicon nitride-based materials 136–41
    zirconia-based materials 149–53
  classification 30–4
  conductivity
    electrical 179–90
    theory 167–79
  dielectric polarization and loss 190–6
  fracture 13–15, 18–29

Ceramic-matrix composites (*cont.*)
  functional materials
    composites for electronics
      158–216
    ferroelectrics, piezoelectrics and
      high-temperature
      superconductors 216–38
  manufacturing processes 34–8
    joining (brazing) 73–95
    machining techniques 53–73
    moulding 41–50
    sintering 50–2
    synthesis and preparation of
      powders and fibres 38–41
  nonstoichiometry and doping in
    oxides 196–204
  physical properties, theory 159–79
  problems and prospects 156–8
  structural materials
    engineering 115–58
    joining (brazing) 73–95
    machining techniques 53–73
    mechanical properties *3*, 95–115
    technology 29–52
    theoretical fundamentals 3–29
  see also Composite materials
Characteristic deformation
  temperature 10–11
Chemical stability, ceramic
  fibre-reinforced composites 270
Chemical vapour impregnation
  process (CVI) xvi
Coal-tar pitches 305
Coke
  density *323*
  yield 305–6
  see also Carbonization
Composite materials
  aftertreatment xvii
  consolidation xv–xvii
  definition x
  finishing xviii
  information sources xix
  market xi, *131*
  moulding xvi
  preparation of starting powders xv
  production processes x, xvi
  properties *3*
  research and development xii, 156–8
  see also Carbon-based composites;
    Ceramic-matrix composites;
    Glass ceramic-based composites

Conductivity
  ceramic-matrix composites 167–79
  current carrier concentration
    180–1
  doping 196–204
  impurity current carriers 181–2
  ionic conduction 184–90
  mobility 182–4
  physical nature of electron
    conduction 180–4
Corrosion properties, ceramic
  fibre-reinforced composites
  269–70
Crack resistance, ceramic-matrix
  composites 97–9, 112–13
Cracking
  ceramic-matrix composites
    bridging between crack edges
      103–5
    composites with continuous
      fibres 107–9
    microcracking and crack
      branching 105–7
    microcracking and
      transformation toughening
      109–10
    micromechanisms of fracture
      25–9
    polymorphic transformations
      100–3
    reorientation of main crack 107
Crystalline structure, refractory
  compounds 5–7
Curie–Weiss law 194

Deformation
  cold 12
  hot 11–12
  warm 12
  see also Characteristic deformation
    temperature
Design in engineering, ceramic-matrix
  composites 115–21
Diamond
  activation volume 14–15
  strength 8, 14–16
Diamond-structure refractory
  compounds, properties 4
Dielectric loss
  equivalent circuit 191, **196**
  physical nature 195–6
Dielectric polarization and loss

dielectric loss 191, 195–6
equivalent circuit 191
physical mechanisms 191–5
polarizability 190–1
Doping
　electrochemical doping and
　　intercalation 203–4
　heterovalent doping and
　　nonstoichiometry of oxides
　　196–204
Dunlop materials
　CB5 *395*
　CB7 *395*
Dynamic moulding
　explosive pressing 47
　gas-dynamic moulding 46
　hydrodynamic pressing 46
　vibroimpact moulding 48–9

ECC-based devices
　capacitors 208–9
　heaters 207–8
　varistors 205–7
　optical transparent ceramic
　　materials 209
　resistors 204–5
　sensors
　　gas composition sensors 212–14
　　posistors 211–12
　　solid-electrolyte oxygen sensors
　　　214–16
　　thermistors 210–11
ECC-based devices, *see* Electronic
　　ceramic composites
Effective volume 125–6
Electret 194–5
Electrical conductivity of ceramics,
　　see Conductivity
Electrochemical doping 203–4
Electron paramagnetic resonance
　　(EPR) 220, 223–4, 229–30,
　　231–3
Electronic ceramic composites (ECC),
　　see Ceramic composites for
　　electronics
Electronics, composites for 158–216
Electronics, *see* Ceramic composites
　　for electronics
Engineering
　ceramic-matrix composites
　　general principles of design
　　　115–21

probabilistic analysis of strength
　　and fracture 121–31
Engines, ceramic 133–4
EPR, *see* Electron paramagnetic
　　resonance
Extrusion, ceramic-matrix
　　composites 42–3

Ferroelectrics 216–29
　actuators 237
　applications 234–6
　oxygen-octahedral structure 219
Filler, definition x
Filtration coefficient 294
Fracture
　ceramic-matrix composites 13–5,
　　18–29, 97–9, 107–9
　probabilistic analysis 121–31
　fracture toughness 21–4, 26
　micromechanics 24–9
Fracture toughness, ceramic-matrix
　　composites 21–4, 26
Friction-purpose carbon–carbon
　　composites 385–99
　braking devices 391–2
　friction-wear characteristics of
　　carbon–carbon composites
　　386–91
　manufacturing technology 392–3
　oxidation protection methods
　　395–9
　properties 394–5
　　Aeolor *394*, *395*
　　Sepcarb A2 *394*, **395**
　　Termar-ADF *394*
Functional materials, ceramic-matrix
　　composites
　composites for electronics 158–216
　ferroelectrics, piezoelectrics and
　　high-temperature
　　superconductors 216–38

Gas-phase synthesis 40
Gas turbine engines 132, 133–4
Glass ceramic-based composites
　　255–85
　ceramic fibre-reinforced 266–72
　chemical stability 270
　corrosion properties 269–70
　glass–ceramic matrices,
　　properties *268*
　mechanical properties 269–70

Glass ceramic-based composites (*cont.*)
  reinforced glass ceramics 266–9
    reinforcing fibres, properties *267*
    structure 270
  metallic fibres and net
      reinforcement 256–65
    interaction at matrix–fibre
        interface 265
    moulding of materials 260–4
    thermodynamic properties 264–5
  particle reinforcement 272–83
    filled glass ceramics 272–5
    microstructure and properties
        279–83
    nonmetallic particles 275–9
  see also Composite materials
Graphite 289
Graphitization 327–30
  cokes **393**
Gravimol **424**, 425–31

Hardness, temperature dependence,
    covalent crystals **9**
High-density carbon–carbon
    composites 400–1
Hot isostatic pressing (HIP) 51–2
Hydrothermal synthesis 40

Injection moulding, ceramic-matrix
    composites 43
Intercalation 203–4
Interstitial solutions, properties 5
Ionic conduction 184–90
  superconductors 187–90
  types 184–7

Karbosil **424**, 425–30
KM-5415 *378*, **380**
Knudsen formula 302
KUP-VM *378*, *379*

Layered-structure ceramic materials
    111
Linear thermal expansion coefficient
    370–1

Machining techniques
  ceramic-matrix composites 53–73
    machinability 54–8
    microcutting 55, *56*
    nonoxides 70–3
    oxide–carbide tool ceramics 60–9

oxide ceramics 58–60
Manufacturing processes
  carbon-based composites **304**,
      330–64
    assembling of multidirectional
        skeletons 344–8
    braiding 340–3
    lay-up 337
    three- and four-dimensional
        composites 352–64
    two-dimensional composites **346**,
        348–52
    weaving of three-dimensional
        structures 337–40
    winding of preforms 331–7
  ceramic-matrix composites **3**, 34–8
    joining (brazing) 73–95
    machining techniques 53–73
    moulding 41–50
    sintering 50–2
    synthesis and preparation of
        powders and fibres 38–41
  friction-purpose materials 392–3
  refractory carbon fibre–carbide
      composites 418–21
Market, structural ceramics xi, *131*,
    132
Matrix, definition ix
Mechanical properties
  ceramic-matrix composites 3, 4,
      95–115
    experimental data 109–15
    crack resistance 97–9
    theoretical prediction 99–109
  mechanical strength of adhesive
      contact 86–90
  quartz ceramics-based materials
      153–6
  silicon carbide-based materials 17,
      141–3
  silicon nitride-based materials **17**,
      18, 136–41, 143–9
  zirconia-based materials 149–53
Metallic fibre reinforced composites
    256–65
  interaction at matrix–fibre
      interface 265
  moulding of materials 260–4
  thermodynamic properties 264–5
Metallic fibres and nets 256–65
Microcracking and crack branching
    105–7

Index

Moulding
  ceramic-matrix composites 41–50
    dynamic moulding 46–7
    elastostatic pressing 45
    extrusion 42–3
    hydrostatic (isostatic) pressing 445
    injection moulding 43
    slip casting 43
    vibroimpact moulding 47–9
  glass ceramic-based composites 260–4
    hot pressing conditions 261–4
    preparation 261

Nasicon 189–90
Near-electrode polarization 193
NMR, *see* Nuclear magnetic resonance
Nonoxide ceramics, machining 70–3
Nonstoichiometry, heterovalent doping 196–204
Nuclear magnetic resonance 227

Optical transparent ceramic materials 209
Organosilicon compounds 146
Oxide ceramics
  electrochemical doping and intercalation 203–4
  heterovalent doping and nonstoichiometry 196–203
  machining 58–60
Oxide–carbide tool ceramics
  machining 60–9
  properties *61–4*, **65**

Partially stabilized zirconia 149, 153
Particle reinforced glass ceramic-based composites 272–83
  filled glass ceramics 272–5
  microstructure and properties 279–83
  nonmetallic particles 275–9
Permeability coefficient 294
Perovskite structure 218, 227
Petroleum pitches **305**, 307
Piezoelectrics 216–29
  applications 234–6
  motors 236–7
  oxygen-octahedral structure 219

piezotransformers 236–7
Pitch-based matrices 304–11
Pitches 304–6
Plasma generators 39
Poisson's ratio, carbon fibre-based composites *298*, 374
Polarizability 190–1
Polaron 183–4
Polymorphic transformations, ceramic-matrix composites 100–3
Porosity
  carbon materials 292–7
  cokes 313
  high-density carbon–carbon composites 400–2
Posistors 211–12
Powder synthesis 38–41
Powder technology, ceramic-matrix composites 36–8
Preform winding 331–7
Pyrocarbon 315–8
PZT 219, 223, 226

Quartz ceramics-based materials
  applications 153–6
  mechanical properties 153–6

R-curve 101–3, 110
Recrystallization temperature *10*
Refractory carbon fibre–carbide composites 418–21
Refractory materials
  carbon fibre–carbide composites 418–21
  crystalline structure 4–7
  slip systems *6*
  temperature dependence of yield strength 8–18
Reinforcing materials, definition x
Resin-based matrices 311–18

Self-propagating high-temperature synthesis (SHS) 39–40
Sepcarb materials
  physicomechanical characteristics *376*
  Sepcarb-500 *413*
  Sepcarb A2 *394*, **395**
Sialons 138–9
Silicon carbide, mechanical properties 17

Silicon carbide-based materials
  applications 4, 141–3
  mechanical properties 141–3, *141*
Silicon nitride-based materials
  applications 4, 136–41
  mechanical properties **17**, 18, *127*, 136–41, *140*
Sintering
  ceramic-matrix composites 50–2
  glass ceramic-based composites 275–8
Slip casting, ceramic-matrix composites 43
Sol–gel process 40
Space Shuttle 155, 386, 431
Strength
  carbon-based composites 300–5
  ceramic-matrix composites *3*, 121–31
  refractory materials 8
Stress–strain curves, carbon fibre-based composites **299**
Structural materials, ceramic-matrix composites
  applications 131–56
  engineering 115–58
    designing principles 115–21
    probabilistic approach 121–31
  formation of composites 38–52
    moulding 41–50
    preparation of powders and fibres 38–41
    sintering 50–2
  joining (brazing) 73–95
    joining methods and brazed products 92–5
    mechanical strength of adhesive contact 86–90
    stresses in brazed joints 90–2
    wetting with metallic melts 76–86
  machining techniques 53–73
    machinability 54–8
    nonoxides 70–3
    oxide–carbide tool ceramics 60–9
    oxide ceramics 58–60
  mechanical properties *3*, *4*, 95–115
    crack resistance 97–9
    experimental data 109–15

  theoretical prediction 99–109
  technology 29–52
    classification of ceramic-matrix composites 30–4
    manufacturing processes **33**, 34–8
  theoretical fundamentals 3–29
    crystalline structure 5–7
    fracture 18–29
    temperature dependence of yield strength 8–18
Superconductors
  high-temperature 229–34, 237–8
  ionic 187–90
Synthesis and preparation of powders and fibres, ceramic-matrix composites 38–41

Termar materials *396*, **398**
Texture, carbon materials 291–2
Thermal protection materials, properties *156*
Thermistors 210–11
Transformation toughening 109–10

Varistors 205–7
VOK-60, grinding 66–7

Weaving of three-dimensional carbon-fibre structures 337–40
Weibull modulus, ceramic-matrix composites 124
Wetting, brazing, metallic melts 76–86
Whisker- and continuous fibre-reinforced materials 112–15
Winding, carbon-based materials 331–7

Yield strength, temperature dependence, refractory compounds 8–18

Zirconia-based materials
  applications 149–53
  mechanical properties 149–53